Monographs in Theoretical Computer Science
An EATCS Series

Springer
Berlin
Heidelberg
New York
Barcelona
Hong Kong
London
Milan
Paris
Singapore
Tokyo

Eike Best
Raymond Devillers
Maciej Koutny

Petri Net Algebra

With 111 Figures

Springer

Authors

Prof. Dr. Eike Best
Fachbereich Informatik
Carl von Ossietzky Universität
Oldenburg
26111 Oldenburg, Germany
Eike.Best@informatik.uni-oldenburg.de

Prof. Dr. Raymond Devillers
Faculté des Sciences
Laboratoire d'Informatique Théorique
Université Libre de Bruxelles
1050 Bruxelles, Belgium
rdevil@ulb.ac.be

Prof. Dr. Maciej Koutny
Department of Computing Science
University of Newcastle
Newcastle upon Tyne NE1 7RU, U.K.
Maciej.Koutny@newcastle.ac.uk

Series Editors

Prof. Dr. Wilfried Brauer
Institut für Informatik,
Technische Universität München
Arcisstraße 21, 80333 München
Germany
Brauer@informatik.tu-muenchen.de

Prof. Dr. Grzegorz Rozenberg
Leiden Institute
of Advanced Computer Science
University of Leiden
Niels Bohrweg 1, 2333 CA Leiden
The Netherlands
rozenber@liacs.nl

Prof. Dr. Arto Salomaa
Turku Centre for Computer Science
Lemminkäisenkatu 14 A, 20520 Turku
Finland
asalomaa@utu.fi

Library of Congress Cataloging-in-Publication Data applied for

Die Deutsche Bibliothek - CIP-Einheitsaufnahme

Best, Eike:
Petri net algebra / Eike Best ; Raymond Devillers ; Maciej Koutny. -
Berlin ; Heidelberg ; New York ; Barcelona ; Hong Kong ; London ;
Milan ; Paris ; Singapore ; Tokyo : Springer, 2001
(Monographs on theoretical computer science)

ACM Computing Classification (1998): F.3.2, F.1.1–2, D.2.2, G.2

ISBN 978-3-642-08677-9

Springer-Verlag Berlin Heidelberg New York,
a member of BertelsmannSpringer Science+Business Media GmbH

Printed in Germany

Cover Design: *design & production* GmbH, Heidelberg

Preface

In modern society services and support provided by computer-based systems have become ubiquitous and indeed have started to fundamentally alter the way people conduct their business. Moreover, it has become apparent that among the great variety of computer technologies available to potential users a crucial role will be played by concurrent systems. The reason is that many commonly occurring phenomena and computer applications are highly concurrent: typical examples include control systems, computer networks, digital hardware, business computing, and multimedia systems. Such systems are characterised by ever increasing complexity, which results when large numbers of concurrently active components interact. This has been recognised and addressed within the computing science community. In particular, several formal models of concurrent systems have been proposed, studied, and applied in practice.

This book brings together two of the most widely used formalisms for describing and analysing concurrent systems: Petri nets and process algebras. On the one hand, process algebras allow one to specify and reason about the design of complex concurrent computing systems by means of algebraic operators corresponding to common programming constructs. Petri nets, on the other hand, provide a graphical representation of such systems and an additional means of verifying their correctness efficiently, as well as a way of expressing properties related to causality and concurrency in system behaviour. The treatment of the structure and semantics of concurrent systems provided by these two types of models is different, and it is thus virtually impossible to take full advantage of their overall strengths when they are used separately.

The idea of combining Petri nets and process algebras can be traced back to the early 1970s. More directly, this book builds on work carried out by its authors within two EU-funded research projects. It presents a step-by-step development of a rigorous framework for the specification and verification of concurrent systems, in which Petri nets are treated as composable objects, and as such are embedded in a general process algebra. Such an algebra is given an automatic Petri net semantics so that net-based verification techniques, based on structural invariants and causal partial orders, can be applied. The book contains full proofs and carefully chosen examples, and describes several possible directions for further research. A unique aspect is that the development of the Petri net algebra is handled so as to allow for further application-oriented extensions and modifications.

The book is primarily aimed at researchers, lecturers, and graduate students interested in studying and applying formal methods for concurrent computing systems. It is self-contained in the sense that no previous knowledge of Petri nets and process algebras is required, although we assume that the reader is familiar with basic concepts and notions of set theory and graph theory.

We would like to express our deepest gratitude to our friends and colleagues with whom we conducted the work presented in this monograph. In particular, we would like to thank Javier Esparza, Hans Fleischhack, Bernd Grahlmann, Jon G. Hall, Richard P. Hopkins, Hanna Klaudel, and Elisabeth Pelz. We would also like to thank Wilfried Brauer for his support and encouragement during the writing of this book, Hans Wössner for his excellent editorial advice, and an anonymous reviewer for several very useful comments and suggestions. Last but not least, we would like to thank our respective families for their unfailing support.

December 2000,

Oldenburg — Eike Best
Bruxelles — Raymond Devillers
Newcastle upon Tyne — Maciej Koutny

This book was typeset with Leslie Lamport's LaTeX document preparation system, which is itself based on Donald Knuth's TeX. To produce diagrams, we used the *graphs* macro package by Frank Drewes.

Contents

1. Introduction

Computing Science is about building systems. Many of such systems, and an increasing number of them, are concurrent or distributed. That may be so, for instance, because more than one processors inhabit a single machine, or because several machines at different locations are made to cooperate with each other, or because dedicated processing elements are built into a single system.

Concurrency theory is sufficiently wide and difficult a subject of Computing Science such that, to this date, no single mathematical formalisation of it, of which there exist many, can claim to have absolute priority over others. Much research has gone into exploring individual concurrency theoretical approaches and exploiting their particular advantages. The state of the art is such that many formalisms have been developed to the extent of being very useful, and actually used, in practice. Other research has gone into attempting to combine the relative advantages of two or more, originally perhaps competing, approaches. Such research, it may be hoped, would lay foundations such that in some not too distant day, a uniform and useful generalised formalism emerges that comprises the most important advantages of the present ones.

This book is about combining two widely known and well studied theories of concurrency: *process algebras* and *Petri nets*. Advantages of process algebras include the following ones:

- They allow the study of connectives directly related to actual programming languages.
- They are compositional by definition. Compositionality refers to the ability of composing larger systems from smaller ones in a structured way.
- They come with a variety of concomitant and derived logics. Such logics may facilitate reasoning about important properties of systems.
- They come with a variety of algebraic laws. Such laws may be used to manipulate systems; in particular, to refine them, or to prove them correct with respect to some specification.

Advantages of Petri nets are the following, amongst others:

- Petri nets sharply distinguish between states and activities (the latter being defined as changes of state). This corresponds to the distinction between places (local states) and transitions (local activities).
- Global states and global activities are not basic notions, but are derived from their local counterparts. This is very intuitive and suits well the description of a distributed system as the sum of its local parts. It also facilitates the description of concurrent behaviour by partially ordered sets.
- While being satisfactorily formal, Petri nets also come with a graphical representation which is easy to grasp and has therefore a wide appeal for practitioners interested in applications.
- By their representation as bipartite graphs, Petri nets have useful links both to graph theory and to linear algebra. Both links can be exploited for the verification of systems.

Trying to achieve a full combination of all the advantages just mentioned would certainly be an ambitious task. This book's aim is more modest in that it only attempts to forge links between a fundamental (but restricted) class of Petri nets and a basic (but again restricted) process algebra. However, the book does attempt to make these links as tight and strong as possible. The core plan of this book is to investigate the structural and behavioural aspects of these two basic models, and its main achievement, perhaps, is the revelation that there is, in fact, an extremely strong equivalence between them.

The basic process algebraic setup we take as our starting point is the Petri Box Calculus (PBC [7]), which in itself is a variant of Robin Milner's Calculus of Communicating Systems (CCS [80, 81]). There are many other incarnations of process algebra [2, 54, 59, 62], some of which will be described in due course in this book. The reason why we prefer CCS as a starting point is that, of all the mentioned ones, it seems to be most directly useful for describing the meaning of a concurrent programming language, which is that by which most concurrent systems are actually implemented.

The basic Petri net theoretic setup we take as our starting point is the place/transition net model [90]. Again, this is but one of the many net-based models that have been investigated [15, 63]. However, it is a fundamental one of which the others are extensions or variations. Also, place/transition nets share with CCS-like process algebras the property of being rather straightforwardly useful for concurrent programming languages [3] (and, more generally, for concurrent algorithms [92]).

The idea to combine a process algebra such as CCS with a Petri net model such as place/transition nets is not new. Perhaps the first author advocating such an approach was Peter Lauer in [71], attributing it there to Brian Randell. In the years between this early memorandum and now, many articles and theses [45], and also several books, have been written with exactly this connection in mind. In [95], Dirk Taubner gives high-level Petri net semantics to a process algebra which is akin to CCS. In [84], Ernst-Rüdiger Olderog uses (place/transition) Petri net semantics of another process algebraic variant as a stepping stone in the design of concurrent systems. In [62], place/transition nets are used as semantics of yet another process algebra.

All these setups subtly differ from each other, and they are also limited in their various individual ways. For instance, recursion is absent in [62], and is treated nondenotationally and limited to guarded cases in [84, 95]. More importantly, perhaps, than these differences is the fact that none of the cited works has demonstrated a very strong equivalence ranging from the set of operators in each framework to the generated behaviour, and holding in full generality.[1] It is the desire for such an equivalence which distinguishes the present work from the other ones mentioned above, and we achieve it at the expense of neglecting other valid aims, such as extensions to high-level nets or to other process-algebraic setups, and also the treatment of substantial case studies.

In the course of searching for the desired strong connections, the authors have experienced several surprises. One germ to such a surprise was the separation, already proposed in PBC, of the CCS synchronisation operator into parallel composition and (the communication part of the) synchronisation. This has eventually led to a general classification of all operators into place-based ones and transition-based ones; note that there needs to be a close connection between process algebras and Petri nets, just in order to formulate such a classification.

An advantage of this classification is that all operators within each class can be treated similarly. An exception to this rule is recursion, which can be seen as belonging to the first class, but needs a more involved treatment than other place-based operators. But, ultimately, all these operators are based on a single net-theoretic operation: transition refinement. (It may seem odd to the reader that place-based operators should be based on a transition-related operation, but the simple explanation is that the environment of a refined transition, which is a set of places, plays an important role in such a definition.) Likewise, all operators in the second class, i.e., transition-based ones, are based on a single operation: relabelling. This is not actually an indigenous Petri net operation; rather, it refers to the transition labelling of a net. It can be defined in a sufficiently powerful way so that all operators of the second class become special instances of relabelling. As a matter of fact, in a formal treatment, we will integrate refinement and relabelling into only one operation, of which everything else is just a particular case.

[1] In one of them, it has in fact been conjectured (see [84], conjecture 3.7.6) that it is impossible to achieve such an equivalence. This conjecture, along with the fact that in our setup we do achieve such an equivalence, shows what far-reaching consequences such subtle and innocent-looking distinctions may have, and what scope there is for fine tuning and adjusting existing theories until, perhaps, an optimal one eventually emerges.

Viewing all operators as incarnations of such a powerful operation leads to a further generalisation. This is so because refining a transition in a net can be viewed as a function from nets to nets. For instance, sequential composition can be viewed as a very simple two-transition net which takes two nets as arguments and creates another one (their sequence) as a result. Nets used in this way as functions from (tuples of) nets to nets will be called Operator Boxes. Usually, in a process algebra, a handful of operators (each corresponding to a particular operator box) is considered. However, because of the generalised setup described above, we may now formulate all our results in terms of a very large class of operator boxes, all of which could be interpreted as process-algebraic operators. It is one of the principal results of this book to state precisely which of such nets can, and which cannot, be viewed as good operators.

The main criterion for this, in fact, will be whether or not the desired (behavioural) equivalence does hold. What we desire is that the entire reachability graph, that is, the set of reachable states, together with the information as to which of them are reachable from others, is exactly the same, whether one formalises it in terms of process algebra or in terms of nets. In order to achieve this equivalence, we will employ a net-independent behavioural semantics of process algebra (so-called SOS semantics, for structured operational semantics [89]) and compare it with the net-induced semantics. The equivalence of both, for a large class of operator boxes, is the central result in Chap. 7 of this book.

The structure of the book is as follows:

Chapter 2 is introductory, focussing, in particular, on the above-mentioned separation between synchronisation and concurrent composition. It also serves to introduce CCS, Petri nets, PBC, and its SOS semantics. Moreover, the penultimate section of Chap. 2 contains an intuitive motivation why CCS and PBC are used for concurrent programming languages. The entire chapter is written in a fairly informal style to make for easy reading. In passing, Chap. 2 introduces some of the basic notions used throughout the book.

In Chap. 3, the behavioural SOS semantics of PBC is defined. The style of giving this semantics is adapted to concurrent evolutions, already anticipating the desired equivalence result.

In Chap. 4, a Petri net semantics of PBC is developed. This semantics is based on a special kind of labelled place/transition nets, which makes them amenable to being composed. Instead of defining separately, for each operator, the rules to follow in order to derive nets corresponding to expressions constructed from it, a general simultaneous refinement and relabelling operation is defined together with a notion of operator box.

In Chap. 5, the treatment of recursive expressions is dealt with using a general fixpoint theory developed in order to solve recursive equations on Petri nets.

In Chap. 6, a general analysis of nets obtained through refinements or recursive equations is conducted, based on the structural notion of S-invariants. Many interesting results are obtained in this way concerning their behaviour (such as safeness, or lack of auto-concurrency), together with the facts that complete evolutions are always finite, and that extending the Petri net semantics to infinite steps would not lead to a substantially increased expressive power.

In Chap. 7, the SOS and Petri net semantics of Chaps. 3, 4, and 5 are taken up more directly. In particular, the equivalence of these two approaches to define the semantics of our algebra is proved formally. Moreover, the proof is done in such a general and generic setting that it will not be necessary to re-evaluate the whole theory if, at a later stage, it is desired to introduce new operators or basic processes, provided some straightforward conditions are fulfilled.

In Chap. 8, an argument is made that PBC, and indeed other process algebras, are instances of the general framework developed in the previous chapter.

In Chap. 9, the applicability of the theory developed so far is finally demonstrated by giving a compositional semantics for a concurrent programming language. The emphasis is not so much on the novelty of this approach, but on the ease with which the translations can be given, and the results proved. The hope is expressed (and some examples are given that may be seen as supporting this view) that thanks to such an ease, it will be possible to use process-algebraic and net-theoretic methods harmoniously side by-side in the verification of important properties of algorithms expressed by concurrent programs.

Chapter 10 summarises the material covered by this book and looks at the possible future directions of research, while the Appendix contains solutions of some selected exercises. Besides references, the book also contains an index which enables the reader to find definitions and places of use of important terms and notations.

This work builds on previous work conducted by two European research projects, DEMON (Esprit Basic Research Action 3148) and CALIBAN (Esprit Basic Research Working Group 6067), see [7, 6, 10, 11, 12, 8, 32, 25, 30, 31, 36, 40, 68, 69, 70]; and, later, within the ARC Project 1032 BAT (Box Algebra with Time). The authors gratefully acknowledge contributions which are due to many discussions with numerous researchers, in particular with Javier Esparza, Hans Fleischhack, Bernd Grahlmann, Jon G. Hall, Richard P. Hopkins, Hanna Klaudel, and Elisabeth Pelz.

2. The Petri Box Calculus

In this chapter we outline a process algebra, whose name – Petri Box Calculus (PBC) – arises from its original Petri net semantics [6, 7]. PBC combines a number of features taken from other process algebras, such as COSY (COncurrent SYstems notation, [62]), CCS (Calculus of Communicating Systems, [80, 81]), SCCS (Synchronous CCS, [81]), CSP (Communicating Sequential Processes, [58]), TCSP (Theoretical CSP, [59]) and ACP (Algebra of Communicating Processes, [2]). However, there are some fundamental differences with each of these process algebras since PBC has been designed with two specific objectives in mind:

- To support a compositional Petri net semantics, together with an equivalent – more syntax-oriented – structured operational semantics (SOS) [89].
- To provide as flexible as possible a basis for developing a compositional semantics of high-level concurrent specification and programming languages with full data and control features.

The discussion in this chapter is largely informal; its aim is to familiarise the reader with the main points and topics that will be discussed to much greater extent, and with considerably more formal substance, in the rest of this book. In particular, we aim at explaining the following issues:

- The operators of PBC. We will concentrate on the sequential composition, parallel composition and synchronisation, as these are the main points of difference between PBC and Milner's CCS, which can be viewed as the model directly inspiring the design of PBC.
- The reasons why these operators differ from their counterparts in other process algebras.
- PBC semantics of synchronisation.
- How PBC can model a concurrent programming language.

It is worth pointing out that although PBC has not been specifically designed as a common generalisation of already existing models (such as COSY, CCS, and CSP), it turns out that all these can be encoded in PBC; we will present suitable translations in Sect. 8.2.

2.1 An Informal Introduction to CCS

A process algebra is usually constructed from a set of basic processes (or process constants) and a set of operators, each operator having a fixed arity indicating the number of its operands. A basic process may always be seen as a nullary operator, just as in logic a constant can be seen as a nullary function. Operators may be applied in the prefix, suffix, infix, or functional mode (with precedence or priority conventions, possibly overridden by the use of parentheses). For instance, the standard basic CCS [80, 81] corresponds to the following syntax:

$$E_{\text{ccs}} ::= \mathsf{nil} \mid a.E_{\text{ccs}} \mid \tau.E_{\text{ccs}} \mid E_{\text{ccs}} + E_{\text{ccs}} \mid X \mid E_{\text{ccs}} \,|\, E_{\text{ccs}} \mid E_{\text{ccs}}[f] \mid E_{\text{ccs}} \backslash a \ . \tag{2.1}$$

This syntax contains a single basic process nil, an infinite family of unary prefix operators parameterised by action names (the so-called prefixing operators, $a.$) together with the silent prefixing unary operator (also in prefix notation, $\tau.$), two binary infix operators (the choice, $+$, and composition, $|$), and two infinite families of unary postfix operators (the restrictions, $\backslash a$, also parameterised by action names, and the relabellings, $[f]$, parameterised by functions f modifying action names). Variables (X) may also be considered as basic processes, but they are associated with defining expressions of the form $X \stackrel{\text{df}}{=} E_{\text{ccs}}$. In this way, variables provide support for different levels of abstraction within a CCS specification, as well as recursion.

Every process has an associated set of behaviours. It is one of the main objectives of formal semantics of process algebras to make this notion precise. As we will see, this may be quite involved, in particular, for recursive expressions, depending on the notion of behaviour one is interested in, and also on the kind of restrictions imposed on recursive expressions. Here, we only describe informally the semantics of CCS expressions.

The basic process nil 'does nothing'. The process $a.E_{\text{ccs}}$ 'does' a and thereafter behaves like E_{ccs}. The process $\tau.E_{\text{ccs}}$ does τ and thereafter behaves like E_{ccs}, the difference being that τ is an internal (silent) action while a is not. The process $E_{\text{ccs}}+F_{\text{ccs}}$ behaves either like E_{ccs} or F_{ccs}. The process $E_{\text{ccs}} \,|\, F_{\text{ccs}}$ behaves both like E_{ccs} and F_{ccs}, together with some further (synchronised) activities which we will describe in Sect. 2.5. The process $E_{\text{ccs}}[f]$ behaves like E_{ccs}, except that function f is applied to all actions. The process $E_{\text{ccs}}\backslash a$ behaves like E_{ccs}, except that the actions a and $\widehat{a}$ may not be executed; here $\widehat{a}$ is the conjugate of a, to be explained later. And, finally, the process X behaves like the process E_{ccs} in its defining equation $X \stackrel{\text{df}}{=} E_{\text{ccs}}$.

We can use this informal semantics to generate behaviours. Consider the CCS expression $a.(b.\mathsf{nil})$ which can 'make an a-move', thereafter 'make a b-move', and then terminate. This is formalised in CCS as the sequence

$$\underbrace{a.(b.\mathsf{nil})}_{E_0} \xrightarrow{a} \underbrace{b.\mathsf{nil}}_{E_1} \xrightarrow{b} \underbrace{\mathsf{nil}}_{E_2}$$

that exhibits an important feature of the CCS treatment of expressions and their behaviours. For E_0, E_1 and E_2 are expressions with completely different *structure*. In other words, the original expression has changed its structure through a behaviour. Thus in CCS, and indeed several other process algebras, the structure and behaviour of process expressions are very closely intertwined. We will later argue that for PBC, because of its strong connections with Petri nets, a different approach is preferred.

Exercise 2.1.1. Guess possible moves of the expression $(a.(b.\mathsf{nil}) + c.\mathsf{nil})$.

2.2 An Informal Introduction to Petri Nets

For the purpose of this section, a Petri net may be viewed as a bipartite directed graph with two kinds of nodes, respectively called *places* and *transitions*. In diagrams, places are represented as circles, and transitions as rectangles (or squares). A place s may be an input of a transition t (if an arc leads from s to t), or an output of a transition t (if an arc leads from t to s), or both an input and output place (if there is an arc from s to t and an arc from t to s). In the latter case, s is called a *side place* (or a *side condition*) of t, and the pair $\{s,t\}$ forms a *side-loop*.

The graph of a Petri net describes the structure of a dynamically evolving system. The *behaviour* of that system is defined with respect to a given starting marking (state) of the graph, called the *initial marking*. In general, a *marking* is a function from the set of places to the set of natural numbers,

$$\mathbf{N} = \{0, 1, 2, \ldots\} ,$$

and if the marking of a place s is n, then we say that s *contains n tokens*. The behaviour of a marked net (i.e., a Petri net with an initial marking) can be defined as a set of execution sequences, which transform markings into other markings through *occurrences of transitions*. The rule for a change of a marking can be described as follows.

Let us consider a marking M of a Petri net N. A transition t is called *enabled at* M if each of its input places contains at least one token. Only an enabled transition t may ever occur which, by definition, is tantamount to the following change of marking: a token is removed from each input place of t, and a token is added to each output place of t; no other place is affected. Notice that no 'negative token' can result from this, since t was enabled. We use the notation

$$M\ [t\rangle\ M' \quad \text{(or } (N,M)\ [t\rangle\ (N,M')\text{, to emphasise the underlying net)}$$

to express that M enables t and that M' arises out of M according to the rule just described; M' is then said to be *directly reachable* from M, and the reflexive transitive closure[1] of this relation gives the overall reachability relation for markings. Notice that if s is a side condition of t, then the rule implies that for t to be enabled, there must be at least one token on s, but that the occurrence of t does not change the marking of s. A marking is called *safe* if it returns 0 or 1 for every place, and a marked net is called *safe* if every marking reachable from the initial one is safe.

Many variants of the basic enabling and occurrence rules may be found in the literature; in particular, such rules may result from the introduction of arc weights, concurrent occurrences and labelled tokens. Some will be introduced and used later in this book.

Figure 2.1 shows an unmarked Petri net, N, and two marked Petri nets, (N, M_0) and (N, M_1). Note that the unmarked net can be interpreted as a marked net whose trivial 'empty' marking assigns zero to each place. Note also that the underlying graphs are the same in all three cases, only the markings are different.

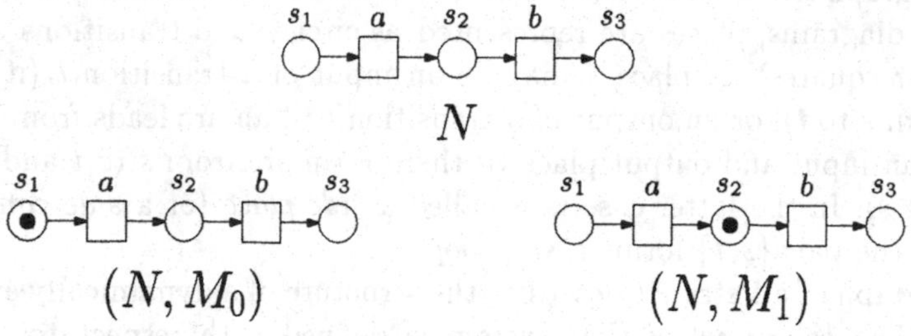

Fig. 2.1. A Petri net N, and the same net with markings M_0 and M_1

Formally, the net N of Fig. 2.1 can be described as a triple (S, T, W) such that S is the set of places, $S = \{s_1, s_2, s_3\}$, T is the set of transitions, $T = \{a, b\}$, and W assigns a number (in this case, only 0 or 1) to every pair (s, t) and (t, s), where $s \in S$ and $t \in T$, depending on whether or not an arrow leads from s to t or from t to s, respectively. For instance, we have $W(s_1, a) = 1$ and $W(s_1, b) = 0$ (where 1 stands for 'an arrow', and 0 stands for 'no arrow'). The marking M_0 in Fig. 2.1 can be described as a function satisfying $M_0(s_1) = 1$, $M_0(s_2) = 0$, and $M_0(s_3) = 0$, and the marking M_1 as a function satisfying $M_1(s_1) = 0$, $M_1(s_2) = 1$, and $M_1(s_3) = 0$. Note that M_0 enables transition a which may therefore occur, and if it does, the resulting marking is M_1. Thus $(N, M_0)\ [a\rangle\ (N, M_1)$.

Intuitively, N describes a sequence of two transitions, a and b, and is related to the CCS expression, $a.(b.\mathsf{nil})$, discussed in Sect. 2.2. But which

[1] The *reflexive transitive closure* of a relation $R \subseteq X \times X$ is the least relation R^* containing R which is reflexive ($(x, x) \in R^*$ for all $x \in X$) and transitive ($(x, y), (y, z) \in R^* \Rightarrow (x, z) \in R^*$ for all $x, y, z \in X$).

of the three nets (if any) would correspond to the semantics of $a.(b.\mathsf{nil})$? The answer should be clear: it is (N, M_0) because only this net has the same *behaviour* as $a.(b.\mathsf{nil})$, where 'behaviour' is determined by the transition rule of Petri nets. The net N by itself has no behaviour (as no transition is enabled), and in the marked net (N, M_1), b can occur but not a, contrary to what is specified by $a.(b.\mathsf{nil})$.

Consider now what happens when a is executed in either model. In terms of CCS, the expression $a.(b.\mathsf{nil})$ is transformed into the expression $b.\mathsf{nil}$. In terms of nets, the marked net (N, M_0) is transformed into the marked net (N, M_1) which has the same underlying net, N. But if we were to define a Petri net corresponding to the expression $b.\mathsf{nil}$, then we would not even think of taking (N, M_1); rather, an obvious choice would be a marked net with only two places and one transition, b, in between (something like (N, M_1) after deleting s_1 and a).

The above example demonstrates that the ways of generating behaviours in net theory and in CCS are different. In the latter, the structure of an expression may change completely, while in the former, only markings change but the structure of a net remains the same through any behaviour. And, indeed, other examples demonstrate this as well. For instance, a CCS expression involving the choice operator, $(a.\mathsf{nil}) + (b.\mathsf{nil})$, can make an a-move and thereafter become the expression nil, forgetting all about the nonoccurring b, as well as a.

This is a fundamental difference which stems from different underlying ideas about systems modelling. In net theory, after a transition has occurred, we may still recover its presence from the static structure of the system, even though it may never again be executed (in some sense, the remains of the past are kept forever); there is thus a sharp distinction between the static and dynamic aspects of a system, captured by the absence or the presence of a marking. In CCS, after an action has occurred, as in the above example, we may safely forget about it precisely because it may never be executed (the past is irreversibly lost), and hence we may safely change the structure of the expression; there is thus a close relation between the static and dynamic aspects of a system.

Exercise 2.2.1. Draw a Petri net corresponding to the CCS expression $(a.(b.\mathsf{nil}) + c.\mathsf{nil})$.

Exercise 2.2.2. Show that your answer to the previous exercise is not the only net corresponding to $(a.(b.\mathsf{nil}) + c.\mathsf{nil})$.

2.3 The Structure and Behaviour of PBC Expressions

Since PBC aims at supporting Petri net semantics, we will devise a behavioural semantics for it which respects the division between the static and dynamic aspects in system modelling found in Petri net theory. It follows from our discussion in the previous section that such a semantics will necessarily differ from that of CCS. The basic idea is to introduce into the syntax of process expressions a device that achieves what is modelled in Petri nets by markings. More precisely, we will define expressions 'without markings' (called *static expressions*) corresponding to unmarked nets, and expressions 'with markings' (called *dynamic expressions*) corresponding to marked nets. Similar to the case of nets, a static expression can always be seen as a dynamic expression with a trivial 'empty marking'. The syntax of static and dynamic expressions will be defined in Sects. 3.1 and 3.4, respectively. Below, we use E and F to denote static expressions, and G and H to denote dynamic expressions.

As an example, consider the static PBC expression $E = a;b$ which corresponds to the unmarked net N of Fig. 2.1. E is actually a simplified expression that, strictly speaking, does not conform to the PBC syntax in Sect. 3.1, according to which we would write $E = \{a\};\{b\}$ instead of $E = a;b$. The brackets around a and b, however, are not relevant for the discussion in this section and will be explained later in Sect. 2.5.

Making an expression dynamic amounts to overbarring and/or underbarring it, or parts of it. In general, overbarring a (sub)expression means that it is currently in its initial state, while underbarring means that it is in its terminal state. For instance, the dynamic PBC expression $\overline{E} = \overline{a;b}$ corresponds to E, but with a marking that lets E be executed from its beginning, i.e., to the 'initial marking' of E. The net corresponding to $\overline{E}$ is the marked net (N, M_0) of Fig. 2.1. Consider next a syntactically legal dynamic PBC expression $G = \underline{a};b$. It has the same underlying expression E as $\overline{E}$, but in a state in which its first action, a, has just been executed, i.e., it shows the instant in which the final state of a has just been reached. Thus, G corresponds to the marked net (N, M_1) of Fig. 2.1.

Overbarring and underbarring are syntactic devices that introduce what could, at first glance, be seen as an unnecessary complication. Consider again expression $E = a;b$ in a state in which its second part, b, is just about to be executed, $G' = a;\overline{b}$. Which marking of N should this dynamic expression correspond to? There is only one reasonable answer to this question, namely, again the marked net (N, M_1) of Fig. 2.1! Thus, in general, we may have a many-to-one relationship between dynamic expressions and marked nets, and we will use the symbol $\equiv$ to relate dynamic expressions which are not necessarily syntactically equal, but are in the same relationship as G and G'. Note that $\overline{E} = \overline{a;b}$ also has a dynamic expression to which it is equivalent but not syntactically equal, namely, $H = \overline{a};b$. Intuitively, $\overline{a;b} \equiv \overline{a};b$ can be viewed as stating that 'the initial state of a sequence is the initial state of the first component', and $\underline{a};b \equiv a;\overline{b}$ as stating that 'the state in which the first component of a sequence has terminated is the same as the state in which the second component may begin its execution'.

We do not regard this many-to-one relationship as incidental, or one to be overcome, perhaps, by a better syntactic device than overbarring and underbarring. Rather, we view it as evidence of a fundamental difference between Petri nets and process algebras; expressions of the latter come with an in-built structure, while Petri nets do not. For instance, $E = a;b$ is immediately recognised as a sequence of a and b. In the example of Fig. 2.1, it so happens that one may also recognise the structure of the nets to be that of a sequence. However, this is due solely to the simplicity of these nets. For a Petri net, it is in general far from obvious how it may be constructed from smaller nets, while it is always clear for an expression in a process algebra how it is constructed from subexpressions.

That an expression of a process algebra always reveals its syntactic structure should be viewed as a great advantage. For example, it is often possible to reason 'by syntactic induction'; to build proof systems 'by syntactic definition' around such an algebra; and last but not least, the availability of operators for the modular construction of systems is usually appreciated by practitioners. On the other hand, one of the disadvantages of such a syntactic view is that some effort has to be invested into the definition of the behaviour of expressions. While this can usually be done inductively, it is still necessary to go through all the operators one by one which sometimes leads to ad hoc (and very disparate) definitions. A nonstructured model such as Petri nets has a clear advantage in this respect; the behaviour of a net (with respect to a marking) is defined by a single rule which covers all the cases. Moreover, Petri nets represent concurrency in systems behaviour in a direct and intuitive way (cf. Sects. 4.2.3 and 6.4.1), and support various analysis techniques based on their graph-theoretic structure, which will also be discussed in this book (cf. Chap. 6).

The PBC attempts to occupy a middle position and combine the advantages of both Petri nets and process algebras. PBC expressions, whether static or dynamic, come with an in-built structure, as do expressions in other process algebras. This has several desirable consequences, such as those mentioned above, even though this book will not cover them to the extent they deserve. On the other hand, the behavioural rules for (dynamic) PBC expressions are in fact the Petri net transition rule in disguise; we make this point very clear in those parts of the book about *consistency results* (in particular, Sect. 7.3.6). While it is still true – as for other process algebras – that every operator has a separate behavioural rule, all of these rules look very much alike. As a matter of fact, the rules are so similar that they allow for far-reaching generalisations, making it possible to add almost arbitrary new operators to PBC as long as they satisfy a few simple conditions. Later, in Chap. 7, we will use the name *box algebra*, rather than PBC, to denote such a generalised (and generic) process algebra.

The need to use the structural equivalence relation $\equiv$ can be viewed as the price that has to be paid for being able to combine the advantages of Petri nets and process algebras in the adopted setting. However, we are more than willing to pay it, since after a thorough analysis of the structural equivalence in Chap. 7, it will turn out that one can provide sufficient conditions for nets to be well-behaved operators. This is clearly a desirable result because it identifies objects possessing a fundamental property (or, at least, a large class of such objects).

Exercise 2.3.1. In Fig. 2.1, the transition b is enabled at marking M_1. Draw and define formally the marked net which is obtained after the occurrence of b. Design dynamic PBC expression(s) corresponding to such a net.

Exercise 2.3.2. Draw a marked net corresponding to the dynamic PBC expression $((a;\overline{b}) \,\square\, c)$.

Exercise 2.3.3. Discuss whether or not $(\overline{a} \,\square\, \overline{b})$ should be allowed as a dynamic PBC expression.

2.4 Sequential Composition

It may be seen from our previous discussion that the CCS expression $a.(b.\mathsf{nil})$ and the PBC expression $a;b$ correspond to each other. In general, CCS allows sequential behaviour to be expressed by a so-called *prefixing*, $a.F$, while PBC contains a true sequential operator, $E;F$, meaning that E is to be executed before F. The main difference is that in CCS the first part of a sequential composition, E, is restricted to $E = a$ or $E = \tau$, and so $E;F$ (or $E.F$) would not, in general, be a valid CCS expression. A fully expressive sequential composition can be simulated in CCS by a combination of prefixing, parallel composition and restriction, in a rather complex manner (see [95]).

An important reason for allowing full sequential composition is that in a majority of imperative languages, it is a basic construct (usually denoted by the semicolon) allowing one to arrange in a sequence any two subprograms, and program modelling is one of the main applications of PBC. Another reason is that the semantics of full sequential composition is no more complicated than that of prefixing, mainly because the nil process is not necessary in PBC. To see why, compare the CCS way of deriving the behaviour of $a.(b.\mathsf{nil})$ and the PBC way of deriving the behaviour of $\overline{a;b}$:

$$\text{In CCS:} \quad a.(b.\mathsf{nil}) \xrightarrow{a} b.\mathsf{nil} \xrightarrow{b} \mathsf{nil}\,.$$

$$\text{In PBC:} \quad \overline{a;b} \equiv \overline{a};b \xrightarrow{a} \underline{a};b \equiv a;\overline{b} \xrightarrow{b} a;\underline{b} \equiv \underline{a;b}\,.$$

In either case an occurrence sequence ab is derived, i.e., the expressions can make an a-move followed by a b-move, but the PBC style of deriving ab is more symmetric – it could be read from back to front as well – while the CCS style is shorter. However, a crucial observation is that in PBC there is no need for an expression such as nil; instead, the fact that 'an execution has ended' is expressed by the derivation of an expression which is underbarred, e.g., $\underline{a;b}$.

In general, if E is an arbitrary static expression, and if we call $\overline{E}$ 'the initial state of E', then we might as well – and will, in fact – call $\underline{E}$ 'the final (or terminal) state of E'. The reader must be warned, however, that this terminology is slightly deceptive, as there may be expressions E for which no behaviour at all can lead from $\overline{E}$ to $\underline{E}$. This may happen if, after starting at $\overline{E}$ a deadlock, or an infinite loop from which it is impossible to escape, can be entered; we will give a suitable example in Sect. 2.7. Thus calling $\underline{E}$ a 'final state' does not imply that this state is reachable from $\overline{E}$.

The *raison d'être* for the CCS process constant nil can thus be appreciated: it provides a syntactic way of describing final states. Given the CCS style of describing sequential behaviour, in which the prefix of an expression is consumed as the expression is executed, there has to be a syntactic 'something' to describe what remains of the expression once the last action has been executed.

Exercise 2.4.1. Guess possible moves of the PBC expression $\overline{(a \,\square\, b); (c \,\square\, d)}$.

2.5 Synchronisation

Although both CCS and PBC are models based upon the idea that in a distributed environment, all complex interactions can be decomposed into primitive binary interactions, they differ in its actual realisation. Let us first recall the meaning of the CCS synchronisation [81].

Suppose that two people, A and B, wish to communicate by exchanging a handshake. In CCS, one could represent the action 'A extends his hand towards B' by a symbol a; the action of 'B extends her hand towards A' by another symbol $\widehat{a}$ (the *conjugate*[2] of a); and the fact that A and B are standing face to face by the symbol $\mid$, yielding the CCS expression $a \mid \widehat{a}$ which is a parallel composition of two simpler CCS expressions. More exactly, we should consider $(a.\mathsf{nil}) \mid (\widehat{a}.\mathsf{nil})$, but we may ignore the nils for simplicity. Any additional semantics of a and $\widehat{a}$ – for instance, whether they happen on a first encounter of A and B – is of no concern at this point because all one wants to describe is the handshake itself. Moreover, the same scheme may be used to describe other (e.g., less symmetrical) kinds of greetings; for instance, a could be interpreted as 'extending one's lips', while $\widehat{a}$ could be interpreted as 'extending one's cheek'.

By definition, the CCS expression a has only one possible activity: an execution of a, which represents the fact that A extends his hand towards B. Similarly, the CCS expression $\widehat{a}$ has only one possible activity: B extends

[2] In CCS, conjugation is usually denoted by overbarring; we use hatting instead, because overbarring has already been used to construct dynamic expressions.

her hand towards A. The expression $a \mid \widehat{a}$, however, has the following possible activities: (i) A may extend his hand and withdraw it, leaving B with the same possibility; (ii) B may extend her hand and withdraw it, then A does the same; and (iii) both A and B extend their hands which results in a handshake. Using the existing Petri net semantics of CCS expressions (e.g., in [44, 95]), these three possible activities could be represented in Petri net terms as shown on the right hand side of Fig. 2.2. The actual handshake – that is, the transition labelled τ – removes the possibility of the further shaking of hands between A and B. The special symbol τ is interpreted as an internal or silent action. In terms of the example, it indicates that the actual handshake is 'known' only to the two people involved, namely A and B, and cannot involve nor be experienced by a third party.

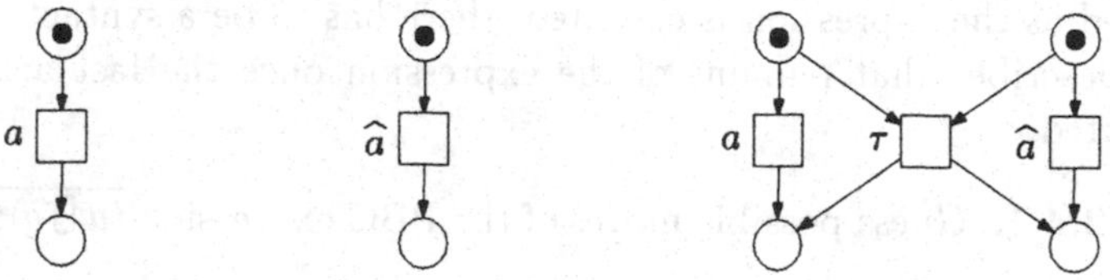

Fig. 2.2. Activity of a (left), activity of $\widehat{a}$ (middle), and activity of $a \mid \widehat{a}$ (right)

Let us clarify at this point some terminology that has been borrowed from the theory of Petri nets but is not standard in CCS. We will call the carriers of activity *transitions* and the entities such as a and $\widehat{a}$ in the above example *actions*. Thus, the CCS expression a has one transition which is also (or more precisely, corresponds to) the action a. Similarly, the expression $\widehat{a}$ has one transition corresponding to $\widehat{a}$. However, the expression $a \mid \widehat{a}$ has three transitions although it syntactically comprises only two actions. It could be said that executing a transition corresponds to an actual activity, normally changing the state of the system, while an action is the way in which this transition is perceived from outside, i.e., the interpretation of the activity; in the Petri net theory this is often formalised through a labelling of transitions by actions. A similar observation can be made for CCS expressions such as $a.(a.\mathsf{nil})$. In the Petri net of this expression, there would be two transitions (just as in the semantics of $a.(b.\mathsf{nil})$ discussed in the previous section), even though the expression contains only a single action (which can be executed twice).

Consider now the question of how one could describe a handshake between three people A, B, and C.[3] In Petri net terms, such an activity could be modelled as depicted schematically by the three-inputs/three-outputs transition

[3] Such *multi-handshakes* are sometimes performed by players of a basketball team, before a game. An example from an entirely different domain is provided by the painting 'The Oath of the Horatii' by Jacques Louis David in the Louvre.

H in Fig. 2.3. It should be clear that such a three-way synchronisation cannot be modelled using directly the two-way handshake mechanism of CCS; the reason is that τ cannot be synchronised with any other action as there is nothing like $\widehat{\tau}$ in CCS. While one might try to simulate a 3-way handshake using a series of 2-way handshakes, PBC proposes instead to extend the CCS framework so that a 3-way (and, in general, n-way) synchronisation could be expressed in a different way. One possibility, which is quite akin to the ACP approach [2], would be to generalise conjugation, e.g., to a relation that allows the grouping of more than just two actions. However, PBC takes a different path, still based on considering primitive synchronisations to be binary in nature, which employs a device of structured actions. Note that although one might think that 3-way handshakes occur rarely in practice, we will argue in Sect. 2.8 (and then in Chap. 9) that n-way synchronisations make perfect sense in the modelling of programming languages, and the way of describing them in PBC is particularly convenient.

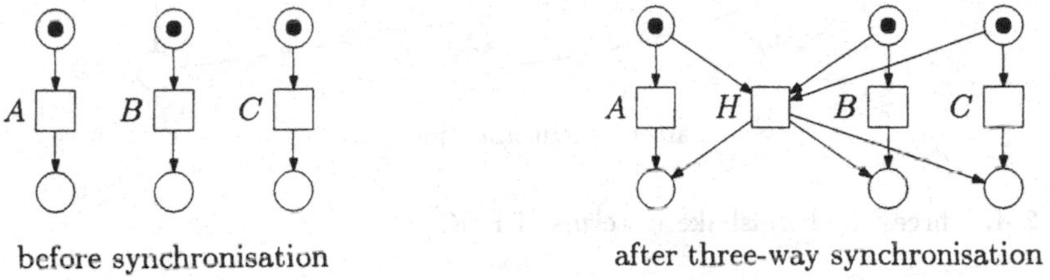

Fig. 2.3. Handshake between three people

To understand multiway synchronisation in PBC, it is perhaps more intuitive to interpret a pair of conjugate CCS actions $(a, \widehat{a})$ in the original way [80], i.e., as denoting two links between two agents that fit together. Then a denotes a communication capability in search, as it were, of a matching $\widehat{a}$; similarly, $\widehat{a}$ denotes a communication capability in search of a matching a. Once a communication link has been established, i.e., the pair $(a, \widehat{a})$ matched, a private synchronisation link between two agents is created with no further externally visible communication capability. The PBC model extends this idea in the sense that the result of a synchronisation does not have to be completely internal or silent, but may contain further communication capabilities linked with other activities. This generalisation is achieved by allowing a transition to correspond to a finite *group* of communication capabilities, such as $\emptyset$, $\{a\}$, $\{\widehat{a}\}$, $\{a, b\}$, $\{\widehat{a}, b\}$, or even $\{a, \widehat{a}, b\}$. Then, for instance, an activity $\{\widehat{a}\}$ may be synchronised with $\{a, b\}$ using a conjugate pair $(a, \widehat{a})$. This results in a synchronised activity which is not silent but rather has the communication capability $\{b\}$, as the pair $(a, \widehat{a})$ is internalised but the b remains untouched; formally, $(\{\widehat{a}\} \cup \{a, b\}) \backslash \{a, \widehat{a}\} = \{b\}$.

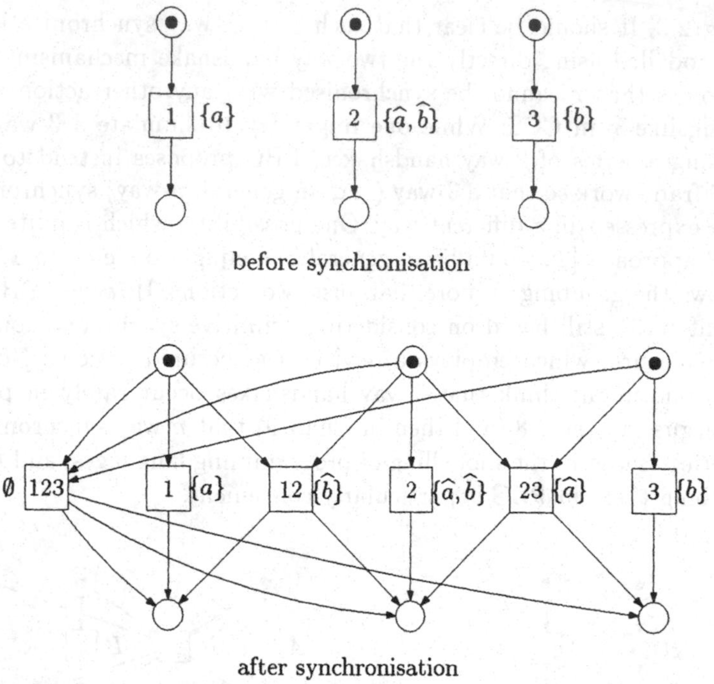

Fig. 2.4. Three-way handshake in terms of PBC

Figure 2.4 shows a Petri net of 3-way PBC handshake with the following possible synchronisations being represented: a denotes 'A shakes hand with B', $\widehat{a}$ denotes 'B shakes hand with A', $\widehat{b}$ denotes 'B shakes hand with C', and b denotes 'C shakes hand with B'. Since B performs $\widehat{a}$ and $\widehat{b}$ simultaneously (which follows from $\{\widehat{a}, \widehat{b}\}$ being the label of a single transition), the resulting transition 123 (in Fig. 2.4) describes a simultaneous handshake between the three people. Moreover, transition 12 describes a 2-way handshake between A and B (it is labelled by $\{\widehat{b}\}$ and so it has a further capability to link with C) and transition 23 describes the handshake only between B and C (it is labelled by $\{\widehat{a}\}$ and so it has a further capability to link with A). Transition 123 can be thought of either as a 3-way synchronisation between 1, 2, and 3, or as a 2-way synchronisation between 1 and 23, or as a 2-way synchronisation between 12 and 3.

There is an asymmetry in the realisation of this 3-way handshake with respect to the way the problem was expressed in terms of A, B, and C, since B does not look exactly like A and C (the action of B is composed of two communication capabilities instead of a single one as is the case for both A and C). Moreover, two extra partial handshakes (12 and 23) are generated if we compare the net with that in Fig. 2.3. This seems to be the price for adopting the idea of building a multiway synchronisation out of primitive binary synchronisations, but we shall later see that PBC – more precisely, its generalisation, the box algebra – may also be equipped with an explicit 'multi-handshake' synchronisation, as is also done in ACP.

In order to implement multiway synchronisation, we still need to discuss whether sets or multisets should be used to describe groups of communication capabilities, and whether the associativity of synchronisation can always be guaranteed. This is also a good place to introduce some basic definitions concerning multisets.

A *multiset* over a given *carrier set* A is a mapping $\mu : A \to \mathbf{N}$, and μ is *finite* if its *support*, i.e., the set $\{a \in A \mid \mu(a) > 0\}$, is finite. The set of finite multisets over A will be denoted by $\mathsf{mult}(A)$.[4] The *cardinality* of $\mu \in \mathsf{mult}(A)$ is defined as the natural number $|\mu| = \sum_{a \in A} \mu(a)$. The operations of multiset *sum*, (non-negative) *difference*, and *multiplication* by a natural number are thus defined, where $a \in A$ and $n \in \mathbf{N}$.

Sum:	$(\mu{+}\mu')(a) = \mu(a) + \mu'(a)$.
Difference:	$(\mu{-}\mu')(a) = \max(\mu(a) - \mu'(a), 0)$.
Multiplication:	$(n{\cdot}\mu)(a) \quad = n \cdot \mu(a)$.

Note that if μ and μ' are finite, then so are $\mu{+}\mu'$, $\mu{-}\mu'$, and $n{\cdot}\mu$. An element a of the carrier set *belongs* to a multiset μ if $\mu(a) > 0$, i.e., if it belongs to the support of μ; we will denote this by $a \in \mu$. A multiset μ is *included* in a multiset μ' if $\mu(a) \leq \mu'(a)$, for all $a \in A$; we will denote this by $\mu \subseteq \mu'$ (notice that if μ' is finite, so is μ). In the examples we will enumerate finite multisets in the usual way; for instance, $\{a, a, b\}$ will denote a multiset such that $\mu(a) = 2$, $\mu(b) = 1$, and $\mu(c) = 0$, for all $c \in A \setminus \{a, b\}$. Finally, we will often call a multiset μ such that $\mu(a) \leq 1$, for all $a \in A$, simply a *set*.

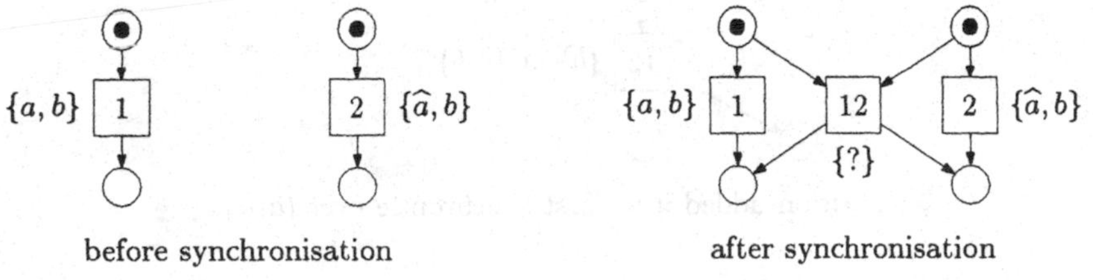

Fig. 2.5. A further example

Returning to multiway synchronisation, let us consider the net on the left of Fig. 2.5. According to the PBC approach, the two transitions 1 and 2 can be synchronised using the conjugate pair $(a, \widehat{a})$. However, what synchronisation

[4] Notice that a marking may be viewed as a multiset over S, but its support may be infinite if the net has infinitely many places (and for reasons that will be explained later, we do not want to exclude infinite nets), so that this notation will not be used for markings.

capability should the resulting transition 12 still possess? In the context of what has already been said, there are only two meaningful answers: either the set $\{b\}$, or the multiset $\{b, b\}$. The PBC model chooses the second alternative, and the next example explains why.

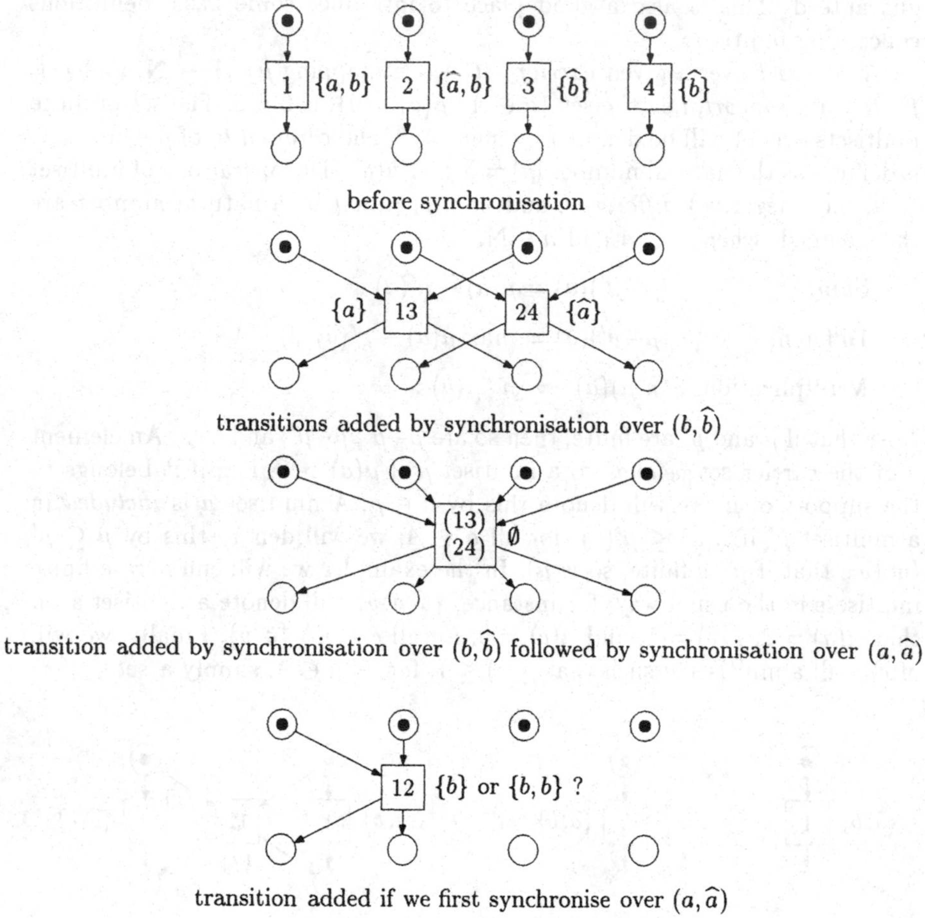

Fig. 2.6. A four-way synchronisation

According to the rules of the game, in Fig. 2.6, transition 1 can be synchronised with transition 3 using $(b, \widehat{b})$, yielding transition 13 with (communication) capability $\{a\}$. Similarly, 2 can be synchronised with 4, yielding transition 24 with capability $\{\widehat{a}\}$. The two new transitions can then be synchronised using $(a, \widehat{a})$, yielding transition (13)(24) with capability $\emptyset$, that is, a silent transition. Note that no multisets were involved nor needed thus far. Now suppose that we first use $(a, \widehat{a})$ in order to synchronise transitions. Then transitions 1 and 2 can be synchronised, yielding transition 12. If this transition has only the communication capability $\{b\}$, then no further synchronisation involving $(b, \widehat{b})$-pairs will yield the four-way synchronisation obtained before. If, however, transition 12 has the communication capability $\{b, b\}$, then one of its b's can be used to synchronise with transition 3, the other to synchronise with transition 4, and a 4-way silent synchronisation can be obtained, as ((12)3)4 or ((12)4)3, where the bracketing indicates (as it did previously) the 2-way CCS-like synchronisations. Since it is highly desirable that the order of synchronisations be irrelevant, this explains why it is preferable to employ multisets of communication capabilities instead of sets.

Hence – and this is formally a gist of the above discussion – PBC is based on a set $\mathsf{A}_{\mathrm{PBC}}$ of primitive actions or *action particles*, ranged over by $a, b, \ldots$, as shown in the above examples. Moreover, it is assumed that there is a conjugation function $\widehat{\ }: \mathsf{A}_{\mathrm{PBC}} \to \mathsf{A}_{\mathrm{PBC}}$ with the properties of *involution*, $\widehat{\widehat{a}} = a$, and *discrimination*, $\widehat{a} \neq a$, for all action particles a. This is the same basic setup as in CCS, except that the conjugation is denoted by the 'hat' $\widehat{\ }$ instead of the 'overbar' $\overline{\ }$, for reasons already explained. However, unlike CCS, PBC allows as a label of a transition any finite multiset over $\mathsf{A}_{\mathrm{PBC}}$. That is, the set of *labels* (or, for more process algebraically inclined readers, *elementary expressions*, or, to emphasise the fact that more than one primitive action may be combined in a single transition, the set of *multiactions*) is defined as $\mathsf{Lab}_{\mathrm{PBC}} = \mathsf{mult}(\mathsf{A}_{\mathrm{PBC}})$.

This specialises to the basic CCS framework as follows. The CCS expressions a and $\widehat{a}$ correspond to the PBC expressions $\{a\}$ and $\{\widehat{a}\}$, respectively, and the CCS expression τ corresponds to the PBC expression $\emptyset$. No PBC multiaction α with $|\alpha| > 1$ has a direct CCS equivalent. Note that $|\alpha|$ is well defined, since, according to our convention of considering only finite multisets (unless stated otherwise), a multiaction is a finite multiset of action particles.

Exercise 2.5.1. Guess how the expression $((\{a\} \,\square\, \{b\})\,;\, \{c\})$, written in the full PBC syntax, could be 'implemented' if only prefixing was available, i.e., if $E; F$ was allowed only under the assumption that E is a multiaction.

2.6 Synchronisation and Parallel Composition

Some of the CCS operators (nil, $+$, $a.$, and $\tau.$) essentially describe how the actions of the various components of an expression are organised in time with respect to each other, without looking at the shape of these components, i.e., they are only concerned with control flow. In CCS terminology, they are often called 'dynamic connectives', but we shall call them 'control flow operators' to avoid confusion with dynamic PBC expressions. Two other operators ($[f]$ and $\backslash a$) take into account the identity of actions performed by the argument (modifying or selectively removing them); they are usually called 'static connectives', but we will call them 'communication interface operators' to avoid confusion with static PBC expressions. The CCS composition operator $\mid$ is special because it both performs a parallel composition of its arguments (a control flow aspect) and adds synchronisations of conjugate pairs of actions (a communication aspect). While this is not harmful, it is valid to ask whether it would perhaps be semantically simpler to separate the control flow and the communication interface aspects of the composition operator.

Another observation is that CCS composition performs synchronisation for all the conjugate pairs. For instance, the CCS expression

$$a.(a.(b.\mathsf{nil})) \mid \widehat{a}.(\widehat{a}.(\widehat{b}.\mathsf{nil}))$$

creates five synchronisations (three of them being actually executable), but the syntax of CCS does not allow one to express synchronisation using only the pair $(b, \widehat{b})$ but not the pair $(a, \widehat{a})$. For reasons that will become clear in Sect. 2.8, and later in Chap. 9, in PBC it is desirable to be able to say precisely which action names are being used for synchronisation. An obvious refinement of the CCS composition operator would be to allow selective synchronisation, as in TCSP [59]. For instance, an operator $\mid_a$ could specify that only $(a, \widehat{a})$ pairs may be used for synchronisation. However, in this case it is difficult to obtain some useful algebraic laws such as associativity of parallel composition. For instance, the expression $(\{b\} \mid_a \{a\}) \mid_b \{\widehat{b}\}$ specifies a synchronisation between the first and the third action, but $\{b\} \mid_a (\{a\} \mid_b \{\widehat{b}\})$ specifies no synchronisation at all. While it is of course possible to live with a restricted form of associativity, PBC proposes a different approach in order to ensure that it holds. The idea is to disentangle synchronisation from parallel composition, regarding the former as a unary rather than binary operation. What is gained by adopting this point of view is a reduction of the complexity of operators and, in particular, a very simple parallel composition operator. Moreover, as we shall see in Chap. 3, one can easily modify synchronisation without having to involve parallel composition as well. The nonassociativity in the last example then arises from the nondistributivity of synchronisation over parallel composition, and not from the nonassociativity of the latter.

The price we shall pay is that we may need to accept some 'unexpected' (or 'unwanted') synchronisations. Consider the PBC expression $\{a\}; \{\widehat{a}\}$ which denotes a transition with communication capability $\{a\}$ followed by a transition with communication capability $\{\widehat{a}\}$, and a corresponding Petri net shown on the left-hand side of Fig. 2.7. Let us then consider a synchronisation expression, E sy a, meaning that E is augmented by all possible (combinations of) binary synchronisations involving pairs $(a, \widehat{a})$ of conjugate basic actions, even if they do not arise from parallel components. The Petri net of $(\{a\}; \{\widehat{a}\})$ sy a would then look like the net shown in the right-hand side of Fig. 2.7. It contains a silent synchronisation transition which is dead from the initial marking, and thus has no impact on the behaviour. Such a transition would not be present in the CCS context where synchronisation is tied to parallel composition. However, Petri nets and our approach open the doors to various interesting generalisations, like that allowing one to start from non-safe markings (even though we shall generally refrain from doing so in this book); then such a dead transition may suddenly become executable, e.g., if we start from the marking doubling the one indicated in Fig. 2.7.

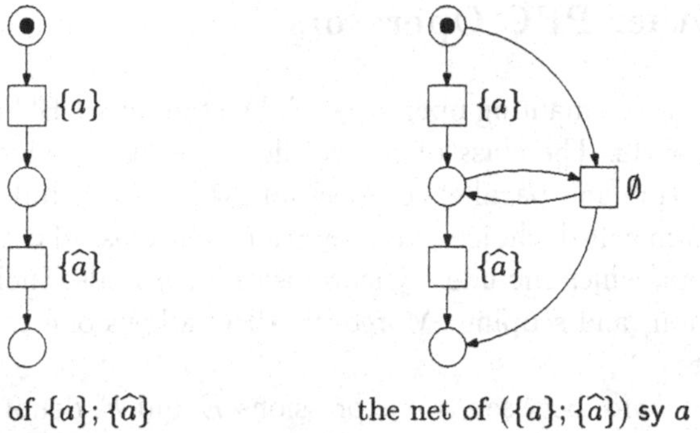

Fig. 2.7. 'Unexpected' synchronisation

Treating synchronisation as a unary operator separated from parallel composition makes it similar to the restriction operator \ of CCS, but playing an opposite role; while synchronisation adds, restriction removes certain transitions. Algebraically, however, the two operators share some interesting properties, as we will see in Chap. 3.

In the Petri net semantics of PBC, the distinction between control flow and communication interface operators manifests itself very clearly; the former (sequence, choice, parallel composition, and iteration) act upon places, whereas the latter (relabelling, synchronisation, restriction, and scoping) act upon transitions. Later, we will use *generalised relabellings* in order to describe communication interface operators, and *transition refinement* as a means of describing control flow operators; in fact, general relabellings and transition refinement will be combined together into a single meta-operator called *net refinement*, which will be introduced in Sect. 4.3.3.

As a last argument in favour of the separation of parallel composition and synchronisation, we observe that parallel composition is quite naturally of a binary nature, whereas synchronisation may be needed in a more general context, like that encountered in the 3-way handshake example, and a unary operator provides much welcomed flexibility here.

Summarising the above discussion, in PBC, we shall use a binary parallel composition construct $E\|F$ which offers simultaneously the behaviours of E and F, and a family of unary synchronisation operators specifying exactly which pairs of conjugate basic actions must be considered for synchronisation, E sy a. Since our actions are multisets of action particles, some care is needed to define exactly which synchronisations are allowed; this will be done, after further discussion, in Sect. 3.2.7.

Exercise 2.6.1. Guess possible moves of $\overline{\left((\{a\}\|\{\widehat{a}\});\{b\} \right)}$ sy a.

2.7 Other PBC Operators

Most of the remaining operators of the standard PBC are akin to their CCS counterparts. The class of control flow operators, which includes sequential composition and parallel composition, will also include alternative composition (often called 'choice') and iteration. The class of communication interface operators, which includes synchronisation, will also include basic relabelling, restriction, and scoping. Moreover, PBC allows one to specify recursive behaviours.

The *choice* between two expressions E and F, denoted by $E \,\square\, F$, specifies behaviour which is, in essence, the union of the behaviours of E and F. That is, any behaviour which is a behaviour of E or F is an acceptable behaviour of $E \,\square\, F$, and no other is. The choice operator, as its counterpart in CCS, allows one to choose between two possible sets of behaviours. The only difference with the corresponding CCS operator, $+$, will be syntactic, the PBC notation following the syntactic convention of Dijkstra's guarded commands [34].

The *basic iteration* operator of PBC obeys the syntax $[E * F * E']$, where E is an initialisation (which may be executed once, at the beginning), F is a body (whose execution may be repeated arbitrarily many times, after initialisation), and E' is a termination (which may be executed at most once, at the end). Later, we will also consider various generalisations of the basic iteration construct.

Recursion employs variables whose behaviour is defined via equations. For example, $X \stackrel{\mathrm{df}}{=} \{a\};X$ specifies the behaviour of X to be an indefinite succession of the executions of $\{a\}$. PBC recursion is considerably more general than in other process algebras, due to the absence of any constraints (elsewhere, for example, recursion is often assumed to be guarded).

Basic relabelling is defined with respect to a function f which consistently renames action particles; $E[f]$ has all the behaviours of E, except that their constituents are relabelled according to f. This operation, as we will show in due course, is the simplest version of a very general mechanism which we will call 'relabelling', and of which all the communication interface operations mentioned so far are special cases. Basic relabelling is theoretically relevant mainly with respect to recursion where it adds the possibility of generating infinitely many action particles from a finite specification. For example, consider the equation $X \stackrel{\mathrm{df}}{=} \{0\};X[\mathit{add1}]$ where *add1* is a relabelling function adding one to each action particle $k \in \mathbf{N}$. The behaviour generated by X is then the infinite sequence $\{0\}\{1\}\{2\}\ldots$

Restriction is defined with respect to an action particle a. For example, E rs a, has all the behaviours of E except those that involve a or $\widehat{a}$. Using restriction, we may give a first example of an expression whose terminal state is not reachable from the initial one. Consider $F = (\{a\}\ \mathsf{rs}\ a)$. Then $\underline{F}$ is not reachable from $\overline{F}$ because, by the definition of restriction, the latter can make no move (i.e., has no nonempty behaviour) at all. However, $\underline{\{a\}\ \mathsf{rs}\ a}$ is a valid dynamic expression.

Scoping is a derived operator which consists of synchronisation followed by restriction (on the same action particle); its importance is in being useful in the modelling of blocks in programming languages, as outlined in the next section, and described more fully in Chap. 9.

The theory we shall develop is very flexible; as a consequence, it will be fairly easy to add new operators to the algebra, but not without some care, as explained in the rest of this book.

Exercise 2.7.1. Guess possible moves of the PBC expressions

$$\overline{\Big(\big((\{a\}\|\{\widehat{a}\});\{b\}\big)\ \mathsf{sy}\ a\Big)}\ \mathsf{rs}\ a$$

$$\overline{\big((\{a\}\|\{\widehat{a}\});\{b\}\big)}\ \mathsf{rs}\ a$$

$$\overline{\Big(\big((\{a\}\|\{\widehat{a}\});\{b\}\big)[f]\Big)}\ \mathsf{rs}\ a\ ,$$

where f maps b to a, $\widehat{b}$ to $\widehat{a}$, and is the identity otherwise.

2.8 Modelling a Concurrent Programming Language

Recall that one of the main motivations behind the design of PBC was to provide a flexible framework for defining semantics of concurrent programming languages. This section considers such an issue informally, but we will return to the subject and examine it thoroughly in Chap. 9. Take, for example, the following fragment of a concurrent program:

...

begin **var** $x : \{0,1\}$;

$[\![\, x{:=}x \oplus 1 \,]\!] \ \|\ [\![\, x{:=}y \,]\!]$

end

...

where y is assumed to be declared by **var** $y : \{0,1\}$ in some outer block, $\oplus$ denotes the addition modulo 2, $\|$ denotes 'shared variable parallelism' (as variable x occurs on both sides of $\|$), and $[\![\ldots]\!]$ delineates an atomic action (i.e., an action whose execution can be considered as instantaneous).

Let us discuss how to construct an appropriate, and as small as possible, Petri net modelling this block. Moreover, such a net should be constructed compositionally from nets derived for its three constituents (i.e., the declaration **var** $x : \{0,1\}$, and the two atomic assignments, $[\![\, x{:=}x \oplus 1 \,]\!]$ and $[\![\, x{:=}y \,]\!]$), and itself be composable (i.e., usable in further compositions with similar nets, e.g., derived for the outer blocks). Using PBC and its Petri net semantics, such a problem can be solved in the following way.

To model assignments and program variables we will use a special kind of action particles in the form of indexed terms x_{vw} and $\widehat{x_{vw}}$, where $v, w \in \{0,1\}$. Each such term denotes a change of the value of the program variable x from v to w, or a test of the value of x, if $v = w$. Then we can translate the two assignments into two PBC expressions:

$$\begin{array}{ll} [\![x{:=}x{\oplus}1]\!] & \rightsquigarrow \{\widehat{x_{01}}\} \,\square\, \{\widehat{x_{10}}\} \\ [\![x{:=}y]\!] & \rightsquigarrow \{\widehat{x_{00}}, \widehat{y_{00}}\} \,\square\, \{\widehat{x_{10}}, \widehat{y_{00}}\} \,\square\, \{\widehat{x_{01}}, \widehat{y_{11}}\} \,\square\, \{\widehat{x_{11}}, \widehat{y_{11}}\} \,. \end{array} \tag{2.2}$$

The first expression means that 'either x could be 0, and then it is changed into 1, or x could be 1, and then it is changed into 0', which is a sound semantics of $[\![x{:=}x{\oplus}1]\!]$ for a variable of type $\{0,1\}$. Unlike the multiactions in the first line, the two-element multiactions in the second PBC expression are not singletons because $[\![x{:=}y]\!]$ involves two variables, x and y. Each multiaction in the second line should be interpreted as denoting two simultaneously executable accesses to these variables, one checking the value of y, the other updating the value of x. For instance, the multiaction $\{\widehat{x_{10}}, \widehat{y_{00}}\}$ denotes the value of y being checked to be 0 and, simultaneously, the value of x being changed from 1 to 0. The expressions for $[\![x{:=}x\oplus 1]\!]$ and $[\![x{:=}y]\!]$ given in (2.2) have corresponding Petri nets with respectively two and four transitions, as shown in Fig. 2.8.

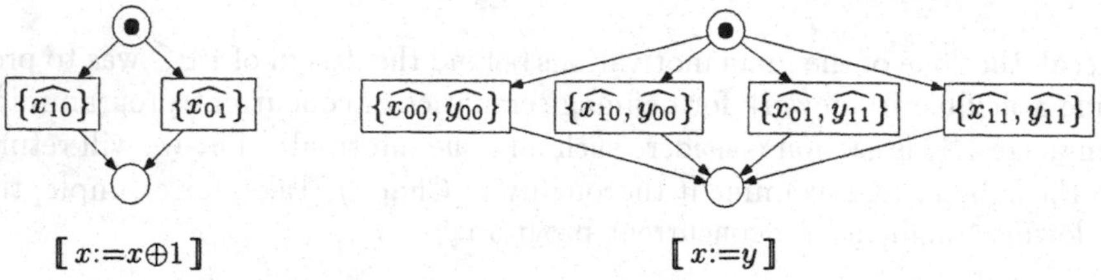

Fig. 2.8. Petri nets representing $[\![x{:=}x{\oplus}1]\!]$ and $[\![x{:=}y]\!]$

Let us now turn to a Petri net corresponding to the declaration of x which will use action particles x_{vw}, where $v, w \in \{0,1\}$. Since each x_{vw} is the conjugate of $\widehat{x_{vw}}$, and so the conjugate of an action particle appearing in (2.2), the net of a variable can be put in parallel with the nets of atomic assignments, and then the nets can be synchronised in order to describe an entire block.

The net describing **var** $x : \{0,1\}$ is shown – in a simplified form – in Fig. 2.9. The simplification is due to the fact that additional transitions are necessary to model the initialisation and destruction of the variable in its block (both features are of no relevance in the present discussion and will be added later, in Sects. 3.3.2 and 4.4.10, and in Chap. 9). Here it is arbitrarily assumed that the current value of x is 1, and thus the net contains a token in the corresponding place.

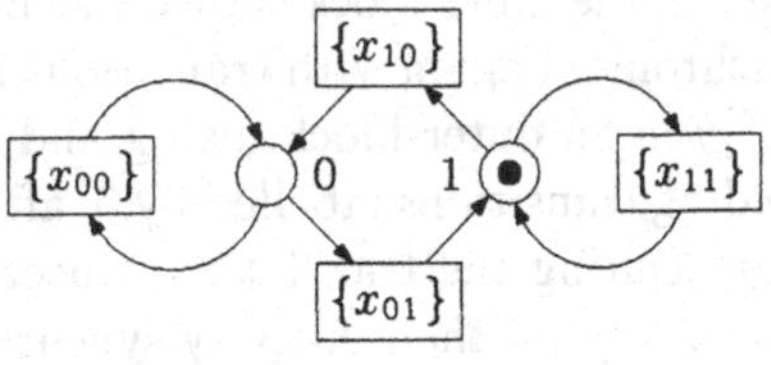

Fig. 2.9. (Part of the) Petri net representing a binary variable x

In order to describe the block structure of the program fragment (with x being a local and y a global variable of the block), the net of **var** $x : \{0,1\}$ needs to be synchronised with the nets of $[\![x{:=}x \oplus 1]\!]$ and $[\![x{:=}y]\!]$. For example, the transition labelled $\{x_{10}\}$ coming from the declaration of x is synchronised with the transition labelled $\{\widehat{x_{10}}, \widehat{y_{00}}\}$ coming from the action $[\![x{:=}y]\!]$ using and consuming the conjugate pair $(x_{10}, \widehat{x_{10}})$ but retaining $\widehat{y_{00}}$ (in fact, as we have already seen, this is built into the definition of synchronisation). The resulting transition has the label $\{\widehat{y_{00}}\}$. After synchronisation, all the transitions with labels containing the x_{uv} and $\widehat{x_{uv}}$ action particles are removed by applying the PBC restriction operator, since the variable x is not known outside its declaring block. The net obtained through synchronisation and restriction (i.e., scoping), is shown in Fig. 2.10.

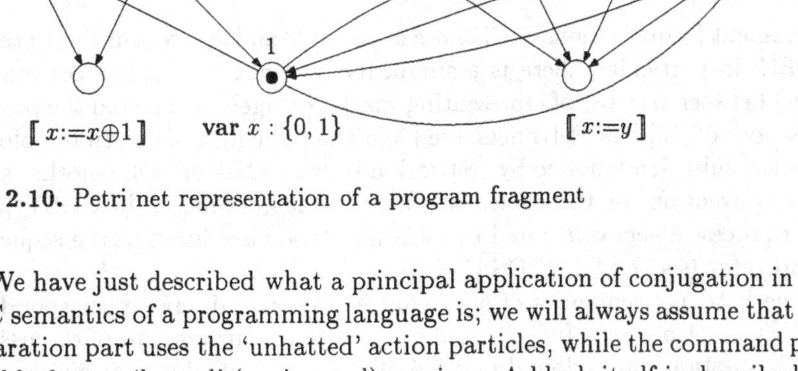

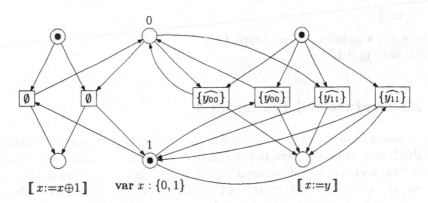

Fig. 2.10. Petri net representation of a program fragment

We have just described what a principal application of conjugation in the PBC semantics of a programming language is; we will always assume that the declaration part uses the 'unhatted' action particles, while the command part of a block uses 'hatted' (conjugated) versions. A block itself is described by putting the command and declaration parts in parallel and then synchronising and restricting (hence scoping) over all variables that are local to it. This will leave only nonlocal accesses to be still visible in the communication interface, i.e., in the transition labels.

Consider now the net of the inner block shown in Fig. 2.10. Transitions labelled $\{\widehat{y_{vv}}\}$ can be synchronised again with transitions labelled $\{y_{vv}\}$ coming from the declaration of y in an outer block, using and consuming conjugate pairs $(y_{vv}, \widehat{y_{vv}})$ and yielding transitions labelled by $\emptyset$, after what is, in essence, a 3-way synchronisation. During the translation process, we could therefore witness an effective application of the multiway synchronisation mechanism, in the modelling of an assignment involving two program variables which are declared in different blocks but accessed in the same atomic assignment.

Exercise 2.8.1. Draw a Petri net translation of the assignment

$$[\![z{:=}y+1]\!] \, ,$$

where y is a variable of type $\{0,1\}$ and z is a variable of type $\{0,1,2\}$.
Hint: See Fig. 2.8.

Exercise 2.8.2. Draw a Petri net translation of the declaration

var $z : \{0,1,2\}$.

Hint: See Fig. 2.9.

Exercise 2.8.3. Draw a Petri net translation of the block

begin **var** $z : \{0,1,2\}$;
$[\![z{:=}y+1]\!]$
end ,

where y is a global variable of type $\{0,1\}$.
Hint: See Fig. 2.10.

2.9 Literature and Background

More recent formalisations of CCS use arguments and devices similar to those of PBC; in particular, there is a similarity (but not a complete correspondence) between our way of representing markings algebraically and the proved transitions of [13]. For Petri nets, see [90]. The basic idea of describing blocks by synchronisation followed by restriction is due to Milner [80]. Another syntactic convention for the choice operator is the comma, as in COSY [62]. Other process algebras inspired by CCS have also introduced a true sequencing operator (see, e.g., LOTOS [75]).

The Petri net semantics of nil ('what is the net – if any – corresponding to nil'?) is not obvious. Indeed, in the literature, a fair amount of discussion on this question can be found (see [44, 95]). Since PBC does not employ nil, we avoid the problems associated with giving it a Petri net semantics.

3. Syntax and Operational Semantics

In this chapter we introduce the syntax and operational semantics of the Petri Box Calculus. We discuss the basic version of PBC, i.e., the part which directly originates from CCS and was informally introduced in the previous chapter, as well as possible extensions. We also present two notions which identify behaviourally equivalent PBC expressions and later, equivalent Petri nets (in Chaps. 4 and 5), and equivalent systems (in Chap. 7). A number of straightforward, yet often important, results are left as an exercise to the reader (but see also the Appendix).

3.1 Standard PBC Syntax

Static expressions. A *(standard) static PBC expression* is a word generated by the syntax

$$\begin{array}{rl|l|l|l|l|} E ::= & \alpha & E\|E & E \,\square\, E & E;E & [E*E*E] \\ & X & E \;\mathsf{sy}\; a & E[f] & E \;\mathsf{rs}\; a & [a:E] \end{array} \quad (3.1)$$

possibly with parentheses, used – if needed – to resolve ambiguities (we shall not assume any precedence order among the operators, nor associativity). In the syntax, α – a constant called *basic action* or *multiaction* – is an element of $\mathsf{Lab}_{\mathsf{PBC}} = \mathsf{mult}(\mathsf{A}_{\mathsf{PBC}})$ (cf. Sect. 2.5 where $\mathsf{A}_{\mathsf{PBC}}$ and $\mathsf{Lab}_{\mathsf{PBC}}$ were first used); a is an element of $\mathsf{A}_{\mathsf{PBC}}$; X belongs to a set $\mathcal{X}$ of predefined *hierarchical* (or recursion) variables – not to be confused with program variables – ranged over by $X, Y, Z, \ldots$, each variable X having associated a unique *defining equation* $X \stackrel{\mathrm{df}}{=} E$; and f is a *relabelling function* from $\mathsf{A}_{\mathsf{PBC}}$ to $\mathsf{A}_{\mathsf{PBC}}$. We shall sometimes require, as in the CCS framework, that such a function preserves conjugates, (i.e., $f(\widehat{a}) = \widehat{f(a)}$, for every action particle $a \in \mathsf{A}_{\mathsf{PBC}}$). Notice that such an f commutes with $\widehat{\ }$ (i.e., $f \circ \widehat{\ } = \widehat{\ } \circ f$), and that $\widehat{\ }$ itself is a relabelling function preserving conjugates.

PBC operators can be split into two groups. The three binary operators, i.e., $\|$ (disjoint parallelism), $\square$ (choice), and ';' (sequence), together with the ternary operator $[\;*\;*\;]$ (iteration with explicit initialisation and termination) are *control flow operators.* The four unary (classes of) operators, i.e., $[f]$ (basic relabelling), sy a (synchronisation), rs a (restriction), and $[a:]$ (scoping) are *communication interface operators*. The *finite* PBC is obtained by excluding iteration and hierarchical variables. The following expressions of finite PBC will serve as running examples:

$$\begin{array}{llll} \text{Example 0:} & E_0 = \emptyset & \text{Example 5:} & E_5 = (a\|b)\,\square\,c \\ \text{Example 1:} & E_1 = \{\widehat{a}, b\} & \text{Example 6:} & E_6 = ((a;b)\,\square\,c)\ \mathsf{rs}\ c \\ \text{Example 2:} & E_2 = (a;b)\,\square\,c & \text{Example 7:} & E_7 = (a\|b)\ \mathsf{rs}\ b \\ \text{Example 3:} & E_3 = a;(b\,\square\,c) & \text{Example 8:} & E_8 = (a\|\widehat{a})\ \mathsf{sy}\ a \\ \text{Example 4:} & E_4 = a\|b & \text{Example 9:} & E_9 = ((a\|\{\widehat{a},\widehat{a}\})\|a)\ \mathsf{sy}\ a\,. \end{array} \tag{3.2}$$

To avoid excessive bracketing, we will often replace, in example expressions (as those above) and the corresponding Petri nets, a singleton multiaction $\{a\} \in \mathsf{Lab}_{\mathrm{PBC}}$ by its only action particle, a. This explains why, in Sect. 2.3, $E = a;b$ was called a 'simplified' expression; according to the convention just adopted, it simplifies the expression $E' = \{a\};\{b\}$.

Dynamic Expressions. A *(standard) dynamic PBC expression* is a word generated by the syntax:

$$\begin{array}{rl} G ::= & \overline{E} \mid \underline{E} \mid G\,\square\,E \mid E\,\square\,G \mid G\ \mathsf{sy}\ a \mid \\ & G[f] \mid [a:G] \mid G;E \mid E;G \mid G\ \mathsf{rs}\ a \mid \\ & G\|G \mid [G*E*E] \mid [E*G*E] \mid [E*E*G] \end{array} \tag{3.3}$$

where E stands for a static expression defined by the syntax (3.1), $a \in \mathsf{A}_{\mathrm{PBC}}$, and $f : \mathsf{A}_{\mathrm{PBC}} \to \mathsf{A}_{\mathrm{PBC}}$ is a relabelling function. As in Chap. 2, dynamic expressions will be ranged over by $G, H, \ldots$, and static expressions by $E, F, \ldots$.

The first two clauses of (3.3) concern a static expression E in its entry (or initial) state, $\overline{E}$, and exit (or terminal, or final) state, $\underline{E}$. Both are valid dynamic expressions, irrespective of whether or not the final state is reachable from the initial state. Note that an expression such as $G;H$ is *not* a dynamic expression. The reason is that sequential composition does not allow both its components to be active at the same time. At this point, the assumption about markings being safe (first pronounced in Sect. 2.6, and made more explicit later on in this book) is already built into the theory. And the shape of dynamic expressions, but not the shape of static expressions, depends on that assumption. The syntax for sequence is therefore split into two clauses. The first clause, $G;E$, means that the first component of a sequential composition is currently active (which includes its entry and exit states), while the second component is currently dormant. The second clause, $E;G$, means that the second component is active and the first is dormant. Similar remarks hold for $\square$ and $[**]$; we require syntactically that only one of the parts of a choice or iteration expression is ever active. Note, however, that parallel composition, $\|$, is allowed – even required – to have *both* of its components active.

$\overline{E}$ is a dynamic expression associated in a canonical way with a static expression E, in the sense that $\overline{E}$ describes the initial state of E. Henceforth, whenever we talk about 'the behaviour of E', we mean the behaviour that is generated by $\overline{E}$, the initial state of E. We will now characterise the behaviour of the example expressions, E_0–E_9, through the behaviour of their initial dynamic counterparts, $\overline{E_0}$–$\overline{E_9}$.

The expression E_0 (more precisely, the dynamic expression $\overline{E_0}$) can do a 'silent' $\emptyset$-move and terminate. E_1 can do a non-silent $\{\widehat{a}, b\}$-move; this move is interpreted as performing synchronously (i.e., inseparably and at the same time) two visible action particles, $\widehat{a}$ and b. E_2 can either do an a-move followed by a b-move and terminate, or a c-move and terminate. E_3 can do an a-move, followed either by a b-move or by a c-move and terminate. E_4 can make a single a-move followed by a b-move and terminate, or a single b-move followed by an a-move and terminate, but also a move consisting of a *concurrent execution* of a and b. E_5 behaves either as expression E_4, or it can make a single c-move and terminate. E_6 can make an a-move followed by a b-move and terminate, but it cannot make a c-move. E_7 can make an a-move but it cannot make a b-move, nor terminate. E_8 can make all the moves of $a\|\widehat{a}$ and, in addition, a silent synchronisation and terminate. The last expression, E_9, can make the same moves as the expression $(a\|\{\widehat{a}, \widehat{a}\})\|a$ and, in addition, three synchronisations: an $\widehat{a}$-move synchronising the left and the middle components of the parallel composition; an $\widehat{a}$-move synchronising the middle and the right-hand side components; and a silent move synchronising all three components.

For a dynamic expression G, $\lfloor G \rfloor$ will denote the underlying static expression, obtained by deleting all the overbars and underbars. Formally, we have $\lfloor \overline{E} \rfloor = \lfloor \underline{E} \rfloor = E$ as well as

$$
\begin{array}{lll}
\lfloor G;E \rfloor = \lfloor G \rfloor ;E & \lfloor G[f] \rfloor = \lfloor G \rfloor [f] & \lfloor [G * E * F] \rfloor = [\lfloor G \rfloor * E * F] \\
\lfloor E;G \rfloor = E; \lfloor G \rfloor & \lfloor G \text{ sy } a \rfloor = \lfloor G \rfloor \text{ sy } a & \lfloor [E * G * F] \rfloor = [E * \lfloor G \rfloor * F] \\
\lfloor G \| H \rfloor = \lfloor G \rfloor \| \lfloor H \rfloor & \lfloor G \text{ rs } a \rfloor = \lfloor G \rfloor \text{ rs } a & \lfloor [E * F * G] \rfloor = [E * F * \lfloor G \rfloor] \\
\lfloor G \,\square\, E \rfloor = \lfloor G \rfloor \,\square\, E & \lfloor [a : G] \rfloor = [a : \lfloor G \rfloor] & \\
\lfloor E \,\square\, G \rfloor = E \,\square\, \lfloor G \rfloor \,. & &
\end{array}
$$

Recursion. PBC employs hierarchical variables which can appear on the left-hand sides as well as on the right-hand sides of defining (recursive) equations, at any position where a multiaction would be allowed. For instance, $X \stackrel{\text{df}}{=} a;X$ is a defining equation for a variable X in which X occurs on the right-hand side, i.e., it is a true recursion; such an X denotes a process performing indefinitely a succession of a-moves, without ever terminating. Recursion may result from more complex definitions, such as

$$X \stackrel{\text{df}}{=} Y \| Z$$

$$Y \stackrel{\text{df}}{=} a; Z$$

$Z \stackrel{\mathrm{df}}{=} b \,\square\, (Y; Z)$.

The above is a *(recursive) system of equations* with one defining equation for each of X, Y, and Z; recall that there is exactly one defining equation per variable.

Not all defining equations necessarily represent true recursion; for instance, in the expression $a; X; (b \,\square\, X)$, with X defined by $X \stackrel{\mathrm{df}}{=} c \| d$, the variable achieves a hierarchical (and more compact) presentation, but does not lead to recursion. In general, a hierarchical variable X is *recursion free* if there is no infinite sequence of (not necessarily distinct) variables $X_0, X_1, X_2, \ldots$ of $\mathcal{X}$ such that $X = X_0$ and, for every $i \geq 0$, X_{i+1} appears on the right-hand side of the equation defining X_i. In this book, we will deal with recursion in its most general form, i.e., we do not introduce any syntactic restrictions on the placement of variables in defining expressions; they may occur at the beginning, at the end, in the middle, even everywhere at once, e.g., as in $X \stackrel{\mathrm{df}}{=} (X \,\square\, X); (X; X)$.

An expression involving hierarchical variables has no meaning by itself; its behaviour depends on the defining equations of the variables it employs. From a strictly formal point of view, these equations should be given beforehand, i.e., before expressions are constructed, to form a context in which the theory is developed. In practice, however, some of the equations may be introduced later, when expressions describing a system are being designed.

Although it is possible to specify recursion using operators with *internalised* context, such as $\mu_{[X_i = E_i | i \in I]} E$, where I is an indexing set, this technique will not be employed here.[1] In any case, it is always possible to transform expressions defined using internalised contexts into equivalent ones conforming to the notation we have adopted, perhaps after renaming some of the variables, e.g., those which have distinct definitions in different internalised contexts. For instance, $\mu_{[X=a;X]}(X; a)$ may be rewritten as $X; a$ with the defining equation $X \stackrel{\mathrm{df}}{=} a; X$. And, if later we needed a process described by

$$\mu_{[X=a;(b\|X)]} \; (a; (X \,\square\, \mu_{[X=b;X]} X)) \;,$$

this could be rewritten as $a; (Y \,\square\, Z)$ with two defining equations

$$Y \stackrel{\mathrm{df}}{=} a; (b \| Y) \quad \text{and} \quad Z \stackrel{\mathrm{df}}{=} b; Z \;.$$

It is also possible to rewrite defining equations to remove (complex) nested applications of PBC operators (nested applications of operators in defining equations will be disallowed, for technical reasons, in Chaps. 5–8); for instance, $Y \stackrel{\mathrm{df}}{=} a; (b \,\square\, Y)$ may be replaced by two equations,

$$Y \stackrel{\mathrm{df}}{=} a; Z \quad \text{and} \quad Z \stackrel{\mathrm{df}}{=} b \,\square\, Y \;.$$

There is a clear connection between the iteration and recursion constructs. Apart from the fact that they both may generate infinite behaviours, $[E *$

[1] But it was used in the prior work on PBC [6, 7, 31].

$F * F']$ has a corresponding (i.e., one with the same behaviour) expression $E; X$ with X being defined by $X \stackrel{\mathrm{df}}{=} (F; X) \,\square\, F'$. However, as we shall see in Chap. 4, iteration will always have a finite Petri net representation (if only E, F, and F' have finite representations), which is not generally the case for recursion. This and the fact that iteration is used explicitly in the syntax of programming languages motivated its introduction as a separate operator.

A number of examples for recursion are given below:

$$
\begin{array}{ll}
\text{Example 10: } X \stackrel{\mathrm{df}}{=} a; (X; b) & \text{Example 14: } X \stackrel{\mathrm{df}}{=} 0; (X[add2]) \\
\text{Example 11: } X \stackrel{\mathrm{df}}{=} (a; (X; b)) \,\square\, c & \text{Example 15: } X \stackrel{\mathrm{df}}{=} (0\|1) \,\square\, (X[add2]) \\
\text{Example 12: } X \stackrel{\mathrm{df}}{=} a\|X & \text{Example 16: } X \stackrel{\mathrm{df}}{=} X \\
\text{Example 13: } X \stackrel{\mathrm{df}}{=} a; X & \text{Example 17: } X \stackrel{\mathrm{df}}{=} a \,\square\, (X; a)\ .
\end{array}
\tag{3.4}
$$

Example 10 defines a process X which can do any number of a-moves (including infinitely many), but never a b-move; it never terminates either. In example 11, X can do whatever it could do in example 10 and, in addition, it can make at any point a single c-move followed by as many b-moves as there have been a-moves before the c-move; its execution terminates after the last possible b-move (or after c, if there was an immediate c-move initially). Example 12 admits a finite or infinite sequence of finitely many concurrent a-moves; it never terminates. Example 13 gives rise to the same behaviour – but not the same net structure – as example 10; note that example 10 is not directly expressible in CCS, but example 13 is. Example 14 is similar to the previous one, but the presence of a relabelling function *add2* (assumed to add two to each action particle $k \in \mathbb{N}$) allows one to generate a behaviour with infinitely many different action particles, i.e., a sequence of moves corresponding to the series of all even natural numbers. Example 15 describes a choice between one of infinitely many concurrent moves, each such move consisting of a pair of consecutive natural numbers, $2{\cdot}n$ and $2{\cdot}n + 1$. After any of these numbers has been executed, only the paired (next odd or previous even) one can be taken. The intuitive meaning of the remaining two examples, 16 and 17, may not be completely clear. The formal semantics developed in this chapter will interpret them as follows: in example 16, X can make no move at all while, in example 17, it can perform any finite succession of a-moves and then terminate, but not an infinite succession of a-moves.

3.2 Structured Operational Semantics

Structured operational semantics (SOS, [89]) is a well-established approach to defining the set of possible moves of process expressions or, more technically speaking, an operational semantics of a process algebra. An SOS consists of a set of axioms and derivation rules from which evolutionary behaviours of (dynamic) expressions can be derived.[2] The adjective 'operational' refers to the fact that the objects being defined are behavioural ones (rather than, e.g., logical ones, in which case one would rather say 'axiomatic'); and 'structured' refers to the fact that a separate set of rules is given for each operator of the syntax, relating the behaviour of an expression to that of its subexpressions (i.e., the definition is based on an inductive scheme).

Let *expr* and *expr'* be two dynamic expressions of a process algebra (recall from Sects. 2.3 and 3.1 that in CCS all expressions are dynamic objects, while in PBC they have a separate syntax). A possible evolution leading from *expr* to *expr'* can be given in the form of a triple

$$expr \xrightarrow{act} expr' \tag{3.5}$$

meaning that in a system state described by *expr*, an action, or a 'move', *act* may be performed,[3] leading to a state described by *expr'* (notice a similarity with the Petri net transition rule in Sect. 2.2). SOS *axioms* specify a set of basic evolutions (3.5) while SOS *derivation rules* take the following form.

$$\frac{evol_1\ ,\ \ldots\ ,\ evol_n}{evol}\ cond$$

Such a rule states that if evolutions $evol_1, \ldots, evol_n$ have already been derived and a boolean condition, *cond*, on the (parameters of) evolutions $evol_i$ is satisfied, then *evol* is a legal evolution. If the condition holds vacuously, i.e., $cond = true$, it is omitted. Usually, and also in this book, the *premise* of the rule, i.e., the $evol_i$'s, are evolutions of subterms of a compound expression whose evolution is described by *evol*. That is, the rules have the form

$$\frac{expr_{i_1} \xrightarrow{act_1} expr'_{i_1}\ ,\ \ldots\ ,\ expr_{i_n} \xrightarrow{act_n} expr'_{i_n}}{op(expr_1, \ldots, expr_m) \xrightarrow{act} op(expr'_1, \ldots, expr'_m)}\ cond \tag{3.6}$$

where *op* is one of the operators of a process algebra, $n \geq 0$, $\{i_1, \ldots, i_n\} \subseteq \{1, \ldots, m\}$, $expr'_i = expr_i$ for $i \notin \{i_1, \ldots, i_n\}$, and *act* is determined by $act_1, \ldots, act_n$ and *cond*. SOS semantics is thus compositional since the derivation rules build the behaviour of a compound expression from the behaviours of its components. Notice that an axiom (basic evolution) may always be viewed as a derivation rule without a premise ($n = 0$) and without a condition ($cond = true$).

For instance, the CCS 'prefixing' axiom $a.E \xrightarrow{a} E$ means that by performing action a, an expression $a.E$ is transformed into expression E, while the CCS derivation 'τ-rule'

[2] With a slight disregard of the abbreviation, we will sometimes use the phrase 'SOS semantics' to mean 'SOS'.

[3] Such a move should not be confused with action particles or multiactions.

$$\frac{E \xrightarrow{a} E' \ , \ F \xrightarrow{\widehat{a}} F'}{E \mid F \xrightarrow{\tau} E' \mid F'} \tag{3.7}$$

means that if two conjugate actions are executable concurrently, then they can always be executed synchronously as a silent action.

3.2.1 The Basic Setup

By specialising the general scheme, we will now define an SOS semantics of dynamic PBC expressions. Any activity associated with a static expression E will then be that of the corresponding *initial* dynamic expression $\overline{E}$. If an activity terminates successfully, then it will lead to the corresponding *terminal* dynamic expression $\underline{E}$.

The SOS axioms and derivation rules defined in this chapter will look much like the CCS prefixing axiom or the derivation τ-rule, except that they will involve expressions decorated by overbars and underbars which indicate currently active (sub)expressions. In Chap. 7, we will see that there is a strong correspondence between the rules we are going to define here, and the transition rule of Petri nets.

Let us now have a closer look at the labels of the evolutions; i.e., the *act* part of (3.5). In CCS, *act* is simply a basic action name, such as a, b, and $\widehat{a}$. This is no longer the case for PBC. To start with, we need to consider multiactions, such as $\emptyset$, $\{a\}$, $\{\widehat{a}\}$, and $\{\widehat{a}, b\}$, rather than individual action particles. Moreover, we have already explained advantages offered, first, by the close relationship of the PBC algebra and Petri nets, and, second, by the separation of concurrent composition and synchronisation. All this points at the possibility of specifying the SOS semantics of PBC in such a way that it explicitly describes *concurrent behaviours*, and we will use *steps* of multiactions — defined as finite multisets of multiactions — to achieve the desired effect. Other devices could be used as well, e.g., labelled partial orders or infinite steps of multiactions. But these would not add that much here, as we shall see in Sects. 6.4 and 7.3.8.

What has been called in this section an 'action', *act*, now becomes an element of the set $\mathsf{mlab} = \mathsf{mult}(\mathsf{Lab}_{\mathsf{PBC}}) = \mathsf{mult}(\mathsf{mult}(\mathsf{A}_{\mathsf{PBC}}))$, ranged over by $\Gamma, \Delta, \ldots$, and will henceforth be referred to as a 'move' or a 'step'.

We will aim at formalising evolutions $G \xrightarrow{\Gamma} H$ such that G is a dynamic PBC expression which can execute a step $\Gamma \in \mathsf{mlab}$ (i.e., execute simultaneously, or concurrently, all the multiactions in Γ) and yield a dynamic expression H. Note that the *empty step*, $\Gamma = \emptyset$, is allowed; it denotes a 'non-action' (or 'inaction'), and thus we will have

$$G \xrightarrow{\emptyset} G \ ,$$

for every expression G. The empty step should not be confused with the singleton step consisting only of the empty multiaction, $\Gamma = \{\emptyset\}$; in general, it will not be the case that

$$G \xrightarrow{\{\emptyset\}} G\,.$$

We shall be interested in a *step sequence* semantics, i.e., in evolutions of the form $G \xrightarrow{\sigma} H$, where $\sigma = \Gamma_1\Gamma_2\ldots\Gamma_n$ – itself called a *step sequence* – is a sequence of steps with $n \geq 0$. This means that there are dynamic expressions $G_0, G_1, \ldots, G_n$ such that $G = G_0$, $H = G_n$, and

$$G_{i-1} \xrightarrow{\Gamma_i} G_i$$

is a valid evolution, for every $1 \leq i \leq n$. The empty step will act as a neutral element in step sequences; for instance, the evolutions

$$G \xrightarrow{\emptyset\Gamma\emptyset\emptyset\Delta\emptyset} H \text{ and } G \xrightarrow{\Gamma\Delta} H$$

will be considered equivalent. Moreover, we shall use a derived relation on dynamic expressions, $\equiv$, defined thus:

$$G \equiv H \quad\Leftrightarrow\quad G \xrightarrow{\emptyset} H\,. \tag{3.8}$$

Dynamic expressions related by $\equiv$ will be called *structurally equivalent* (for technical reasons, in Chap. 7, we will extend $\equiv$ to static expressions; see also Exc. 3.2.51 in this chapter). The following set of *inaction rules* captures the main properties of empty steps, and hence also of the $\equiv$-relation.

(*INACTION*)				
	IN1	$G \xrightarrow{\emptyset} G$	ILN	$\dfrac{G \xrightarrow{\emptyset\Gamma} H}{G \xrightarrow{\Gamma} H}$
	IN2	$\dfrac{G \xrightarrow{\emptyset} H}{H \xrightarrow{\emptyset} G}$	IRN	$\dfrac{G \xrightarrow{\Gamma\emptyset} H}{G \xrightarrow{\Gamma} H}$

IN1 (INaction rule 1) implies that $\equiv$ is reflexive, and IN2 implies that it is symmetric. Each of the other two rules, ILN (Inaction is Left Neutral) and IRN (Inaction is Right Neutral), implies the transitivity of $\equiv$ (it suffices to take $\Gamma = \emptyset$). Thus $\equiv$ is an equivalence relation in the usual sense.[4] For a dynamic expression G, we will denote by $[G]_\equiv$ the equivalence class of $\equiv$ containing G. Moreover, $[G\rangle$ will denote the least set of expressions containing G such that if $H \in [G\rangle$ and $H \xrightarrow{\Gamma} C$, for some $\Gamma \in \mathsf{mlab}$, then $C \in [G\rangle$.

[4] A relation $R \subseteq X \times X$ is an *equivalence relation* if it is reflexive, transitive, and symmetric ($(x, y) \in R \Rightarrow (y, x) \in R$ for all $x, y \in X$). An *equivalence class* of R is a maximal nonempty set $X' \subseteq X$ such that $X' \times X' \subseteq R$.

Exercise 3.2.1. Show that $G \xrightarrow{\varepsilon} G$ where ε denotes the empty step sequence.

We may now further explain why the symbol '$\widehat{\ }$' instead of '$\overline{\ }$' was used for conjugation. Even though consistent bracketing would never lead to an ambiguity between overbarring and hatting, using the same symbol could be confusing. For instance, $\overline{\{a\}}$, representing the initial state of $\{a\}$, may be distinguished from the conjugate multiaction $\{\overline{a}\}$, but the distinction is subtle and it would vanish if we used a simplified notation which omits parentheses in singleton multiactions. Hence it is better to use the notations $\overline{\{a\}}$ and $\{\widehat{a}\}$ since then the simplified expressions, $\overline{a}$ and $\widehat{a}$, are clearly distinct.

3.2.2 Equivalence Notions

The question of whether two concurrent systems can be regarded as behaviourally equivalent will recur in one guise or another throughout the entire book. For instance, to conclude that parallel composition is commutative, we should be able to assert that, in particular, $\overline{a\|b}$ and $\overline{b\|a}$ are equivalent expressions. But what does the phrase 'equivalent expressions' actually mean? As it is often the case, there are several reasonable answers and one needs to look closely at the problem of choosing a sound behavioural equivalence for PBC expressions.

A first candidate could be the relation $\equiv$. Indeed, it is an equivalence relation, and a behavioural one since, due to ILN and IRN, if

$$G' \equiv G \xrightarrow{\sigma} H \equiv H',$$

then $G \xrightarrow{\sigma} H'$, $G' \xrightarrow{\sigma} H$, and $G' \xrightarrow{\sigma} H'$, for every step sequence σ. But $\equiv$ is too demanding and restrictive a relation for our purposes; intuitively, one may think of $\equiv$ as relating dynamic expressions which represent different views of the same system state[5] (cf. the discussion in Sects. 2.3 and 2.4). For instance, $\overline{a\|b} \not\equiv \overline{b\|a}$, since no empty move leading from $\overline{a\|b}$ to $\overline{b\|a}$ can be derived using the rules presented in this chapter (and the Petri nets associated with $\overline{a\|b}$ and $\overline{b\|a}$ in Chap. 4 will be isomorphic but not exactly the same). Yet the equivalence of $\overline{a\|b}$ and $\overline{b\|a}$ is required for parallel composition to be commutative.

Moreover, it would not be right to base the equivalence solely on the equality of step sequences generated by dynamic expressions. Consider, for instance, an expression $\overline{\mathsf{stop}}$, where stop is a hierarchical variable with the defining equation $\mathsf{stop} \stackrel{\mathrm{df}}{=} \mathsf{stop}$. Such an expression is only able to perform a looping empty move, $\overline{\mathsf{stop}} \xrightarrow{\emptyset} \overline{\mathsf{stop}}$, and other similar moves (see Exc. 3.2.45), without ever reaching a terminal state. As a result, $\overline{a;a}$ and $\overline{(a;a)\,\square\,(a;\mathsf{stop})}$ will generate the same sequences of nonempty moves: $\{\{a\}\}$ and $\{\{a\}\}\{\{a\}\}$;

[5] We shall later see that $\equiv$-equivalent expressions always correspond to the same Petri net (cf. Thm. 7.3.3).

but it would not be correct to consider these two expressions equivalent since the latter may be blocked after the first execution of a (if the right branch of the choice is followed), while this may not happen for the former. Hence, it is necessary to take into account the *branching structure* of sequences of moves.

Transition Systems. A well-established way of representing the branching structure of dynamic systems is to use (labelled) transition systems. A *transition system* is a quadruple $\mathsf{ts} = (V, L, A, v_{in})$ consisting of a set V of *states*, a set L of *arc labels*, a set $A \subseteq V \times L \times V$ of *arcs*,[6] and an *initial* state v_{in}. Being essentially model independent, transition systems are often the preferred tool to compare different semantics and system models. In this book, we will associate a transition system both with a PBC expression evolving according to the rules of SOS semantics, and with the Petri net corresponding to that expression evolving according to the transition occurrence rule. It will then be possible to state that these semantics are equivalent, through a suitable equivalence defined at the transition system level.

A possible way to associate a transition system with a dynamic expression G could be to take $\mathsf{ts} = ([G\rangle, \mathsf{mlab}, A, G)$, where

$$A = \{(H, \Gamma, C) \in [G\rangle \times \mathsf{mlab} \times [G\rangle \mid H \xrightarrow{\Gamma} C\}.$$

An alternative would be $\mathsf{ts}' = (\mathcal{G}, \mathsf{mlab}, \mathcal{A}, [G]_{\equiv})$, where

$$\mathcal{G} = \{[H]_{\equiv} \mid H \in [G\rangle\} \text{ and } \mathcal{A} = \{([H]_{\equiv}, \Gamma, [C]_{\equiv}) \mid (H, \Gamma, C) \in A\},$$

since $\equiv$-equivalent expressions have the same behaviours, being essentially two different views of the same system state. However, neither ts nor ts' would capture all the subtle properties of the PBC model. Indeed, using either of them, one would conclude that $\overline{\mathsf{stop}}$ and $\underline{\mathsf{stop}}$ are equivalent expressions, since they only allow empty moves. But this would create a worrying situation since $\overline{\mathsf{stop}}; a$ and $\underline{\mathsf{stop}}; a$ are certainly not equivalent; as we shall see in Sect. 3.2.6, the latter allows an a-move while the former only allows empty moves. What we in fact need is not a behavioural equivalence, but more exactly a behavioural *congruence*, i.e., a behavioural equivalence, $\sim$, which is preserved by every PBC operator op:

$$G_1 \sim H_1 \wedge \ \dots \ \wedge G_n \sim H_n \implies op(G_1, \dots, G_n) \sim op(H_1, \dots, H_n).$$

The last example suggests that in order to define a behavioural congruence, it is necessary to distinguish terminal expressions from non-terminal ones. However, this is still not enough, since $\underline{a}$ and $\underline{b}$ are both terminal expressions allowing only empty moves, yet $[c * \underline{a} * c]$ and $[c * \underline{b} * c]$ are certainly not equivalent; as we shall see in Sect. 3.2.14, the former allows one to perform a series of a-moves (possibly terminated by a c-move), while the latter allows a

[6] In the literature, such an arc is usually called a 'transition'; however, we have already reserved this name for another purpose.

series of b-moves (again, possibly terminated by a c-move). Hence, it is necessary to take into account the fact that some (but not all) PBC operators switch, in the behavioural sense, terminal subexpressions to the corresponding initial ones. And, by a symmetric argument, it is necessary to take into account the fact that some of the operators switch initial subexpressions to the corresponding terminal ones.

It turns out that distinguishing initial and terminal expressions from other dynamic expressions, and switching between initial and terminal expressions, may be captured by the following simple device. When constructing a transition system associated with an expression (but *only* for that purpose and *never* for its subexpression), we shall artificially augment the action set $\mathsf{Lab}_{\mathrm{PBC}}$ by two special actions, **redo** and **skip**, and add the following two axioms.

$$(SKIP - REDO) \quad \boxed{\text{SKP} \quad \overline{E} \xrightarrow{\{\mathsf{skip}\}} \underline{E} \qquad \text{RDO} \quad \underline{E} \xrightarrow{\{\mathsf{redo}\}} \overline{E}}$$

The initial and terminal expressions will be distinguished from the other ones in that only they will allow **skip**- and **redo**-moves. We will use $\mathsf{mlab}_{\mathsf{sr}} = \mathsf{mult}(\mathsf{Lab}_{\mathrm{PBC}} \uplus \{\mathsf{skip}, \mathsf{redo}\})$ to denote the set of augmented move labels, and $[G\rangle_{\mathsf{sr}}$ to denote the least set of expressions containing a dynamic expression G such that if $H \in [G\rangle_{\mathsf{sr}}$ and $H \xrightarrow{\Gamma} C$, for some $\Gamma \in \mathsf{mlab}_{\mathsf{sr}}$, then $C \in [G\rangle_{\mathsf{sr}}$. Then, with each dynamic expression G, we shall associate two transition systems. The first one, called the *non-reduced transition system* of G, is defined as $\mathsf{ts}_G^{\mathrm{nr}} = ([G\rangle_{\mathsf{sr}}, \mathsf{mlab}_{\mathsf{sr}}, A_{\mathsf{sr}}, G)$, where

$$A_{\mathsf{sr}} = \{(H, \Gamma, C) \in [G\rangle_{\mathsf{sr}} \times \mathsf{mlab}_{\mathsf{sr}} \times [G\rangle_{\mathsf{sr}} \mid H \xrightarrow{\Gamma} C\} .$$

The second transition system, called the *transition system* of G, is defined as $\mathsf{ts}_G = (\mathcal{G}_{\mathsf{sr}}, \mathsf{mlab}_{\mathsf{sr}}, \mathcal{A}_{\mathsf{sr}}, [G]_{\equiv})$, where

$$\mathcal{G}_{\mathsf{sr}} = \{[H]_{\equiv} \mid H \in [G\rangle_{\mathsf{sr}}\} \text{ and } \mathcal{A}_{\mathsf{sr}} = \{([H]_{\equiv}, \Gamma, [C]_{\equiv}) \mid (H, \Gamma, C) \in A_{\mathsf{sr}}\} .$$

Note that since we do not exclude the empty moves in the definition of $\mathcal{A}_{\mathsf{sr}}$, all the nodes of ts_G have self-loops labelled by $\emptyset$ which will not be shown in the diagrams. Note also that $\mathsf{ts}_G^{\mathrm{nr}}$ is non-reduced in the sense that distinct $\equiv$-equivalent expressions in $[G\rangle_{\mathsf{sr}}$, which in ts_G are collapsed into a single node, are represented by different nodes.

Finally, with each static expression E, we shall associate two transition systems, $\mathsf{ts}_E = \mathsf{ts}_{\overline{E}}$ and $\mathsf{ts}_E^{\mathrm{nr}} = \mathsf{ts}_{\overline{E}}^{\mathrm{nr}}$.

Isomorphism and Strong Equivalence. The strongest notion of behavioural equivalence that can be defined for transition systems is isomorphism. Two transition systems over possibly different sets of arc labels L and L', $\mathsf{ts} = (V, L, A, v_{\mathrm{in}})$, and $\mathsf{ts}' = (V', L', A', v'_{\mathrm{in}})$, are *isomorphic* if there is a bijection[7] $isom : V \to V'$ such that $isom(v_{\mathrm{in}}) = v'_{\mathrm{in}}$ and

[7] A mapping $f : X \to Y$ is a *bijection* if it is surjective ($f(X) = Y$) and injective ($f(x) = f(y) \Rightarrow x = y$ for all $x, y \in X$).

$A' = \{(isom(v), l, isom(w)) \mid (v, l, w) \in A\}$.

An equivalence based on the isomorphism of non-reduced transition systems of PBC expressions would be uselessly restrictive since. For instance, a variable X with the defining equation $X \stackrel{\text{df}}{=} a\|b$, would 'artificially' increase the $\equiv$-class of $\overline{a\|b}$ (by $\overline{X}$, see Sect. 3.2.15), adding an extra node to the non-reduced transition system of $\overline{a\|b}$, but it would not affect the non-reduced transition system of $\overline{b\|a}$. Yet one would expect thesc two expressions to be equivalent.[8] This leads to an alternative definition. Two PBC expressions, G and H (being both either static or dynamic), are *ts-isomorphic*if their transition systems, ts_G and ts_H, are isomorphic. We denote this by $G \cong H$.

Isomorphism is not the only way to define a sound behavioural congruence. In this book, we shall also use what in CCS terminology is called strong equivalence (weak equivalences and congruences, obtained by essentially ignoring the silent $\{\emptyset\}$-moves can also be defined, but we shall not treat them here). Two transition systems over possibly different sets of labels, $\mathsf{ts} = (V, L, A, v_{\text{in}})$ and $\mathsf{ts}' = (V', L', A', v'_{\text{in}})$, are *strongly equivalent* if there is a relation $\mathcal{SB} \subseteq V \times V'$, itself called a *strong bisimulation*, such that $(v_{\text{in}}, v'_{\text{in}}) \in \mathcal{SB}$, and if $(v, v') \in \mathcal{SB}$, then

$$\begin{array}{lcll} (v, l, w) \in A & \Longrightarrow & \exists w' \in V' : & (v', l, w') \in A' \wedge (w, w') \in \mathcal{SB} \\ (v', l, w') \in A' & \Longrightarrow & \exists w \in V \ : & (v, l, w) \in A \ \ \wedge (w, w') \in \mathcal{SB} . \end{array}$$

Figure 3.1 shows two strongly equivalent non-isomorphic transition systems. Notice that in a graphical representation of transition systems, initial states are represented by small black rings, while other nodes by small black circles.

Exercise 3.2.2. Find a strong bisimulation for these transition systems.

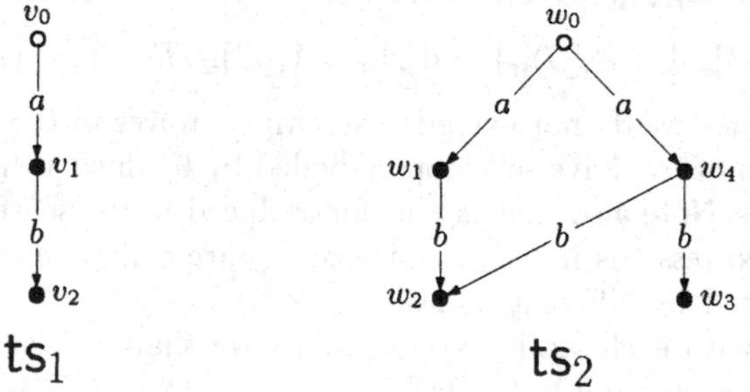

Fig. 3.1. Strongly equivalent transition systems (v_0 and w_0 are the initial states)

Two PBC expressions, G and H (being both either static or dynamic), are *strongly equivalent* if $\mathsf{ts}^{\text{nr}}_G$ and $\mathsf{ts}^{\text{nr}}_H$ are strongly equivalent transition systems. We will denote this by $G \approx H$. Here, we used non-reduced transition

[8] We will return to this example in Sect. 3.2.15.

systems, but this is irrelevant since it is easy to see that, for all expressions G and H, $\mathsf{ts}_G^{\mathrm{nr}}$ and $\mathsf{ts}_H^{\mathrm{nr}}$ are strongly equivalent if and only if ts_G and ts_H are strongly equivalent; moreover, for every expression G, $\mathsf{ts}_G^{\mathrm{nr}}$ and ts_G are strongly equivalent. It may also be observed that

$$G \equiv H \;\Rightarrow\; G \cong H \;\Rightarrow\; G \approx H\ .$$

Moreover, the two implications cannot be reversed, which will be proved by some of the examples and exercises presented later on. The reader will also be able to check that $\equiv$, $\cong$, and $\approx$ are indeed congruences for the PBC operators.

In the rest of this chapter, we will specify the structured operational semantics for the standard PBC model. We will present not just the rules, but also motivating remarks and hints for possible extensions. We will proceed through the PBC syntax in the order given by (3.1), except that we treat the two infinite constructs, iteration and recursion, at the end. It is important to stress that no derivation considered below is allowed to involve the two special actions, skip and redo, since their only role is in the construction of transition systems, and in the resulting definitions of behavioural equivalences.

At the end of this chapter, Fig. 3.5 depicts reduced transition systems for some of the example expressions considered in this chapter.

3.2.3 Elementary Actions

Let us examine the semantics of a basic action expression $\alpha \in \mathsf{Lab}_{\mathrm{PBC}}$. Intuitively, α should be allowed to execute two steps, namely the empty step $\emptyset$ (as specified by IN1), and $\{\alpha\}$, the step consisting of a single occurrence of α. At this point, a major difference between CCS and PBC semantics may again be observed. In CCS, the latter evolution would lead to the terminal expression nil, $\alpha \xrightarrow{\alpha} \mathsf{nil}$, from which no further activities can ensue. This, however, is not appropriate for PBC because a static expression is supposed to represent the structure of a corresponding net. An evolution represents the change of its state, i.e., of net marking, and not a change in the structure of the net, nor consequently, in the structure of the expression. In the case of the basic action, α, we consider two dynamic expressions, $\overline{\alpha}$ and $\underline{\alpha}$, respectively denoting the initial and final state of α, and the operational semantics of α is given by the following *action rule*.

$$(OP_\alpha) \quad \boxed{\mathrm{AR} \qquad \overline{\alpha} \xrightarrow{\{\alpha\}} \underline{\alpha}}$$

AR specifies that α corresponds to an atomic transition (i.e., a transition in the Petri net sense). For instance, using the $(INACTION)$ and (OP_α) rules, we may derive the following evolutions for the first two running examples, $E_0 = \emptyset$ and $E_1 = \{\widehat{a}, b\}$, starting from their initial states:

$$\overline{\emptyset} \xrightarrow{\emptyset} \overline{\emptyset} \qquad \text{IN1}$$

$$\overline{\emptyset} \xrightarrow{\{\emptyset\}} \underline{\emptyset} \qquad \text{AR}$$

$$\overline{\{\widehat{a},b\}} \xrightarrow{\emptyset} \overline{\{\widehat{a},b\}} \qquad \text{IN1}$$

$$\overline{\{\widehat{a},b\}} \xrightarrow{\{\{\widehat{a},b\}\}} \underline{\{\widehat{a},b\}} \qquad \text{AR}\,.$$

Notice the difference between the first and second evolution. The former maintains the initial state of E_0 and denotes inaction, while the latter transforms the initial state of E_0 into its terminal state and denotes an execution of a silent transition.

3.2.4 Parallel Composition

The following are SOS rules for parallel composition.

$$(OP_{\|}) \quad \begin{array}{|ll|} \hline \text{IPAR1} & \overline{E\|F} \xrightarrow{\emptyset} \overline{E}\|\overline{F} \\ & \\ \text{PAR} & \dfrac{G \xrightarrow{\Gamma} G'\ ,\ H \xrightarrow{\Delta} H'}{G\|H \xrightarrow{\Gamma+\Delta} G'\|H'} \\ & \\ \text{IPAR2} & \underline{E}\|\underline{F} \xrightarrow{\emptyset} \underline{E\|F} \\ \hline \end{array}$$

In the above, G, G', H, and H' are dynamic expressions, and $\Gamma{+}\Delta$ is the sum of multisets Γ and Δ. $(OP_{\|})$ comprises two inaction rules (axioms), IPAR1 and IPAR2, and one derivation rule, PAR. The latter means that if G can make a step Γ to become G', and H can make a step Δ to become H', then $G\|H$ can make a step $\Gamma{+}\Delta$, by performing concurrently all the component multiactions of Γ and Δ, to become $G'\|H'$, i.e.,

$$\begin{array}{ll} \text{from} & G \xrightarrow{\Gamma} G' \text{ and } H \xrightarrow{\Delta} H' \\ \text{infer} & G\|H \xrightarrow{\Gamma+\Delta} G'\|H'\,. \end{array}$$

PAR is a derivation rule which allows one to deduce a move of a compound expression from the moves of its subexpressions. The latter may be actual nonempty steps involving AR of the previous section, but they may also be given by inaction rules. Thus it would be slightly misleading to call PAR an 'action rule'.

Let us apply the $(OP_{\|})$ rules to the example expression $E_4 = a\|b$ whose initial and final states are given by $\overline{a\|b}$ and $\underline{a\|b}$, respectively. One would expect to be able to infer the move

$$\overline{a\|b} \xrightarrow{\{\{a\},\{b\}\}} \underline{a\|b}\,. \qquad (3.9)$$

Note that we use the convention of omitting brackets around a and b in expressions, but not in the executed steps; allowing the latter could easily lead to a confusion. In the full notation, the move we expect to be able to infer is

$$\overline{\{a\}\|\{b\}} \xrightarrow{\{\{a\},\{b\}\}} \underline{\{a\}\|\{b\}} .$$

Let us see how this derivation can be deduced. From IPAR1, applied with $E = a$ and $F = b$ (note that we reverted to using abbreviated expressions), we may derive

$$\overline{a\|b} \xrightarrow{\emptyset} \overline{a}\|\overline{b} .$$

Now the overbars are located separately on the two sides of the $\|$ operator, and so we may use AR twice, once for each side of $\|$, yielding

$$\overline{a} \xrightarrow{\{\{a\}\}} \underline{a} \quad \text{and} \quad \overline{b} \xrightarrow{\{\{b\}\}} \underline{b} .$$

Hence it is possible to instantiate PAR with $G = \overline{a}$, $\Gamma = \{\{a\}\}$, $G' = \underline{a}$, $H = \overline{b}$, $\Delta = \{\{b\}\}$, and $H' = \underline{b}$. The rule can be applied and, after observing that $\{\{a\}\} + \{\{b\}\} = \{\{a\},\{b\}\}$, we obtain

$$\overline{a}\|\overline{b} \xrightarrow{\{\{a\},\{b\}\}} \underline{a}\|\underline{b} .$$

Finally, IPAR2 yields $\underline{a}\|\underline{b} \xrightarrow{\emptyset} \underline{a\|b}$, and the whole inference looks as follows:

$$\begin{array}{lll} \overline{a\|b} & \xrightarrow{\emptyset} \overline{a}\|\overline{b} & \text{IPAR1} \\ & \xrightarrow{\{\{a\},\{b\}\}} \underline{a}\|\underline{b} & \text{AR (twice), PAR} \\ & \xrightarrow{\emptyset} \underline{a\|b} & \text{IPAR2} . \end{array}$$

Thus we obtained the following derivation

$$\overline{a\|b} \xrightarrow{\emptyset\{\{a\},\{b\}\}\emptyset} \underline{a\|b}$$

which, together with the inaction rules ILN and IRN, means that (3.9) is indeed a valid move. Since the last step is a routine application of the inaction rules, we have omitted it here, and in other examples presented in this chapter.

As already explained, the inaction rule $\overline{E\|F} \xrightarrow{\emptyset} \overline{E}\|\overline{F}$ should not be interpreted as denoting anything actually happening in the behavioural sense. Rather, it describes two different views of the same system state; the difference between $\overline{E\|F}$ and $\overline{E}\|\overline{F}$ is that the former specifies an initial state of $E\|F$ while the latter specifies the parallel composition of two separate initial states of respectively E and F, and the inaction rule says that these two views are equivalent. From IN2 in the $(INACTION)$ rules, it follows that we also have a symmetric evolution

$\overline{E}\|\overline{F} \xrightarrow{\emptyset} \overline{E\|F}$.

The reason we have chosen IPAR1 as a basic rule instead of its symmetric counterpart is that, when deriving behaviours, we usually use IPAR1 rather than its inverse. A similar remark holds for IPAR2, and other inaction rules introduced later on.

Let us apply the $(OP_{\|})$ rules again to the same example expression. Starting from the expression $\overline{a\|b}$, the nonmaximal step $\{\{a\}\}$, by which only the left-hand side but not the right-hand side of the expression is executed, should also be possible. Indeed, this is the case, as can be seen below:

$$\begin{array}{llllll} \overline{a\|b} & \xrightarrow{\emptyset} \overline{a}\|\overline{b} & \text{as before} & & & \\ & \xrightarrow{\{\{a\}\}} \underline{a}\|\overline{b} & \text{by PAR,} & \text{since} & \text{by AR} & \overline{a} \xrightarrow{\{\{a\}\}} \underline{a} \\ & & & & \text{by IN1} & \overline{b} \xrightarrow{\emptyset} \overline{b} \\ & & & & \text{and} & \{\{a\}\}+\emptyset = \{\{a\}\} . \end{array}$$

The dynamic expression $\underline{a}\|\overline{b}$ produced in this derivation is not of the form $\overline{E}$ or $\underline{E}$, nor is $\equiv$-equivalent to any such expression. That is, it describes an intermediate state of $a\|b$ which is neither initial nor final.

Exercise 3.2.3. Show that

$$\begin{array}{llll} E\|F & \cong F\|E & G\|H & \cong H\|G \\ E\|(F\|E') & \cong (E\|F)\|E' & G\|(H\|G') & \cong (G\|H)\|G' \end{array}$$

which means that parallel composition is *commutative* and *associative*.

The proof of the results formulated in the last exercise, as well as other results establishing equivalence between various types of expressions stated later in this section, cannot be carried out formally until the whole SOS semantics of PBC has been introduced and, in particular, the treatment of recursion explained. We therefore suggest that the reader returns to this, and other, exercises after reaching the end of Sect. 3.2. The first property formulated in the last exercise (i.e., the commutativity of parallel composition) is discussed in the Appendix.

Exercise 3.2.4. Show that the following derivation rules for parallel composition can be deduced from PAR.

$$\text{PARL}\ \frac{G \xrightarrow{\Gamma} G'}{G\|H \xrightarrow{\Gamma} G'\|H} \qquad \text{PARR}\ \frac{H \xrightarrow{\Gamma} H'}{G\|H \xrightarrow{\Gamma} G\|H'}$$

In other words, show that individual behaviours of the two components of a parallel composition are always possible.

Note: By saying that PARL and PARR are 'deduced' rules, we mean that any move obtained using PARL or PARR can also be inferred using PAR and other, already defined, rules.

3.2.5 Choice Composition

The remaining control flow operators of the standard PBC, i.e., choice, sequence, and iteration, all follow the pattern expounded in the previous section; their operational semantics consists of some inaction rules and some derivation rules. The following are the rules for the choice between two expressions.

$$
(OP_{\square}) \quad
\begin{array}{|llll|}
\hline
\text{IC1L} & \overline{E \,\square\, F} \xrightarrow{\emptyset} \overline{E} \,\square\, F & \text{IC1R} & \overline{E \,\square\, F} \xrightarrow{\emptyset} E \,\square\, \overline{F} \\[1ex]
\text{CL} & \dfrac{G \xrightarrow{\Gamma} G'}{G \,\square\, F \xrightarrow{\Gamma} G' \,\square\, F} & \text{CR} & \dfrac{H \xrightarrow{\Delta} H'}{E \,\square\, H \xrightarrow{\Delta} E \,\square\, H'} \\[1ex]
\text{IC2L} & \underline{E} \,\square\, F \xrightarrow{\emptyset} \underline{E \,\square\, F} & \text{IC2R} & E \,\square\, \underline{F} \xrightarrow{\emptyset} \underline{E \,\square\, F} \\
\hline
\end{array}
$$

We apply the rules $(OP_{\square})$ to the example expression $E_5 = (a\|b) \,\square\, c$. One would expect that for the dynamic expression $\overline{E_5} = \overline{(a\|b) \,\square\, c}$, the following steps (amongst others) are possible:

$\{\{a\},\{b\}\}$ leading to the final state

$\{\{a\}\}$ leading to a nonfinal state

$\{\{c\}\}$ leading to the final state.

Also, one would expect that $\{\{a\},\{c\}\}$, for instance, is not a valid step. Below we show that the first step is indeed possible:

$$
\begin{array}{lll}
\overline{(a\|b) \,\square\, c} & \xrightarrow{\emptyset} \ \overline{(a\|b)} \,\square\, c & \text{IC1L} \\
 & \xrightarrow{\{\{a\},\{b\}\}} \ \underline{(a\|b)} \,\square\, c & \text{CL, PAR, AR (twice),} \\
 & & \text{IPAR1, IPAR2, ILN, IRN} \\
 & \xrightarrow{\emptyset} \ \underline{(a\|b) \,\square\, c} & \text{IC2L}\ .
\end{array}
$$

Exercise 3.2.5. Do the same for the steps $\{\{a\}\}$ and $\{\{c\}\}$, and convince yourself that the step $\{\{a\},\{c\}\}$ is impossible.

Exercise 3.2.6. Show that

$$
\begin{array}{ll}
E \,\square\, F \cong F \,\square\, E & \underline{G} \,\square\, E \cong E \,\square\, \underline{G} \\
E \,\square\, (F \,\square\, E') \cong (E \,\square\, F) \,\square\, E' & \underline{G} \,\square\, (E \,\square\, F) \cong (\underline{G} \,\square\, E) \,\square\, F \\
E \,\square\, (\underline{G} \,\square\, F) \cong (E \,\square\, \underline{G}) \,\square\, F & E \,\square\, (F \,\square\, \underline{G}) \cong (E \,\square\, F) \,\square\, \underline{G}
\end{array}
$$

which means that choice composition is *commutative* and *associative*.

Exercise 3.2.7. Show that

$$E \Box E \approx E \qquad G \Box \lfloor G \rfloor \approx G \qquad \alpha \Box \alpha \cong \alpha$$

which means that choice composition is *idempotent*.

Exercise 3.2.8. Show that $E \Box E \cong E$ does not in general hold.
Hint: Notice that $E = \lfloor G \rfloor$ does not necessarily imply $G \Box E \equiv E \Box G$.

3.2.6 Sequential Composition

The following rules describe operational semantics of the sequential composition of two expressions.

$(OP_;)$				
	IS1	$\overline{E;F} \xrightarrow{\emptyset} \overline{E};F$	SL	$\dfrac{G \xrightarrow{\Gamma} G'}{G;F \xrightarrow{\Gamma} G';F}$
	IS2	$\underline{E};F \xrightarrow{\emptyset} E;\overline{F}$		
	IS3	$E;\underline{F} \xrightarrow{\emptyset} \underline{E;F}$	SR	$\dfrac{H \xrightarrow{\Delta} H'}{E;H \xrightarrow{\Delta} E;H'}$

We apply these rules to the example expression $E_2 = (a;b) \Box c$. One would expect that $\overline{(a;b) \Box c}$ can make an $\{\{a\}\}$-step followed by a $\{\{b\}\}$-step, and reach its final state. The following derivation shows that such a behaviour is indeed valid:

$$\begin{array}{lll}
\overline{(a;b) \Box c} & \xrightarrow{\emptyset} \overline{(a;b)} \Box c & \text{IC1L} \\
& \xrightarrow{\emptyset} (\overline{a};b) \Box c & \text{IS1, CL} \\
& \xrightarrow{\{\{a\}\}} (\underline{a};b) \Box c & \text{AR, SL, CL} \\
& \xrightarrow{\emptyset} (a;\overline{b}) \Box c & \text{IS2, CL} \\
& \xrightarrow{\{\{b\}\}} (a;\underline{b}) \Box c & \text{AR, SR, CL} \\
& \xrightarrow{\emptyset} \underline{(a;b)} \Box c & \text{IS3, CL} \\
& \xrightarrow{\emptyset} \underline{(a;b) \Box c} & \text{IC2L} .
\end{array}$$

Note that in the second line, we used not just IS1, but also CL. In this case, CL was used not to describe an action, but to capture the fact that the inaction of the left-hand part of the choice propagates to the topmost level of the whole expression. Another derivation shows that the example expression $E_3 = a;(b \Box c)$ can also make an $\{\{a\}\}$-step followed by a $\{\{b\}\}$-step.

$$\begin{array}{lll}
\overline{a;(b \Box c)} & \xrightarrow{\{\{a\}\}} a;(\overline{b} \Box c) & \text{IS1, AR, SL, IS2, IC1L, SR, ILN, IRN (twice)} \\
& \xrightarrow{\{\{b\}\}} \underline{a;(b \Box c)} & \text{AR, CL, SR, IC2L, SR, IS3, IRN (twice)} .
\end{array}$$

Exercise 3.2.9. Show that

$$E;(F;E') \cong (E;F);E' \qquad G;(E;F) \cong (G;E);F$$

$$E;(G;F) \cong (E;G);F \qquad E;(F;G) \cong (E;F);G$$

which means that sequential composition is *associative*. Moreover, show that it is *not* commutative.
Note: A solution to this exercise is provided in the Appendix.

There is a strong similarity between PARL, CL, and SL (and also between their 'R' versions). This suggests that the inaction rules (IPAR1, IC1L, IS1, etc.), rather than the derivation rules, are chiefly responsible for discriminating the semantics of the control flow operators. Quite the opposite holds for the communication interface operators, which will be described in Sects. 3.2.7–3.2.13; their derivation rules differ while their inaction rules have always the same shape.

3.2.7 Synchronisation

The synchronisation, $E \;\mathsf{sy}\; a$, of an expression E with respect to an action particle a adds behaviour to that already offered by E. We will describe three possible ways of defining such an operator, distinguished by the amount of the added behaviour.

3.2.8 Standard PBC Synchronisation

We first cast the scheme described in Sect. 2.5 in terms of structured operational semantics. Since the scheme works inductively, allowing new synchronisations to be created from the existing (in particular, basic) ones, the derivation rule for synchronisation has two constituent rules, as shown below.

(OP_{sy})

ISY1 $$\overline{E \;\mathsf{sy}\; a} \xrightarrow{\emptyset} \overline{E} \;\mathsf{sy}\; a$$

SY1 $$\frac{G \xrightarrow{\Gamma} G'}{G \;\mathsf{sy}\; a \xrightarrow{\Gamma} G' \;\mathsf{sy}\; a}$$

SY2 $$\frac{G \;\mathsf{sy}\; a \xrightarrow{\{\alpha+\{a\}\}+\{\beta+\{\widehat{a}\}\}+\Gamma} G' \;\mathsf{sy}\; a}{G \;\mathsf{sy}\; a \xrightarrow{\{\alpha+\beta\}+\Gamma} G' \;\mathsf{sy}\; a}$$

ISY2 $$\underline{E} \;\mathsf{sy}\; a \xrightarrow{\emptyset} \underline{E \;\mathsf{sy}\; a}$$

The first derivation rule, SY1, states that G sy a incorporates all the steps possible for G, i.e., G sy a has at least as much behaviour as G. The second derivation rule, SY2, states that if G sy a can make a move to G' sy a involving two multiactions, one containing a and the other containing $\widehat{a}$, then G sy a can also make a move in which these two multiactions are combined into a single one, less the synchronising pair $(a, \widehat{a})$. SY2 can be applied repeatedly, provided that conjugate pairs are suitably distributed, because the same pairs of dynamic expressions appear above and below the horizontal line. The (OP_{sy}) rules generalise the CCS rule for parallel composition. In particular, SY2 plays the same role as the 'τ-rule' (3.7); to see this, set $\alpha = \emptyset = \beta$ and $\Gamma = \emptyset$, and treat the empty multiaction as τ.

We apply the (OP_{sy}) rules to the example $E_9 = ((a \| \{\widehat{a}, \widehat{a}\}) \| a)$ sy a, showing that the initial dynamic expression $\overline{E_9}$ can make a $\{\emptyset\}$-move synchronising all of its parallel components and terminate. As this derivation is more involved than those given previously, we split it into a number of stages.

$$
\begin{aligned}
\overline{E_9} &\xrightarrow{\ \emptyset\ } \overline{((a \| \{\widehat{a}, \widehat{a}\}) \| a)} \ \mathsf{sy}\ a && \text{ISY1} \\
&\xrightarrow{\{\emptyset\}} \underline{((a \| \{\widehat{a}, \widehat{a}\}) \| a)} \ \mathsf{sy}\ a && \text{(see below)} \\
&\xrightarrow{\ \emptyset\ } \underline{((a \| \{\widehat{a}, \widehat{a}\}) \| a) \ \mathsf{sy}\ a} && \text{ISY2}\,.
\end{aligned}
\tag{3.10}
$$

The second line of this derivation incorporates one application of SY1, two applications of SY2, several applications of all parts of $(OP_{\|})$, and three applications of AR. This combination of rule applications is decomposed as follows. First, by proceeding similarly as in Sect. 3.2.4, the following may be derived:

$$
\overline{(a \| \{\widehat{a}, \widehat{a}\}) \| a} \xrightarrow{\{\{a\},\{\widehat{a},\widehat{a}\},\{a\}\}} \underline{(a \| \{\widehat{a}, \widehat{a}\}) \| a}\,.
$$

Using SY1, the following can then be deduced:

$$
\overline{((a \| \{\widehat{a}, \widehat{a}\}) \| a)} \ \mathsf{sy}\ a \xrightarrow{\{\{a\},\{\widehat{a},\widehat{a}\},\{a\}\}} \underline{((a \| \{\widehat{a}, \widehat{a}\}) \| a)} \ \mathsf{sy}\ a\,.
$$

We may rewrite this as

$$
\overline{((a \| \{\widehat{a}, \widehat{a}\}) \| a)} \ \mathsf{sy}\ a \xrightarrow{\{\emptyset+\{a\}\}+\{\{\widehat{a}\}+\{\widehat{a}\}\}+\{\{a\}\}} \underline{((a \| \{\widehat{a}, \widehat{a}\}) \| a)} \ \mathsf{sy}\ a\,.
$$

By setting $\alpha = \emptyset$, $\beta = \{\widehat{a}\}$, and $\Gamma = \{\{a\}\}$, this matches the premise of SY2. We may now use this rule to deduce

$$
\overline{((a \| \{\widehat{a}, \widehat{a}\}) \| a)} \ \mathsf{sy}\ a \xrightarrow{\{\emptyset+\{\widehat{a}\}\}+\{\{a\}\}} \underline{((a \| \{\widehat{a}, \widehat{a}\}) \| a)} \ \mathsf{sy}\ a\,.
$$

This may be rewritten as

$$
\overline{((a \| \{\widehat{a}, \widehat{a}\}) \| a)} \ \mathsf{sy}\ a \xrightarrow{\{\emptyset+\{a\}\}+\{\emptyset+\{\widehat{a}\}\}+\emptyset} \underline{((a \| \{\widehat{a}, \widehat{a}\}) \| a)} \ \mathsf{sy}\ a\,.
$$

By setting $\alpha = \emptyset = \beta$ and $\Gamma = \emptyset$, this again matches the premise of SY2. The rule can be used once more, yielding

$$\overline{((a\|\{\widehat{a},\widehat{a}\})\|a)\ \mathsf{sy}\ a} \xrightarrow{\{\emptyset+\emptyset\}+\emptyset} \underline{((a\|\{\widehat{a},\widehat{a}\})\|a)}\ \mathsf{sy}\ a\ .$$

We have therefore shown the second line of (3.10) since the step $\{\emptyset+\emptyset\}+\emptyset$ is just another way of writing $\{\emptyset\}$. Thus, the fact that the initial dynamic expression $\overline{E_9}$ can execute the step $\{\emptyset\}$ and terminate, has been formally deduced from the rules of the operational semantics. Example E_8 can be treated similarly, except that it only involves a single application of ISY2.

Exercise 3.2.10. Show that

$$\begin{array}{ll}
E\ \mathsf{sy}\ a\ \mathsf{sy}\ b \cong E\ \mathsf{sy}\ b\ \mathsf{sy}\ a & G\ \mathsf{sy}\ a\ \mathsf{sy}\ b \cong G\ \mathsf{sy}\ b\ \mathsf{sy}\ a \\
E\ \mathsf{sy}\ a\ \mathsf{sy}\ a \cong E\ \mathsf{sy}\ a & G\ \mathsf{sy}\ a\ \mathsf{sy}\ a \cong G\ \mathsf{sy}\ a \\
E\ \mathsf{sy}\ \widehat{a} \qquad \cong E\ \mathsf{sy}\ a & G\ \mathsf{sy}\ \widehat{a} \qquad \cong G\ \mathsf{sy}\ a
\end{array}$$

which means that synchronisation is *commutative*, *idempotent*, and *insensitive to conjugation*.

Exercise 3.2.11. Show that

$$\begin{array}{llll}
(E;F) & \mathsf{sy}\ a & \cong & (E\ \mathsf{sy}\ a);(F\ \mathsf{sy}\ a) \\
(G;E) & \mathsf{sy}\ a & \cong & (G\ \mathsf{sy}\ a);(E\ \mathsf{sy}\ a) \\
(E;G) & \mathsf{sy}\ a & \cong & (E\ \mathsf{sy}\ a);(G\ \mathsf{sy}\ a) \\[1ex]
(E\,\square\,F) & \mathsf{sy}\ a & \cong & (E\ \mathsf{sy}\ a)\,\square\,(F\ \mathsf{sy}\ a) \\
(G\,\square\,E) & \mathsf{sy}\ a & \cong & (G\ \mathsf{sy}\ a)\,\square\,(E\ \mathsf{sy}\ a) \\
(E\,\square\,G) & \mathsf{sy}\ a & \cong & (E\ \mathsf{sy}\ a)\,\square\,(G\ \mathsf{sy}\ a) \\
(E\|F) & \mathsf{sy}\ a & \cong & ((E\ \mathsf{sy}\ a)\|(F\ \mathsf{sy}\ a))\ \mathsf{sy}\ a \\
(G\|H) & \mathsf{sy}\ a & \cong & ((G\ \mathsf{sy}\ a)\|(H\ \mathsf{sy}\ a))\ \mathsf{sy}\ a
\end{array}$$

which means that synchronisation *distributes* over sequence and choice, but only *propagates* through parallel composition (see Exc. 3.2.42).

Exercise 3.2.12. Show that SY1 and SY2 are together equivalent to the following single rule:

$$\text{SY12}\quad \frac{G \xrightarrow{\Gamma+\sum_{i=1}^{n}\sum_{j=1}^{m_i}\{\alpha_{ij}+h_{ij}\cdot\{a\}+k_{ij}\cdot\{\widehat{a}\}\}} G'}{G\ \mathsf{sy}\ a \xrightarrow{\Gamma+\sum_{i=1}^{n}\{\sum_{j=1}^{m_i}\alpha_{ij}\}} G'\ \mathsf{sy}\ a}\quad n \geq 0 \wedge \forall i\ :\ C_i$$

where C_i is

$$m_i \geq 2 \wedge \left(\prod_{j=1}^{m_i}(h_{ij}+k_{ij}) > 0\right) \wedge \left(\sum_{j=1}^{m_i} h_{ij} = \sum_{j=1}^{m_i} k_{ij} = m_i - 1\right)\ .$$

Note: By saying that the rules SY1 and SY2 are 'equivalent' to SY12, we mean that any move obtained using SY1 and SY2 can be inferred using SY12 instead, and vice versa. Since this exercise is more complicated than the previous ones, and other exercises of the same kind will be proposed later on, we discuss it in detail in the Appendix.

We will now explain how SY12 could be used to derive the move for $\overline{E_9}$ shown in (3.10). First, recall that the following may be derived using the action and parallel composition rules:

$$\overline{(a\|\{\widehat{a},\widehat{a}\})\|a} \xrightarrow{\{\{a\},\{\widehat{a},\widehat{a}\},\{a\}\}} \underline{(a\|\{\widehat{a},\widehat{a}\})\|a}\ .$$

One can easily check that we can then apply the rule SY12 instantiated thus: $\Gamma = \emptyset$, $n = 1$, $m_1 = 3$, $\alpha_{11} = \alpha_{12} = \alpha_{13} = \emptyset$, $h_{11} = h_{13} = 1$, $h_{12} = k_{11} = k_{13} = 0$, and $k_{12} = 2$. This, in turn, yields

$$\overline{((a\|\{\widehat{a},\widehat{a}\})\|a)\ \mathsf{sy}\ a} \xrightarrow{\emptyset+\{\emptyset\}} \underline{((a\|\{\widehat{a},\widehat{a}\})\|a)\ \mathsf{sy}\ a}\ .$$

Exercise 3.2.13. Find out which properties of the synchronisation mechanism defined by (OP_{sy}) would no longer hold if conjugation was not involutive (i.e., $\widehat{\widehat{a}} = a$ could not be assumed).
Convince yourself that leaving out the discriminative property of conjugation ($\widehat{a} \neq a$) would not invalidate any of the results stated in this section, and that SY12 could be simplified if $\widehat{a} = a$.

3.2.9 Auto-synchronisation and Multilink-synchronisation

Since SY1 and SY2 are not the only rules generalising CCS synchronisation, there is a nonobvious decision involved in defining the exact semantics of PBC synchronisation. For instance, it may be noticed that the rules given in Sect. 3.2.8 exclude what might be called *silent auto-synchronisation.* To see this, consider the expression $\{a,\widehat{a}\}\ \mathsf{sy}\ a$. One might wonder whether the $\underline{\{\emptyset\}}$-move is possible from the corresponding initial dynamic expression $\overline{\{a,\widehat{a}\}\ \mathsf{sy}\ a}$. But, after some experimenting, it should be clear that such a move is not allowed, because SY1 and SY2 only admit handshake communication between distinct subexpressions. It is, however, possible to allow auto-synchronisation without destroying the incremental multiway synchronisation scheme defined by (OP_{sy}). Consider the following enlarged set of rules.

(OP'_{sy})

ISY1, SY1, SY2, ISY2

SY3 $$\frac{G\ \mathsf{sy}\ a \xrightarrow{\{\alpha+\{a,\widehat{a}\}\}+\Gamma} G'\ \mathsf{sy}\ a}{G\ \mathsf{sy}\ a \xrightarrow{\{\alpha\}+\Gamma} G'\ \mathsf{sy}\ a}$$

The rule SY3 allows additional synchronisations; in particular, the following derivation is now possible:

$$\overline{\{a,\widehat{a}\}\ \mathsf{sy}\ a} \xrightarrow{\{\emptyset\}} \underline{\{a,\widehat{a}\}\ \mathsf{sy}\ a}\ .$$

Exercise 3.2.14. Verify this.

Since we can combine SY1, SY2 and SY3, it is possible to realise what might be called *multilink-synchronisation*, i.e., synchronisation combining more than one conjugate link between two subexpressions. This idea is illustrated by the derivation

$$\overline{(\{a,a\}\|\{\widehat{a},\widehat{a}\})\ \mathsf{sy}\ a} \xrightarrow{\{\emptyset\}} \underline{(\{a,a\}\|\{\widehat{a},\widehat{a}\})\ \mathsf{sy}\ a}$$

which holds by SY1, SY2 and SY3 (but not by SY1 and SY2 alone), since

$$\overline{(\{a,a\}\|\{\widehat{a},\widehat{a}\})\ \mathsf{sy}\ a} \xrightarrow{\{\{a,a\},\{\widehat{a},\widehat{a}\}\}} \underline{(\{a,a\}\|\{\widehat{a},\widehat{a}\})\ \mathsf{sy}\ a} \qquad \text{SY1}$$

$$\overline{(\{a,a\}\|\{\widehat{a},\widehat{a}\})\ \mathsf{sy}\ a} \xrightarrow{\{\{a,\widehat{a}\}\}} \underline{(\{a,a\}\|\{\widehat{a},\widehat{a}\})\ \mathsf{sy}\ a} \qquad \text{SY2}$$

$$\overline{(\{a,a\}\|\{\widehat{a},\widehat{a}\})\ \mathsf{sy}\ a} \xrightarrow{\{\emptyset\}} \underline{(\{a,a\}\|\{\widehat{a},\widehat{a}\})\ \mathsf{sy}\ a} \qquad \text{SY3}\ .$$

Thus, the standard PBC synchronisation can be described as a one-link multiway synchronisation, while its extension described in this section is a generalisation allowing multiple links as well.

Exercise 3.2.15. Verify that $\overline{(\{a,a\}\|\{\widehat{a},\widehat{a}\})\ \mathsf{sy}\ a}$ cannot make the $\{\emptyset\}$-step when using (OP_{sy}).

Exercise 3.2.16. Show that the commutativity, idempotence, and insensitivity to conjugation of synchronisation, as well as the distributivity properties in Exc. 3.2.11, are all preserved by the addition of SY3.

Exercise 3.2.17. Show that SY1, SY2, and SY3 are together equivalent to the following single rule:

$$\text{SY123} \quad \frac{G \xrightarrow{\Gamma+\sum_{i=1}^{n}\sum_{j=1}^{m_i}\{\alpha_{ij}+h_{ij}\cdot\{a\}+k_{ij}\cdot\{\widehat{a}\}\}} G'}{G\ \mathsf{sy}\ a \xrightarrow{\Gamma+\sum_{i=1}^{n}\{\sum_{j=1}^{m_i}\alpha_{ij}\}} G'\ \mathsf{sy}\ a} \quad n \geq 0 \wedge \forall i : \mathcal{C}_i$$

where $\mathcal{C}_i$ is

$$m_i \geq 1 \wedge \left(\prod_{j=1}^{m_i}(h_{ij}+k_{ij}) > 0\right) \wedge \left(\sum_{j=1}^{m_i} h_{ij} = \sum_{j=1}^{m_i} k_{ij} \geq m_i - 1\right)\ .$$

Exercise 3.2.18. Design a derivation rule which allows multilink- but not auto-synchronisations.

3.2.10 Step-synchronisation

We finally introduce the third synchronisation scheme defined by the following set of rules.

$$(OP''_{\mathsf{sy}}) \quad \boxed{\begin{array}{ll} \multicolumn{2}{c}{\text{ISY1, SY1, SY3, ISY2}} \\ \text{ST} & \dfrac{G \xrightarrow{\{\alpha\}+\{\beta\}+\Gamma} G'}{G \xrightarrow{\{\alpha+\beta\}+\Gamma} G'} \end{array}}$$

The rule ST can be applied to any dynamic expression, not just to a synchronised one; it means that two (or more, through its repeated use) concurrent multiactions may always be combined into a single one, offering simultaneously what they offered separately. For instance, using ST we may infer

$$\overline{a\|b} \xrightarrow{\{\{a,b\}\}} \underline{a\|b}$$

using as the premise another derivation, obtained in Sect. 3.2.4:

$$\overline{a\|b} \xrightarrow{\{\{a\},\{b\}\}} \underline{a\|b}\,.$$

Exercise 3.2.19. Show that ST could be replaced by

$$\boxed{\dfrac{G \xrightarrow{\Gamma+\sum_{i=1}^{n}\sum_{j=1}^{m_i}\{\alpha_{ij}\}} G'}{G \xrightarrow{\Gamma+\sum_{i=1}^{n}\{\sum_{j=1}^{m_i}\alpha_{ij}\}} G'} \quad n \geq 0 \wedge \forall i: m_i \geq 1}$$

Convince yourself that replacing $m_i \geq 1$ by $m_i \geq 0$ would have a significant impact on the operational semantics of expressions (essentially, empty steps would then be treated as silent moves).
Hint: Notice that $\Gamma+\sum_{j=1}^{0}\{\alpha_j\} = \Gamma+\emptyset = \Gamma$ while $\Gamma+\{\sum_{j=1}^{0}\alpha_j\} = \Gamma+\{\emptyset\}$.

From (OP''_{sy}), we can derive all the standard synchronisations, together with auto-synchronisations and multilink-synchronisations, without the need for SY2. For instance, the derivations

$$\overline{(a\|\widehat{a})\ \mathsf{sy}\ a} \xrightarrow{\{\emptyset\}} \underline{(a\|\widehat{a})\ \mathsf{sy}\ a}$$

$$\overline{\{a,\widehat{a}\}\ \mathsf{sy}\ a} \xrightarrow{\{\emptyset\}} \underline{\{a,\widehat{a}\}\ \mathsf{sy}\ a}$$

$$\overline{(\{a,a\}\|\{\widehat{a},\widehat{a}\})\ \mathsf{sy}\ a} \xrightarrow{\{\emptyset\}} \underline{(\{a,a\}\|\{\widehat{a},\widehat{a}\})\ \mathsf{sy}\ a}$$

are all possible with ST, SY1, and SY3, as one may easily verify. The synchronisation rule described in this section is close to that of 'synchronous

CCS' (SCCS [81]) and Esterel [14]. It is the strongest possible synchronisation, since it allows arbitrary concurrent transitions to be combined into a single one.

Exercise 3.2.20. Show that SY2 is a derived rule given (OP''_{sy}).

Exercise 3.2.21. Show that the addition of ST preserves the property of being a congruence for the relations $\equiv$, $\cong$, and $\approx$. More precisely, show that, for every $\sim \in \{\equiv, \cong, \approx\}$, $G \sim H$ without ST implies $G \sim H$ with ST. Show that the reverse implication does not in general hold.
Hint: Consider $a\|b$ and $(a\|b) \,\square\, \{a, b\}$.

Exercise 3.2.22. Show that the commutativity, idempotence, and insensitivity to conjugation of synchronisation, as well as the distributivity properties in Exc. 3.2.11, are all preserved by the addition of ST.

In PBC, we shall (in a way, arbitrarily) choose the (OP_{sy}) rules as they appear to be the 'smallest' generalisation of the CCS synchronisation. However, we shall discuss properties of the other two synchronisation schemes as well.

3.2.11 Basic Relabelling

Like CCS, PBC allows one to uniformly modify all the actions offered by an expression. Since the actions in PBC are multisets of action particles, we shall assume that one can take a relabelling for the latter and then apply it element-wise to multiactions.

Let f be a relabelling function for action particles, $f\colon \mathsf{A_{PBC}} \to \mathsf{A_{PBC}}$. The relabelling of an expression E with respect to f is the expression $E[f]$ which has the same behaviour as E except that all the executed actions are changed according to f. Formally, we first lift f to a function $f : \mathsf{Lab_{PBC}} \to \mathsf{Lab_{PBC}}$ through the formula

$$f(\alpha) = \sum_{a\in\alpha} \alpha(a) \cdot \{f(a)\} ,$$

where the sum $(\sum)$ and the multiplication by a natural number $(\cdot)$ have their usual multiset meaning; in particular, $\alpha = \sum_{a\in\alpha} \alpha(a) \cdot \{a\}$. Next we lift f to a function $f : \mathsf{mult}(\mathsf{Lab_{PBC}}) \to \mathsf{mult}(\mathsf{Lab_{PBC}})$, in the following way:

$$f(\Gamma) = \sum_{\alpha\in\Gamma} \Gamma(\alpha) \cdot \{f(\alpha)\} .$$

We then define the following operational semantics for the basic relabelling.

IR1	$\overline{E[f]} \xrightarrow{\emptyset} \overline{E}[f]$	RR	$\dfrac{G \xrightarrow{\Gamma} G'}{G[f] \xrightarrow{f(\Gamma)} G'[f]}$
IR2	$\underline{E}[f] \xrightarrow{\emptyset} \underline{E[f]}$		

For example, using the above rules, one can show that $\{b,b\};\{c\}$ has the same behaviour as $(\{a,b\};\{c\})[f_{a\mapsto b}]$, where $f_{a\mapsto b}$ is a relabelling function mapping a to b, and leaving other action particles unchanged.

Exercise 3.2.23. Show that

$$E[f][g] \cong E[g\circ f] \qquad G[f][g] \cong G[g\circ f]$$

which means that successive relabellings can be *collapsed.*

Exercise 3.2.24. Show that, if *id* is the identity relabelling, then

$$E[id] \cong E \qquad G[id] \cong G$$

which means that $[id]$ is a *neutral* operator.

Exercise 3.2.25. Show that

$$\begin{array}{ll}
(E;F)[f] \cong (E[f]);(F[f]) & (E \,\Box\, F)[f] \cong (E[f]) \,\Box\, (F[f]) \\
(G;E)[f] \cong (G[f]);(E[f]) & (G \,\Box\, E)[f] \cong (G[f]) \,\Box\, (E[f]) \\
(E;G)[f] \cong (E[f]);(G[f]) & (E \,\Box\, G)[f] \cong (E[f]) \,\Box\, (G[f]) \\
(E\|F)[f] \cong (E[f])\|(F[f]) & (G\|H)[f] \cong (G[f])\|(H[f])
\end{array}$$

which means that relabelling *distributes* over sequence, choice and parallel composition (see also Exc. 3.2.42).

Exercise 3.2.26. For each of the three synchronisation schemes considered above, show that if f is a bijective relabelling preserving conjugates, then

$$(E \;\mathsf{sy}\; a)[f] \cong (E[f]) \;\mathsf{sy}\; f(a) \qquad (G \;\mathsf{sy}\; a)[f] \cong (G[f]) \;\mathsf{sy}\; f(a)$$

which may be interpreted as a kind of *distributivity.* Discuss the significance of each of the assumptions made about f.

3.2.12 Restriction

Restriction with respect to an action particle a means that all the steps of a given expression are allowed except those including a or its conjugate, $\widehat{a}$. Thus, unlike synchronisation which may add new moves to those already present, restriction can suppress some, as specified by the following rules.

$$(OP_{\mathsf{rs}}) \quad \begin{array}{|lcc|}
\hline
\text{IRS1} & \overline{E \;\mathsf{rs}\; a} \xrightarrow{\;\emptyset\;} \overline{E} \;\mathsf{rs}\; a & \\
\text{RS} & \dfrac{G \xrightarrow{\;\Gamma\;} G'}{G \;\mathsf{rs}\; a \xrightarrow{\;\Gamma\;} G' \;\mathsf{rs}\; a} & \forall \alpha\in\Gamma \,:\, a,\widehat{a}\notin\alpha \\
\text{IRS2} & \underline{E} \;\mathsf{rs}\; a \xrightarrow{\;\emptyset\;} \underline{E \;\mathsf{rs}\; a} & \\
\hline
\end{array}$$

Exercise 3.2.27. Verify that $E_6 \cong a; b$.

Exercise 3.2.28. Describe the essential difference between $E_7 = (a \| b)\ \mathsf{rs}\ b$ and $E'_7 = a$.

Exercise 3.2.29. Show that

$$\begin{array}{llll} E\ \mathsf{rs}\ a\ \mathsf{rs}\ b \cong E\ \mathsf{rs}\ b\ \mathsf{rs}\ a & & G\ \mathsf{rs}\ a\ \mathsf{rs}\ b \cong G\ \mathsf{rs}\ b\ \mathsf{rs}\ a \\ E\ \mathsf{rs}\ a\ \mathsf{rs}\ a \cong E\ \mathsf{rs}\ a & & G\ \mathsf{rs}\ a\ \mathsf{rs}\ a \cong G\ \mathsf{rs}\ a \\ E\ \mathsf{rs}\ \widehat{a} \quad\ \ \cong E\ \mathsf{rs}\ a & & G\ \mathsf{rs}\ \widehat{a} \quad\ \ \cong G\ \mathsf{rs}\ a \end{array}$$

which means that restriction is *commutative*, *idempotent*, and *insensitive to conjugation*.

Exercise 3.2.30. Show that

$$\begin{array}{lll} (E; F)\ \mathsf{rs}\ a & \cong & (E\ \mathsf{rs}\ a); (F\ \mathsf{rs}\ a) \\ (G; E)\ \mathsf{rs}\ a & \cong & (G\ \mathsf{rs}\ a); (E\ \mathsf{rs}\ a) \\ (E; G)\ \mathsf{rs}\ a & \cong & (E\ \mathsf{rs}\ a); (G\ \mathsf{rs}\ a) \\[1ex] (E \,\Box\, F)\ \mathsf{rs}\ a & \cong & (E\ \mathsf{rs}\ a) \,\Box\, (F\ \mathsf{rs}\ a) \\ (G \,\Box\, E)\ \mathsf{rs}\ a & \cong & (G\ \mathsf{rs}\ a) \,\Box\, (E\ \mathsf{rs}\ a) \\ (E \,\Box\, G)\ \mathsf{rs}\ a & \cong & (E\ \mathsf{rs}\ a) \,\Box\, (G\ \mathsf{rs}\ a) \\[1ex] (E \| F)\ \mathsf{rs}\ a & \cong & (E\ \mathsf{rs}\ a) \| (F\ \mathsf{rs}\ a) \\ (G \| H)\ \mathsf{rs}\ a & \cong & (G\ \mathsf{rs}\ a) \| (H\ \mathsf{rs}\ a) \end{array}$$

which means that restriction *distributes* over sequence, choice and parallel composition (see also Exc. 3.2.42).

Exercise 3.2.31. For each of the three synchronisation schemes considered above, show that if $a \notin \{b, \widehat{b}\}$, then

$$(E\ \mathsf{sy}\ a)\ \mathsf{rs}\ b \cong (E\ \mathsf{rs}\ b)\ \mathsf{sy}\ a \qquad (G\ \mathsf{sy}\ a)\ \mathsf{rs}\ b \cong (G\ \mathsf{rs}\ b)\ \mathsf{sy}\ a$$

which means that noninterfering synchronisation and restriction *commute*.

Exercise 3.2.32. For each of the three synchronisation schemes considered above, show that if f is a bijective relabelling preserving conjugates, then

$$(E\ \mathsf{rs}\ a)[f] \cong E[f]\ \mathsf{rs}\ f(a) \qquad (G\ \mathsf{rs}\ a)[f] \cong G[f]\ \mathsf{rs}\ f(a)$$

which may be viewed as a form of *distributivity*.

3.2.13 Scoping

Scoping – combining synchronisation and restriction over the same action particle – is a derived operator defined by $[a : E] = (E\ \mathsf{sy}\ a)\ \mathsf{rs}\ a$. The explicit scoping operator has been introduced because it is very useful, as already mentioned in Sect. 2.8, in the modelling of the block structure found in high-level programming languages. Although its semantics is implied by that of restriction and synchronisation, we provide for it a separate set of rules:

$$(OP_{[:]})\quad \begin{array}{lll} \text{ISC1} & \overline{[a:E]} \xrightarrow{\emptyset} [a:\overline{E}] & \\ \text{SC} & \dfrac{G \text{ sy } a \xrightarrow{\Gamma} G' \text{ sy } a}{[a:G] \xrightarrow{\Gamma} [a:G']} & \forall \alpha \in \Gamma : a, \widehat{a} \notin \alpha \\ \text{ISC2} & [a:\underline{E}] \xrightarrow{\emptyset} \underline{[a:E]} & \end{array}$$

The rules combine those for synchronisation and restriction. There is also a direct derivation rule for scoping but it relies, of course, on the chosen semantics of synchronisation.

Exercise 3.2.33. Show that with the (OP_{sy}) rules, SC is equivalent to

$$\dfrac{G \xrightarrow{\sum_{i=1}^{n} \sum_{j=1}^{m_i} \{\alpha_{ij}\}} G'}{[a:G] \xrightarrow{\sum_{i=1}^{n} \{\sum_{j=1}^{m_i} (\alpha_{ij} - \alpha_{ij}(a)\cdot\{a\} - \alpha_{ij}(\widehat{a})\cdot\{\widehat{a}\})\}} [a:G']} \quad n \geq 0 \wedge \mathcal{C}$$

where the condition $\mathcal{C}$ is given by

$$\mathcal{C} = \forall i \ : \ \left(\sum_{j=1}^{m_i} \alpha_{ij}(a) = \sum_{j=1}^{m_i} \alpha_{ij}(\widehat{a}) \right) \ \wedge \ cond_i$$

and each of the conditions $cond_i$ is defined thus:

$$cond_i \ = \ \begin{pmatrix} m_i = 1 \\ \wedge \ \alpha_{i1}(a) = 0 \end{pmatrix} \ \vee \ \begin{pmatrix} m_i \geq 2 \\ \wedge \ \prod_{j=1}^{m_i} (\alpha_{ij}(a) + \alpha_{ij}(\widehat{a})) > 0 \\ \wedge \ \sum_{j=1}^{m_i} \alpha_{ij}(a) = m_i - 1 \end{pmatrix} .$$

The main condition in Exc. 3.2.33 means that, in the i-th group of multiactions, the number of a's is exactly counterbalanced by the number of $\widehat{a}$'s, so that they will disappear and allow the group to pass through an implicit restriction, provided that the a's and $\widehat{a}$'s all take part in a synchronisation. The latter is captured by $cond_i$ which expresses a linking condition between the multiactions α_{ij} in the i-th group; either the group has a single multiaction without an a or an $\widehat{a}$ and is then kept intact, or it is a group with at least two multiactions, each multiaction comprises at least one a or $\widehat{a}$, and the total number of the a's (and so also of the $\widehat{a}$'s) is equal to the minimum number of links (induced by conjugate $(a, \widehat{a})$ pairs) needed to connect m_i multiactions.

Exercise 3.2.34. Show that with the (OP'_{sy}) rules, SC is equivalent to the same rule as in Exc. 3.2.33, except that, for every i,

$$cond_i = \begin{pmatrix} m_i = 1 \\ \wedge\ \alpha_{i1}(a) = 0 \end{pmatrix} \vee \begin{pmatrix} m_i \geq 1 \\ \wedge\ \prod_{j=1}^{m_i}(\alpha_{ij}(a) + \alpha_{ij}(\widehat{a})) > 0 \\ \wedge\ \sum_{j=1}^{m_i} \alpha_{ij}(a) \geq m_i - 1 \end{pmatrix}.$$

Exercise 3.2.35. Show that with the (OP''_{sy}) rules, SC is equivalent to the same rule as in Exc. 3.2.33, except that $cond_i = (m_i \geq 1)$.

The very simple form of $cond_i$ in the last exercise results from the fact that here we allow any grouping of steps, and any number of auto-synchronisations as well as multiple synchronisation links, so that the only true constraint is the balancing condition, $\sum_{j=1}^{m_i} \alpha_{ij}(a) = \sum_{j=1}^{m_i} \alpha_{ij}(\widehat{a})$.

Exercise 3.2.36. Show that

$$[a : E] \cong (E\ \mathsf{sy}\ a)\ \mathsf{rs}\ a \qquad [a : G] \cong (G\ \mathsf{sy}\ a)\ \mathsf{rs}\ a$$

which means that scoping is indeed a *derived* operator.

Exercise 3.2.37. Show that the $(OP_{[:]})$ rules can be deduced from those for synchronisation and restriction, assuming that $[a : E] = (E\ \mathsf{sy}\ a)\ \mathsf{rs}\ a$ and $[a : G] = (G\ \mathsf{sy}\ a)\ \mathsf{rs}\ a$.

Exercise 3.2.38. Show that

$$[a : [b : E]] \cong [b : [a : E]] \qquad [a : [b : G]] \cong [b : [a : G]]$$

$$[a : [a : E]] \cong [a : E] \qquad [a : [a : G]] \cong [a : G]$$

$$[\widehat{a} : E] \cong [a : E] \qquad [\widehat{a} : G] \cong [a : G]$$

which means that scoping is *commutative*, *idempotent*, and *insensitive to conjugation*.

Exercise 3.2.39. Show that

$$[a : (E; F)] \cong [a : E]; [a : F] \qquad [a : (E \,\square\, F)] \cong [a : E] \,\square\, [a : F]$$

$$[a : (E; G)] \cong [a : E]; [a : G] \qquad [a : (E \,\square\, G)] \cong [a : E] \,\square\, [a : G]$$

$$[a : (G; E)] \cong [a : G]; [a : E] \qquad [a : (G \,\square\, E)] \cong [a : G] \,\square\, [a : E]$$

which means that scoping *distributes* over sequence and choice (see also Exc. 3.2.42). Show that similar property does not hold for parallel composition.

Exercise 3.2.40. Show that if $a \notin \{b, \widehat{b}\}$, then

$$[a : E]\ \mathsf{rs}\ b \ \cong\ [a : (E\ \mathsf{rs}\ b)] \qquad [a : G]\ \mathsf{rs}\ b \ \cong\ [a : (G\ \mathsf{rs}\ b)]$$

$$[a : E]\ \mathsf{sy}\ b \ \cong\ [a : (E\ \mathsf{sy}\ b)] \qquad [a : E]\ \mathsf{sy}\ b \ \cong\ [a : (E\ \mathsf{sy}\ b)]$$

which means scoping *commutes* with noninterfering synchronisation and restriction.

Exercise 3.2.41. Show that if f is a bijective relabelling preserving conjugates, then

$$[a : E][f] \cong [f(a) : (E[f])] \qquad [a : G][f] \cong [f(a) : (G[f])]$$

which may be viewed as a form of *distributivity.*

3.2.14 Iteration

None of the operators described thus far can specify repetitive behaviours. PBC provides two ways of specifying such behaviours: *iteration* (treated in its simplest form here, and in generalised forms later, in Sect. 3.3.1) and *recursion.* The iteration of an expression F with an initialisation E and a termination E' is denoted syntactically by $[E * F * E']$, thus defining a ternary operator. Such a construct specifies the execution of E, followed (if E does not deadlock before reaching its end and does not enter an infinite loop) by an arbitrary number of executions of F (including zero and infinitely many times), possibly followed (if none of the executions of F deadlocks or loops indefinitely, and F is not executed indefinitely) by the execution of E'. Providing a loop with nonempty initialisation and termination has some semantic advantages, as will be discussed later, and also corresponds to common programming practice. For example, with the syntax defined in Chap. 9, consider the loop

$$[\![x{:=}0]\!] \, ; \, \mathbf{do} \, [\![x * x - 16 \neq 0]\!] \, ; \, [\![x{:=}x{+}1]\!] \, \mathbf{od}$$

which specifies a linearly ascending search for a non-negative integer root of the polynomial $h(x) = x^2 - 16$. The loop has an explicit initialisation $[\![x{:=}0]\!]$ and an implicit termination, namely the test for zero: $[\![x * x - 16 = 0]\!]$. Hence it loosely corresponds to the expression

$$\Big[[\![x{:=}0]\!] * \Big([\![x * x - 16 \neq 0]\!] \, ; \, [\![x{:=}x + 1]\!] \Big) * [\![x * x - 16 = 0]\!] \Big] \, .$$

The operational semantics of iteration with initialisation and termination are given in Fig. 3.2. The rules follow an already established pattern, and so their meaning should by now be clear even without the help of an example. The rule that effects repetition is IIT2b, without it $[E * F * E']$ would have the same behaviour as $E; (F; E')$.

Exercise 3.2.42. Show that

$$[E * F * E'] \; \mathsf{sy} \; a \cong [(E \; \mathsf{sy} \; a) * (F \; \mathsf{sy} \; a) * (E' \; \mathsf{sy} \; a)]$$

$$[E * F * E'] \; \mathsf{rs} \; a \cong [(E \; \mathsf{rs} \; a) * (F \; \mathsf{rs} \; a) * (E' \; \mathsf{rs} \; a)]$$

$$[a : [E * F * E']] \cong [[a : E] * [a : F] * [a : E']]$$

and that the same holds for the corresponding dynamic expressions, which means that all communication interface operators *distribute* over the ternary iteration.

$$
(OP_{[**]}) \quad
\begin{array}{|ll|}
\hline
\text{IIT1} & \overline{[E * F * E']} \xrightarrow{\emptyset} [\overline{E} * F * E'] \\[2ex]
\text{IIT2a} & [\underline{E} * F * E'] \xrightarrow{\emptyset} [E * \overline{F} * E'] \\[2ex]
\text{IIT2b} & [E * \underline{F} * E'] \xrightarrow{\emptyset} [E * \overline{F} * E'] \\[2ex]
\text{IIT2c} & [E * \underline{F} * E'] \xrightarrow{\emptyset} [E * F * \overline{E'}] \\[2ex]
\text{IIT3} & [E * F * \underline{E'}] \xrightarrow{\emptyset} \underline{[E * F * E']} \\[2ex]
\text{IT1} & \dfrac{G \xrightarrow{\Gamma} G'}{[G * F * E'] \xrightarrow{\Gamma} [G' * F * E']} \\[3ex]
\text{IT2} & \dfrac{G \xrightarrow{\Gamma} G'}{[E * G * E'] \xrightarrow{\Gamma} [E * G' * E']} \\[3ex]
\text{IT3} & \dfrac{G \xrightarrow{\Gamma} G'}{[E * F * G] \xrightarrow{\Gamma} [E * F * G']} \\
\hline
\end{array}
$$

Fig. 3.2. SOS-rules for iteration

3.2.15 Recursion

The operational semantics of a variable X with a defining equation $X \stackrel{\text{df}}{=} E$ is given by

$$
(OP_{\text{rec}}) \quad \boxed{\text{IREC} \quad \mathcal{C}[X] \xrightarrow{\emptyset} \mathcal{C}[E]}
$$

where $\mathcal{C}[.]$ is a valid dynamic context[9] – a dynamic expression with a 'hole' – so that $\mathcal{C}[X]$ and $\mathcal{C}[E]$ are valid dynamic expressions. IREC is the recursion *unfolding* rule which, in particular, implies that

$$
X \xrightarrow{\emptyset} E \quad \text{and} \quad \overline{X} \xrightarrow{\emptyset} \overline{E}\,.
$$

Recursion does not have any derivation rule because, by the unfolding rule IREC, E can always replace X and depending on the structure of E, other rules can be applied.

Several interesting derivations might still be obtained if we replaced IREC by the following two simpler recursion rules used in an earlier version of the theory (see [69]):

---

[9] Formally, a *dynamic context* is defined by introducing a special variable, denoted by '.', so that $\mathcal{C}[.]$ is simply a dynamic expression with (possibly) this extra variable. Then $\mathcal{C}[E]$ is obtained by replacing each '.' by E; it may be checked that if E is a valid static expression, then $\mathcal{C}[E]$ is a valid dynamic expression.

$$\overline{X} \xrightarrow{\emptyset} \overline{E} \quad \text{and} \quad \underline{X} \xrightarrow{\emptyset} \underline{E}\,.$$

But then, crucially, the structural equivalence relation on expressions would be strictly weaker. For example, consider the following two derivations which are possible with the two simpler rules:

$$\overline{X \square [a : (b; a)]} \xrightarrow{\emptyset} X \square \overline{[a : (b; a)]} \xrightarrow{\{\{b\}\}} X \square [a : (\underline{b}; a)]$$

$$\begin{aligned}\overline{X \square [a : (b; a)]} &\xrightarrow{\emptyset} \overline{X} \square [a : (b; a)] \xrightarrow{\emptyset} \overline{E} \square [a : (b; a)] \\ &\xrightarrow{\emptyset} \overline{E \square [a : (b; a)]} \xrightarrow{\emptyset} E \square \overline{[a : (b; a)]} \\ &\xrightarrow{\{\{b\}\}} E \square [a : (\underline{b}; a)]\,.\end{aligned}$$

The resulting expressions, $X \square [a : (\underline{b}; a)]$ and $E \square [a : (\underline{b}; a)]$, have the same corresponding Petri nets, yet

$$X \square [a : (\underline{b}; a)] \equiv E \square [a : (\underline{b}; a)] \tag{3.11}$$

would not hold if we replaced IREC by the two simpler rules (unless, of course, $E = X$). The consequences would be profound since in such a case the transition systems generated by $\overline{X \square [a : (b; a)]}$ and the corresponding Petri net would not be isomorphic (only strongly bisimilar), rendering invalid our main consistency result stated in Chap. 7. However, with IREC (3.11) does hold which explains why we introduced this more powerful unfolding rule.

Let us consider example 12, where X is defined by $X \stackrel{\mathrm{df}}{=} a\|X$. One would expect that $\overline{X}$ can make an $\{\{a\},\{a\}\}$-step. To derive it, we need to unfold X twice, as shown in the following.

$$\begin{aligned}\overline{X} &\xrightarrow{\emptyset} \overline{a\|X} && \text{IREC (using the context } \mathcal{C}[.] = \overline{.}\text{)} \\ &\xrightarrow{\emptyset} \overline{a}\|\overline{X} && \text{IPAR1} \\ &\xrightarrow{\emptyset} \overline{a}\|\overline{(a\|X)} && \text{IREC (using the context } \mathcal{C}[.] = \overline{a}\|\overline{.}\text{)} \\ &\xrightarrow{\emptyset} \overline{a}\|(\overline{a}\|\overline{X}) && \text{IPAR1, IN1, PAR} \\ &\xrightarrow{\{\{a\},\{a\}\}} \underline{a}\|(\underline{a}\|\overline{X}) && \text{AR (twice), IN1, PAR (twice)}\,.\end{aligned}$$

Note that neither this, nor any other derivation for $\overline{X}$, can ever eliminate a subterm of the form $\overline{(\ldots)}$. This is indicative of the fact that this expression never terminates.

Exercise 3.2.43. Show that $\overline{X}$, where $X \stackrel{\mathrm{df}}{=} a \square (X; a)$ is defined in example 17, can execute a step sequence $\{\{a\}\}\{\{a\}\}\ldots\{\{a\}\}$ of any finite length (and terminate), but not an infinite sequence of $\{\{a\}\}$-steps.

Exercise 3.2.44. Show that $\overline{X}$, where X is defined in example 14, can generate an infinite step sequence of consecutive even natural numbers.

Exercise 3.2.45. Show that, with the defining equation $\mathsf{stop} \stackrel{\mathrm{df}}{=} \mathsf{stop}$, the variable stop behaves as stated in Sect. 3.2.2. That is, only looping empty moves are ever possible, unless there are also equations of the form $X \stackrel{\mathrm{df}}{=} \mathsf{stop}$, in which case there are also empty moves,

$$\overline{X} \xrightarrow{\emptyset} \overline{\mathsf{stop}} \quad \text{and} \quad \underline{X} \xrightarrow{\emptyset} \underline{\mathsf{stop}} ,$$

which preserve the property of being an initial or terminal expression.

Exercise 3.2.46. Show that if $a \in \alpha$ or $\widehat{a} \in \alpha$, then

$$\mathsf{stop} \cong \alpha \;\mathsf{rs}\; a \cong [a : \alpha] \cong (\alpha; E) \;\mathsf{rs}\; a \cong [a : (\alpha; E)] .$$

Exercise 3.2.47. Show that $[E * F * E'] \approx (E; X)$ where $X \stackrel{\mathrm{df}}{=} E' \,\square\, (F; X)$.

We may now explain precisely why we decided to base the $\cong$-equivalence on ts instead of $\mathsf{ts}^{\mathrm{nr}}$. Assume that $\mathcal{X} = \{X\}$ and that the defining equation for the only variable is $X \stackrel{\mathrm{df}}{=} a \| b$. Then $\mathsf{ts}_{\overline{a\|b}}$ and $\mathsf{ts}_{\overline{b\|a}}$ are isomorphic transition systems, yet $\mathsf{ts}^{\mathrm{nr}}_{\overline{a\|b}}$ and $\mathsf{ts}^{\mathrm{nr}}_{\overline{b\|a}}$ are not isomorphic since the former has eight nodes and the latter six.

Exercise 3.2.48. Verify this.

Exercise 3.2.49. Show that the derivation relation is closed in the domain of dynamic PBC expressions, i.e., if D is a dynamic expression and $D \xrightarrow{\Gamma} H$, then H is also a dynamic PBC expression.

Exercise 3.2.50. Show that the initial state of an expression cannot be re-entered (by executing a non-redo-action), nor can the terminal state be escaped from (by executing a non-skip-action), i.e., show that if $\Gamma \neq \emptyset$ and

$$D \xrightarrow{\Gamma} H ,$$

then neither $D \not\equiv \underline{E}$ nor $H \not\equiv \overline{E}$, for any static expression E.

Exercise 3.2.51. The structural equivalence relation $\equiv$ can be extended to static expressions (as will be shown in Chap. 7) by stipulating that in the domain of static boxes, it is the smallest equivalence relation such that, for each variable X defined by $X \stackrel{\mathrm{df}}{=} E$, and each valid static context $\mathcal{C}[.]$,

$$\mathcal{C}[X] \xrightarrow{\emptyset} \mathcal{C}[E] .$$

Show that if E and F are static expressions, then $E \equiv F$ if and only if $\overline{E} \equiv \overline{F}$ (as well as $E \equiv F$ if and only if $\underline{E} \equiv \underline{F}$).

3.3 Extensions

In this section we discuss three kinds of extensions of the PBC syntax: generalised iterative constructs in Sect. 3.3.1, data variables in Sect. 3.3.2, and generalised control flow and communication interface operators motivated by the algebraic properties of basic operators, in Sects. 3.3.3 and 3.3.4.

3.3.1 Generalised Iterations

An iteration $[E*F*E']$ is mainly based on its looping part, F. By comparison, its initialisation, E, and termination, E', might be seen as marginal. We now consider what happens if either of them, or both, are omitted. That is, we will discuss the following alternative loop constructs:

$$\begin{array}{ll} [E * F\rangle & \text{initialisation } E\text{, iteration } F\text{, no termination} \\ \langle F * E'] & \text{iteration } F\text{, termination } E'\text{, no initialisation} \\ \langle F\rangle & \text{iteration } F \text{ only}\,. \end{array}$$

The four variants of iteration form a hierarchy: $[E * F\rangle$ could be treated as $E;\langle F\rangle$, $\langle F*E']$ as $\langle F\rangle;E'$, and $[E*F*E']$ as any of $E;\langle F*E']$, $[E*F\rangle;E'$ and $E;\langle F\rangle;E'$. Intuitively, $\langle F * E']$ corresponds to a hierarchical variable X with the defining equation $X \overset{\mathrm{df}}{=} (F;X) \,\square\, E'$, but there is a semantical problem with this construction which will soon be discussed. The loop $[E * F\rangle$ has no exact corresponding recursive expression. One might try $(E;X)$, with X being defined by $X \overset{\mathrm{df}}{=} (F;X) \,\square\, \emptyset$, but this would introduce an extra silent action, thus making $(E;X)$ equivalent to $[E*F*\emptyset]$, rather than $[E*F\rangle$, which could be undesirable, depending on the intended use. The last loop construct, $\langle F\rangle$, being the most general one, suffers from the problems troubling both $\langle F * E']$ and $[E * F\rangle$.

The operational semantics of the three alternative iteration constructs might be defined by first extending the set of dynamic expressions with

$$[G * E\rangle\ ,\ [E * G\rangle\ ,\ \langle G * E]\ ,\ \langle E * G]\ ,\ \text{ and } \langle G\rangle\ ,$$

and then adding derivation rules shown in Fig. 3.3.

While the derivation rules for generalised iterations seem to be obtained in a natural way from those given for the ternary iteration operator, they exhibit a surprising (and highly unwanted) behaviour when considered, for example, in the context of the choice operator. Indeed, one may easily check that the following derivations are allowed by the rules $(OP_{\langle *]})$:

$$\begin{array}{lllllll} \overline{\langle a * b] \,\square\, c} & \xrightarrow{\emptyset} & \overline{\langle a * b]} \,\square\, c & \xrightarrow{\emptyset} & \langle \overline{a} * b] \,\square\, c & \xrightarrow{\{\{a\}\}} & \langle \underline{a} * b] \,\square\, c \\ & \xrightarrow{\emptyset} & \langle \overline{a} * b] \,\square\, c & \xrightarrow{\emptyset} & \overline{\langle a * b]} \,\square\, c & \xrightarrow{\emptyset} & \overline{\langle a * b] \,\square\, c} \\ & \xrightarrow{\emptyset} & \langle a * b] \,\square\, \overline{c} & \xrightarrow{\{\{c\}\}} & \langle a * b] \,\square\, \underline{c} & \xrightarrow{\emptyset} & \underline{\langle a * b] \,\square\, c}\,. \end{array}$$

$(OP_{[*\rangle})$

IIT11	$\overline{[E * F\rangle} \xrightarrow{\emptyset} [\overline{E} * F\rangle$
IIT12a	$[\underline{E} * F\rangle \xrightarrow{\emptyset} [E * \overline{F}\rangle$
IIT12b	$[E * \underline{F}\rangle \xrightarrow{\emptyset} [E * \overline{F}\rangle$
IIT13	$[E * \underline{F}\rangle \xrightarrow{\emptyset} \underline{[E * F\rangle}$
IT11	$\dfrac{G \xrightarrow{\Gamma} G'}{[G * F\rangle \xrightarrow{\Gamma} [G' * F\rangle}$
IT12	$\dfrac{G \xrightarrow{\Gamma} G'}{[E * G\rangle \xrightarrow{\Gamma} [E * G'\rangle}$

$(OP_{\langle *]})$

IIT21	$\overline{\langle E * F]} \xrightarrow{\emptyset} \langle \overline{E} * F]$
IIT22a	$\langle \underline{E} * F] \xrightarrow{\emptyset} \langle \overline{E} * F]$
IIT22b	$\langle \underline{E} * F] \xrightarrow{\emptyset} \langle E * \overline{F}]$
IIT23	$\langle E * \underline{F}] \xrightarrow{\emptyset} \underline{\langle E * F]}$
IT21	$\dfrac{G \xrightarrow{\Gamma} G'}{\langle G * F] \xrightarrow{\Gamma} \langle G' * F]}$
IT22	$\dfrac{G \xrightarrow{\Gamma} G'}{\langle E * G] \xrightarrow{\Gamma} \langle E * G']}$

$(OP_{\langle\rangle})$

IIT31	$\overline{\langle E\rangle} \xrightarrow{\emptyset} \langle \overline{E}\rangle$
IIT32	$\langle \underline{E}\rangle \xrightarrow{\emptyset} \langle \overline{E}\rangle$
IIT33	$\langle \underline{E}\rangle \xrightarrow{\emptyset} \underline{\langle E\rangle}$
IT31	$\dfrac{G \xrightarrow{\Gamma} G'}{\langle G\rangle \xrightarrow{\Gamma} \langle G'\rangle}$

Fig. 3.3. SOS-rules for generalised iterations

That is, it is possible to start performing a loop (by executing one or more a's) and afterwards leave it (without performing the terminal b) and enter the right branch of the choice. (A similar example can be constructed when such a loop is nested inside an enclosing loop, even if the outer loop is the standard one, with initialisation and termination; in Chap. 4, we shall see that this corresponds to a very specific feature of the Petri net translation.) Whereas such a scenario may be seen as an interesting application of SOS-rules, it is counterintuitive as it does not correspond to any behaviour one would expect from a choice construct. Hence it seems to be a good idea either not to adopt the rules $(OP_{\langle *]})$ and the associated loop structure, or – if one adopts them – to make sure that such generalised loops do not occur, in particular, at the start of an enclosing choice. The latter may be enforced by adding contextual clauses, or by modifying the syntax of the language in order to define 'safe' subexpressions allowed in loops or choices (this technique will be exploited in Sect. 8.1.3 and in Chap. 9).

A similarly undesired behaviour may be observed of the first generalised loop construct:

$$
\begin{aligned}
\overline{[a * b\rangle \,\Box\, c} &\xrightarrow{\emptyset} [a * b\rangle \,\Box\, \overline{c} \xrightarrow{\{\{c\}\}} [a * b\rangle \,\Box\, \underline{c} \xrightarrow{\emptyset} \underline{[a * b\rangle \,\Box\, c} \\
&\xrightarrow{\emptyset} \underline{[a * b\rangle} \,\Box\, c \xrightarrow{\emptyset} [a * \underline{b}\rangle \,\Box\, c \xrightarrow{\emptyset} [a * \overline{b}\rangle \,\Box\, c \\
&\xrightarrow{\{\{b\}\}} [a * \underline{b}] \,\Box\, c\,.
\end{aligned}
$$

Here it is possible to execute the right branch of the choice and then enter the loop, without performing the initial a. Again, this is not a behaviour one would expect of a choice construct, and our conclusion is the same as before, i.e., one should restrict the usage of such a generalised loop construct; in particular, one should disallow it at the end of an enclosing choice.

The operational semantics of the third generalised loop is very simple, but this simplicity is offset by a combination of the problematic features exhibited by the two previous loop constructs. For instance, $(OP_{\langle\rangle})$ allow the following derivation.

$$
\begin{aligned}
\overline{\langle a\rangle \,\Box\, c} &\xrightarrow{\emptyset} \overline{\langle a\rangle} \,\Box\, c \xrightarrow{\emptyset} \langle \overline{a}\rangle \,\Box\, c \xrightarrow{\{\{a\}\}} \langle \underline{a}\rangle \,\Box\, c \xrightarrow{\emptyset} \langle \overline{a}\rangle \,\Box\, c \\
&\xrightarrow{\emptyset} \overline{\langle a\rangle} \,\Box\, c \xrightarrow{\emptyset} \overline{\langle a\rangle \,\Box\, c} \xrightarrow{\emptyset} \langle a\rangle \,\Box\, \overline{c} \xrightarrow{\{\{c\}\}} \langle a\rangle \,\Box\, \underline{c} \\
&\xrightarrow{\emptyset} \underline{\langle a\rangle \,\Box\, c} \xrightarrow{\emptyset} \underline{\langle a\rangle} \,\Box\, c \xrightarrow{\emptyset} \langle \underline{a}\rangle \,\Box\, c \xrightarrow{\emptyset} \langle \overline{a}\rangle \,\Box\, c \\
&\xrightarrow{\{\{a\}\}} \langle \underline{a}\rangle \,\Box\, c \xrightarrow{\emptyset} \underline{\langle a\rangle} \,\Box\, c \xrightarrow{\emptyset} \underline{\langle a\rangle \,\Box\, c}\,.
\end{aligned}
$$

That is, it is possible to start performing the loop (by executing one or more a's), then enter the other branch of the choice and finally re-enter the loop.

The above examples provide a justification for the decision to include in the basic version of PBC only the least general iteration, which is completely unproblematic when combined in any way with the remaining operators. Other loop operators may be used only when it is 'safe'; in particular, their usage should be constrained in the presence of enclosing choices and loops. We will discuss this issue extensively in Sect. 8.1.3.

Exercise 3.3.1. Show that all communication interface operators distribute over generalised iterations (cf. Exc. 3.2.42).

3.3.2 Data Variables

To model data variables of a programming language, we shall need an iterationlike construct (see the net of the declaration **var** $x : \{0, 1\}$ in Fig. 2.9), but somewhat different from that already introduced. In the formal treatment, instead of introducing another family of operators modelling data, we shall simply add another set of basic processes (constants).

Let x be a program variable of type $Type(x)$ introduced by its declaration, and $\bullet$ be a special symbol not in $Type(x)$, interpreted as an 'undefined' value.

For such a variable, the set of action particles $\mathsf{A}_{\mathrm{PBC}}$ contains the symbols x_{kl} and $\widehat{x_{kl}}$ for all $k, l \in \mathit{Type}(x) \cup \{\bullet\}$; moreover, x_{kl} and $\widehat{x_{kl}}$ are conjugate action particles. Recall from Sect. 2.8 that x_{uv} represents the change of the value of x from u to v. Similarly, $x_{\bullet u}$ represents the initialisation of x to the value u, and $x_{u\bullet}$ represents the 'destruction' of x with the current value u. The behaviour of variables will be modelled by the unhatted symbols; those hatted are reserved for the modelling of program statements.

The behaviour of x will be represented by a looplike basic process expression $[x_{()}]$. Intuitively, it is composed of an initialisation, a core looping behaviour (but more complex than that of iteration) and a termination. Such an expression can first execute $x_{\bullet u}$, for any value u in $\mathit{Type}(x)$, then a sequence of zero or more executions of x_{uv}, where $u, v \in \mathit{Type}(x)$, and possibly terminate by executing $x_{u\bullet}$. Each such execution is carried out under a restriction that, if x_{ku} and x_{vl} are two consecutive actions, then $u = v$.

For the basic expression $[x_{()}]$ – modelling the behaviour of data variable x – we shall use a set of rules similar to those introduced for iteration. The operational semantics is defined as follows, where $u, v \in \mathit{Type}(x)$:

$$(OP_{[()]})\quad \boxed{\begin{array}{lll} \mathrm{DAT1} & [x_{()}] \xrightarrow{\{\{x_{\bullet u}\}\}} [\overline{x_{(u)}}] \\ \mathrm{DAT2} & [\overline{x_{(u)}}] \xrightarrow{\{\{x_{uv}\}\}} [\overline{x_{(v)}}] \\ \mathrm{DAT3} & [\overline{x_{(u)}}] \xrightarrow{\{\{x_{u\bullet}\}\}} \underline{[x_{()}]} \end{array}}$$

In the above, $[\overline{x_{(u)}}]$ is a *constant* (not a compound expression) representing x with the current value u. We illustrate the rules $(OP_{[()]})$ on a binary variable x with $\mathit{Type}(x) = \{0, 1\}$:

$$\begin{array}{lll} \overline{[x_{()}]} & \xrightarrow{\{\{x_{\bullet 0}\}\}} [\overline{x_{(0)}}] & \mathrm{DAT1} \\ & \xrightarrow{\{\{x_{01}\}\}} [\overline{x_{(1)}}] & \mathrm{DAT2} \\ & \xrightarrow{\{\{x_{11}\}\}} [\overline{x_{(1)}}] & \mathrm{DAT2} \\ & \xrightarrow{\{\{x_{1\bullet}\}\}} \underline{[x_{()}]} & \mathrm{DAT3}\ . \end{array}$$

Notice how the constants representing variable x change according to the actions being executed; moreover, at the end, the same expression is obtained as that at the start (except for the overbar and underbar, respectively).

The rules $(OP_{[()]})$ would very much resemble those for iteration if, in the latter, the inaction rules were replaced by an explicit identification of $\equiv$-equivalent expressions. For instance, DAT2 could be seen as a compact representation of the following derivation 'inferred' from the iteration rules:

$$\begin{array}{lll} [\overline{x_{(u)}}] & \xrightarrow{\{\{x_{uv}\}\}} [x_{(v)}] & \sim \text{IT2} \\ & \xrightarrow{\emptyset} [\overline{x_{(v)}}] & \sim \text{IIT2b} . \end{array}$$

Basic processes similar to $[x_{()}]$ and $[\overline{x_{(u)}}]$ may also be devised for programming data structures other than variables (where reading and writing are the only basic operations), like stacks (where the basic operations are push and pop), or FIFO queues (where the basic operations are add and delete). The latter will be used in Chap. 9 to model communication channels.

3.3.3 Generalised Control Flow Operators

We extend the three binary control flow operators by allowing any number of arguments, indexed by the elements of a nonempty set I:

$$\|_{i\in I} E_i \quad , \quad \Box_{i\in I} E_i \quad , \quad \text{and} \quad ;_{i\in I} E_i \; . \tag{3.12}$$

The above constructs are motivated by the associativity of the corresponding binary operators (Excs. 3.2.3, 3.2.6, and 3.2.9). For the indexed sequence operator, I must be a finite or infinite sequence; for the indexed choice and parallel composition operators, I need not be ordered since the corresponding binary operators are also commutative (Excs. 3.2.3 and 3.2.6). When I is finite and contains at least two elements, the expressions in (3.12) are equivalent (in the sense of the $\cong$-relation) to multiple applications of their binary counterparts; for example, $\|_{i\in\{0,1,2\}} E_i$ is equivalent to $E_0\|(E_1\|E_2)$. When I is infinite, there are equivalent definitions in terms of a system of recursive equations. If we assume that I is the set of natural numbers, then the three constructs in (3.12) are equivalent to the variable X_0 which is evaluated in the context of an infinite system of equations defined respectively by

$$\begin{aligned} X_i &\stackrel{\text{df}}{=} E_i \| X_{i+1} \\ X_i &\stackrel{\text{df}}{=} E_i \,\Box\, X_{i+1} \\ X_i &\stackrel{\text{df}}{=} E_i ; X_{i+1} \; . \end{aligned} \tag{3.13}$$

No such translations exist if I is uncountable.[10] However, even then we will be able to provide the indexed choice and parallel composition with a Petri net as well as operational semantics, as we shall see in Chaps. 4 and 7, respectively.

Given the indexed choice operator in (3.12), the construct used to represent a data variable x might be redefined as

$$[x_{()}] = \Box_{u\in \mathit{Type}(x)} (x_{\bullet u} ; X_u)$$

with the following defining equation for each hierarchical variable X_u:

$$X_u \stackrel{\text{df}}{=} \left(\Box_{v\in \mathit{Type}(x)} (x_{uv} ; X_v)\right) \Box\, x_{u\bullet} \; .$$

[10] A set is *countable* if it is finite or has the cardinality of the set of natural numbers.

However, the Petri net semantics derived for such a system of equations (using the technique developed in Chaps. 5 and 7) would then be infinite even for a variable with a finite type.

Exercise 3.3.2. Design a syntax for dynamic expressions corresponding to the indexed control flow operators, and then propose SOS-rules capturing their intended meaning.

Exercise 3.3.3. Show that

$$\|_{i\in I}(\|_{j\in J_i} E_j) \cong \|_{j\in J} E_j$$
$$\square_{i\in I}(\square_{j\in J_i} E_j) \cong \square_{j\in J} E_j \,,$$

where $\{J_i\}_{i\in I}$ is a family of mutually disjoint indexing sets, and $J = \bigcup_{i\in I} J_i$. Moreover, show that if I is a sequence and each J_i is a finite sequence (except perhaps the last one, if I is finite), then

$$;_{i\in I}\,(;_{j\in J_i}\; E_j) \cong ;_{j\in J}\; E_j \,,$$

where J is the sequence induced in the obvious way by I and the J_i's.

3.3.4 Generalised Communication Interface Operators

Given that synchronisation, restriction, and scoping are commutative and idempotent (see Excs. 3.2.10, 3.2.29, and 3.2.38), one may introduce operators E sy A, E rs A, and $[A : E]$, where $A \subseteq \mathrm{A_{PBC}}$ is a set of action particles. The idea here is to apply the corresponding unary operations for all the action particles in A, in any order (due to commutativity) and without worrying about repetitions (idempotence means that considering multisets instead of sets of action particles would add no extra power). For a finite set A, this would simply be a way of compacting the notation, but by allowing infinite sets A, the expressiveness is strictly increased. We will later need the extended communication interface operators to deal with program variables. But if all such variables have finite domains, we will only need the extended operators with finite sets A. The SOS rules capturing the intended meaning of the extended synchronisation, restriction and scoping are shown in Fig. 3.4.

Exercise 3.3.4. Design derivation rules for extended versions of the other two synchronisation schemes discussed earlier in this chapter.

Exercise 3.3.5. Design derivation rules for extended scoping, assuming the other two synchronisation schemes discussed earlier in this chapter.

Exercise 3.3.6. Show that, for every static or dynamic expression C,

IGSY1
$$\overline{E \,\mathsf{sy}\, A} \xrightarrow{\emptyset} \overline{E} \,\mathsf{sy}\, A$$

GSY1
$$\frac{G \xrightarrow{\;\Gamma\;} G'}{G \,\mathsf{sy}\, A \xrightarrow{\;\Gamma\;} G' \,\mathsf{sy}\, A}$$

GSY2
$$\frac{G \,\mathsf{sy}\, A \xrightarrow{\{\alpha+\{a\}\}+\{\beta+\{\widehat{a}\}\}+\Gamma} G' \,\mathsf{sy}\, A}{G \,\mathsf{sy}\, A \xrightarrow{\{\alpha+\beta\}+\Gamma} G' \,\mathsf{sy}\, A.} \quad a \in A$$

IGSY2
$$\underline{E} \,\mathsf{sy}\, A \xrightarrow{\emptyset} \underline{E \,\mathsf{sy}\, A}$$

IGRS1
$$\overline{E \,\mathsf{rs}\, A} \xrightarrow{\emptyset} \overline{E} \,\mathsf{rs}\, A$$

GRS
$$\frac{G \xrightarrow{\;\Gamma\;} G'}{G \,\mathsf{rs}\, A \xrightarrow{\;\Gamma\;} G' \,\mathsf{rs}\, A} \quad \forall a \in A, \alpha \in \Gamma \;:\; a, \widehat{a} \notin \alpha$$

IGRS2
$$\underline{E} \,\mathsf{rs}\, A \xrightarrow{\emptyset} \underline{E \,\mathsf{rs}\, A}$$

IGSC1
$$\overline{[A : E]} \xrightarrow{\emptyset} [A : \overline{E}]$$

GSC
$$\frac{G \,\mathsf{sy}\, A \xrightarrow{\;\Gamma\;} G' \,\mathsf{sy}\, A}{[A : G] \xrightarrow{\;\Gamma\;} [A : G']} \quad \forall a \in A, \alpha \in \Gamma \;:\; a, \widehat{a} \notin \alpha$$

IGSC2
$$[A : \underline{E}] \xrightarrow{\emptyset} \underline{[A : E]}$$

Fig. 3.4. SOS-rules for extended communication interface operators

$$C \,\mathsf{sy}\, a \cong C \,\mathsf{sy}\, \{a\} \qquad (C \,\mathsf{sy}\, A) \,\mathsf{sy}\, B \cong (C \,\mathsf{sy}\, B) \,\mathsf{sy}\, A$$
$$C \,\mathsf{sy}\, A \cong C \,\mathsf{sy}\, (A \cup \widehat{A}) \qquad (C \,\mathsf{sy}\, A) \,\mathsf{sy}\, B \cong C \,\mathsf{sy}\, (A \cup B)$$

$$C \,\mathsf{rs}\, a \cong C \,\mathsf{rs}\, \{a\} \qquad (C \,\mathsf{rs}\, A) \,\mathsf{rs}\, B \cong (C \,\mathsf{rs}\, B) \,\mathsf{rs}\, A$$
$$C \,\mathsf{rs}\, A \cong C \,\mathsf{rs}\, (A \cup \widehat{A}) \qquad (C \,\mathsf{rs}\, A) \,\mathsf{rs}\, B \cong C \,\mathsf{rs}\, (A \cup B)$$

$$[a : C] \cong [\{a\} : C] \qquad [A : [B : C]] \cong [B : [A : C]]$$
$$[A : C] \cong [(A \cup \widehat{A}) : C] \qquad [A : [B : C]] \cong [(A \cup B) : C]$$

where $\widehat{A} = \{\widehat{a} \mid a \in A\}$. Show also that other properties of the basic communication interface operators carry over to the more general setting.

3.4 Extended PBC Syntax

Having presented the standard PBC and a number of its possible extensions, we now give a syntax which incorporates some of them.

An *(extended) static PBC expression* is a word generated by the syntax

$$\begin{array}{rl|l|l|l|l|l}
E ::= & \alpha & E \;\mathsf{sy}\; a & E \;\mathsf{sy}\; A & E \| E & [E * E * E] & \\
 & X & E \;\mathsf{rs}\; a & E \;\mathsf{rs}\; A & E \,\square\, E & E[f] & \\
 & [x_{()}] & [a : E] & [A : E] & E ; E & &
\end{array} \tag{3.14}$$

where $A \subseteq \mathsf{A}_{\mathsf{PBC}}$ is a set of action particles and x is a program variable; the meaning of the remaining items is the same as in (3.1). The *(extended) dynamic PBC expressions* are defined by the next syntax, where E is a static expression given by (3.14) and $u \in \mathit{Type}(x)$.

$$\begin{array}{rl|l|l|l|l|}
G ::= & \overline{E} & \underline{E} & G \| G & G \;\mathsf{sy}\; a & G \;\mathsf{sy}\; A \\
 & [\overline{x_{(u)}}] & G ; E & E ; G & G \;\mathsf{rs}\; a & G \;\mathsf{rs}\; A \\
 & G[f] & G \,\square\, E & E \,\square\, G & [a : G] & [A : G] \\
 & [G * E * E] & [E * G * E] & [E * E * G] \,. & &
\end{array} \tag{3.15}$$

The extensions have been chosen with the view of providing a suitable model for a concurrent programming language, in Chap. 9. But they are still not sufficient; in Chap. 8, we shall further extend the syntax of PBC to include two generalised iteration operators: $[E * F\rangle$ and $\langle F * E]$ in the static case, and $[G * E\rangle$, $[E * G\rangle$, $\langle G * E]$, and $\langle E * G]$ in the dynamic case, but with restrictions on their usage, as explained before. The second of these two loops will be crucial in providing the semantics of guarded while-loops in Chap. 9.

3.5 Examples of Transition Systems

The SOS-rules may be used to construct transition systems of static and dynamic PBC expressions. This is illustrated in Fig. 3.5 for some of the running examples. We consider transition systems up to isomorphism, and so $\mathsf{ts}_{E_{10}} = \mathsf{ts}_{E_{13}}$ means that the two transition systems are isomorphic. Notice that the initial states are here characterised by the presence of an outgoing skip and an incoming redo transition, since all the considered expressions are static ones. Recall that we do not represent the $\emptyset$-moves which label self-loops adjacent to all the nodes and do not appear elsewhere.

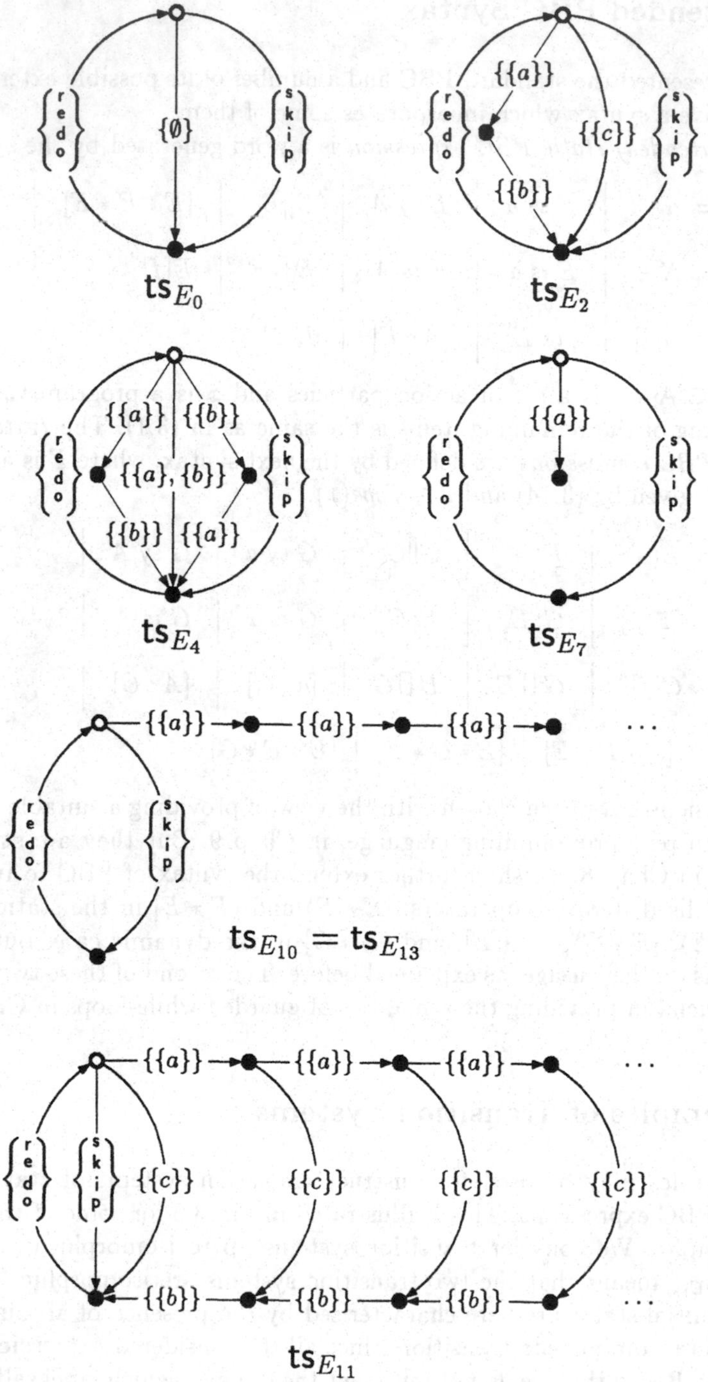

Fig. 3.5. Transition systems generated by some of the example expressions

3.6 Literature and Background

The technique of decorating expressions with overbars and underbars has been developed independently of the 'proved transitions' semantics of Boudol and Castellani [13], but bears certain resemblance to it.

It is, in principle, possible to consider dynamic expressions of the form $\overline{\overline{E}}$, $\underline{\underline{E}}$, $\underline{\overline{E}}$, ..., corresponding to multiple 'activations' of an expression, which is similar to auto-concurrent transitions in non-safe Petri nets. However, since we shall focus on the safe case, such expressions will not be considered beyond this chapter.

If $\overline{\overline{E}}$ was to be admitted as an initially marked expression, not only would one have to change the syntax of dynamic expressions, but also the 'unexpected' transitions in its Petri net semantics, such as those discussed in Sect. 2.6 (see Fig. 2.7), would become relevant. Indeed, consider the expression

$$\overline{\overline{(\{a\};\{\widehat{a}\})\ \mathsf{sy}\ a}}\ .$$

Such an expression could use one of the initial overbars to execute $\{a\}$, and then use the other initial overbar to execute $\{a\}$ again, in synchronisation with $\{\widehat{a}\}$, as shown below:

$$\begin{aligned}\overline{\overline{(\{a\};\{\widehat{a}\})\ \mathsf{sy}\ a}} &\xrightarrow{\emptyset} (\overline{\overline{\{a\}}};\{\widehat{a}\})\ \mathsf{sy}\ a \xrightarrow{\{\{a\}\}} (\underline{\overline{\{a\}}};\{\widehat{a}\})\ \mathsf{sy}\ a\\ &\xrightarrow{\emptyset} (\overline{\{a\}};\overline{\{\widehat{a}\}})\ \mathsf{sy}\ a \xrightarrow{\{\emptyset\}} (\underline{\{a\}};\underline{\{\widehat{a}\}})\ \mathsf{sy}\ a\ .\end{aligned}$$

Hence the expression could execute the step sequence $\{\{a\}\}\{\emptyset\}$, which is clearly not executable by $\overline{(\{a\};\{\widehat{a}\})\ \mathsf{sy}\ a}$. Note that the same step sequence can be generated by the transition rule of Petri nets in Fig. 2.7, provided there are two tokens on the initial place instead of one.

The multilink synchronisation scheme has been discussed in [29], [30], [40], and [41], while the single link scheme is adopted from [7].

Other extensions of the standard PBC to model, e.g., communication channels, may be found in [10, 39, 74], and some will be introduced in Chap. 9. We shall not give a full list of such extensions here; however, we would like to stress that they rely on the high flexibility of the theory and, consequently, on the ease of extending not only the syntax and the informal semantics of the algebra, but also its formal operational and Petri net semantics. This flexibility will be a constant concern in the development of the theory presented in this book, leading in particular to a generic box algebra introduced in Chap. 7.

The **redo** action bears some resemblance to the δ action of LOTOS ([75]), which also allows us to distinguish a successful termination from an unsuccessful one. More recently, the **skip** and **redo** actions were used in the context of the modelling of workflow processes by Petri Nets [1].

4. Petri Net Semantics

Petri nets have long been provided with different kinds of coherent semantics (e.g., [90, 83, 15]); in particular, concurrency semantics, such as trace [77], step [42], process [48, 5], and partial word semantics [49, 94, 100]. Therefore, a natural idea to get a fully fledged concurrency semantics for a process expression is to associate with it a Petri net. This technique has already been exploited for various process algebras [13, 19, 47, 44, 84, 95], but in many cases only a fragment of a model has been successfully translated. Here we shall show how to accomplish this task not only in full generality, but also entirely compositionally, thanks to a careful choice of the operators and very general mechanisms introduced to combine nets. In order to obtain a compositional translation of PBC expressions into Petri nets, we shall define for the latter operators corresponding to those introduced for the process algebra; the translation itself will then be a homomorphism. This chapter shows how to do this for the finite PBC and for iteration; in Chap. 5, we will turn to the translation of recursion variables. To avoid (or at least minimise) the risk of having to redesign a large part of the theory whenever a small change needs to be made to the algebra (for example, due to the introduction of a new operator), we shall develop a general mechanism to define net operators based on an auxiliary notion of net refinement.

The structure of the chapter is as follows. First, we introduce labelled Petri nets which will be used to model PBC expressions. Then we define the operation of net refinement, and in the third part of this chapter, we provide a compositional translation from (static as well as dynamic) PBC expressions to Petri nets. The presentation in this chapter is self-contained; all the relevant concepts will be formally defined, even those belonging to the elementary Petri net theory.

4.1 Compositionality and Nets

In this book, we aim to obtain a model in which it is possible to apply both Petri net and process algebra specific analysis techniques. An example of the former is an S-invariant-based structural analysis of nets, and of the latter a structural operational semantics together with related behavioural equivalences. In order to take full advantage of both kinds of techniques, we need a sound and expressive framework for composing Petri nets. We decided right from the start to base our approach to compositionality on transition refinement, i.e., we envisaged that each operator on nets would be based on a Petri net Ω whose transitions $t_1, t_2, \ldots$ are refined by the corresponding nets $\Sigma_1, \Sigma_2, \ldots$ in the process of forming a new net $\Omega(\Sigma_1, \Sigma_2, \ldots)$.

However, being only able to compose nets is clearly not enough. For our scheme to make sense, we also need an additional property, namely that of preserving semantic properties. For instance, to be able to use S-invariant analysis techniques, it ought to be the case that $\Omega(\Sigma_1, \Sigma_2, \ldots)$ is covered by S-invariants provided that the nets $\Sigma_1, \Sigma_2, \ldots$ were. Similarly, $\Omega(\Sigma_1, \Sigma_2, \ldots)$ should admit a structured operational semantics whenever $\Sigma_1, \Sigma_2, \ldots$ admitted one. Clearly, whether or not properties like these hold will strongly depend on the properties of the net Ω. In an ideal world, one would be looking for a single class of net operators (or, equivalently, a single class of nets Ω defining them) which would guarantee the preservation of all the relevant semantic properties. Indeed, as we shall see later, the net operators corresponding to PBC operators described in the previous two chapters do preserve S-coverability and admit structured operational semantics. However, when one tries to generalise the framework to possibly the widest class of net operators, it turns out that S-coverability and SOS semantics are orthogonal issues.

We therefore decided to carry out two essentially independent studies of compositionality in Petri net theory, *S-compositionality* and *SOS-compositionality*, the former being oriented towards the S-invariant analysis, and the latter towards the structured operational semantics. They will give rise to different sets of net operators, although their intersection is nonempty and will include, in particular, all net operators corresponding to the PBC operators introduced in the previous chapter. Despite being different, S-compositionality and SOS-compositionality do share a significant number of definitions and properties. We shall therefore introduce a third framework in which we will deal with a number of notions and results common to both kinds of compositionality; in particular, we will be able to distinguish those (labelled) nets that are easily composable with one another. These considerations will be shown in Sects. 4.2 and 4.3 of this chapter, and will then be applied to the standard PBC, in Sect. 4.4 and Chap. 5. The diagram shown in Fig. 4.1, where each arrow points from a more general framework to a less general one, summarises the structure of this and the next four chapters. In each case, we will specify exactly what are the nets Ω used to define net operators and the domain of nets to which such operators can be applied. The model of compositionality introduced in this chapter will be very general and therefore the technical results, however highly relevant, will be relatively straightforward.

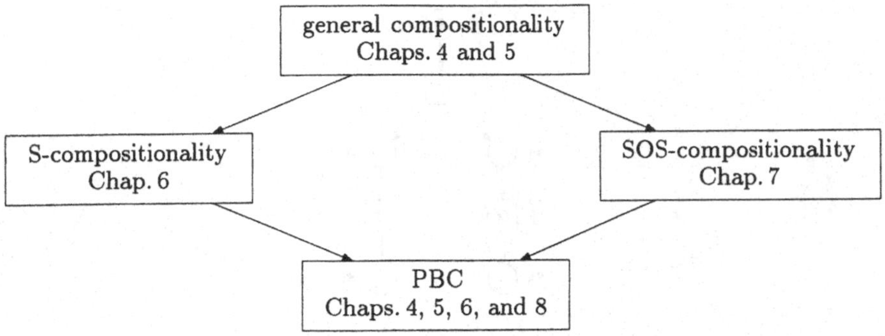

Fig. 4.1. Relating different frameworks that support compositionality

4.2 Labelled Nets and Boxes

In this section we delineate several classes of Petri nets whose interfaces are expressed by the labelling of places and transitions. Such nets will serve both as basic semantic objects of interest (plain boxes defined in Sect. 4.2.5), and as patterns (or functions) capable of creating new plain boxes (operator boxes defined in Sect. 4.3.1).

4.2.1 An Example

In view of what we have said about the operational semantics of PBC, a possible model for the dynamic expression $\alpha; (\alpha \| \alpha)$ might look like the (safe) net depicted in Fig. 4.2. Looking at this diagram, it is clear that the action name α cannot be used to identify transitions, since there are three of them, all corresponding to the same action α. Hence, we have to employ Petri nets with transitions being *labelled* by action names. For Petri nets defining net operators, we shall also use transition labelling, but with more complicated labels, called *(generalised) relabellings*, in order to allow combining together transitions coming from the composed nets. And, since action names may be treated as a special kind of (constant) general relabellings, the latter will be used in full generality.

Now, if we look closer at the net in Fig. 4.2, we may identify three different kinds of places. The place s_0 is special in the sense that it contains a token corresponding to the expression in its initial state. And, by symmetry, the places s_3 and s_4, when holding one token each, characterise its terminal state. The two remaining places, s_1 and s_2, may be considered as internal; they contribute to intermediate markings corresponding to intermediate dynamic expressions. The different roles of places may be captured by a suitable labelling mechanism, with three possible values identifying the three kinds of places. As a result, although static expressions will correspond to nets with empty markings, the labelling of places will allow one to associate with any such expression markings corresponding to its entry and exit states.

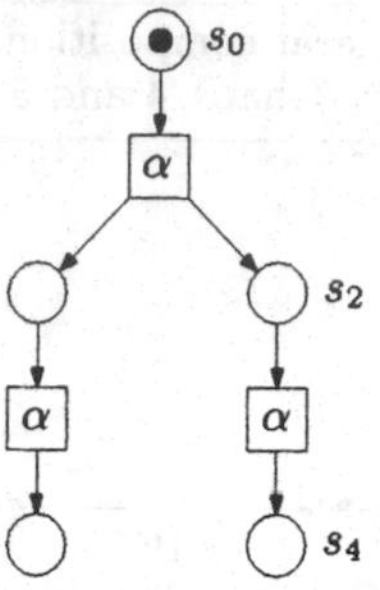

Fig. 4.2. A net corresponding to $\overline{\alpha;(\alpha\|\alpha)}$

4.2.2 Actions and Relabellings

We assume a set Lab of *actions* to be given, which at this point is an arbitrary set. Later, in Sect. 4.4 – when we employ the tools developed in the first part of this chapter to define a Petri net semantics of PBC – we shall take Lab to be the set $\mathsf{Lab}_{\mathrm{PBC}}$ of PBC multiactions. Intuitively, an element $\alpha \in \mathsf{Lab}$ denotes some interface activity. The notation 'Lab' has been chosen because actions will serve as transition labels; when executing a transition we shall be able to consider the corresponding action.

In order to obtain a concise and uniform translation from expressions to nets, we shall consider (generalised) relabellings which will specify interface changes applied to nets involved in the translation. Formally, a *relabelling* ϱ is a relation

$$\varrho \subseteq \Big(\mathsf{mult}(\mathsf{Lab})\Big) \times \mathsf{Lab} \tag{4.1}$$

such that $(\emptyset, \alpha) \in \varrho$ if and only if $\varrho = \{(\emptyset, \alpha)\}$. An intuition behind a pair (Γ, α) belonging to ϱ is that it allows any (finite) group of transitions whose labels match the argument, i.e., the multiset of actions Γ, to be combined (glued) into a new transition labelled α. Since Γ is a multiset – an unordered object – the order of transitions in such a group does not matter, and we immediately obtain a kind of simple commutativity of the operation described by ϱ. The reason why the empty multiset is normally excluded from being a Γ is that it would be matched by the empty group of transitions, thus allowing one to create transitions 'out of nothing'; in technical terms this would lead to isolated transitions in the definition of net refinement and we will later find good reasons to exclude such transitions. Three specific relabellings are of particular importance. The *identity* relabelling,

$$\varrho_{\mathrm{id}} = \Big\{(\{\alpha\}, \alpha) \,\Big|\, \alpha \in \mathsf{Lab}\Big\}$$

captures the 'keep things as they are' interface (non)change. A *constant* relabelling,

$$\varrho_\alpha = \{(\emptyset, \alpha)\}$$

where α is an action in Lab, can be identified with α itself, so that we may consider the set of actions Lab to be embedded in the set of all relabellings. A relabelling which is not constant will be called *transformational.* A *restriction*

$$\varrho_{\mathsf{Lab}'} = \Big\{(\{\alpha\}, \alpha) \;\Big|\; \alpha \in \mathsf{Lab}'\Big\}$$

filters out transitions labelled by actions not belonging to a set $\mathsf{Lab}' \subseteq \mathsf{Lab}$.

4.2.3 Labelled Nets

The class of Petri nets we shall work with are labelled nets. Not surprisingly, such a broad term has several different meanings in the Petri net theory, and so it must be defined precisely. In this book, a *labelled net* will be a tuple

$$\Sigma = (S, T, W, \lambda, M) \tag{4.2}$$

such that: S and T are disjoint sets of respectively *places* and *transitions*; W is a *weight function* from the set $(S\times T)\cup(T\times S)$ to the set of natural numbers $\mathbf{N}$; λ is a *labelling* function for places and transitions such that $\lambda(s) \in \{\mathsf{e}, \mathsf{i}, \mathsf{x}\}$, for every place $s \in S$, and $\lambda(t)$ is a relabelling of the form (4.1), for every transition $t \in T$; and M is a *marking*, i.e., a mapping assigning a natural number to each place $s \in S$ (thus it is a finite or infinite multiset of places). This generalises the net model we considered in Sect. 2.2.

We adopt the standard rules about representing nets as directed graphs, viz. places are represented as circles, transitions as rectangles (or squares), the flow relation generated by W is indicated by arcs annotated with the corresponding weights, and markings are shown by placing tokens within circles. In the diagrams, place labels will be put near (or inside) the corresponding circles, and transition labels inside the corresponding rectangles. As usual, the zero weight arcs will be omitted and the unit weight arcs (or unitary arcs) will be left as plain arcs, i.e., unannotated. To avoid ambiguity, we will sometime decorate the various components of Σ with the index Σ; thus, T_Σ will denote the set of transitions of Σ, etc.

We shall not require that the nets be finite; indeed, we shall need infinite nets to model recursion in Chap. 5. This means, however, that some care will have to be taken at various points in the development of the theory. Although a labelled net Σ always has a marking M_Σ, we found it convenient to call Σ *marked* if there is at least one place s such that $M_\Sigma(s) > 0$; otherwise Σ will be called *unmarked*. Figure 4.3 shows the graph of a marked labelled net $\Sigma_0 = (S_0, T_0, W_0, \lambda_0, M_0)$ defined as follows (note that we represent the functions W_0, λ_0, and M_0 as relations).

$$
\begin{aligned}
S_0 &= \{s_0, s_1, s_2, s_3\} \\
T_0 &= \{t_0, t_1, t_2\} \\
W_0 &= \Big((TS \cup ST) \times \{1\}\Big) \cup \Big(\big((S \times T) \setminus ST \cup (T \times S) \setminus TS\big) \times \{0\}\Big) \\
\lambda_0 &= \{(s_0, \mathsf{e}), (s_1, \mathsf{i}), (s_2, \mathsf{x}), (s_3, \mathsf{e}), (t_0, \alpha), (t_1, \beta), (t_2, \alpha)\} \\
M_0 &= \{(s_0, 1), (s_1, 0), (s_2, 0), (s_3, 1)\}
\end{aligned}
$$

where

$$TS = \{(t_0, s_1), (t_1, s_2), (t_2, s_3)\} \text{ and } ST = \{(s_0, t_0), (s_1, t_1), (s_3, t_2)\} .$$

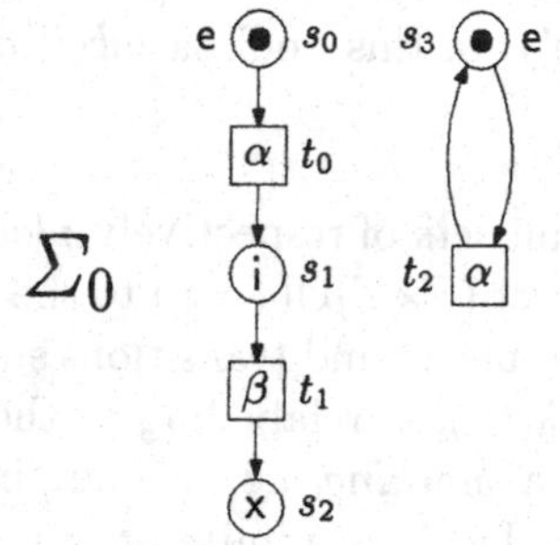

Fig. 4.3. A labelled net

Step Sequence Semantics. We employ a finite step sequence semantics for a labelled net $\Sigma = (S, T, W, \lambda, M)$, in order to capture concurrency in the behaviour of the system modelled by Σ. A finite multiset of transitions $U \in \mathsf{mult}(T)$, called a *step*, is *enabled* by Σ if for every place $s \in S$,

$$M(s) \geq \sum_{t \in U} U(t) \cdot W(s, t)$$

i.e., if there are enough tokens in each place to satisfy simultaneously the needs of all the transitions of U, respecting their multiplicity. We denote this by $\Sigma[U\rangle$, or $M[U\rangle$ if the net is understood from the context. An enabled step U can be *executed* leading to a new marking M' defined, for every place $s \in S$, by

$$M'(s) = M(s) + \sum_{t \in U} U(t) \cdot \Big(W(t, s) - W(s, t)\Big) .$$

Depending on the context, we will denote this by $M\ [U\rangle\ M'$ or $\Sigma\ [U\rangle\ \Theta$, where Θ is the labelled net (S, T, W, λ, M'). The above captures formally, and generalises to steps rather than single transition occurrences, the definitions given in Sect. 2.2.

As it happens, all the steps we shall really need in the net semantics of PBC expressions are sets of transitions rather than true multisets. However, the latter are required for the general labelled net framework we have adopted. Although we use the term 'step' to refer both to a finite multiset of transitions here, and to a finite multiset of multiactions in Chap. 3, it will always be clear from the context which interpretation is meant.

Transition labelling may be extended to steps, through the formula

$$\lambda(U) = \sum_{t \in U} U(t) \cdot \left\{\lambda(t)\right\} \in \mathsf{mult}(\mathsf{Lab}) .$$

In particular, we will denote $\Sigma\ [\Gamma\rangle_{\mathsf{lab}}\ \Theta$ whenever there is a multiset of transitions U such that $\Sigma\ [U\rangle\ \Theta$and $\Gamma = \lambda(U)$. This allows one to translate various behavioural notions defined in terms of multisets of transitions into notions based on multisets of transition labels (or *labelled steps*).

A *finite step sequence* of Σ is a sequence $\sigma = U_1 \ldots U_k$ of steps for which there are labelled nets $\Sigma_0, \ldots, \Sigma_k$ such that $\Sigma = \Sigma_0$ and $\Sigma_{i-1}\ [U_i\rangle\ \Sigma_i$, for every $1 \leq i \leq k$. Depending on the context, this shall be denoted by $\Sigma\ [\sigma\rangle\ \Sigma_k$ or $M_\Sigma\ [\sigma\rangle\ M_{\Sigma_k}$. Moreover, the marking M_{Σ_k} will be called *reachable* from M_Σ, and denoted $M_{\Sigma_k} \in [M_\Sigma\rangle$; and Σ_k *derivable* from Σ, and denoted $\Sigma_k \in [\Sigma\rangle$.

As in Sect. 3.2.1, empty steps are always enabled (which follows easily from the above formulae), and can be ignored when one considers a step sequence. The only difference is that here the empty step always relates a net to itself: $\Sigma\ [\emptyset\rangle\ \Theta \Leftrightarrow \Sigma = \Theta$. A net exhibits *auto-concurrency* if it enables a step sequence which is not a sequence of sets of transitions (i.e., one of the steps in the sequence is a true multiset). The label of a step may be a multiset rather than a set, even if the net exhibits no auto-concurrency, since different concurrent transitions may have the same label.

A step is another word for *concurrent occurrence of transitions*. For the marked net Σ_0 shown in Fig. 4.3, transitions t_0 and t_2 are concurrently enabled and so $\{t_0, t_2\}$ is an enabled step. After this step has been executed, transitions t_1 and t_2 become concurrently enabled and hence $\{t_0, t_2\}\{t_1, t_2\}$ is a step sequence of Σ_0. A step does not need to be maximal; for instance, $\{t_0\}$ is also a step of Σ_0, and $\{t_0\}\{t_1\}\{t_2\}\{t_2\}$ and $\{t_0, t_2\}\{t_1\}\{t_2\}$ are step sequences. In terms of labelled steps, $\{t_0, t_2\}\{t_1\}\{t_2\}$ corresponds to $\{\alpha, \alpha\}\{\beta\}\{\alpha\}$. Different step sequences may correspond to the same labelled step sequence; for example, both $\{t_0\}\{t_2\}\{t_1\}$ and $\{t_2\}\{t_0\}\{t_1\}$ correspond to $\{\alpha\}\{\alpha\}\{\beta\}$.

Related Notions and Notations. If the labelling of a place s in a labelled net Σ is e, then s is an *entry* place, if i, then s is an *internal* place, and if x, then s is an *exit* place. By convention, ${}^\circ\Sigma$, Σ°, and $\ddot{\Sigma}$ denote respectively the entry, exit, and internal places of Σ. For every place (transition) x, we use ${}^\bullet x$ to denote its pre-set which is the set of all transitions (places) y such that there is an arc from y to x, that is, $W(y, x) > 0$. The post-set $x^\bullet$ is defined in a similar way, and comprises all y such that $W(x,y) > 0$. The pre- and post-set notation extends to sets R of places and transitions through $^\bullet R = \bigcup\{^\bullet r \mid r \in R\}$ and $R^\bullet = \bigcup\{r^\bullet \mid r \in R\}$. In what follows, all nets are assumed to be T-*restricted* which means that the pre- and post-sets of each transition are nonempty. However, no assumption of that kind is made for places. For the labelled net of Fig. 4.3, we have $^\circ\Sigma_0 = \{s_0, s_3\}$, $\Sigma_0^\circ = \{s_2\}$, $^\bullet s_0 = \emptyset$, $s_0^\bullet = \{t_0\}$, and $\{s_0, s_1\}^\bullet = \{t_0, t_1\} = {}^\bullet\{s_1, s_2\}$.

A labelled net Σ is *finite* if both S and T are, *simple* if W always returns 0 or 1, and *pure* if for all transitions $t \in T$, $^\bullet t \cap t^\bullet = \emptyset$. If Σ is not pure then there are $s \in S$ and $t \in T$ such that $W(s,t) \cdot W(t,s) > 0$. In such a case, the pair $\{s,t\}$ will be called a *side-loop*, and s will be called a *side place* of t. The net in Fig. 4.3 is finite and simple but not pure as it contains a side-loop, $\{s_3, t_2\}$.

A marking M of Σ is *safe* if $M(S) \subseteq \{0,1\}$, i.e., if for every $s \in S$, $M(s) = 0$ or $M(s) = 1$. A safe marking can and will often be identified with the set of places to which it assigns 1. A marking is *clean* if it is not a proper super-multiset of $^\circ\Sigma$ nor Σ°, i.e., if $^\circ\Sigma \subseteq M$ or $\Sigma^\circ \subseteq M$ implies $^\circ\Sigma = M$ or $\Sigma^\circ = M$, respectively. It is k-*clean* if it is not a proper super-multiset of $k \cdot {}^\circ\Sigma$ nor $k \cdot \Sigma^\circ$, i.e., if $k \cdot {}^\circ\Sigma \subseteq M$ or $k \cdot \Sigma^\circ \subseteq M$ implies $M = k \cdot {}^\circ\Sigma$ or $M = k \cdot \Sigma^\circ$, respectively (so that cleanness amounts to 1-cleanness).

We will call $^\circ\Sigma$ the *entry marking* of Σ, and Σ° the *exit marking*; note that both are safe and clean. The marking of the net in Fig. 4.3 is safe and clean. It would cease to be clean if we added a token to it, even if this new token was put on one of the entry places because the corresponding multiset of tokens would be strictly greater than $^\circ\Sigma_0$.

A marked net is *(k-)clean* if all its reachable markings are (k-)clean, *safe* if all its reachable markings are safe, k-*bounded* if no reachable marking puts more than k tokens on any place (so that 1-boundedness is the same as safeness), and *bounded* if there is k such that it is k-bounded.

A labelled net Σ is ex-*restricted* if there is at least one entry and at least one exit place, $^\circ\Sigma \neq \emptyset \neq \Sigma^\circ$. Σ is *e-directed* (*x-directed*) if the entry (respectively, exit) places are free from incoming (respectively, outgoing) arcs, $^\bullet(^\circ\Sigma) = \emptyset$ (respectively, $(\Sigma^\circ)^\bullet = \emptyset$). Moreover, Σ is *ex-directed* if it is both e-directed and x-directed. The next property is more of a behavioural nature: we will say that Σ is *ex-exclusive* if, for every marking M reachable from M_Σ or $^\circ\Sigma$ or Σ°, it is the case that $M \cap {}^\circ\Sigma = \emptyset$ or $M \cap \Sigma^\circ = \emptyset$, i.e., it is not possible to mark simultaneously an entry and an exit place. The labelled net Σ_0 in Fig. 4.3 is ex-restricted and x-directed, but it is neither e-directed nor ex-exclusive.

A transition $t \in T$ is *ex-skewed* if $^\bullet t \cap {}^\circ\Sigma \neq \emptyset \neq {}^\bullet t \cap \Sigma^\circ$ or $t^\bullet \cap {}^\circ\Sigma \neq \emptyset \neq t^\bullet \cap \Sigma^\circ$. Note that a labelled net which contains an ex-skewed transition t is ex-restricted but not ex-directed, and that if t can be executed, then the net is not ex-exclusive.

Let ind_Σ be the symmetric, not necessarily transitive, relation on the transitions of Σ, defined by

$$\mathsf{ind}_\Sigma = \left\{(t,u) \in T \times T \;\middle|\; ({}^\bullet t \cup t^\bullet) \cap ({}^\bullet u \cup u^\bullet) = \emptyset\right\}.$$

It is called the *independence relation* because two transitions related by ind_Σ have no impact on their respective environments. If they are both enabled separately, then they are enabled simultaneously (as a step). Conversely, if Σ is safe (which will later be our exclusive case of interest), then it can be shown that any two transitions occurring in the same step are independent. Notice that T-restrictedness implies the irreflexivity of the independence relation, i.e., $(t,t) \notin \mathsf{ind}_\Sigma$, for every $t \in T$.

We will use three direct ways of modifying the marking of a labelled net $\Sigma = (S,T,W,\lambda,M)$. First, we define $\lfloor \Sigma \rfloor$ as $(S,T,W,\lambda,\emptyset)$. Typically, this operation is used when Σ is marked; it corresponds to erasing all tokens, and also to erasing all over- and underbars in a dynamic expression which yields a static expression (cf. Sects. 2.3 and 3.1). Second, we define $\overline{\Sigma}$ and $\underline{\Sigma}$ as, respectively, $(S,T,W,\lambda,{}^\circ\Sigma)$ and $(S,T,W,\lambda,\Sigma^\circ)$. These operations are typically applied if Σ is unmarked; they correspond to placing one token on each entry place (respectively,one token on each exit place), and also to the operations of overbarring (underbarring) a whole static expression to obtain an initial (respectively, terminal) dynamic expression. It should be stressed that $\lfloor . \rfloor$, $\overline{(.)}$, and $\underline{(.)}$ are syntactic operations having nothing to do with derivability (or reachability) in the sense of the step sequence semantics.

One might ask why in the definition of the step sequence semantics was the step U was assumed finite, even though we do allow infinite nets. The question is relevant because the difference between the adopted semantics and that which would allow infinite steps is a profound one, as the example of Fig. 4.4 shows: under the finite step sequence rule t is never enabled, but this is not the case if we allow infinite steps and execute $\{t_1, t_2, \ldots\}$ first.

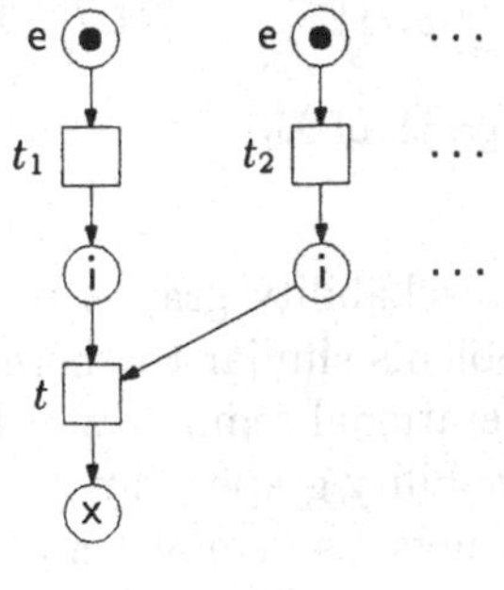

Fig. 4.4. An infinite net allowing an infinite step

A possible answer could be to say that in the 'real world' an infinite step is not feasible since it would require an infinite set of processors or other computing devices to execute. However, we are not entirely happy with such an answer. One of the reasons is that a similar argument could be used to dismiss any work on infinite data types since there will never be enough memory to represent, say, the set of all integers. We therefore decided not to give any justification at all. Simply, the finiteness of steps reflects the prevailing attitudes in the field of concurrency theory, and seriously simplifies the general setup (see [5]). However, we will investigate the consequences of allowing infinite steps in Chap. 6, showing how the theory developed for the finite case needs to be changed or re-evaluated.

4.2.4 Equivalence Notions

As for PBC expressions, various behavioural equivalences may also be defined for labelled nets. It may first be observed that the whole set of step sequences of a labelled net may be specified by defining its *step reachability graph*, whose nodes are all the reachable markings (or equivalently, all the derivable nets) and whose arcs are labelled with steps which transform one marking into another. For example, Fig. 4.5 represents the step reachability graph of the net Σ_0 shown in Fig. 4.3; the empty steps are left implicit as in the graphical representation of the transition systems of PBC expressions in Chap. 3 (the same will be true of the transition systems of labelled nets introduced next). The arc labels may be transition steps (as in Fig. 4.5) or labelled steps. Any finite path in such a graph starting at the initial node, denoted by a small ring rather than a black circle, specifies a legal step sequence of the marked net Σ_0, and vice versa.

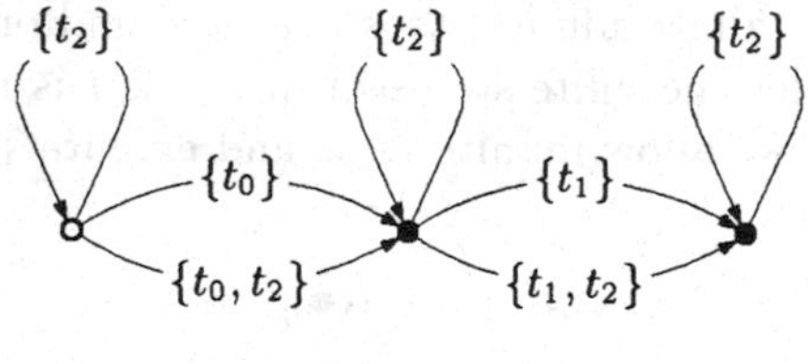

Fig. 4.5. Step reachability graph of Σ_0

Using (labelled) step reachability graphs to represent the behaviour of labelled nets leads to problems similar to those encountered in Sect. 3.2.2, when we discussed the operational semantics of PBC expressions. In particular, isomorphism of reachability graphs is not preserved by, e.g., sequential and choice composition of nets, as demonstrated in Figs. 4.6 and 4.7. (Both net operations are performed according to the definitions given later in this chapter.) We address this problem by introducing a device similar to that applied in the case of PBC expressions. Recall that the idea there was to augment the behaviour with two auxiliary actions, skip and redo, which could transform an expression in its initial state into its terminal state, and vice

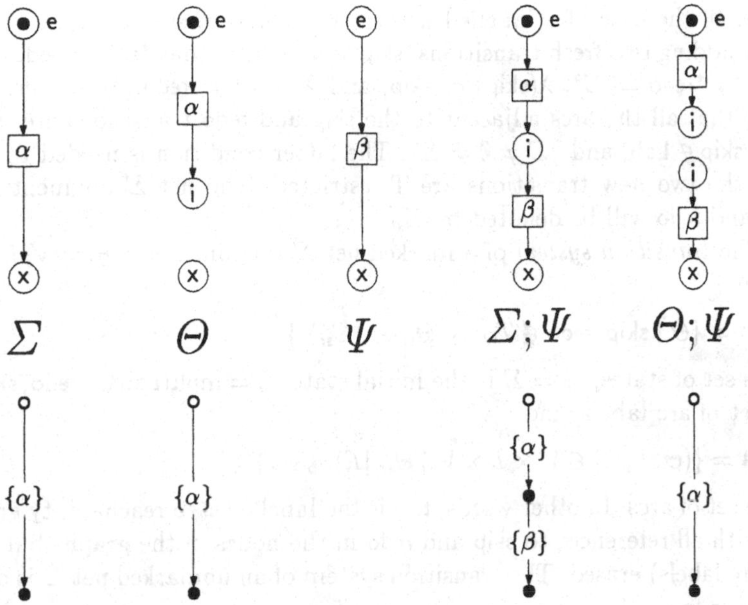

Fig. 4.6. Five nets and the corresponding (labelled) step reachability graphs, demonstrating that isomorphism of reachability graphs is not preserved by sequential composition

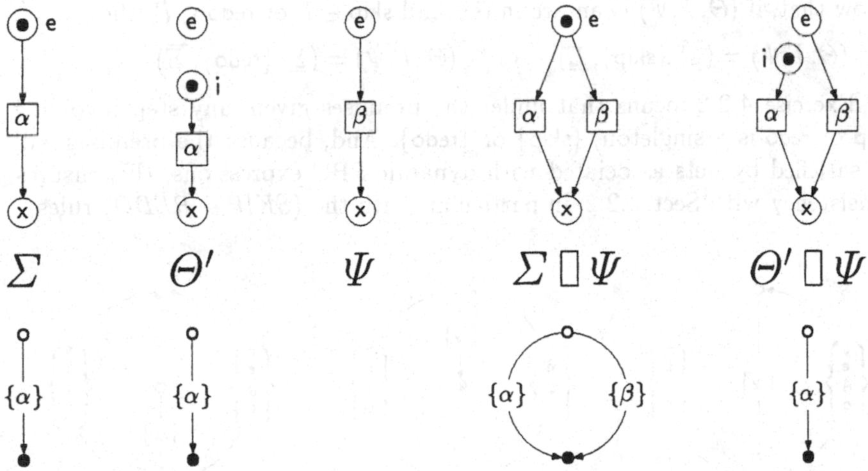

Fig. 4.7. Five nets and the corresponding (labelled) step reachability graphs demonstrating that isomorphism of reachability graphs is not preserved by choice composition

versa. In the case of a labelled net Σ, we achieve a similar effect by artificially adding two fresh transitions, skip and redo, so that ${}^\bullet\mathsf{skip} = \mathsf{redo}^\bullet = {}^\circ\Sigma$, $\mathsf{skip}^\bullet = {}^\bullet\mathsf{redo} = \Sigma^\circ$, $\lambda(\mathsf{skip}) = \mathsf{skip}$, and $\lambda(\mathsf{redo}) = \mathsf{redo}$. Moreover, we assume that all the arcs adjacent to the skip and redo transitions are unitary, $\mathsf{redo}, \mathsf{skip} \notin \mathsf{Lab}$, and ${}^\circ\Sigma \neq \emptyset \neq \Sigma^\circ$. The latter condition is needed to ensure that the two new transitions are T-restricted. The net Σ augmented with skip and redo will be denoted by Σ_{sr}.

The *transition system* of a marked net Σ is defined as $\mathsf{ts}_\Sigma = (V, L, A, v_0)$ where

$$V = \{\Theta \mid \mathsf{skip}, \mathsf{redo} \notin T_\Theta \;\wedge\; \Theta_{\mathsf{sr}} \in [\Sigma_{\mathsf{sr}}\rangle\,\}$$

is the set of states, $v_0 = \Sigma$ is the initial state, $L = \mathsf{mult}(\mathsf{Lab} \cup \{\mathsf{redo}, \mathsf{skip}\})$ is the set of arc labels, and

$$A = \{(\Theta, \Gamma, \Psi) \in V \times L \times V \mid \Theta_{\mathsf{sr}}\, [\Gamma\rangle_{\mathsf{lab}}\, \Psi_{\mathsf{sr}}\}$$

is the set of arcs. In other words, ts_Σ is the labelled step reachability graph of Σ_{sr} with all references to skip and redo in the nodes of the graph (but not in the arc labels) erased. The transition system of an unmarked net Σ is defined as $\mathsf{ts}_\Sigma = \mathsf{ts}_{\overline{\Sigma}}$.

Exercise 4.2.1. Determine sufficient conditions for Σ_{sr} to be safe and clean. What would be the answer if Σ was ex-directed?

Exercise 4.2.2. Let Σ an ex-restricted labelled net such that Σ_{sr} is clean. Show that, if (Θ, Γ, Ψ) is an arc in ts_Σ and $\mathsf{skip} \in \Gamma$ or $\mathsf{redo} \in \Gamma$, then

$$(\Theta, \Gamma, \Psi) = (\overline{\Sigma}, \{\mathsf{skip}\}, \underline{\Sigma}) \quad \text{or} \quad (\Theta, \Gamma, \Psi) = (\underline{\Sigma}, \{\mathsf{redo}\}, \overline{\Sigma})\,.$$

Exercise 4.2.2 means that under the premises given, any step involving skip or redo is a singleton, {skip} or {redo}. And, because the premises will be satisfied by nets associated with dynamic PBC expressions, this ensures consistency with Sect. 3.2.2; in particular, with the (*SKIP* – *REDO*) rules.

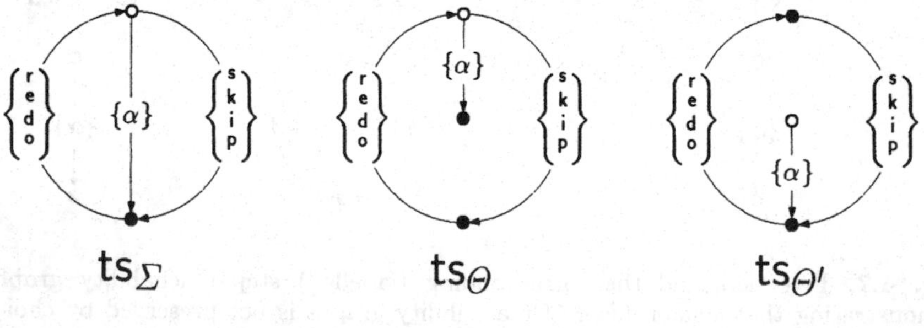

Fig. 4.8. Transition systems of labelled nets

Figure 4.8 shows that augmenting labelled nets with the redo and skip transitions allows one to discriminate between the nets Σ, Θ, and Θ' depicted in Figs. 4.6 and 4.7. Intuitively, skip and redo allow for distinguishing the entry and exit markings, and accounting for the possibility that a net left through the exit marking may later be re-entered through the entry marking.

Let Σ and Θ be two labelled nets which are either both unmarked, or both marked. The nets are *ts-isomorphic*, $\Sigma \cong \Theta$, if ts_Σ and ts_Θ are isomorphic transition systems, and *strongly equivalent*, $\Sigma \approx \Theta$, if ts_Σ and ts_Θ are strongly equivalent transition systems. Notice that there is no need to consider an isomorphism up to the empty moves, as in Sect. 3.2.2, since here $\Sigma\,[\emptyset\rangle\,\Sigma' \Leftrightarrow \Sigma = \Sigma'$. Figure 4.9 shows two strongly equivalent labelled nets;moreover, the reader may check that in Fig. 4.6, $\Theta \cong \Theta;\Psi$. The two equivalence relations are unaffected by adding or dropping transitions which may never be executed in the augmented nets (that is, augmented with skip and redo).

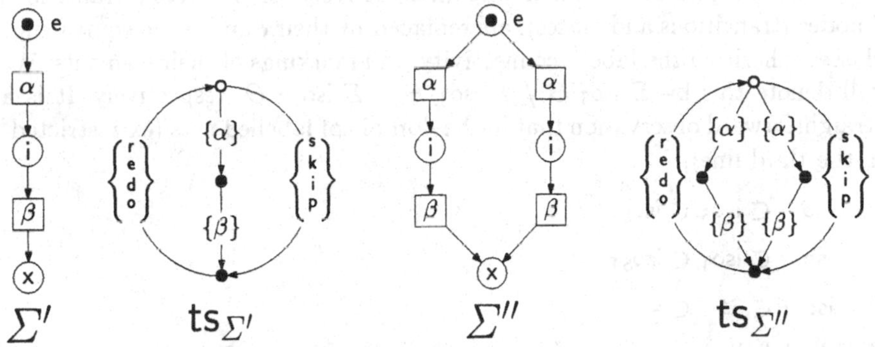

Fig. 4.9. Two strongly equivalent (but not ts-isomorphic) labelled nets

What we have done is not the only way to define sound equivalence notions in the domain of labelled nets. In particular, it may be observed that ts-isomorphism and strong equivalence are behavioural notions, hence it may be difficult to check if two nets are indeed equivalent. Now, since we are working with nets, it is also possible to define equivalence notions based on the structure of nets rather than on their behaviour. But, again, there are different ways to do so.

Arguably the strongest structural equivalence, other than equality, is net isomorphism. Two labelled nets, Σ and Θ, are *isomorphic* if there is a bijective mapping $isom : S_\Sigma \cup T_\Sigma \to S_\Theta \cup T_\Theta$ such that for all $s \in S_\Sigma$ and $t \in T_\Sigma$,

$$
\begin{array}{ll}
isom(s) \in S_\Theta & isom(t) \in T_\Theta \\
\lambda_\Theta(isom(s)) = \lambda_\Sigma(s) & \lambda_\Theta(isom(t)) = \lambda_\Sigma(t) \\
W_\Theta(isom(s), isom(t)) = W_\Sigma(s,t) & W_\Theta(isom(t), isom(s)) = W_\Sigma(t,s) \\
M_\Theta(isom(s)) = M_\Sigma(s)\ . &
\end{array}
$$

We will denote this by Σ iso Θ and call *isom* an isomorphism for Σ and Θ.

Weaker equivalences are obtained by allowing the two nets to differ only by duplicate places and/or transitions. Two places, s and s', *duplicate* each other in a labelled net Σ if $\lambda_\Sigma(s) = \lambda_\Sigma(s')$, $M_\Sigma(s) = M_\Sigma(s')$, and for every transition t, $W_\Sigma(s,t) = W_\Sigma(s',t)$ and $W_\Sigma(t,s) = W_\Sigma(t,s')$. Then, in any evolution of the net, the two places do not add anything with respect to each other. Similarly, two transitions, t and t', *duplicate* each other if $\lambda_\Sigma(t) = \lambda_\Sigma(t')$, and for every place s, $W_\Sigma(s,t) = W_\Sigma(s,t')$ and $W_\Sigma(t,s) = W_\Sigma(t',s)$. Then, in any label-based evolution of the net, the two transitions do not add anything with respect to each other. Clearly, the relation of *duplication* is an equivalence relation for places and transitions; thus it is possible to replace places and/or transitions by their duplication equivalence classes, obtaining nets with essentially the same behaviour. Two labelled nets, Σ and Θ, will be called *place-duplicating / transition-duplicating / node-duplicating* if they lead to isomorphic labelled nets when, respectively, their places / transitions / nodes (transitions and places), are replaced by their duplication equivalence classes inheriting the labels, connectivity and markings of their elements. We will denote this by $\Sigma\ \mathsf{iso_S}\ \Theta$ / $\Sigma\ \mathsf{iso_T}\ \Theta$ / $\Sigma\ \mathsf{iso_{ST}}\ \Theta$, respectively. It is a straightforward observation that in the domain of labelled nets (ex-restricted, in the third line),

$$\mathsf{iso} \subset \mathsf{iso_S} \subset \mathsf{iso_{ST}}$$

$$\mathsf{iso} \subset \mathsf{iso_T} \subset \mathsf{iso_{ST}}$$

$$\mathsf{iso_{ST}} \subset \cong \subset \approx .$$

For the labelled nets in Fig. 4.10, we have that $\Sigma_1\ \mathsf{iso_T}\ \Sigma_2\ \mathsf{iso_{ST}}\ \Sigma_3\ \mathsf{iso_S}\ \Sigma_1$, and that Σ_4 is not equivalent to any of the other three nets.

Exercise 4.2.3. Verify this.

Exercise 4.2.4. Show that $\Sigma\ \mathsf{iso_{ST}}\ \Theta$ if and only if there are relations $\omega_S \subseteq S_\Sigma \times S_\Theta$ and $\omega_T \subseteq T_\Sigma \times T_\Theta$ such that the following hold:

- $S_\Sigma \cup T_\Sigma = \{x \mid \exists y : (x,y) \in \omega_S \cup \omega_T\}$.
- $S_\Theta \cup T_\Theta = \{y \mid \exists x : (x,y) \in \omega_S \cup \omega_T\}$.
- $W_\Sigma(s,t) = W_\Theta(r,w)$, for $(s,r) \in \omega_S$ and $(t,w) \in \omega_T$.
- $W_\Sigma(t,s) = W_\Theta(w,r)$, for $(s,r) \in \omega_S$ and $(t,w) \in \omega_T$.
- $M_\Sigma(s) = M_\Theta(r)$, for $(s,r) \in \omega_S$.
- $\lambda_\Sigma(x) = \lambda_\Theta(y)$, for $(x,y) \in \omega_S \cup \omega_T$.

Find similar characterisations for $\mathsf{iso_T}$ and $\mathsf{iso_S}$.

4.2.5 Boxes

A *box* is an ex-restricted labelled net Σ. We require Σ to be ex-restricted to ensure that the operation of net refinement is well defined. We will later see how the theory depends on this assumption, cf. Sects. 4.3.7 and 5.4.5. There

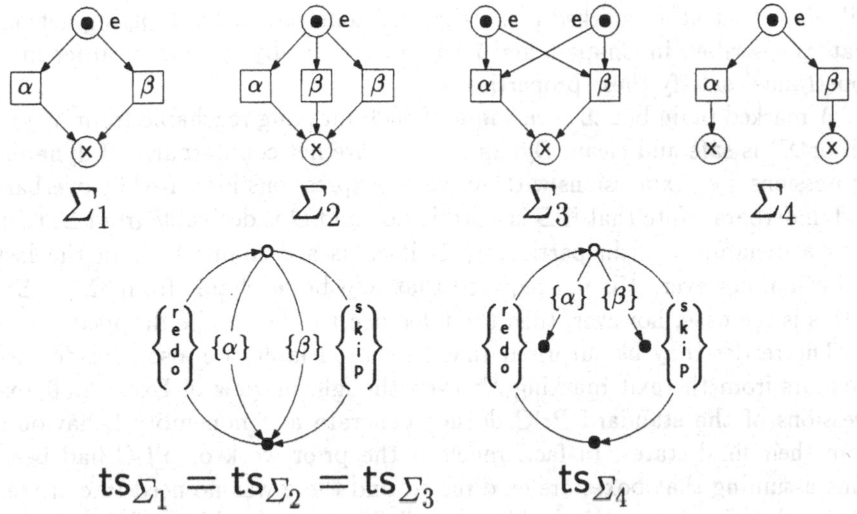

Fig. 4.10. Four labelled nets and their transition systems

are two main classes of boxes which we will be interested in, viz. plain boxes and operator boxes.

Plain boxes – defined in this section – are basic semantical objects of interest. They form the Petri net domain upon which various operators are defined, just as expressions form the domain of a process algebra upon which operators are defined. When giving a Petri net semantics of a process algebra, such as PBC or CCS, we will associate a plain box with every expression.

Operator boxes – defined in Sect. 4.3.1 – define ways of constructing new plain boxes out of given ones. When translating a process algebra into Petri nets, we will aim at associating an operator box with every operator of the process algebra. It is one of the characteristic features of our approach that the same class of nets – boxes – serve to describe two seemingly very different objects, namely the elements and the operators of the semantical domain. However, plain boxes and operator boxes will be distinguished by some additional properties.

Plain Boxes. A box Σ is *plain* if for each transition $t \in T_\Sigma$, the label $\lambda_\Sigma(t)$ is a constant relabelling. Hence, by our convention which identifies constant relabellings with actions, every transition label in a plain box is an action in Lab.

Static and Dynamic Boxes. We now introduce, using simple behavioural properties, two main classes of plain boxes.

An unmarked plain box Σ is *static* if each marking reachable from ${}^\circ\Sigma$ or Σ° is safe and clean. Static boxes are net counterparts of static expressions, i.e., expressions without overbars and underbars. The safeness and cleanness conditions delineate the class of boxes which will be considered. Later, we will show that all unmarked boxes that are constructed by applying net operators described in Chaps. 6 and 7 will automatically (or under rather mild constraints) satisfy these properties.

A marked plain box Σ is *dynamic* if each marking reachable from M_Σ or ${}^\circ\Sigma$ or Σ° is safe and clean. Dynamic boxes are net counterparts of dynamic expressions, i.e., expressions with active subexpressions indicated by overbars and underbars. Note that if Σ is a static box and Θ is derivable from $\overline{\Sigma}$, then Θ is a dynamic box; in particular, $\overline{\Sigma}$ itself is a dynamic box. In the last definition, however, it is not required that M_Σ be reachable from ${}^\circ\Sigma$ or Σ°. If this is the case, however, then the reference to M_Σ can be dropped.

The reader may be surprised that here (and later) we also consider behaviours from the exit marking Σ° even though, in view of Exc. 3.2.50, expressions of the standard PBC do not generate any nonempty behaviours from their final states. In fact, much of the prior work on PBC had been done assuming that boxes are ex-directed and there was no need to consider reachability from the exit markings at all. However, in this book we do not make such an assumption which means, on the one hand, that the formal treatment is more involved and, on the other hand, we will at the end be able to incorporate into the formal framework generalised iteration constructs. This, in turn, will significantly enhance the modelling of the concurrent programming language in Chap. 9.

We shall mainly be interested in dynamic boxes whose markings are reachable from the entry markings, but when combining reachable dynamic boxes together it may sometimes happen that the result does not satisfy this property, much in the same way as some syntactically valid dynamic PBC expressions may not be reached from their corresponding initial expression (for instance, $(a;\overline{b})$ rs a is not reachable from $\overline{(a;b)\ \mathsf{rs}\ a}$). Moreover, requiring that M_Σ be reachable from ${}^\circ\Sigma$ in the definition of a dynamic box could be hard to check in the general case. An objection could be that determining whether all the reachable markings are safe and clean may also be hard, especially for infinite boxes. However, we shall see in Chap. 6 that a simple structural argument may be used to prove that all the constructed boxes have the required property, and in Chap. 7 that a careful behavioural analysis leads to the same conclusion.

When a box Σ is marked (like $\overline{\Sigma}$, because of ex-restrictedness), its reachable markings are always nonempty (due to T-restrictedness). On the other hand, the empty marking of a box has no successor markings except itself. Thus, the distinction between static and dynamic boxes is invariant over behaviour. Moreover, if Σ is a static box, then both $\overline{\Sigma}$ and $\underline{\Sigma}$ are dynamic boxes, and if Σ is a dynamic box, then $\lfloor\Sigma\rfloor$ is a static box.

We further distinguish two special classes of boxes, called the *entry* and *exit* boxes, which comprise all the dynamic boxes Σ such that M_Σ is, respectively, equal to ${}^\circ\Sigma$ and Σ°. The sets of static, entry, and exit boxes are in bijection with each other; the transformations associating with a static box

Σ the entry box $\overline{\Sigma}$ and the exit box $\underline{\Sigma}$ are bijections from the set of static boxes to the set of entry and exit boxes, respectively. The set of dynamic boxes properly contains all the entry and exit boxes, and is disjoint with the set of static boxes (on account of the fact that dynamic boxes are marked, and static boxes unmarked). The sets of plain static, dynamic, entry and exit boxes will, respectively, be denoted by $\mathsf{Box}^{\mathsf{s}}$, $\mathsf{Box}^{\mathsf{d}}$, $\mathsf{Box}^{\mathsf{e}}$ and $\mathsf{Box}^{\mathsf{x}}$, and the whole set of plain boxes will be denoted by Box.

Proposition 4.2.1. Let Σ be a dynamic box and U a step enabled by Σ.

(1) Every labelled net derivable from Σ is a dynamic box.
(2) U is a set of mutually independent transitions: $U \times U \subseteq \mathsf{ind}_\Sigma \cup id_{T_\Sigma}$.[1]
(3) All the arcs adjacent to the transitions in U are unitary:

$$W_\Sigma(U \times S_\Sigma) \cup W_\Sigma(S_\Sigma \times U) \subseteq \{0,1\}.$$

Proof. (1) Suppose that Θ is derivable from a dynamic box Σ. Θ is marked since Σ is T-restricted. Clearly, all the markings reachable from M_Θ or ${}^\circ\Theta$ or Θ° are safe and clean since those reachable from M_Σ and ${}^\circ\Sigma$ and Σ° are safe and clean and ${}^\circ\Theta = {}^\circ\Sigma$ and $\Theta^\circ = \Sigma^\circ$. Hence Θ is a dynamic box.

(2,3) Follow directly from the safeness of the markings before and after the execution of step U. □

Exercise 4.2.5. Show that a plain box Σ is dynamic (static) if and only if Σ_{sr} is a dynamic (respectively, static) box.

It is the static boxes Σ (or, equivalently, the entry boxes $\overline{\Sigma}$) which will provide a denotational semantics of static PBC expressions. Two behavioural conditions were imposed on the markings M reachable from the entry or exit marking of a static box Σ. First, we require M to be safe in order to ensure that the semantics of Σ is as simple as possible (in particular, that it does not exhibit auto-concurrency), and so we can directly use a partial order semantics of Petri nets in the style of Mazurkiewicz [77] (see also Sect. 7.3.8). The second condition, that M is always a clean marking, is a consequence of the first condition and our decision to use iterative constructs in the algebra of nets (see Sect. 4.4.9). Moreover, it may be observed that when a net is augmented with the **redo** and **skip** transitions, cleanness implies that executing **redo** leads to the entry marking ${}^\circ\Sigma$ and the net remains safe (and clean). It can also be seen that if $\overline{\Sigma}$ was not clean, then $\overline{\Sigma}_{\mathsf{sr}}$ would not be safe either (nor clean). We will return to this issue in Sect. 4.3.7.

In what follows, we also need to be able to represent markings reachable, e.g., from the entry ones. For this reason we introduced dynamic boxes which, by definition, include all marked nets Θ such that Θ is derivable from $\overline{\Sigma}$, for some static box Σ. Dynamic boxes will provide a denotational semantics for dynamic PBC expressions.

[1] id_{T_Σ} denotes the *identity relation* for the set of transitions T_Σ. In general, $id_X = \{(x,x) \mid x \in X\}$ for every set X.

4.3 Net Refinement

The basis for providing plain boxes with an algebraic structure, and thus for defining a compositional net semantics of PBC expressions, will be a *simultaneous refinement and relabelling meta-operator* (*net refinement*, for short). It captures a general mechanism combining transition refinement and interface change defined by (general) relabellings. Net refinement is defined for a simple box Ω which serves as a pattern for gluing together a set of plain boxes $\boldsymbol{\Sigma}$ (one plain box per transition in Ω) along their entry and exit interfaces. Relabellings annotating the transitions of Ω specify interface changes applied to the boxes in $\boldsymbol{\Sigma}$.

4.3.1 Operator Boxes

An *operator box* is a simple box Ω such that all relabellings annotating its transitions are transformational. Let $\boldsymbol{\Sigma} : T_\Omega \to \mathsf{Box}$ be a function from the transitions of Ω to plain boxes. We will refer to $\boldsymbol{\Sigma}$ as an *Ω-tuple* and denote $\boldsymbol{\Sigma}(v)$ by Σ_v, for every $v \in T_\Omega$. When the set of transitions of Ω is finite, we will assume that $T_\Omega = \{v_1, \ldots, v_n\}$ is an arbitrary but fixed ordering imposed on the set of transitions, and then denote $\boldsymbol{\Sigma} = (\Sigma_{v_1}, \ldots, \Sigma_{v_n})$ or $\boldsymbol{\Sigma} = (\Sigma_1, \ldots, \Sigma_n)$, i.e., we will treat $\boldsymbol{\Sigma}$ as a vector of plain boxes. We shall not require that the boxes in $\boldsymbol{\Sigma}$ be distinct, and the same plain box may be used to refine different transitions of Ω.

As the transitions of an operator box will be refined (replaced) by possibly complex plain boxes, it seems reasonable to consider that their execution may take some time or, indeed, may even last indefinitely (if the subsystem represented by a refining box can deadlock or progress endlessly). This feature may be captured by a special kind of extended markings. A *complex marking* of an operator box Ω is a pair $\mathcal{M} = (M, Q)$ composed of a standard marking M and a finite multiset Q of (engaged[2]) transitions of Ω. M will be called the *real*, and Q the *imaginary* part of $\mathcal{M}$. A standard marking M may, then be identified with the complex marking $(M, \emptyset)$. The enabling and execution rules are extended to the complex markings thus.

A step U is *enabled* at $\mathcal{M}$ if it is enabled at M; we will denote this by $\mathcal{M}\ [U\rangle$. An enabled U can be (fully) *executed*, $\mathcal{M}\ [U\rangle\mathcal{M}'$, yielding a marking $\mathcal{M}' = (M', Q)$ such that, for every $s \in S$,

$$M'(s) \;=\; M(s) + \sum_{t \in U} U(t) \cdot \Big(W(t,s) - W(s,t)\Big) \,.$$

An enabled step U may also be *engaged*, $\mathcal{M}\ [U^+\rangle\mathcal{M}'$, yielding a marking $\mathcal{M}' = (M', Q + U)$ such that, for every $s \in S$,

[2] Complex markings are related to the ST-idea used to characterise equivalences on nets preserved by refinements (see [23, 97, 99]).

$$M'(s) = M(s) - \sum_{t\in U} U(t) \cdot W(s,t) .$$

An engaged step $U \subseteq Q$ may always be *completed*, $\mathcal{M}\ [U^{-}\rangle \mathcal{M}'$, yielding a marking $\mathcal{M}' = (M', Q - U)$ such that, for every $s \in S$,

$$M'(s) = M(s) + \sum_{t\in U} U(t) \cdot W(t,s) .$$

Bringing the above three cases together, we will in general, denote

$$\mathcal{M}\ [U : V^{+} : Y^{-}\rangle\ (M', Q')$$

if $Y \subseteq Q$ and $Q' = Q + V - Y$ and, for every $s \in S$,

$$M(s) \geq \sum_{t\in U+V} \Big(U(t) + V(t)\Big) \cdot W(s,t)$$

$$M'(s) = M(s) + \sum_{t\in U+Y} \Big(U(t) + Y(t)\Big) \cdot W(t,s) - \sum_{t\in U+V} \Big(U(t) + V(t)\Big) \cdot W(s,t) .$$

We will denote by $[\mathcal{M}\rangle$ the set of complex markings reachable from $\mathcal{M}$.

The notions of safeness, k-boundedness and cleanness, can be extended in a natural way to complex markings. A complex marking $\mathcal{M} = (M, Q)$ is *k-bounded* (or *safe*, for $k = 1$) if M is k-bounded (respectively, safe), and it is *k-clean* (or *clean*, for $k = 1$) if M is k-clean (respectively, clean).

Exercise 4.3.1. Show that if Σ is marked with a real marking M and $\mathcal{M} = (M, \emptyset)$, then

- $M' \in [M\rangle$ if and only if $(M', \emptyset) \in [\mathcal{M}\rangle$.
- Σ is k-bounded if and only if all markings in $[\mathcal{M}\rangle$ are k-bounded.
- Σ is k-clean if and only if all markings in $[\mathcal{M}\rangle$ are k-clean.

Let Ω be an operator box with a complex marking $\mathcal{M} = (M, Q)$. The composition operation specified by Ω will be applicable to every Ω-tuple $\boldsymbol{\Sigma}$ of plain boxes provided Σ_v is marked if and only if $v \in Q$.[3] This implies, in particular, that only finitely many Σ_v's are marked. However, for reasons that will be explained in Chap. 7, we will sometimes allow infinite Q, under the restriction that, for every $s \in S_\Omega$, there are finitely many v's in ${}^\bullet s \cap Q$ with Σ_v having a marked exit place, as well as finitely many v's in $s^\bullet \cap Q$ with Σ_v having a marked entry place, i.e., for every $s \in S_\Omega$,

$$\left|\left\{ v \in Q \,\middle|\, \begin{pmatrix} M_{\Sigma_v}(\Sigma_v^\circ) \neq \{0\} \\ \wedge\, v \in {}^\bullet s \end{pmatrix} \vee \begin{pmatrix} M_{\Sigma_v}({}^\circ\Sigma_v) \neq \{0\} \\ \wedge\, v \in s^\bullet \end{pmatrix} \right\}\right| < \infty . \quad (4.3)$$

Clearly, a finite Q fulfills this requirement.

[3] In Chap. 6 it will also be required that, for every $v \in Q$, Σ_v is $Q(v)$-bounded and $Q(v)$-clean, and remains so after adding redo and skip. For the static and dynamic boxes Σ_v, these extra conditions are always fulfilled.

4.3.2 Intuition Behind Net Refinement

The operation of net refinement is carried out by taking Ω and $\boldsymbol{\Sigma}$, and creating a new labelled net, $\Omega(\boldsymbol{\Sigma})$, by first applying to every Σ_v of $\boldsymbol{\Sigma}$ an interface change specified by the relabelling $\lambda_\Omega(v)$ of the corresponding transition v, and then using the result to refine each transition v of Ω. Although the construction of $\Omega(\boldsymbol{\Sigma})$ will be formalised in a single definition, we find it easier to describe separately the interface change operation (phase 1 below) and transition refinement (phase 2).

Phase 1 – Interface Change. Here, relabellings $\varrho_v = \lambda_\Omega(v)$ are applied to the boxes Σ_v. As a result, we obtain an Ω-tuple of intermediate boxes, $\boldsymbol{\Theta} = \{\Theta_v \mid v \in T_\Omega\}$ which is subsequently fed into the second phase (transition refinement).

Fig. 4.11. Two refining boxes

Applying relabelling ϱ_v to Σ_v affects only its transitions (and so indirectly the weight function). Basically, all places together with their markings and labels are retained, but the transition set is reconstructed completely in the following way. Suppose t, u, and w are transitions in Σ_v such that the multiset $\{\lambda_{\Sigma_v}(t), \lambda_{\Sigma_v}(u), \lambda_{\Sigma_v}(w)\}$ belongs to the domain of the relabelling ϱ_v, for example, $(\{\lambda_{\Sigma_v}(t), \lambda_{\Sigma_v}(u), \lambda_{\Sigma_v}(w)\}, \alpha) \in \varrho_v$. Then transitions t, u, and w are synchronised (glued) to yield a composite transition labelled α. The connectivity of the new transition is inherited from the connectivities of transitions t, u, and w, i.e., it simply acquires the sum of the arc weights leading to and from the three transitions.

Taking as an example the pair of boxes shown in Fig. 4.11 where t, y, u, and w are transition names and α, β, γ, and δ are transition labels, and two relabellings,

$$\varrho_1 = \varrho_{\text{id}} \ \ (\text{for } \Sigma_1) \qquad \text{and} \qquad \varrho_2 = \varrho_{\text{id}} \cup \{(\{\alpha, \delta\}, \varepsilon)\} \ \ (\text{for } \Sigma_2)\,,$$

we obtain two intermediate nets shown in Fig. 4.12 (where place and transition names are omitted). In Θ_1, the transition labelled β has been created by transition t and the pair $(\{\beta\}, \beta) \in \varrho_1 = \varrho_{\text{id}}$. Similarly, the transition labelled γ has been created by transition y and $(\{\gamma\}, \gamma) \in \varrho_1$. In Θ_2, the transition labelled α (or δ) has been created by transition u (respectively, w)

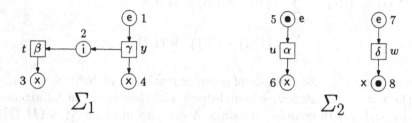

and $(\{\alpha\},\alpha) \in \varrho_2$ (respectively, $(\{\delta\},\delta) \in \varrho_2$). The transition labelled ε has been created by transitions u and w and the pair $(\{\alpha,\delta\},\varepsilon) \in \varrho_2$.

Fig. 4.12. Interface change using ϱ_1 (yielding Θ_1 from Σ_1) and ϱ_2 (yielding Θ_2 from Σ_2)

In order to uniquely identify newly created transitions, we shall use labelled trees as transition names and, in doing so, also record the way in which transitions were formed. Later, it will become apparent that using trees as the names of transitions (and places) is a powerful device which allows one to express the result of several simultaneous and successive refinements. In the treatment of recursion we will even need to consider infinite successions of refinements, and thus (possibly) infinite trees, but for now all trees may be assumed finite.

Using labelled trees as place and transition names is a versatile and unifying device facilitating the definition of all the components of a refined/relabelled net; this will be further justified in the next section, as well as in Chap. 5. For instance, the ε-labelled transition in Fig. 4.12 will have the name shown on the left-hand side of Fig. 4.13, provided that v_2 is the transition in Ω which drives the relabelling of Σ_2 into Θ_2. Notice also that the u and w may themselves be possibly complex trees appended to the (v_2,ε)-labelled root, if Σ_2 has been obtained through previous refinements. The presence of ε in the label of the root is due to the fact that ϱ_2 could have had more than one label associated to the same multiset of actions. For instance, if $(\{\alpha,\delta\},\varepsilon') \in \varrho_2$, then it would be necessary to distinguish a new transition formed by u and w with the label ε from another one, also formed by u and w, but with label ε'.

Phase 2 – Transition Refinement. We now use Ω and the Ω-tuple $\boldsymbol{\Theta}$ from phase 1 to construct the final net, $\Omega(\boldsymbol{\Sigma})$. This time, the exit and entry places of nets in $\boldsymbol{\Theta}$ (which are the same as those in $\boldsymbol{\Sigma}$), rather than transitions, are joined together using pre- and post-places of the transitions in Ω they are meant to refine. The resulting interface places will connect (or glue) together the various (copies of the interiors of the) Θ_i's replacing the corresponding v_i's. We must use copies since the same Σ may be used to refine several transitions in Ω, and only the interiors, i.e., the nets without the entry/exit

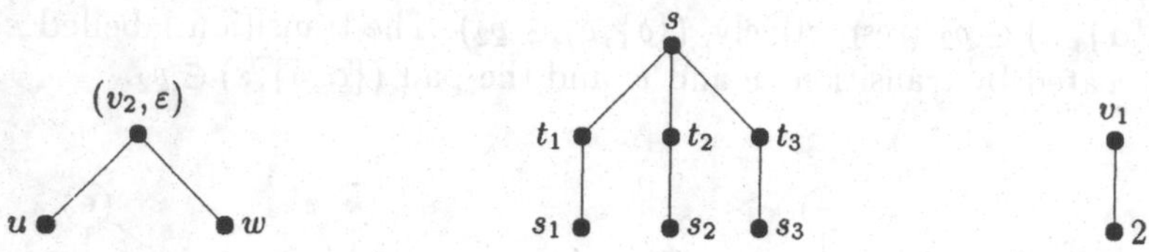

Fig. 4.13. Trees used as place and transition names

places are copied since the remaining places are used to construct the gluing interface.

The new (gluing) places are formed as follows. Suppose that s is a place in Ω such that ${}^\bullet s = \{t_1\}$ and $s^\bullet = \{t_2, t_3\}$. Then, for all possible combinations of three places, s_1, s_2, and s_3, such that s_1 is an exit place of Θ_1, and s_2 and s_3 are entry places of, respectively, Θ_2 and Θ_3, we construct a new place. The connectivity and marking of the new place are inherited from the places s_1, s_2, and s_3, while its label is the same as that of s. Again, the names of newly constructed nodes will be expressed through labelled trees recording the way in which they have been obtained; in the case just considered the new place will be the tree shown in the middle of Fig. 4.13.

Finally, all the transitions as well as internal places of the nets $\boldsymbol{\Theta}$ are carried forward to the final net (in our example of Fig. 4.12, there is only one such place, in Θ_1), with the same labels, connectivity and markings. This is the copying operation of the interiors. The copying of internal places may be achieved by using suitably labelled trees (very simple in this case) to identify the copied elements; for instance, the tree shown on the right of Fig. 4.13 may be used to identify a copy of the internal place 2 of Θ_1 (or of Σ_1 since they have the same places)corresponding to the refinement of v_1. Note that, again, 2 could be a possibly complex tree if Σ_1 was itself derived via refinement. It will not be necessary to transform in the same way the names of transitions of the Θ_i's since this has already been done in phase 1; the roots of the trees constructed there are labelled by pairs like (v_2, ε), the first component of which identifies a particular copy.

Continuing our example, and taking Ω as in Fig. 4.14 with the complex marking $\mathcal{M} = (\emptyset, \{v_2\})$ (which accords with the definition since Σ_1 is unmarked while Σ_2 is marked), we obtain, by applying phase 2 construction to $\boldsymbol{\Theta} = (\Theta_1, \Theta_2)$, the net displayed in Fig. 4.15. For example, the marked entry place of $\Omega(\boldsymbol{\Sigma})$ results from combining the marked entry place of Θ_2 with the only entry place of Θ_1 on account of place 9 of Ω, keeping in mind that the latter has two output transitions, one into which Θ_1 is refined and another one into which Θ_2 is refined (see Fig. 4.16).

There is some degree of behavioural conformity between $\Omega(\boldsymbol{\Sigma})$ and the constituent nets. For instance, the transition labelled α is enabled in $\Omega(\boldsymbol{\Sigma})$ which is due to the fact that the corresponding α-labelled transition is also

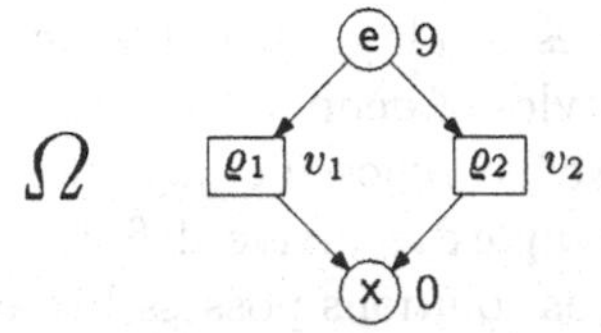

Fig. 4.14. An operator box

enabled in Σ_2. Similarly, the δ-labelled transitions are backward enabled in both $\Omega(\boldsymbol{\Sigma})$ and Σ_2. However, it would be wrong to always draw a reverse conclusion. For instance, if we moved the token from place 8 to place 7 in Σ_2, then the γ-labelled transition in $\Omega(\boldsymbol{\Sigma})$ would become enabled, without a corresponding enabling of the γ-labelled transition in Σ_1. This situation will be analysed in detail in Chap. 7 where, perhaps surprisingly, we will find that it has strong connections with the inaction rules used in the treatment of PBC expressions in Chap. 3.

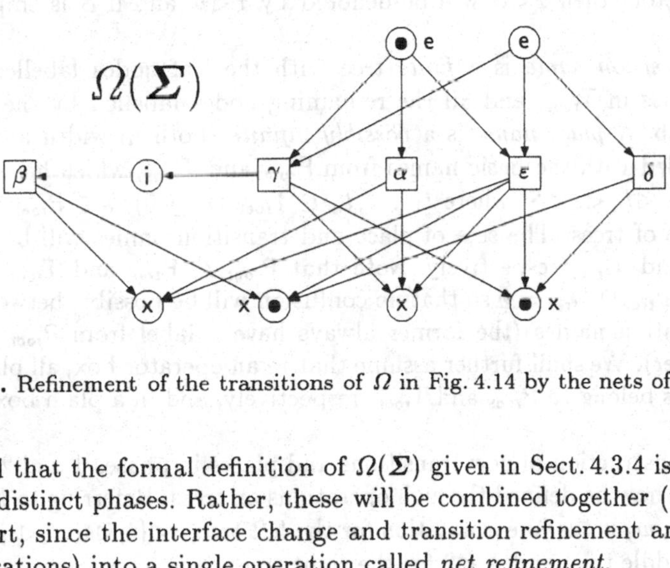

Fig. 4.15. Refinement of the transitions of Ω in Fig. 4.14 by the nets of Fig. 4.11

Recall that the formal definition of $\Omega(\boldsymbol{\Sigma})$ given in Sect. 4.3.4 is not split into two distinct phases. Rather, these will be combined together (with very little effort, since the interface change and transition refinement are orthogonal operations) into a single operation called *net refinement.*

4.3.3 Place and Transition Names

As far as net refinement is concerned, the names (identities) of newly constructed transitions and places are irrelevant, provided that we always choose them fresh. However, in our approach to solving recursive definitions on boxes, it is the *names* of places and transitions which play a crucial role since we use them to define an inclusion order on the domain of labelled nets. This could be contrasted with an alternative approach to defining such an order through injective net mappings (in such a case nets are defined up to isomorphism, and the names of places and transitions do not matter). We will not compare the two styles of defining inclusion orders. Rather, we shall demonstrate that the one we have chosen enjoys two important properties: (i) one can treat arbitrarily complex recursive definitions on boxes; and (ii) the boxes which are generated as solutions possess behavioural properties which match exactly those enjoyed by the corresponding recursive expressions in the associated process algebra. A key device used to obtain such results is the use of labelled (nonempty) trees as place and transition names (we define such labelled trees up to isomorphism; only the structure of the tree and the labels of the nodes do matter), as already indicated in the previous section.

We shall assume that there are two disjoint nonempty sets of basic *place* and *transition names*, $\mathsf{P_{root}}$ and $\mathsf{T_{root}}$; each name $x \in \mathsf{P_{root}} \cup \mathsf{T_{root}}$ can be viewed as a special tree with a single x-labelled node (the root). We shall employ a simple linear notation to denote labelled trees, whereby an expression $x \lhd S$ such that x is some element as above and S is a (possibly infinite) multiset of trees, denotes a tree where all trees of the multiset S are appended (with their multiplicity) to the x-labelled root. Moreover, if $S = \{p\}$ is a singleton, then $x \lhd S$ will be denoted by $x \lhd p$, and if S is empty, then $x \lhd S = x$.

A *transition name* is a *finite* tree with the leaf nodes labelled by the basic names in $\mathsf{T_{root}}$, and all the remaining nodes labelled by the pairs in $\mathsf{T_{root}} \times \mathsf{Lab}$. A *place name* is a *possibly infinite* (both in width and depth) tree labelled with the basic names from $\mathsf{P_{root}}$ and $\mathsf{T_{root}}$, which has the form $t_1 \lhd t_2 \lhd \ldots \lhd t_n \lhd s \lhd S$, where $t_1, \ldots, t_n \in \mathsf{T_{root}}$ $(n \geq 0)$, $s \in \mathsf{P_{root}}$ and S is a multiset of trees. The sets of place and transition names will be denoted by $\mathsf{P_{tree}}$ and $\mathsf{T_{tree}}$, respectively. Note that $\mathsf{P_{root}} \subset \mathsf{P_{tree}}$ and $\mathsf{T_{root}} \subset \mathsf{T_{tree}}$; moreover $\mathsf{P_{tree}} \cap \mathsf{T_{tree}} = \emptyset$ so that no confusion will be possible between place and transition names (the former always have a label from $\mathsf{P_{root}}$ and the latter never). We shall further assume that in an operator box, all places and transitions belong to $\mathsf{P_{root}}$ and $\mathsf{T_{root}}$, respectively, and in a plain box to $\mathsf{P_{tree}}$ and $\mathsf{T_{tree}}$.

The construction of new transitions and places in phases 1 and 2, respectively, can now be defined through operations on trees. Referring to Fig. 4.13, $(v_2, \varepsilon) \lhd \{u, w\}$ is the linear notation for the left tree, $s \lhd \{t_1 \lhd s_1, t_2 \lhd s_2, t_3 \lhd s_3\}$ for the middle tree, and $v_1 \lhd 2$ for the right tree.

Example Continued Figure 4.16 presents a complete representation of the net $\Omega(\Sigma)$ in Fig. 4.15 which served as our running example, by showing explicitly all the transition and place names. For instance, the tree name of the only marked entry place means that this place is the result of combining places 1 (through the refinement of transition v_1) and 5 (through the refinement of transition v_2) on account of place 9 of Ω. The name of the internal place means that this place has been carried from the refining net (where its name was 2) into the refined net via transition v_1. The name of the leftmost transition means that this transition has been created from t by refining v_1

and has the label β. The name of the ε-labelled transition means that it has been created from two transitions, u and w, by refining v_2, and that it has ε as the label.

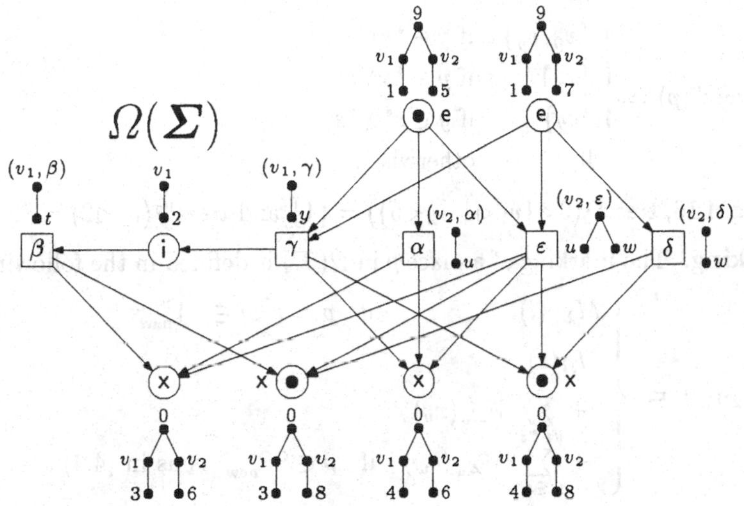

Fig. 4.16. Place names and transition names of the net of Fig. 4.15

4.3.4 Formal Definition of Net Refinement

Under the assumptions made about the operator box Ω and Ω-tuple of plain boxes $\boldsymbol{\Sigma}$, the result of a simultaneous substitution of boxes $\boldsymbol{\Sigma}$ for the transitions in Ω is a labelled net $\Omega(\boldsymbol{\Sigma})$ whose components are defined as follows.

Places. The set of places of $\Omega(\boldsymbol{\Sigma})$ is defined as the (disjoint) union

$$S_{\Omega(\boldsymbol{\Sigma})} = \bigcup_{v \in T_\Omega} \mathsf{ST}^v_{\mathsf{new}} \cup \bigcup_{s \in S_\Omega} \mathsf{SP}^s_{\mathsf{new}}$$

where, for every $v \in T_\Omega$ and $s \in S_\Omega$,

$$\mathsf{ST}^v_{\mathsf{new}} = \{v \lhd i \mid i \in \ddot{\Sigma}_v\}$$

$$\mathsf{SP}^s_{\mathsf{new}} = \{s \lhd (\{v \lhd x_v\}_{v \in {}^\bullet s} + \{w \lhd e_w\}_{w \in s^\bullet}) \mid x_v \in \Sigma_v^\circ \wedge e_w \in {}^\circ\Sigma_w\}\,.$$

If s is an *isolated* place, ${}^\bullet s = s^\bullet = \emptyset$, then by the definition of the tree-appending operation, $\mathsf{SP}^s_{\mathsf{new}} = \{s\}$. The multiset $\{v \lhd x_v\}_{v \in {}^\bullet s} + \{w \lhd e_w\}_{w \in s^\bullet}$ in the definition of a place

$$p = s \lhd \Big(\{v \lhd x_v\}_{v \in {}^\bullet s} + \{w \lhd e_w\}_{w \in s^\bullet}\Big) \in \mathsf{SP}^s_{\mathsf{new}} \tag{4.4}$$

is in fact a set, possibly infinite, because even in case there is a side-loop between s and $v = w$, $x_v \neq e_w$ since x_v is an exit place and e_w is an entry place of Σ_v. The following notation will be useful in manipulating the tree names upon which a newly constructed place is based.

Let y be a transition in Ω. Then for a place $p = v \lhd i$ in $\mathsf{ST}^v_{\mathsf{new}}$, we define $\mathsf{trees}^y(p) = \{i\}$ if $v = y$, and $\mathsf{trees}^y(p) = \emptyset$ if $v \neq y$; moreover, for a place $p \in \mathsf{SP}^s_{\mathsf{new}}$ as in (4.4), we define

$$\mathsf{trees}^y(p) = \begin{cases} \{x_y, e_y\} & \text{if } y \in {}^\bullet s \cap s^\bullet \\ \{x_y\} & \text{if } y \in {}^\bullet s \setminus s^\bullet \\ \{e_y\} & \text{if } y \in s^\bullet \setminus {}^\bullet s \\ \emptyset & \text{otherwise .} \end{cases}$$

In Fig. 4.16, $\mathsf{trees}^{v_1}(9 \lhd \{v_1 \lhd 1, v_2 \lhd 5\}) = \{1\}$ and $\mathsf{trees}^{v_2}(v_1 \lhd 2) = \emptyset$.

Marking. The marking of a place p in $\Omega(\boldsymbol{\Sigma})$ is defined in the following way:

$$M_{\Omega(\boldsymbol{\Sigma})}(p) = \begin{cases} M_{\Sigma_v}(i) & \text{if } \ p = v \lhd i \in \mathsf{ST}^v_{\mathsf{new}} \\ M_\Omega(s) & \\ \quad + \sum_{v \in {}^\bullet s} M_{\Sigma_v}(x_v) & \\ \quad + \sum_{w \in s^\bullet} M_{\Sigma_w}(e_w) & \text{if } \ p \in \mathsf{SP}^s_{\mathsf{new}} \ \text{ is as in (4.4) .} \end{cases} \tag{4.5}$$

Notice that if $s \in S_\Omega$ is an isolated place, then $M_{\Omega(\boldsymbol{\Sigma})}(s) = M_\Omega(s)$. Moreover, $M_{\Omega(\boldsymbol{\Sigma})}(p)$ is a well-defined natural number due to the assumption (4.3).

Transitions. The set of transitions of $\Omega(\boldsymbol{\Sigma})$ is defined as the (disjoint) union

$$T_{\Omega(\boldsymbol{\Sigma})} = \bigcup_{v \in T_\Omega} \mathsf{T}^v_{\mathsf{new}}$$

where, for every $v \in T_\Omega$,

$$\mathsf{T}^v_{\mathsf{new}} = \{(v, \alpha) \lhd R \mid R \in \mathsf{mult}(T_{\Sigma_v}) \wedge (\lambda_{\Sigma_v}(R), \alpha) \in \lambda_\Omega(v)\} .$$

The multiset R in $(v, \alpha) \lhd R$ will never be empty since no pair in $\lambda_{\Sigma_v}(v)$ has the empty multiset as its left argument. We will discuss in the next section why R is allowed to be a true multiset rather than simply a set.

Similarly as for places, we will denote by $\mathsf{trees}(u)$ the multiset of transitions R upon which a newly constructed transition $u = (v, \alpha) \lhd R \in \mathsf{T}^v_{\mathsf{new}}$ is based. In Fig. 4.16, $\mathsf{trees}((v_1, \beta) \lhd t) = \{t\}$ and $\mathsf{trees}((v_2, \varepsilon) \lhd \{u, w\}) = \{u, w\}$.

Labelling. The label of a place or transition x in $\Omega(\boldsymbol{\Sigma})$ is defined in the following way:

$$\lambda_{\Omega(\boldsymbol{\Sigma})}(x) = \begin{cases} \mathsf{i} & \text{if } \ x \in \mathsf{ST}^v_{\mathsf{new}} \\ \lambda_\Omega(s) & \text{if } \ x \in \mathsf{SP}^s_{\mathsf{new}} \\ \alpha & \text{if } \ x = (v, \alpha) \lhd R \in \mathsf{T}^v_{\mathsf{new}} . \end{cases} \tag{4.6}$$

Weight Function. For a place p in $S_{\Omega(\Sigma)}$ and transition u in $\mathsf{T}^v_{\mathsf{new}}$, the values of the weight function are given by

$$W_{\Omega(\Sigma)}(p,u) = \sum_{z\in \mathsf{trees}^v(p)} \sum_{t\in \mathsf{trees}(u)} W_{\Sigma_v}(z,t)\cdot \mathsf{trees}(u)(t)\ , \tag{4.7}$$

$$W_{\Omega(\Sigma)}(u,p) = \sum_{z\in \mathsf{trees}^v(p)} \sum_{t\in \mathsf{trees}(u)} W_{\Sigma_v}(t,z)\cdot \mathsf{trees}(u)(t)\ . \tag{4.8}$$

Note that $\mathsf{trees}(u)(t)$ denotes the number of occurrences (or multiplicity) of t in the multiset $\mathsf{trees}(u)$.

The above definition of $\Omega(\Sigma)$ follows the construction described informally in Sect. 4.3.2; in particular, the trees created as place and transition names of $\Omega(\Sigma)$ obey the constraints formulated in Sect. 4.3.3.

4.3.5 Remarks on Net Refinement

One might ask why we allowed multisets of transitions to be combined, rather than simply sets. Such a question is justified since any true multiset of transitions R will lead to some non-unitary arcs, and so in the safe nets framework, to a dead transition which could be dropped. Moreover, if R was a set, then the two formulae, (4.7) and (4.8), for the weight function would simplify to

$$W_{\Omega(\Sigma)}(p,u) = \sum_{z\in \mathsf{trees}^v(p)} \sum_{t\in \mathsf{trees}(u)} W_{\Sigma_v}(z,t)\ ,$$

$$W_{\Omega(\Sigma)}(u,p) = \sum_{z\in \mathsf{trees}^v(p)} \sum_{t\in \mathsf{trees}(u)} W_{\Sigma_v}(t,z)\ .$$

The answer is twofold. First, as already stated, our aim is to develop a model which would not need a major redesign if we wanted to extend it, say, by allowing other initial markings (for instance, a net corresponding to a generalised PBC expression $\overline{\overline{\alpha;\alpha}}$ could have two tokens in the entry place). Then our argument about transitions with non-unitary arcs being dead would no longer hold, and it seems reasonable to define the refinement in a fully general way, while possibly dropping the dead transitions afterwards, when it is convenient or desirable for other reasons (cf. the discussion in Sect. 5.3.2).

Another, more subtle, argument arises in the context of the structural equivalence $\mathsf{iso_T}$ defined in Sect. 4.2.4. The nets Σ_1 and Σ_2 in Fig. 4.17 are $\mathsf{iso_T}$-equivalent, since they only differ by duplicate transitions, and hence will always have the same behaviour; more precisely, they have isomorphic transition systems. However, their synchronisations (as defined in Sect. 4.4.6) through the rule using sets would not produce nets which are $\mathsf{iso_T}$-equivalent; the synchronisation of Σ_1 with respect to the action particle a has the same structure as Σ_1 itself, while the synchronisation of Σ_2 gives rise to a new (nonduplicate) transition (arising from the synchronisation of t_1 and t_2). This problem does not occur when one adopts the multiset rule (but at a cost of having potentially infinitely many synchronisations as discussed in Sect. 4.4.6). Hence we shall use the multiset-based refinement, even though it may introduce transitions which are clearly dead in the framework based on safe nets.

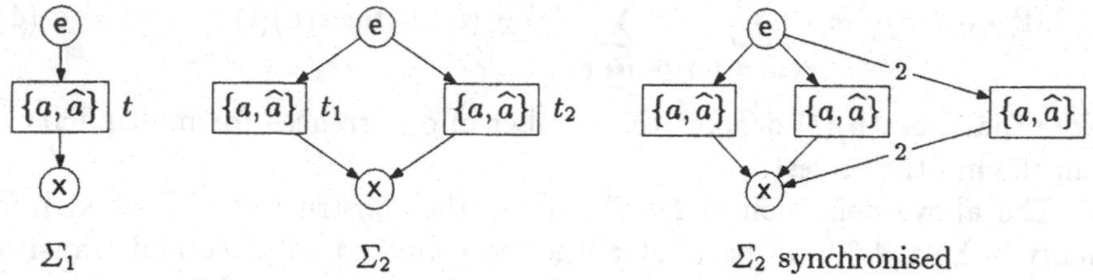

Fig. 4.17. Problem with sets w.r.t. multisets for grouping transitions

There is another subtle point concerning arc weights. Since, by the (intended) safeness of boxes, no transition which is joined with a place by a non-unitary arc can ever be enabled, one could be tempted to postulate that all boxes should be simple (i.e., only weights 0 and 1 be allowed). This could be achieved by redefining the synchronisation in phase 1 above in such a way that only sets of mutually independent transitions be allowed to combine in order to yield new transitions. The resulting model would be simpler, most of the results in Chaps. 5–7 would go through without any change, and the counterexample of Fig. 4.17 would no longer work since in the synchronisation of Σ_2 no new transitions would be added. However, the modified net refinement would fail to induce a monotonic mapping with respect to the inclusion order on labelled nets, rendering our treatment of recursion in Sect. 5.2.1 no longer valid. For instance, in Fig. 4.18, it is the case that Σ_1 is included in Σ_2, but this is no longer true after a synchronisation with the 'independent set' rule, since such a synchronisation does not add any new transition to Σ_2.

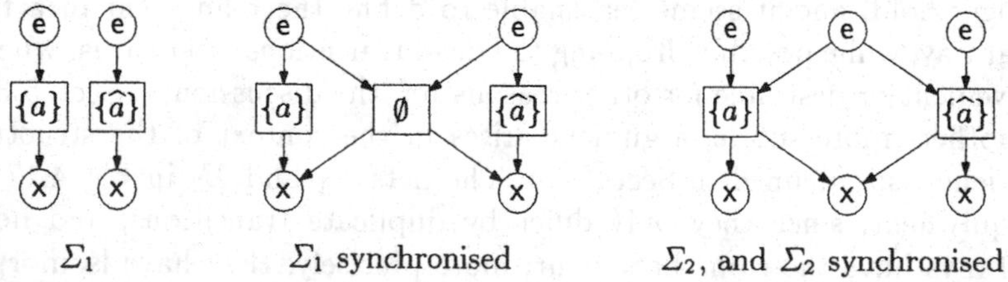

Fig. 4.18. Synchronisation based on independent sets of transitions is not monotonic w.r.t. inclusion order

To summarise, we decided to adopt the present definition but still reserve the right, whenever it is needed, to delete all (or some) dead transitions in the resulting net, since this leads to $\cong$-equivalent nets anyway (see Sect. 4.2.4).

4.3.6 Properties

Up to now, net refinement operation has been treated as purely syntactic operation in the sense that we did not provide any results about the structure nor behaviour of the resulting nets. These issues will be our main concern in Chaps. 6 and 7. Before that, we shall gain some intuition by looking at how the static and dynamic PBC expressions may be translated into plain boxes. Again, the formal results showing how this translation works and why it is equivalent to the SOS semantics developed in Chap. 3 will be given later, in Chaps. 7 and 8. At this point, it is not even certain that, if we start from static boxes, the result of net refinement will be a static box. Indeed, as it will turn out, some extra conditions need to be satisfied for this result to hold. What we can say, however, is that the mechanism we have described is compositional since the definition respects the structure of the nets being composed.

Proposition 4.3.1. The labelled net $\Omega(\boldsymbol{\Sigma})$ defined in Sect. 4.3.4 is a plain box which is unmarked if and only if the marking of the operator box Ω is $\mathcal{M} = (\emptyset, \emptyset)$ (and so each plain box Σ_v is unmarked).

Proof. We first observe that,

$$\forall v \in T_\Omega \; \forall q \in S_{\Sigma_v} \; \exists p \in S_{\Omega(\boldsymbol{\Sigma})} \; : \; q \in \mathsf{trees}^v(p) \, . \tag{4.9}$$

Indeed, if $q \in \ddot{\Sigma}_v$, then we can take $p = v \triangleleft q \in \mathsf{SP}^v_{\mathsf{new}}$. If $q \in {}^\circ\Sigma_v$, then we take any $s \in {}^\bullet v$ (which can be done since Ω is T-restricted) and after that any $p = s \triangleleft \{\ldots, v \triangleleft q, \ldots\} \in \mathsf{SP}^s_{\mathsf{new}}$ will satisfy our requirement (at least one such p exists since all the boxes in $\boldsymbol{\Sigma}$ are ex-restricted). If $q \in \Sigma_v^\circ$, we proceed similarly.

Having shown (4.9), we next observe that

$$\forall s \in S_\Omega \; : \; \mathsf{SP}^s_{\mathsf{new}} \neq \emptyset \tag{4.10}$$

which follows directly from the definition (for an isolated s), or from the fact that all the boxes in $\boldsymbol{\Sigma}$ are ex-restricted (for a non-isolated s). Indeed, in the latter case, we can choose, for each $w \in s^\bullet$ (if any) an entry place $e_w \in {}^\circ\Sigma_w$, and for each $v \in {}^\bullet s$ (if any) an exit place $x_v \in \Sigma_v^\circ$. Then $s \triangleleft (\{v \triangleleft x_v\}_{v \in {}^\bullet s} + \{w \triangleleft e_w\}_{w \in s^\bullet})$ is a place in $\mathsf{SP}^s_{\mathsf{new}}$. Thus (4.10) holds.

We now proceed with the proof proper. It is clear that $\Omega(\boldsymbol{\Sigma})$ is a plain labelled net; in particular, place markings are all natural numbers thanks to the fact that only finitely many nets Σ_v may contribute to the marking of a place in $\Omega(\boldsymbol{\Sigma})$ (see (4.3)). The T-restrictedness and ex-restrictedness of

$\Omega(\boldsymbol{\Sigma})$ (needed to make it a plain box) are inherited from the corresponding properties of the components, as shown below.

$\Omega(\boldsymbol{\Sigma})$ is T-restricted. Let $u = (v, \alpha) \lhd R \in T_{\Omega(\boldsymbol{\Sigma})}$. To show ${}^\bullet u \neq \emptyset$, we take any $t \in R$ and $q \in {}^\bullet t$ (the former is possible since $R \neq \emptyset$, and the latter since Σ_v is T-restricted); by (4.7) and (4.8) it suffices to find $p \in S_{\Omega(\boldsymbol{\Sigma})}$ such that $q \in \mathsf{trees}^v(p)$. This, however, follows immediately from (4.9). To show $u^\bullet \neq \emptyset$, we proceed similarly.

$\Omega(\boldsymbol{\Sigma})$ is ex-restricted. Follows immediately from the ex-restrictedness of Ω and (4.10).

The second part of the proposition follows directly from the definition of the marking of $\Omega(\boldsymbol{\Sigma})$ together with (4.9) and (4.10). □

Proposition 4.3.2. If Ω and Σ_v, for every $v \in ({}^\circ\Omega)^\bullet$, are all e-directed boxes, then so is $\Omega(\boldsymbol{\Sigma})$. Similarly, if Ω and Σ_v, for every $v \in {}^\bullet(\Omega^\circ)$, are all x-directed boxes, then so is $\Omega(\boldsymbol{\Sigma})$.

Proof. Since Ω is e-directed, if p is an entry place of $\Omega(\boldsymbol{\Sigma})$, then it has the form $p = s \lhd \{w \lhd e_w\}_{w \in s^\bullet}$. Moreover, since each Σ_w is e-directed, we have $W_{\Sigma_w}(t, e_w) = 0$, for all $w \in s^\bullet$ and $t \in T_{\Sigma_w}$. Hence, by (4.8), $W_{\Omega(\boldsymbol{\Sigma})}(u, p) = 0$, for every $u \in T_{\Omega(\boldsymbol{\Sigma})}$. Thus $\Omega(\boldsymbol{\Sigma})$ is e-directed. The proof for x-directedness is similar. □

Corollary 4.3.1. If Ω and Σ_v, for every $v \in T_\Omega$, are all ex-directed boxes, then so is $\Omega(\boldsymbol{\Sigma})$. □

Proposition 4.3.3. $\lfloor \Omega(\boldsymbol{\Sigma}) \rfloor = \lfloor \Omega \rfloor (\lfloor \boldsymbol{\Sigma} \rfloor)$ where $\lfloor \boldsymbol{\Sigma} \rfloor$ is $\boldsymbol{\Sigma}$ with each Σ_v changed to $\lfloor \Sigma_v \rfloor$.

Proof. Follows from the definition of $M_{\Omega(\boldsymbol{\Sigma})}$. Notice that the condition (4.3) for applying a refinement is trivially fulfilled for $\lfloor \Omega \rfloor$ and $\lfloor \boldsymbol{\Sigma} \rfloor$. □

Exercise 4.3.2. Show that if neither Ω nor any of the Σ_v's contains an ex-skewed transition, then neither does $\Omega(\boldsymbol{\Sigma})$.

Isomorphism is preserved through net refinement. In what follows, we consider two instances of net refinement, $\Omega(\boldsymbol{\Sigma})$ and $\Omega'(\boldsymbol{\Sigma}')$.

Exercise 4.3.3. Let $\sim \in \{ \mathsf{iso}, \mathsf{iso_T} \}$ and *isom* be an isomorphism from Ω to Ω' such that $\Sigma_v \sim \Sigma'_{isom(v)}$, for every $v \in T_\Omega$. Show that $\Omega(\boldsymbol{\Sigma}) \sim \Omega'(\boldsymbol{\Sigma}')$.

Exercise 4.3.4. Let Ω $\mathsf{iso_S}$ Ω' and *isom* be an isomorphism from Ω_0 to Ω'_0, where Ω_0 and Ω'_0 are obtained from, respectively, Ω and Ω' by replacing all places by their duplication equivalence classes. Moreover, let $\sim \in \{ \mathsf{iso_S}, \mathsf{iso_{ST}} \}$ and $\Sigma_v \sim \Sigma'_{isom(v)}$, for every $v \in T_\Omega$.
Show that $\Omega(\boldsymbol{\Sigma}) \sim \Omega'(\boldsymbol{\Sigma}')$.

We do not consider $\mathsf{iso_T}$- nor $\mathsf{iso_{ST}}$-equivalence for operator boxes, since these two equivalences do not preserve the arity of net operators.

Exercise 4.3.5. Let Ω' be obtained from Ω by replacing all transitions by their duplication equivalence classes and $\Sigma_v = \Sigma'_{[v]}$, for every $v \in T_\Omega$ (where $[v]$ is the duplication equivalence class of v). Show that it is not always the case that $\Omega(\boldsymbol{\Sigma})$ $\mathsf{iso_{ST}}$ $\Omega'(\boldsymbol{\Sigma}')$.

4.3.7 Discussion

T-restrictedness and ex-restrictedness are closely related by the refinement operation. To appreciate this point, let us consider the nets shown in Fig. 4.19. While Ω and Σ_2 are well formed, Σ_1 has no exit place and hence is not ex-restricted. The refinement may be extended to this case, since the definitions we gave do not rely on this condition, but the result of the refinement, $\Omega(\boldsymbol{\Sigma})$, where $\boldsymbol{\Sigma} = (\Sigma_1, \Sigma_2)$, is not T-restricted since the place s of Ω does not give rise to any place in the refined net (all places with a root labelled s should have a leaf labelled by an exit place of Σ_1, and there is none). It may also be noticed that, while all the original nets are safe and clean, the result is not bounded, even if we start from the empty marking. This highlights the importance of T-restrictedness and ex-restrictedness in ensuring the desirable behaviour of the compositionally constructed nets. The significance of ex-directedness will be discussed when we consider Petri net translation of generalised iterations, in Sect. 4.4.13.

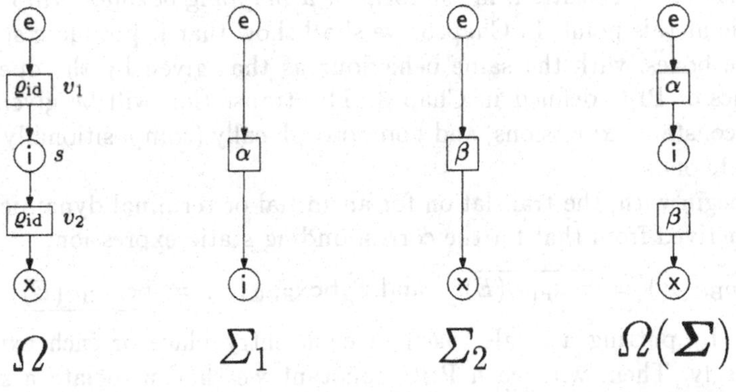

Fig. 4.19. The connection between T-restrictedness and ex-restrictedness

It may also be asked why, while we tried to be as general as possible in defining the refinement mechanism, we decided to only consider simple nets for the role of operator boxes. This stems from the fact that there is no unique natural generalisation to the non-simple case; besides, at the end, operator boxes with arcs weights greater than 1 are not really interesting, since we shall develop our theory only for safe nets. We can easily think of at least two. Indeed, if we look at the definition of places and weights in a refined net, it is possible to cope with non-unitary arcs in Ω by keeping the set of places as it is now and modifying the weight formulae (4.7) and (4.8), by incorporating the weights $W_\Omega(s,v)$ and $W_\Omega(v,s)$ as multiplicative factors, for the transition $u \in \mathsf{T}^v_{\mathsf{new}}$ and place $p \in \mathsf{SP}^s_{\mathsf{new}}$. Alternatively, one could keep the weight formulae intact and modify the definition of $\mathsf{SP}^s_{\mathsf{new}}$ by introducing as many $v \lhd x_v$'s (and $w \lhd e_w$'s) as indicated by the weight of the corresponding arc from v to s (respectively, from s to w). The results would have about the same behaviour, but there might also be some structural differences. We illustrate the latter point using an example.

The refinement shown in Fig. 4.20 yields, with the first rule, two place trees with 2-labelled roots. But it yields three such place trees with the second rule. The behaviours of the two refined nets (considered from their entry markings) are very similar, but the transitions labelled β and γ are independent in the first case, and dependent in the second one (but they may still be executed concurrently). In [6, 25, 31], the first rule was used with good results, but choosing it was simply a matter of taste. Since non-simple operator boxes will not be used in the rest of this book, we shall not elaborate further on this subject.

4.4 Petri Net Semantics of PBC

We will now employ net refinement to associate plain boxes with PBC expressions. The translation, in the form of a mapping $\mathsf{box}_{\mathsf{PBC}}$, will be purely syntactic at this point. In Chap. 8, we shall show that it produces static and dynamic boxes, with the same behaviour as that given by the operational semantics of PBC defined in Chap. 3. The translation will be given explicitly for constant expressions, and homomorphically (compositionally) for the composite ones.

To begin with, the translation for an initial or terminal dynamic expression is derived from that for the corresponding static expression:

$$\mathsf{box}_{\mathsf{PBC}}(\overline{E}) = \overline{\mathsf{box}_{\mathsf{PBC}}(E)} \quad \text{and} \quad \mathsf{box}_{\mathsf{PBC}}(\underline{E}) = \underline{\mathsf{box}_{\mathsf{PBC}}(E)} \ .$$

That is, by putting a single token in each entry place or each exit place, respectively. Then, with each PBC constant we shall associate a static or dynamic box, with each (remaining) PBC operator op an operator box Ω_{op} with the same arity, and define, for the allowed combinations of static and dynamic PBC expressions:

$$\mathsf{box}_{\mathsf{PBC}}\Big(op(H_1,\ldots,H_n)\Big) = \Omega_{op}\Big(\mathsf{box}_{\mathsf{PBC}}(H_1),\ldots,\mathsf{box}_{\mathsf{PBC}}(H_n)\Big) \ .$$

Each H_i above is a static or dynamic expression such that $op(H_1,\ldots,H_n)$ conforms to the syntax given in Sects. 3.1 and 3.4, and Ω_{op} is equipped with a suitable complex marking $\mathcal{M} = (\emptyset, Q)$ such that the operation of net refinement is well defined. The latter means that in each case it suffices to describe an unmarked operator box Ω_{op} only.

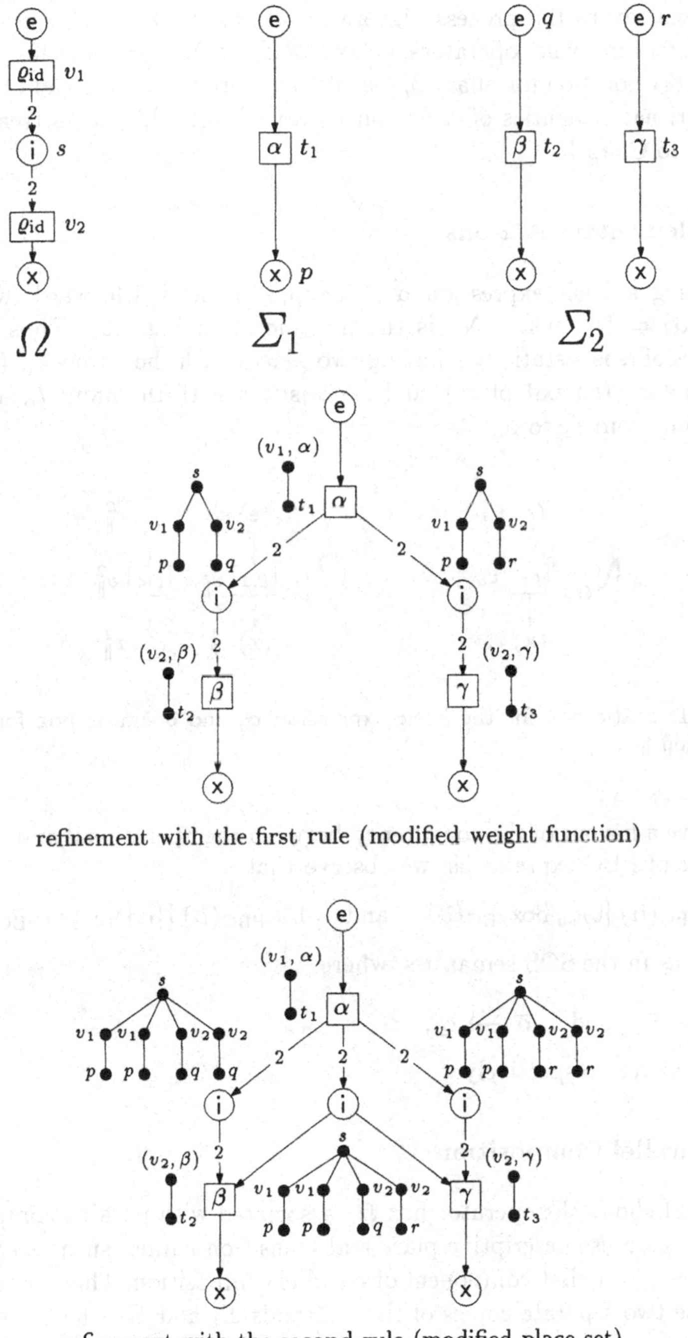
Ω
Σ1
Σ2
refinement with the first rule (modified weight function)
refinement with the second rule (modified place set)

Fig. 4.20. Two versions of refinement for a non-simple operator box (only relevant place and transition trees are shown)

By analogy to the process algebra notation, we shall often use an infix notation for binary net operators, i.e., write Σ_1 *op* Σ_2 instead of $\Omega_{op}(\Sigma_1, \Sigma_2)$, and a suffix notation for unary operators, i.e., write Σ *op* instead of $\Omega_{op}(\Sigma)$. The Petri net semantics of recursion is more involved, and its treatment is deferred to Chap. 5.

4.4.1 Elementary Actions

Translating a basic expression $\alpha \in \mathsf{Lab}_{\mathrm{PBC}}$ is straightforward; we define $\mathsf{box}_{\mathrm{PBC}}(\alpha) = \mathsf{N}_\alpha$, where N_α is the net shown in Fig. 4.21. Thus, the net semantics of α is a static box having two places with the names e_α (an entry place) and x_α (an exit place), and a transition with the name t_α and label α,[4] leading from e_α to x_α.

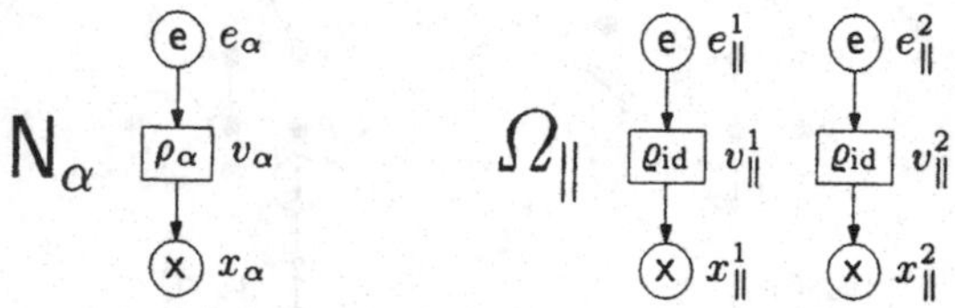

Fig. 4.21. Static box for the basic expression α, and operator box for parallel composition $\|$

To give a flavour of the consistency between the operational and Petri net semantics of PBC expressions, we observe that

$$\mathsf{box}_{\mathrm{PBC}}(\overline{\alpha})\ [\emptyset\rangle_{\mathsf{lab}}\ \mathsf{box}_{\mathrm{PBC}}(\overline{\alpha}) \quad \text{and} \quad \mathsf{box}_{\mathrm{PBC}}(\overline{\alpha})\ [\{\alpha\}\rangle_{\mathsf{lab}}\ \mathsf{box}_{\mathrm{PBC}}(\underline{\alpha})\ ,$$

similarly as in the SOS semantics, where

$$\overline{\alpha} \xrightarrow{\emptyset} \overline{\alpha} \quad \text{and} \quad \overline{\alpha} \xrightarrow{\{\alpha\}} \underline{\alpha}\ ,$$

by IN1 and AR, respectively.

4.4.2 Parallel Composition

Figure 4.21 shows the operator box $\Omega_{\|}$ associated with parallel composition. Here, we again use descriptive place and transition names, such as $e^1_{\|}$ for the entry place of the first component of parallel composition. This operator box will create two separate copies of the operands Σ_1 and Σ_2 which are refined into $v^1_{\|}$ and $v^2_{\|}$, respectively, even if Σ_1 is the same as Σ_2.

An important observation to be made at this point is that

$$\mathsf{box}_{\mathrm{PBC}}(\overline{\alpha\|\beta}) = \mathsf{box}_{\mathrm{PBC}}(\overline{\alpha}\|\overline{\beta})$$

[4] Recall that ϱ_α and α have been identified.

in the sense of net equality, not only net isomorphism. And, at the same time,

$$\overline{\alpha \| \beta} \xrightarrow{\emptyset} \overline{\alpha} \| \overline{\beta} \,,$$

by the inaction rule IPAR1. We shall see later that this is not an isolated case, but an instance of a more general fact. As already explained, the inaction rules should not be interpreted as denoting anything actually happening in the semantic sense, but rather as describing two different views of the same system, and the $\emptyset$-labelled arrow as denoting a change from one point of view to another. This is further illustrated in Fig. 4.22. The dashed lines indicate that the net $\mathsf{box}_{\mathrm{PBC}}(\overline{\alpha \| \beta})$, which corresponds to the left-hand side of IPAR1, is considered as a single entity, whereas the net $\mathsf{box}_{\mathrm{PBC}}(\overline{\alpha} \| \overline{\beta})$, which corresponds to the right-hand side of the rule, is considered as a disjoint union of two nets. Yet the two nets are the same. In a similar way, the difference between $\overline{\alpha \| \beta}$ and $\overline{\alpha} \| \overline{\beta}$ is that the first expression specifies an initial state of $\alpha \| \beta$ while the second specifies two separate initial states of, respectively, α and β, and the inaction rule says that these two views are equivalent.

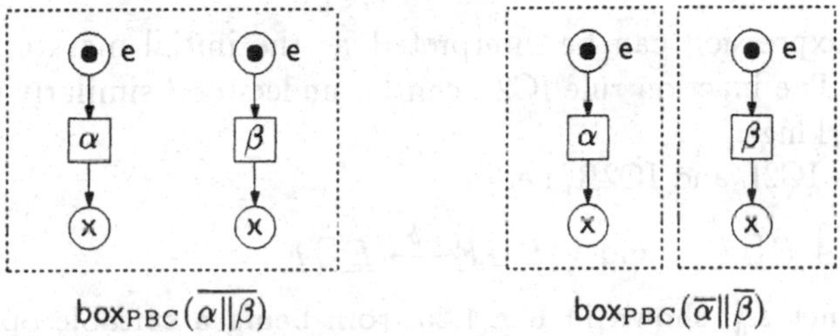

Fig. 4.22. Two different views of the same system (with respect to $\|$)

Exercise 4.4.1. Show that

$$\Sigma \| \Theta \ \mathsf{iso}\ \Theta \| \Sigma \quad \text{and} \quad \Sigma \| (\Theta \| \Psi) \ \mathsf{iso}\ (\Sigma \| \Theta) \| \Psi \,.$$

Note: In this and other exercises, it is assumed that Σ, Θ and Ψ are plain boxes.

The first property formulated in the above exercise (i.e., the commutativity of parallel composition) is discussed in the Appendix.

4.4.3 Choice Composition

The operator box associated with choice composition is given in Fig. 4.23. The naming of places and transitions follows the now well-established convention.

An instance of the inaction rule IC1L, $\overline{\alpha \,\square\, \beta} \xrightarrow{\emptyset} \overline{\alpha} \,\square\, \beta$, can be understood in Petri net terms as shown in Fig. 4.24. The rule says that the initial marking

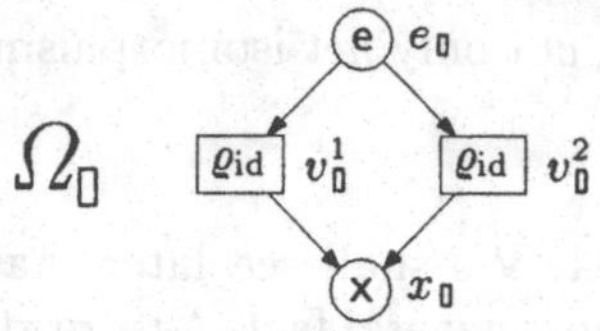

Fig. 4.23. Operator box for choice composition $\Box$

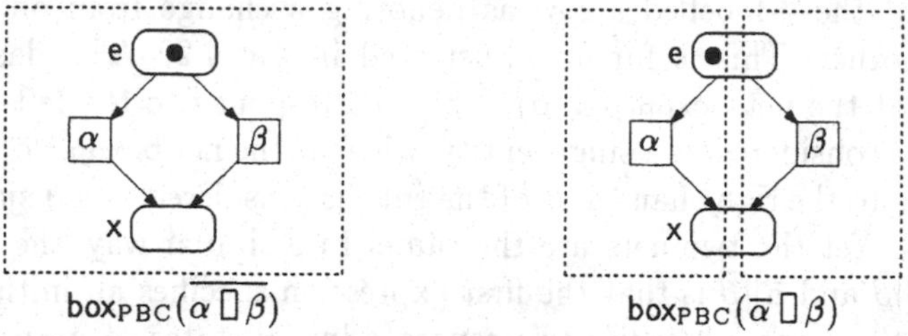

Fig. 4.24. Two different views of the same system (with respect to $\Box$)

of a choice expression can be interpreted as the initial marking of its left constituent. The inaction rule IC2L can be understood similarly in terms of terminal markings.

The rules IC2L and IC2R, i.e.,

$$\underline{E}\,\Box\,F \xrightarrow{\emptyset} \underline{E\,\Box\,F} \quad \text{and} \quad E\,\Box\,\underline{F} \xrightarrow{\emptyset} \underline{E\,\Box\,F}\,,$$

exclude the net $\Omega'_{\Box}$ shown in Fig. 4.25 from being a suitable operator box for choice. The reason is that in $\Omega'_{\Box}\,(\mathsf{N}_\alpha, \mathsf{N}_\beta)$, the marking obtained after an execution of the α-labelled transition cannot be interpreted as being the same as the marking obtained after an execution of the β-labelled transition, and that neither of the two markings is terminal. But, by the inaction rules IC2L and IC2R, the branches of a choice construct should rejoin upon termination.

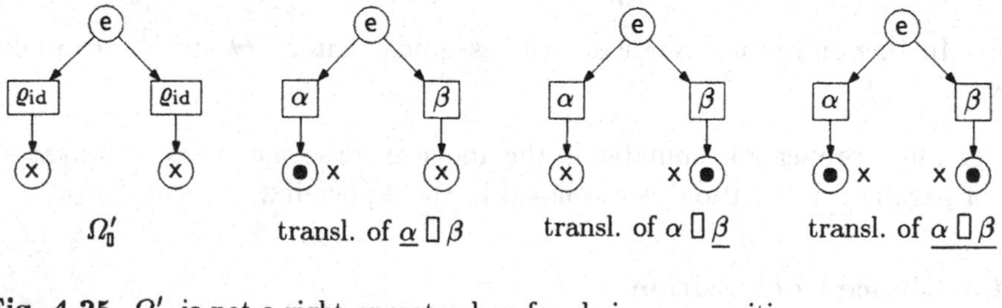

Fig. 4.25. $\Omega'_{\Box}$ is not a right operator box for choice composition

Exercise 4.4.2. Show that

$$\Sigma\,\Box\,\Theta \text{ iso } \Theta\,\Box\,\Sigma \quad \text{and} \quad \Sigma\,\Box\,(\Theta\,\Box\,\Psi) \text{ iso } (\Sigma\,\Box\,\Theta)\,\Box\,\Psi\,.$$

4.4.4 Sequential Composition

The operator box $\Omega_;$ associated with sequential composition is shown in Fig. 4.26.

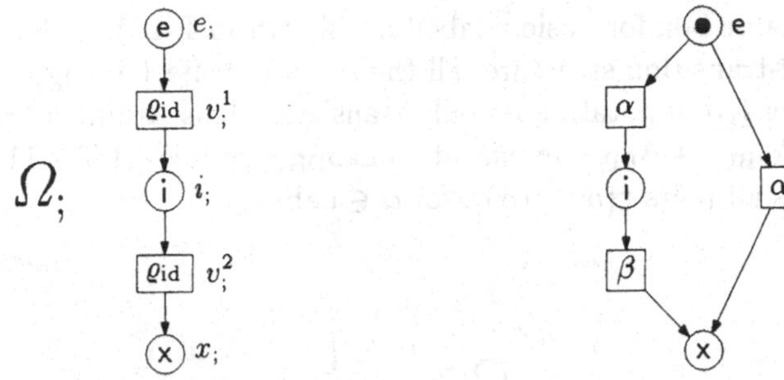

Fig. 4.26. Operator box for sequential composition, and the box of $\overline{(\alpha;\beta)\,\square\,\alpha}$

It is instructive to examine the Petri net corresponding to $\overline{(\alpha;\beta)\,\square\,\alpha}$, shown on the right of Fig. 4.26. Recall that in Sect. 3.2.6 we argued that such an expression can execute the step sequence $\{\alpha\}\{\beta\}$. The same behaviour is also possible for the Petri net, by executing the left α-labelled transition first, and then the β-labelled transition.

The inaction rule IS2 is illustrated in Fig. 4.27 for its instance, $\underline{\alpha};\beta \xrightarrow{\emptyset} \alpha;\overline{\beta}$. Notice that the token on the internal place can be interpreted equally well as the terminal marking of the upper box enclosed in a dashed rectangle, and as the initial marking of the lower box.

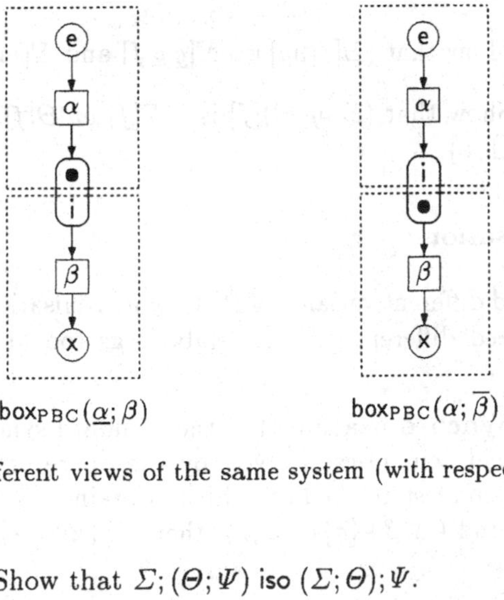

$\mathsf{box}_{\mathsf{PBC}}(\underline{\alpha};\beta)$ $\mathsf{box}_{\mathsf{PBC}}(\alpha;\overline{\beta})$

Fig. 4.27. Two different views of the same system (with respect to ';')

Exercise 4.4.3. Show that $\Sigma;(\Theta;\Psi)$ iso $(\Sigma;\Theta);\Psi$.

The above property (i.e., the associativity of sequential composition) is discussed in the Appendix.

4.4.5 Basic Relabelling

The operator box for basic relabelling, shown in Fig. 4.28, has a simple two-place-one-transition structure, all the semantic effect being contained in the relabelling $\varrho_{[f]}$ annotating its only transition. It is assumed that f is a function $f : \mathsf{A}_{\mathrm{PBC}} \to \mathsf{A}_{\mathrm{PBC}}$, extended to $\mathsf{Lab}_{\mathrm{PBC}}$ as in Sect. 3.2.11, and that $\varrho_{[f]}$ comprises all pairs $(\{\alpha\}, f(\alpha))$, for $\alpha \in \mathsf{Lab}_{\mathrm{PBC}}$.

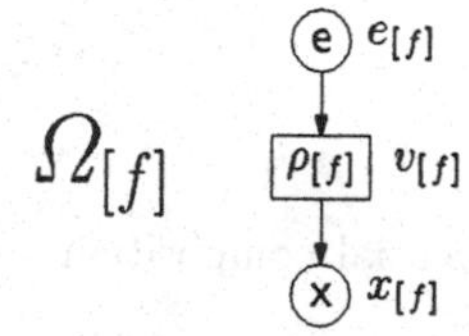

Fig. 4.28. Operator box for basic relabelling

The net semantics of basic relabelling may be compared with its operational semantics, where the semantic effect is contained in the derivation rule, the inaction rules being straightforward. Such an observation is a general one. Control flow operators tend to have simple derivation rules and more complex inaction rules; they also have simple transition relabellings (namely, the identity relabelling) but more complex operator boxes. By contrast, communication interface operators tend to have more complex derivation rules and simple inaction rules.

Exercise 4.4.4. Show that $(\Sigma[f])[g]$ iso $\Sigma[g \circ f]$ and $\Sigma[id]$ iso Σ.

Exercise 4.4.5. Show that $(\Sigma\ op\ \Theta)[f]$ iso $(\Sigma[f]\ op\ \Theta[f])$, for each binary operator $op \in \{\|, \square, ;\}$.

4.4.6 Synchronisation

Since we introduced different variants of PBC synchronisation, in Sects. 3.2.8–3.2.10, we shall need different general relabellings too in order to capture them.

Standard PBC Synchronisation. For the standard synchronisation rules (OP_{sy}), the Petri net semantics is given by the operator box in Fig. 4.29, where $\varrho_{\mathsf{sy}\,a}$ is the smallest relabelling which contains ϱ_{id} and such that, if $(\Gamma, \alpha+\{a\}) \in \varrho_{\mathsf{sy}\,a}$ and $(\Delta, \beta+\{\widehat{a}\}) \in \varrho_{\mathsf{sy}\,a}$, then $(\Gamma+\Delta, \alpha+\beta) \in \varrho_{\mathsf{sy}\,a}$.

Fig. 4.29. Operator box for the standard PBC synchronisation, and the box of $(a\|\widehat{a})$ sy a

We illustrate this synchronisation operator in Fig. 4.29 by giving the translation for $(a\|\widehat{a}\,)$ sy a. The middle transition arises from the second part of the definition of $\varrho_{\mathsf{sy}\ a}$ and

$$(\{\{a\}\},\{a\}),(\{\{\widehat{a}\}\},\{\widehat{a}\}) \in \varrho_{\mathsf{id}} \subseteq \varrho_{\mathsf{sy}\ a}\ .$$

The translation is correct because only the three pairs,

$$(\{\{a\}\},\{a\}) \quad , \quad (\{\{\widehat{a}\}\},\{\widehat{a}\}) \quad , \quad \text{and} \quad (\{\{a\},\{\widehat{a}\}\},\emptyset)$$

in $\varrho_{\mathsf{sy}\ a}$, are relevant as only their first arguments can be matched by the labels of some group of transitions of $\mathsf{box}_{\mathsf{PBC}}(a\|\widehat{a}\,)$.

To formally analyse the different variants of PBC synchronisation, we will use the notion of a linkage graph. Throughout the rest of this section, we will assume that $a \in \mathsf{A}_{\mathsf{PBC}}$ is a fixed action particle used to synchronise transitions. Let

$$(\Gamma,\alpha) = (\{\alpha_1,\ldots,\alpha_n\},\alpha) \in \mathsf{mult}(\mathsf{Lab}_{\mathsf{PBC}}) \times \mathsf{Lab}_{\mathsf{PBC}}$$

be such that $n \geq 1$. A *linkage graph* of (Γ,α) is a (not necessarily unique) finite labelled directed graph LG, in which multiple arcs as well as multiple self-loops are allowed, with n nodes $\xi_1,\ldots,\xi_n$ and m arcs, such that

$$\alpha = \sum_{i=1}^{n} \alpha_i - m\cdot\{a,\widehat{a}\}$$

and each node ξ_i is labelled by α_i. Moreover, for every $i \leq n$, $\alpha_i(a) \geq \mathsf{deg}^+(\xi_i)$ and $\alpha_i(\widehat{a}\,) \geq \mathsf{deg}^-(\xi_i)$, where $\mathsf{deg}^+(\xi_i)$ is the number of arcs outgoing from ξ_i and $\mathsf{deg}^-(\xi_i)$ is the number of arcs incoming to ξ_i. Figure 4.30 shows a possible linkage graph with four arcs for the pair

$$(\Gamma,\alpha) = (\,\{\{a,a,a,a,b\},\{a,\widehat{a},\widehat{a}\},\{\widehat{a},\widehat{a}\}\}\ ,\ \{a,b\}\,)\,.$$

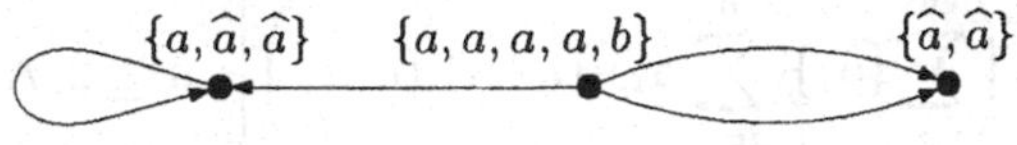

Fig. 4.30. A linkage graph

Exercise 4.4.6. Show that $\varrho_{\mathsf{sy}\,a}$ is the set of all pairs

$$(\Gamma, \alpha) \in \mathsf{mult}(\mathsf{Lab}_{\mathsf{PBC}}) \times \mathsf{Lab}_{\mathsf{PBC}}$$

such that $|\Gamma| \geq 1$ and there is a connected linkage graph for (Γ, α) with exactly $|\Gamma| - 1$ arcs (which means that the undirected version of the graph is connected and acyclic).

We can lift the notion of a linkage graph to synchronisation transitions in the following way. A *linkage graph* for a transition $t = (v, \alpha) \triangleleft R$ in a box obtained from some net Σ through synchronisation on a, is a graph LG with the nodes labelled exactly by the multiset of transitions R and such that, if we replace each transition label u by $\lambda_\Sigma(u)$, then the resulting labelled graph is a linkage graph for $(\lambda_\Sigma(R), \alpha)$. Figure 4.31 shows a (unique) linkage graph for the transition ttt of $\mathsf{box}_{\mathsf{PBC}}(\{a, \widehat{a}, b\})\ \mathsf{sy}\ a = \mathsf{box}_{\mathsf{PBC}}(\{a, \widehat{a}, b\}\ \mathsf{sy}\ a)$.

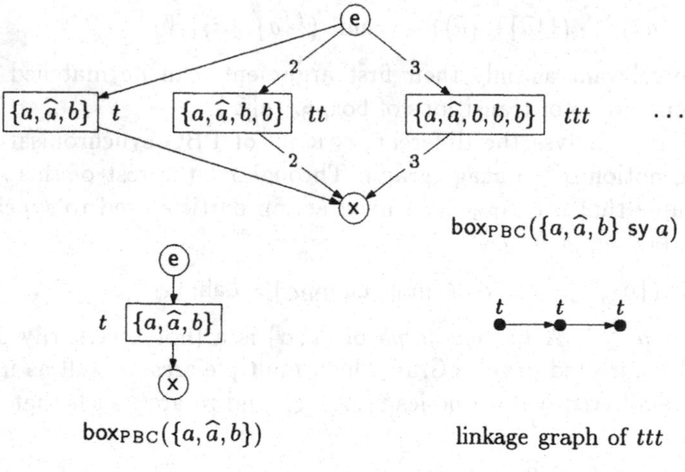

Fig. 4.31. Infinite synchronisation

Exercise 4.4.7. Show that $t \in T_{\Sigma\ \mathsf{sy}\ a}$ if and only if $t = (v_{\mathsf{sy}\,a}, \alpha) \triangleleft R$, where R is a finite multiset of transitions in T_Σ, and there is a connected linkage graph for t with exactly $|R| - 1$ arcs.

Exercise 4.4.8. Show that

$$\varrho_{\mathsf{sy}\,a} = \varrho_{\mathsf{id}} \cup \left\{ \left(\sum_{i=1}^{n} \{\alpha_i\}, \sum_{i=1}^{n} \alpha_i - (n-1) \cdot \{a, \widehat{a}\} \right) \middle|\ n \geq 2 \wedge \mathcal{C} \right\},$$

where $\mathcal{C} = \left(\prod_{i=1}^{n} \left(\alpha_i(a) + \alpha_i(\widehat{a}) \right) > 0 \right) \wedge \left(\sum_{i=1}^{n} \alpha_i(a) \geq n-1 \leq \sum_{i=1}^{n} \alpha_i(\widehat{a}) \right)$.

Exercise 4.4.9. Show that $((\Sigma \text{ sy } a) \text{ sy } b)$ iso $((\Sigma \text{ sy } b) \text{ sy } a)$ and that $(\Sigma \text{ sy } a)$ iso $(\Sigma \text{ sy } \widehat{a})$.

Exercise 4.4.10. Show that $((\Sigma \text{ sy } a) \text{ sy } a)$ iso_T $(\Sigma \text{ sy } a)$.

The idempotence result in the last exercise cannot use iso, since a second application of synchronisation adds again the same group of synchronised transitions, though with different names.

Exercise 4.4.11. Show that $(\Sigma \text{ sy } a)[f]$ iso $(\Sigma[f]) \text{ sy } f(a)$ provided that f is a bijective relabelling which preserves conjugates.

Exercise 4.4.12. Determine whether the properties of the synchronisation operator similar to those formulated in Exc. 3.2.11 hold also in the box domain.

The definition of $\varrho_{\text{sy } a}$, while faithfully reflecting the idea behind CCS synchronisation when restricted to empty and singleton multiactions, has at the same time some drawbacks. For instance, even if we start from a finite net, the result of synchronisation may be infinite, as illustrated by the net modelling the expression $\{a, \widehat{a}, b\}$ sy a, shown in Fig. 4.31. The next result provides a complete characterisation of all such cases.

Theorem 4.4.1. Let Σ be a plain box with finitely many transitions. Then Σ sy a has infinitely many transitions if and only if one of the following holds:

(a) There is $v \in T_\Sigma$ such that $\lambda_\Sigma(v)(a) > 0$ and $\lambda_\Sigma(v)(\widehat{a}) > 0$.
(b) There are $u, w \in \Sigma$ such that $\lambda_\Sigma(u)(a) > 1$ and $\lambda_\Sigma(w)(\widehat{a}) > 1$.

Proof. ($\Longrightarrow$) Suppose that neither (a) nor (b) is satisfied. Since T_Σ is finite, there is $n \geq 0$ such that

$$n = \max\{\lambda_\Sigma(t)(a), \lambda_\Sigma(t)(\widehat{a}) \mid t \in T_\Sigma\} .$$

Without loss of generality, we may assume that $n = \lambda_\Sigma(t)(a)$, for some $t \in T_\Sigma$. From the definition of net refinement and the property formulated as Exc. 4.4.7, it follows that for each transition u of Σ sy a there is at least one acyclic connected linkage graph LG (and that any such linkage graph determines a unique transition). It therefore follows that, for every node ξ of LG,

- $\mathsf{deg}^+(\xi)$ is bounded by the number of the a's in the label of the corresponding transition and $\mathsf{deg}^-(\xi)$ is bounded by the number of the $\widehat{a}$'s in the label of the corresponding transition (from the definition of a linkage graph for a synchronisation transition),
- $\mathsf{deg}^-(\xi) \cdot \mathsf{deg}^+(\xi) = 0$ (since (a) does not hold),
- $\mathsf{deg}^-(\xi) + \mathsf{deg}^+(\xi) \leq n$ (by the choice of n and the previous property),

- $\deg^-(\xi) \leq 1$ (since either $n \leq 1$ and the property is trivial, or $n \geq 2$ but (b) does not hold and we know that there is a transition t with $\lambda_\Sigma(t)(a) = n > 1$).

Hence, since LG is connected, it has a starlike shape with at most $n+1$ nodes. Thus, for every $u \in T_{\Sigma \text{ sy } a}$, $|\mathsf{trees}(u)| \leq n+1$, and so $|T_{\Sigma \text{ sy } a}| \leq \sum_{m=1}^{n+1} |T_\Sigma|^m < \infty$.

($\Longleftarrow$) Suppose that (a) holds. Then, for every $n > 0$, there is a transition $w_n = (v_{\mathsf{sy}\,a}, \alpha) \lhd n{\cdot}\{v\}$ in Σ sy a, for which there is a linkage graph of the form[5]

$$\underbrace{v \to v \to v \to v \to \cdots \to v}_{n \text{ times } v}\ .$$

Suppose now that (b) holds. Then, for every $n > 0$, there is a transition $w^n = (v_{\mathsf{sy}\,a}, \alpha) \lhd n{\cdot}\{u, w\}$ in Σ sy a, for which there is a linkage graph of the form

$$\underbrace{u \to w \leftarrow u \to \cdots w \leftarrow u \to w}_{n \text{ times } u \text{ and } n \text{ times } w}\ .$$

In either case Σ sy a has infinitely many transitions. □

It may be observed that almost all transitions generated by synchronisation from a finite set of transitions are dead if the resulting net is safe but, as explained before, we prefer to retain them all, at least in the current theoretical development. Having said that, in many practical situations, like those described in Chap. 9, it is the case that synchronisation never generates infinite sets of new transitions (cf. Exc. 9.2.1 and Cor. 9.5.3).

Auto-synchronisation and Multilink-synchronisation. To model the rules (OP'_{sy}), we simply change the relabelling of the operator box $\Omega_{\mathsf{sy}\,a}$. The result is shown in Fig. 4.32, where $\varrho_{\mathsf{sy}'\,a}$ is the smallest relabelling containing $\varrho_{\mathsf{sy}\,a}$ and such that, if $(\Gamma, \alpha + \{a, \widehat{a}\}) \in \varrho_{\mathsf{sy}'\,a}$, then $(\Gamma, \alpha) \in \varrho_{\mathsf{sy}'\,a}$.

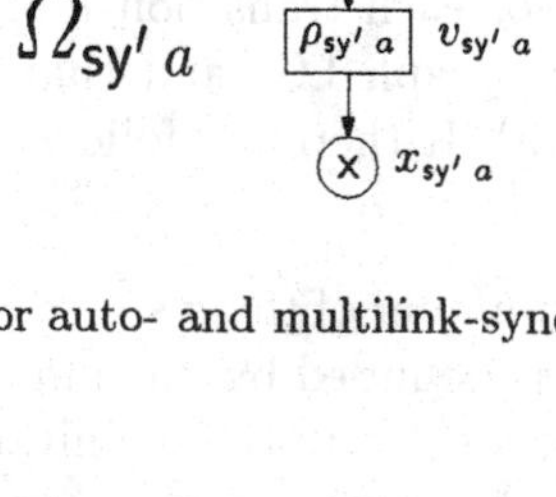

Fig. 4.32. Operator box for auto- and multilink-synchronisation

[5] Note that, for all $n > 0$ and $s \in S_{\Sigma \text{ sy } a}$, $W_{\Sigma \text{ sy } a}(s, w_n) = n \cdot W_{\Sigma \text{ sy } a}(s, w_1)$ so, by the T-restrictedness of Σ sy a, the transitions $w_1, w_2, \ldots$ do not duplicate each other. The same is true of the transitions $w^1, w^2, \ldots$ constructed for case (b).

Exercise 4.4.13. Show that $\varrho_{\mathsf{sy}'\,a}$ is the set of all pairs

$$(\Gamma,\alpha) \in \mathsf{mult}(\mathsf{Lab_{PBC}}) \times \mathsf{Lab_{PBC}}$$

such that $|\Gamma| \geq 1$ and there is a connected linkage graph for (Γ,α).

Exercise 4.4.14. Show that

$$\varrho_{\mathsf{sy}'\,a} \;=\; \varrho_{\mathrm{id}} \,\cup\, \left\{ \left(\sum_{i=1}^{n}\{\alpha_i\}, \sum_{i=1}^{n}\alpha_i - m\cdot\{a,\widehat{a}\} \right) \middle|\; m \geq n-1 \geq 0 \wedge \mathcal{C} \right\},$$

where $\mathcal{C} \;=\; \left(\prod_{i=1}^{n}\left(\alpha_i(a)+\alpha_i(\widehat{a})\right) > 0\right) \wedge \left(\sum_{i=1}^{n}\alpha_i(a) \geq m \leq \sum_{i=1}^{n}\alpha_i(\widehat{a})\right).$

Exercise 4.4.15. Show that $((\Sigma\ \mathsf{sy}'\ a)\ \mathsf{sy}'\ b)\ \mathsf{iso}\ ((\Sigma\ \mathsf{sy}'\ b)\ \mathsf{sy}'\ a)$ and $(\Sigma\ \mathsf{sy}'\ a)\ \mathsf{iso}\ (\Sigma\ \mathsf{sy}'\ \widehat{a})$.

Exercise 4.4.16. Show that $((\Sigma\ \mathsf{sy}'\ a)\ \mathsf{sy}'\ a)\ \mathsf{iso_T}\ (\Sigma\ \mathsf{sy}'\ a)$.

Exercise 4.4.17. Show that $(\Sigma\ \mathsf{sy}'\ a)[f]\ \mathsf{iso}\ (\Sigma[f])\ \mathsf{sy}'\ f(a)$ provided that f is a bijective relabelling which preserves conjugates.

Exercise 4.4.18. Design a relabelling that would correspond to a synchronisation operator allowing multi- but not auto-synchronisations.

Step-synchronisation. If we were to be faithful to the idea from Sect. 3.2.10 that each concurrent activity ought to correspond to a transition, whatever the marking, it would be necessary to work with 'saturated' nets containing a transition corresponding to every possible step of transitions (i.e., with the sum of the labels – thus implying that there is a summation operation for labels, as it is the case here – and the sum of the connectivities of every finite multiset[6] of transitions), for the translation of the basic expressions as well as for the nets constructed through box operators. Thus we would almost always work with infinite nets, which would not be practical. For instance, Fig. 4.33 shows the saturated version of the translation of the PBC expression $E = a$. The definition of net refinement would also have to be modified. Afterwards, once an initial marking has been chosen, it would of course be possible to drop all dead transitions, but the situation would be uncomfortable anyway.

Another, in our view more satisfactory, way of addressing the problem is to modify the execution semantics for Petri nets. Instead of considering the standard step sequence semantics, one may accept that if a net can perform a multiset of transitions concurrently, it may also perform them simultaneously, which allows one to consider behaviours constituted of sequences of multisets of multisets of transitions, and when filtered through the labelling, this gives sequences of multisets of labels through the formula

[6] Limiting the grouping to sets, or to sets of independent transitions, would not be satisfactory, for reasons explained in Sect. 4.3.5.

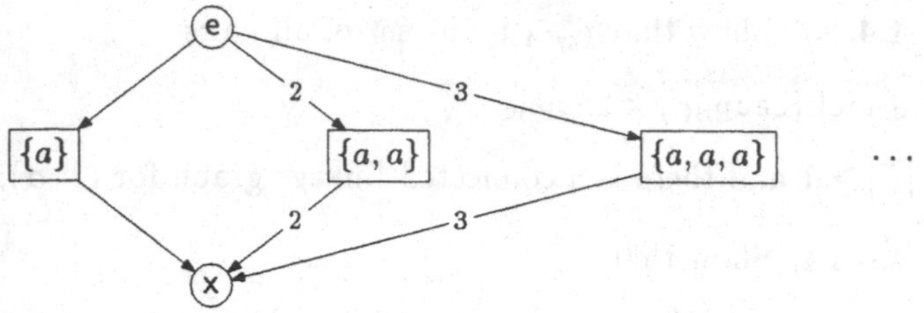

Fig. 4.33. A saturated version of $\mathsf{box}_{\mathsf{PBC}}(a)$

$$\lambda\left(\sum_{i=1}^{n}\left\{\sum_{j=1}^{m_i}\{t_{ij}\}\right\}\right) = \sum_{i=1}^{n}\left\{\sum_{j=1}^{m_i}\lambda(t_{ij})\right\} .$$

This corresponds to the grouping allowed by the rule ST, implying that if actions may be executed concurrently, then they may also be executed simultaneously. Of course, this would have impact on transition systems associated with labelled nets and, consequently, on the derived equivalence notions. In particular, one would be able to define an interesting structural equivalence such that not only would it be possible to arbitrarily drop or add duplicate transitions (or places), but also transitions which are equivalent (w.r.t. labelling and connectivity) to the step of at least two other transitions. The result would have exactly the same (modified) transition systems as the original net, whatever the initial marking. Let us call **step** the resulting *(step-)equivalence*, i.e., Σ **step** Ψ if the nets Σ and Ψ are isomorphic after all duplicate and step transitions have been dropped. Interestingly, this is related to the finiteness of synchronisation.

Finiteness of Synchronisation. The basic idea arises from the observation that, in the example shown in Fig. 4.34, almost all transitions are equivalent to steps of more elementary ones, in the sense that they have the label and connectivity identical to the sum of the labels and connectivities of the components of the corresponding step; for instance, ttt^m is equivalent to the step $\{t^m, tt^m\}$ as well as $\{t^m, t^m, t^m\}$, and tt^l is equivalent to the step $\{t^l, t^m\}$ (where the m means the maximal saturation of the conjugate $(a, \widehat{a})$ pairs and the l means the minimal, or least, saturation). In other words, in each case there is a corresponding linkage graph which is a disjoint union of linkage graphs of other transitions. We shall show that this situation is general. In the following, a transition t is called *elementary* if all its linkage graphs are connected.

Proposition 4.4.1. If T_Σ is finite, then there are finitely many elementary transitions in Σ sy$'$ a.

Proof. Suppose that the set T_{elem} of the elementary transitions of Σ sy$'$ a is infinite. Below, for every finite multiset R of transitions of Σ, we denote

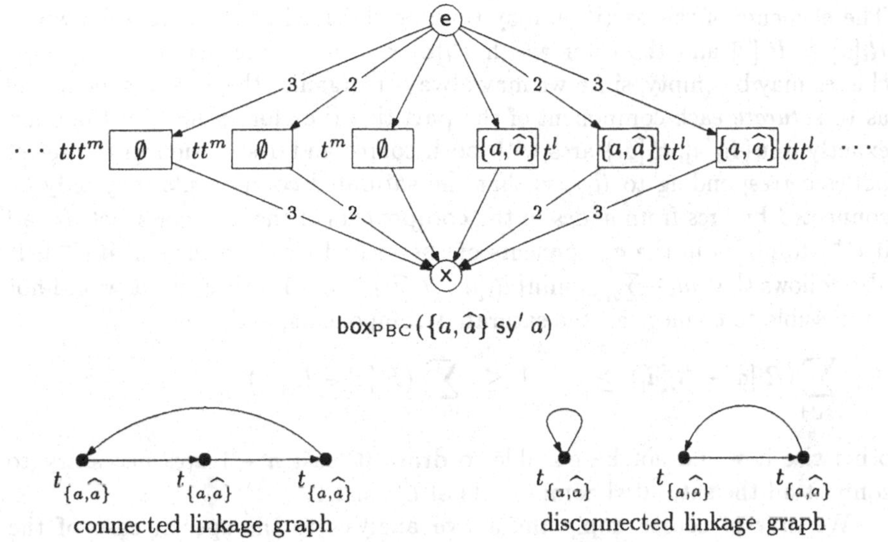

Fig. 4.34. Synchronisation with auto- and multilinks, and two linkage graphs for ttt^m (the second one is composed of linkage graphs for t^m and tt^m)

$$R[a] = \sum_{u \in R} R(u) \cdot \lambda_\Sigma(u)(a) \quad \text{and} \quad R[\widehat{a}] = \sum_{u \in R} R(u) \cdot \lambda_\Sigma(u)(\widehat{a}) .$$

A transition $t = (v_{\mathsf{sy}'\,a}, \alpha) \lhd R_t \in T_{\Sigma\ \mathsf{sy}'\ a}$ may have more than one linkage graph, but all such graphs have the same nodes (up to isomorphism, since their labelling multiset is R_t), and the same number m_t of arcs (note that $m_t = R_t[a] - \alpha(a) = R_t[\widehat{a}] - \alpha(\widehat{a})$); moreover, no two distinct transitions may have linkage graphs with the same nodes (even up to isomorphism). It also follows that $m_t \leq \min\{R_t[a], R_t[\widehat{a}]\}$, so that a single R_t, or a finite family of the R_t's, may only contribute finitely many elements to T_{elem} and may be ignored. In particular, we may retain for further consideration only non-singleton R_t's.

Let $t = (v_{\mathsf{sy}'\,a}, \alpha) \lhd R_t \in T_{elem}$ be such that $|R_t| \geq 2$. Then, for every (nontrivial) partitioning $\mathcal{P} = \{R_1, \ldots, R_n\}$ of R_t such that $R_t = \sum_{i=1}^{n} R_i$, $n \geq 2$, and $R_i \neq \emptyset$ (for each $i \leq n$), we have that

$$m_t > \sum_{i=1}^{n} \min\{R_i[a], R_i[\widehat{a}]\} .$$

Otherwise, the arcs of a linkage graph for t could easily (since we may find a decomposition $m_t = \sum_{i=1}^{n} m_i$ with $m_i \leq \min\{R_i[a], R_i[\widehat{a}]\}$ for $i = 1, \ldots, n$) be reorganised in such a way that each subset of nodes corresponding to a component of the partition would be isolated from the other ones, contradicting t being elementary. It must also be the case that $R_i[a] \neq R_i[\widehat{a}]$ (for each $i \leq n$), otherwise a straightforward reorganisation of the arcs would isolate the corresponding subset of nodes, again contradicting t being elementary.

The elements of the partition may thus be classified as those R_i's for which $R_i[a] > R_i[\widehat{a}]$ and those for which $R_i[a] < R_i[\widehat{a}]$. Moreover, none of these classes may be empty, since we may always reorganise the arcs in such a way as to *saturate* each component of the partition (i.e., for each $i \leq n$ there are exactly $\min\{R_i[a], R_i[\widehat{a}]\}$ arcs with both source and destination in the set of nodes corresponding to R_i), so that the saturated components may only be connected by arcs from nodes in the components of the first class (let us call it $\mathcal{C}^+$) to nodes in the components of the second class (let us call it $\mathcal{C}^-$). It also follows that $m_t - \sum_{i=1}^{n} \min\{R_i[a], R_i[\widehat{a}]\} \geq n-1$, otherwise it would not be possible to connect all the saturated components; and

$$\sum_{i \in \mathcal{C}^+} (R_i[a] - R_i[\widehat{a}]) \geq n - 1 \leq \sum_{i \in \mathcal{C}^-} (R_i[\widehat{a}] - R_i[a]) ,$$

otherwise it would not be possible to draw at least $n-1$ arcs necessary to connect all the saturated components of $\mathcal{C}^+$ and $\mathcal{C}^-$.

We may of course apply the above analysis to the special case of the (unique) partitioning of the non-singleton R_t onto singleton components: $R_t = \sum_{i=1}^{n} \{u_i\}$. As a result, we may infer that no $u \in R_t$ is such that $\lambda_\Sigma(u)(a) = \lambda_\Sigma(u)(\widehat{a})$, and the above classification only depends on the characteristics of individual transitions in R_t: $\mathcal{C}^+$ is composed of nodes labelled by transitions u such that $\lambda_\Sigma(u)(a) > \lambda_\Sigma(u)(\widehat{a})$, and $\mathcal{C}^-$ of nodes labelled by transitions u such that $\lambda_\Sigma(u)(a) < \lambda_\Sigma(u)(\widehat{a})$.

We now observe that since T_Σ is finite and for every finite multiset R of transitions of T_Σ, there can be only finitely many transitions t of $T_{\Sigma \textsf{ sy}' a}$ such that $t = (v_{\textsf{sy}' a}, \alpha) \lhd R$, there is no upper bound on the size of the multisets R_t. Hence, since T_Σ is finite, from the above observation about us being able to draw at least $n-1$ arcs between the saturated components (here singletons), there are $u, v \in T_\Sigma$ and $t_1, t_2, \ldots \in T_{elem}$ such that

$$k = \lambda_\Sigma(u)(a) - \lambda_\Sigma(u)(\widehat{a}) > 0$$
$$l = \lambda_\Sigma(v)(\widehat{a}) - \lambda_\Sigma(v)(a) > 0$$
$$1 < R_{t_1}(u) < R_{t_2}(u) < \ldots$$
$$1 < R_{t_1}(v) < R_{t_2}(v) < \ldots .$$

Hence there is t_j such that $|R_{t_j}| > k+l$, $R_{t_j}(v) \geq k$, and $R_{t_j}(u) \geq l$. Let V be the multiset $k \cdot \{v\} + l \cdot \{u\}$. If we now consider a (nontrivial) partitioning of R_{t_j} with V as one of its components, it is clear that $V[a] = V[\widehat{a}]$, contradicting one of the properties obtained earlier in this proof, and hence contradicting $t_j \in T_{elem}$. □

Corollary 4.4.1. If Σ is finite, then there is a finite representative in the step-equivalence class of Σ sy' a.

Proof. Immediate from Prop. 4.4.1. □

It is remarkable that the property we have just proved does not hold for the standard version of the synchronisation operator, as shown by the example in Fig. 4.31; none of the transitions of $\mathsf{box}_{\mathrm{PBC}}(\{a, \widehat{a}, b\}\ \mathsf{sy}\ a)$ is equivalent to a step of other transitions since their labels all have exactly one a and exactly one $\widehat{a}$.

Exercise 4.4.19. Adapt the definition of a linkage graph, and the properties of the various forms of synchronisation, to the case when the conjugation is not discriminative, i.e., if $\widehat{a} \neq a$ cannot be assumed. Determine also what properties would be lost if the conjugation was not involutive, i.e., if $\widehat{\widehat{a}} \neq a$ could not be assumed.

4.4.7 Restriction

The net semantics of a restricted expression is almost self-explanatory; the net Σ rs a is simply Σ where all transitions whose labels carry at least one a or $\widehat{a}$ have been erased. This corresponds to the operator box $\Omega_{\mathsf{rs}\,a}$ shown in Fig. 4.35, where $\varrho_{\mathsf{rs}\,a} = \varrho_{\mathsf{mult}(\mathrm{A}_{\mathrm{PBC}} \setminus \{a, \widehat{a}\})}$ is the restriction on the multiactions without any a or $\widehat{a}$, i.e., $\varrho_{\mathsf{rs}\,a}$ comprises all pairs $(\{\alpha\}, \alpha)$ such that $\alpha \in \mathsf{Lab}_{\mathrm{PBC}}$ and $a, \widehat{a} \notin \alpha$.

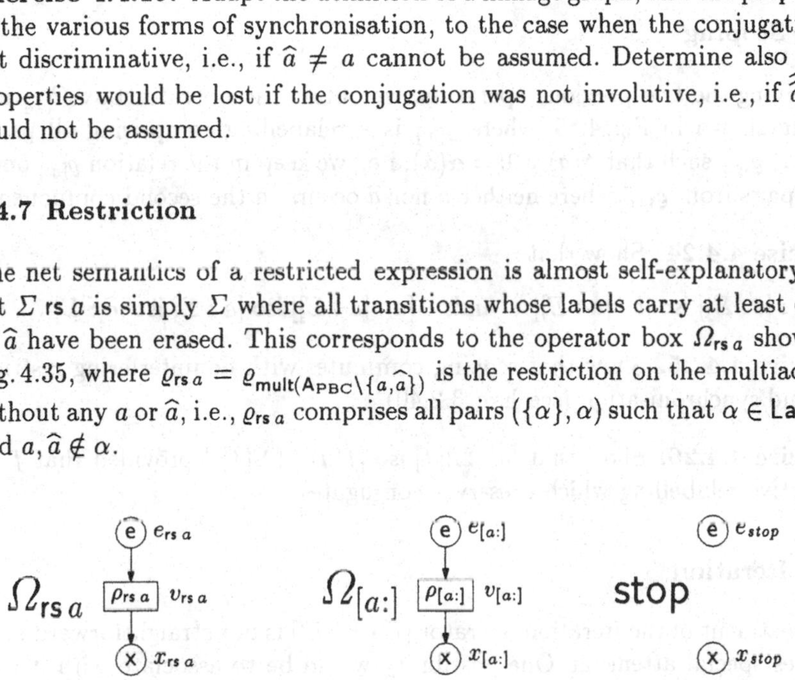

Fig. 4.35. Operator boxes for restriction and scoping, and the simplest ex-box

Applying restriction may create boxes without any transitions. Boxes whose transition and internal place sets are empty will be called ex-*boxes* (see Fig. 4.35). They correspond, for instance, to the **stop** expression of LOTOS [75], which deadlocks without reaching the terminated state, and to the stop expression of PBC (which we may assume to be present in the syntax, since it can be expressed by the recursive equation stop $\stackrel{\mathrm{df}}{=}$ stop as in Sect. 3.2.2, or by $\{a\}$ rs a as in Exc. 3.2.46).

Exercise 4.4.20. Show that $((\Sigma\ \mathsf{rs}\ a)\ \mathsf{rs}\ b)$ iso $((\Sigma\ \mathsf{rs}\ b)\ \mathsf{rs}\ a)$ and that $((\Sigma\ \mathsf{rs}\ a)\ \mathsf{rs}\ a)$ iso $(\Sigma\ \mathsf{rs}\ a)$ iso $(\Sigma\ \mathsf{rs}\ \widehat{a}\,)$.

Exercise 4.4.21. Show that Σ iso $(\Sigma \,\Box\, \mathsf{stop})$ iso $(\mathsf{stop} \,\Box\, \Sigma)$ which means that stop is a neutral element for choice.

Exercise 4.4.22. Show that restriction distributes over the control flow operators, and commutes with noninterfering synchronisation (see Excs. 3.2.30 and 3.2.31).

Exercise 4.4.23. Show that $(\Sigma \text{ rs } a)[f]$ iso $(\Sigma[f] \text{ rs } f(a))$ provided that f is a bijective relabelling which preserves conjugates.

4.4.8 Scoping

Combining the two previous operators, we obtain the operator box $\Omega_{[a:]}$ for scoping shown in Fig. 4.35, where $\varrho_{[a:]}$ is a relabelling comprising all pairs $(\Gamma, \alpha) \in \varrho_{\mathsf{sy}\, a}$ such that $\alpha(a) = 0 = \alpha(\widehat{a})$, i.e., we keep in the relation $\varrho_{[a:]}$ only those pairs from $\varrho_{\mathsf{sy}\, a}$ where neither a nor $\widehat{a}$ occurs in the second component.

Exercise 4.4.24. Show that

$$[a : [b : \Sigma]] \text{ iso } [b : [a : \Sigma]] \quad \text{and} \quad [a : [a : \Sigma]] \text{ iso } [a : \Sigma] \text{ iso } [\widehat{a} : \Sigma] .$$

Exercise 4.4.25. Show that scoping commutes with noninterfering restriction and synchronisation (see Exc. 3.2.40).

Exercise 4.4.26. Show that $[a : \Sigma][f]$ iso $[f(a) : (\Sigma[f])]$ provided that f is a bijective relabelling which preserves conjugates.

4.4.9 Iteration

The treatment of the iteration operator $[E * F * E']$ is not straightforward and deserves special attention. One possibility would be to associate with it the operator box $\Omega_{[*]}$ shown in Fig. 4.36. This box first performs the behaviour represented by (i.e., refining) $v^1_{[*]}$ which corresponds to E, then zero, one or more (possibly infinitely many) times the behaviour represented by $v^2_{[*]}$ which corresponds to F, and finally (if entered) the behaviour represented by $v^3_{[*]}$ which corresponds to E'. However, $\Omega_{[*]}$ does not always preserve the safeness of the operand boxes, as can be seen by taking the PBC expression $[\alpha * (\beta \| \alpha) * \alpha]$, whose translation is shown in Fig. 4.37.

It may be observed that, if we start from the entry marking, after the execution of t_1 and t_2, the place s_{12} will accumulate two tokens. And, as one can see, the net is 2-bounded, but not safe. On the level of executions, the sequential as well as step behaviour of the net is as expected, but if we consider partial order semantics, there are unwanted dependencies. Thus the operator $\Omega_{[*]}$ is not entirely satisfactory. A careful analysis of the reasons of the non-safeness phenomenon (see [25, 31], and Chap. 6) shows that

- The situation is never worse than that exhibited by the example in Fig. 4.37, i.e., the boxes obtained for the PBC expressions with the operator $\Omega_{[*]}$ are always 2-bounded.

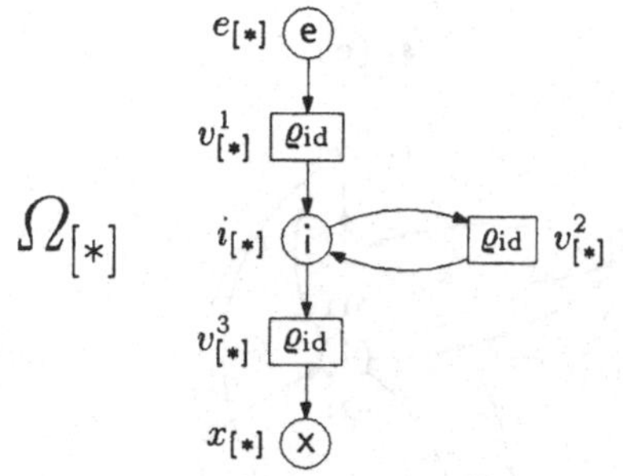

Fig. 4.36. Operator box for iteration (2-bounded version)

- The non-safeness is 'free of self-concurrency' in the sense that each transition always has at least one safe pre-place and at least one safe post-place.
- The origin of non-safeness can be traced to the fact that the operator $\Omega_{[*]}$ has a side-loop (between $i_{[*]}$ and $v^2_{[*]}$) and that the operand replacing $v^2_{[*]}$ has disjoint, independently starting and terminating subnets, for example, like that modelling $\beta\|\alpha$; in other words, the operand net is not ex-exclusive.

If we want to allow non-ex-exclusive operands in the looping part of iteration,[7] the only way to guarantee safeness is to get rid of the side-loop. This leads to the operator box $\Omega_{[**]}$ shown in Fig. 4.38, and the translation for a static PBC expression $[E * F * E']$ is then given by

$$\mathsf{box}_{\mathrm{PBC}}\Big([E * F * E']\Big) = \Omega_{[**]}\begin{pmatrix}\mathsf{box}_{\mathrm{PBC}}(E), \mathsf{box}_{\mathrm{PBC}}(F), \mathsf{box}_{\mathrm{PBC}}(E') \\ \mathsf{box}_{\mathrm{PBC}}(E), \mathsf{box}_{\mathrm{PBC}}(F), \mathsf{box}_{\mathrm{PBC}}(E')\end{pmatrix}$$

where the ordering of the transitions of $\Omega_{[**]}$ is $v^{10}_{[**]}, v^{20}_{[**]}, v^{30}_{[**]}, v^{11}_{[**]}, v^{21}_{[**]}, v^{31}_{[**]}$. It may be noticed that one of the transitions $v^{10}_{[**]}$ and $v^{11}_{[**]}$ is not strictly necessary for obtaining the desired behaviour, but we keep both just to maintain the symmetry of the operator box. Strictly speaking, the operator $\Omega_{[**]}$ has the arity of 6, but we turned it into a ternary one by restricting the allowable combinations of its operands.

The situation is more complex for the translation of dynamic iterative expressions. For example, there are two different translations for $[E * G * E']$, namely, with $F = \lfloor G \rfloor$:

$$\Omega_{[**]}\begin{pmatrix}\mathsf{box}_{\mathrm{PBC}}(E), \mathsf{box}_{\mathrm{PBC}}(G), \mathsf{box}_{\mathrm{PBC}}(E') \\ \mathsf{box}_{\mathrm{PBC}}(E), \mathsf{box}_{\mathrm{PBC}}(F), \mathsf{box}_{\mathrm{PBC}}(E')\end{pmatrix}$$

and

$$\Omega_{[**]}\begin{pmatrix}\mathsf{box}_{\mathrm{PBC}}(E), \mathsf{box}_{\mathrm{PBC}}(F), \mathsf{box}_{\mathrm{PBC}}(E') \\ \mathsf{box}_{\mathrm{PBC}}(E), \mathsf{box}_{\mathrm{PBC}}(G), \mathsf{box}_{\mathrm{PBC}}(E')\end{pmatrix}.$$

They are isomorphic (by symmetry), but not exactly the same in the compositional context; in particular, the inaction rule IIT2a

[7] In Chap. 8, we will show how to guarantee ex-exclusiveness syntactically.

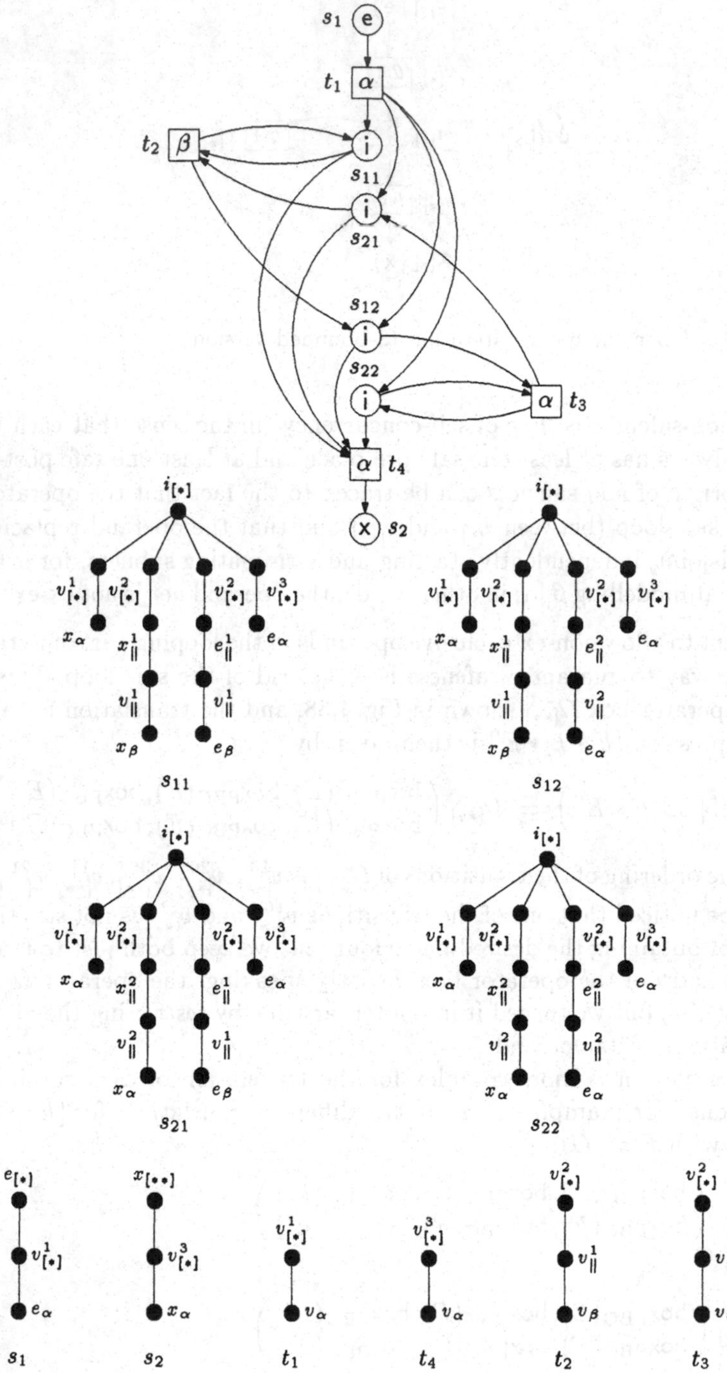

Fig. 4.37. Translation of $[\alpha * (\beta \| \alpha) * \alpha]$ obtained using $\Omega_{[*]}$

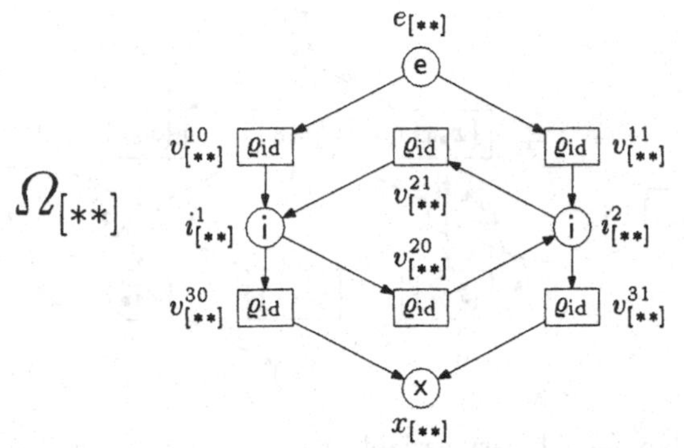

Fig. 4.38. Operator box for iteration (safe version)

$$[\underline{E} * F * E'] \xrightarrow{\emptyset} [E * \overline{F} * E']$$

will not correspond to

$$\Omega_{[**]}\begin{pmatrix} \mathsf{box_{PBC}}(\underline{E}), & \mathsf{box_{PBC}}(F) \\ \mathsf{box_{PBC}}(E'), & \mathsf{box_{PBC}}(E) \\ \mathsf{box_{PBC}}(F), & \mathsf{box_{PBC}}(E') \end{pmatrix} = \Omega_{[**]}\begin{pmatrix} \mathsf{box_{PBC}}(E), & \mathsf{box_{PBC}}(F) \\ \mathsf{box_{PBC}}(E'), & \mathsf{box_{PBC}}(E) \\ \mathsf{box_{PBC}}(\overline{F}), & \mathsf{box_{PBC}}(E') \end{pmatrix}$$

but rather to

$$\Omega_{[**]}\begin{pmatrix} \mathsf{box_{PBC}}(\underline{E}), & \mathsf{box_{PBC}}(F) \\ \mathsf{box_{PBC}}(E'), & \mathsf{box_{PBC}}(E) \\ \mathsf{box_{PBC}}(F), & \mathsf{box_{PBC}}(E') \end{pmatrix} = \Omega_{[**]}\begin{pmatrix} \mathsf{box_{PBC}}(E), & \mathsf{box_{PBC}}(\overline{F}) \\ \mathsf{box_{PBC}}(E'), & \mathsf{box_{PBC}}(E) \\ \mathsf{box_{PBC}}(F), & \mathsf{box_{PBC}}(E') \end{pmatrix}.$$

However, this is more acceptable than non-safeness and we shall disregard this (minor) problem. Another solution is to modify the syntax of iteration by providing an explicit annotation indicating which copy of any of the three operand nets is currently marked, as shown in Chap. 8. But, in situations where non-safeness does not occur, one may use the simpler operator $\Omega_{[*]}$ instead of the safe, but more complex and perhaps cumbersome, operator $\Omega_{[**]}$.

Exercise 4.4.27. Show that

$$[\Sigma * \Theta * \Psi] \text{ rs } a \ \text{ iso } \ [(\Sigma \text{ rs } a) * (\Theta \text{ rs } a) * (\Psi \text{ rs } a)] .$$

4.4.10 Data Variables

Since the looplike data expressions are PBC constants, we shall need to provide an explicit translation for them. The translation is exemplified in Fig. 4.39 for a binary variable x with $Type(x) = \{0, 1\}$.

For a program variable x, $\mathsf{box_{PBC}}([x_{()}]) = \mathsf{N}_{[x_{()}]} = (S_x, T_x, W_x, \lambda_x, \emptyset)$ is the associated box whose places and transitions are

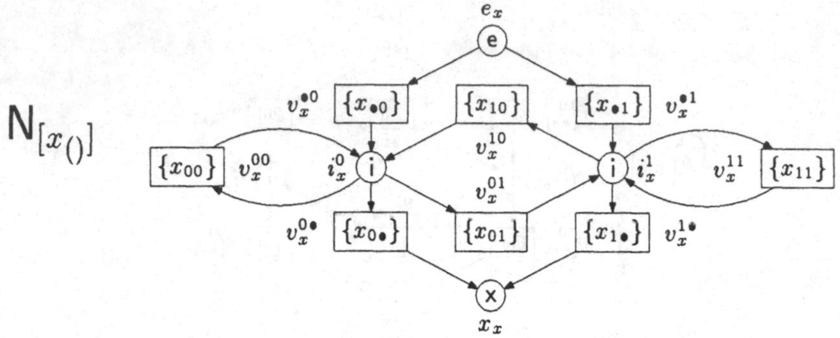

Fig. 4.39. Static box for a binary variable x

$$S_x = \{e_x, x_x\} \cup \{i_x^u \mid u \in \mathit{Type}(x)\}$$

$$T_x = \{v_x^{\bullet u}, v_x^{u\bullet} \mid u \in \mathit{Type}(x)\} \cup \{v_x^{uv} \mid u, v \in \mathit{Type}(x)\} ;$$

the weight function W_x returns 1 for all pairs

$$(e_x, v_x^{\bullet u}) , (v_x^{\bullet, u}, i_x^u) , (i_x^u, v_x^{uv}) , (v_x^{uv}, i_x^v) , (i_x^u, v_x^{u\bullet}) , (v_x^{u\bullet}, x_x)$$

where $u, v \in \mathit{Type}(x)$, and 0 otherwise; and the label function is defined, for all values $u, v \in \mathit{Type}(x)$, by

$$\lambda_x(e_x) = \mathsf{e} \qquad \lambda_x(x_x) = \mathsf{x} \qquad \lambda_x(i_x^u) = \mathsf{i}$$
$$\lambda_x(v_x^{\bullet u}) = \{x_{\bullet u}\} \qquad \lambda_x(v_x^{u\bullet}) = \{x_{u\bullet}\} \qquad \lambda_x(v_x^{uv}) = \{x_{uv}\} .$$

Moreover, for every $u \in \mathit{Type}(x)$, $\mathsf{box}_{\mathrm{PBC}}([\overline{x_{(u)}}])$ is the same as $\mathsf{box}_{\mathrm{PBC}}([x_{()}])$, but with a single token in the place i_x^u. Note that

$$\overline{[x_{()}]} \xrightarrow{\{\{x_{\bullet u}\}\}} [\overline{x_{(u)}}]$$
$$[\overline{x_{(u)}}] \xrightarrow{\{\{x_{uv}\}\}} [\overline{x_{(v)}}]$$
$$[\overline{x_{(u)}}] \xrightarrow{\{\{x_{u\bullet}\}\}} \underline{[x_{()}]}$$

are the operational semantics moves corresponding to the transitions of $\mathsf{N}_{[x_{()}]}$.

4.4.11 Generalised Control Flow Operators

Operator boxes for generalised control flow operators are shown in Fig. 4.40, where it is assumed that $I = \{i_1, i_2, \ldots\}$ is an infinite set of natural numbers satisfying $i_1 < i_2 < \ldots$ and $J = \{i_1, i_2, \ldots, i_n\}$ is a nonempty finite set of natural numbers satisfying $i_1 < i_2 < \ldots < i_n$.

The operator box for generalised choice is rather simple; there are as many transitions labelled ϱ_{id} as there are indices in I (for the finite set J, the operator $\Omega_{\square J}$ would be as that shown in Fig. 4.40 with all transitions $v_{\square J}^{i_j}$, for

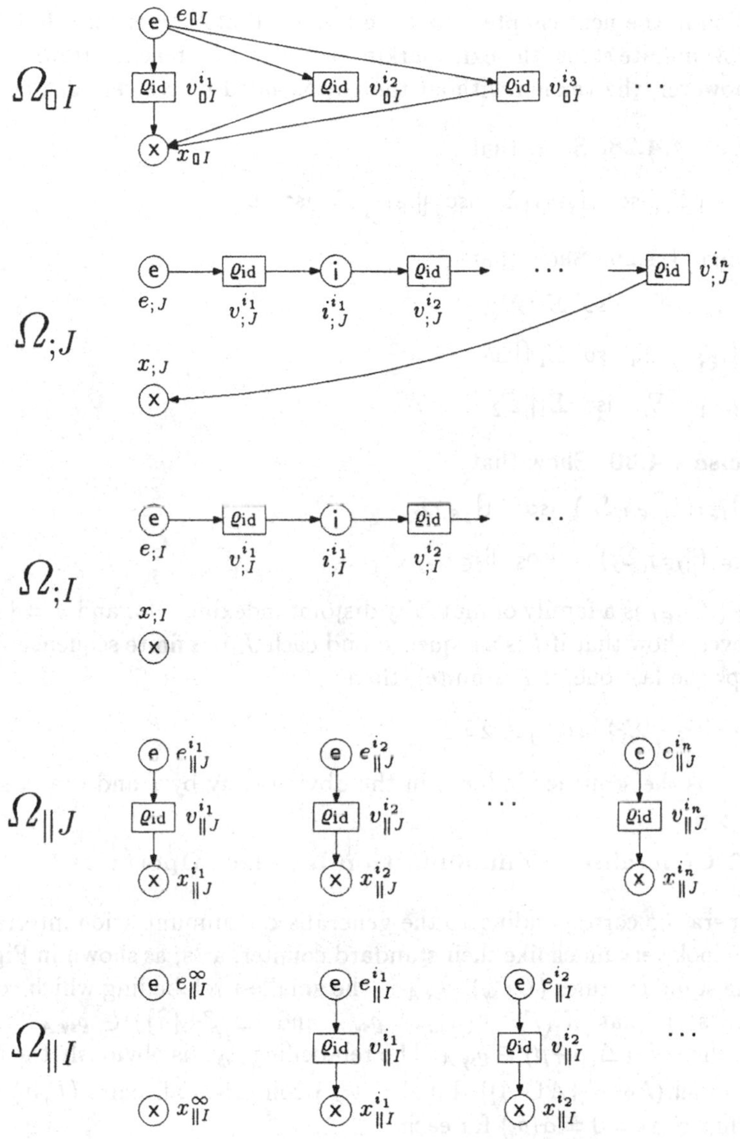

Fig. 4.40. Operator boxes for generalised control flow operators

$j > n$, deleted). The operator box for generalised sequence takes two slightly different forms, depending on whether the index set is finite or not. If it is infinite, the isolated exit place is necessary to fulfill the ex-restrictedness; in particular, this means that the exit marking is unreachable from the entry one. The operator box for generalised parallel composition again takes two slightly different forms, depending on whether the index is finite or not. If it is infinite, the isolated entry and exit places may for the time being seem unnecessary; their role will be explained when we consider the treatment of recursion in the next chapter. Here we remark that they ensure that, even if we allow infinite steps, the exit marking will never be reached from the entry one; however, the version without the two isolated places can also be used.

Exercise 4.4.28. Show that

$$;_{i\in\{1\}} \Sigma_i \text{ iso } \square_{i\in\{1\}}\Sigma_i \text{ iso } \|_{i\in\{1\}}\Sigma_i \text{ iso } \Sigma_1 .$$

Exercise 4.4.29. Show that

$$;_{i\in\{1,2\}} \Sigma_i \quad \text{iso } \Sigma_1 ; \Sigma_2$$

$$\square_{i\in\{1,2\}}\Sigma_i \quad \text{iso } \Sigma_1 \,\square\, \Sigma_2$$

$$\|_{i\in\{1,2\}}\Sigma_i \quad \text{iso } \Sigma_1 \| \Sigma_2 .$$

Exercise 4.4.30. Show that

$$\square_{i\in I}(\,\square_{j\in J_i}\Sigma_j) \text{ iso } \quad \square_{j\in J}\Sigma_j$$

$$\|_{i\in I}(\|_{j\in J_i}\Sigma_j) \quad \text{iso}_S \ \|_{j\in J}\Sigma_j ,$$

where $\{J_i\}_{i\in I}$ is a family of mutually disjoint indexing sets, and $J = \bigcup_{i\in I} J_i$. Moreover, show that if I is a sequence and each J_i is a finite sequence (except perhaps the last one, if I is finite), then

$$;_{i\in I} (;_{j\in J_i} \Sigma_j) \text{ iso } ;_{j\in J} \Sigma_j ,$$

where J is the sequence induced in the obvious way by I and the J_i's.

4.4.12 Generalised Communication Interface Operators

Box operators corresponding to the generalised communication interface operators look very much like their standard counterparts, as shown in Fig. 4.41. For the standard rules (OP_{sy}), $\varrho_{\mathsf{sy}\,A}$ is the smallest relabelling which contains ϱ_{id} and such that, if $(\Gamma, \alpha+\{a\}) \in \varrho_{\mathsf{sy}\,A}$ and $(\Delta, \beta+\{\widehat{a}\}) \in \varrho_{\mathsf{sy}\,A}$ for some $a \in A$, then $(\Gamma+\Delta, \alpha+\beta) \in \varrho_{\mathsf{sy}\,A}$. The relabelling $\varrho_{\mathsf{rs}\,A}$ is obviously the restriction on $\mathsf{mult}(\mathsf{A_{PBC}} \backslash (A \cup \widehat{A}))$. Finally, $\varrho_{[A:]}$ comprises all pairs $(\Gamma, \alpha) \in \varrho_{\mathsf{sy}\,A}$ such that $\alpha(a) = 0 = \alpha(\widehat{a}\,)$ for each $a \in A$.

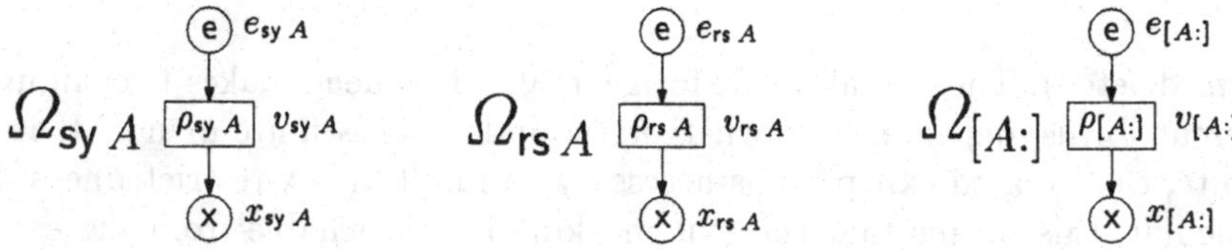

Fig. 4.41. Operator boxes for generalised communication interface operators

Exercise 4.4.31. Define a relabelling $\varrho'_{\mathsf{sy}\,A}$ corresponding to the operational semantics rules (OP'_{sy}).

Exercise 4.4.32. Define a relabelling allowing multi- but not auto-synchronisations.

Exercise 4.4.33. Show that

$$\begin{array}{llllll}
\Sigma \;\mathsf{sy}\; a & \mathsf{iso} & \Sigma \;\mathsf{sy}\; \{a\} & (\Sigma \;\mathsf{sy}\; A) \;\mathsf{sy}\; B & \mathsf{iso} & (\Sigma \;\mathsf{sy}\; B) \;\mathsf{sy}\; A \\
\Sigma \;\mathsf{sy}\; A & \mathsf{iso} & \Sigma \;\mathsf{sy}\; (A \cup \widehat{A}) & (\Sigma \;\mathsf{sy}\; A) \;\mathsf{sy}\; B & \mathsf{iso}_{\mathrm{T}} & \Sigma \;\mathsf{sy}\; (A \cup B) \\
 & & & & & \\
\Sigma \;\mathsf{rs}\; a & \mathsf{iso} & \Sigma \;\mathsf{rs}\; \{a\} & (\Sigma \;\mathsf{rs}\; A) \;\mathsf{rs}\; B & \mathsf{iso} & (\Sigma \;\mathsf{rs}\; B) \;\mathsf{rs}\; A \\
\Sigma \;\mathsf{rs}\; A & \mathsf{iso} & \Sigma \;\mathsf{rs}\; (A \cup \widehat{A}) & (\Sigma \;\mathsf{rs}\; A) \;\mathsf{rs}\; B & \mathsf{iso} & \Sigma \;\mathsf{rs}\; (A \cup B) \\
 & & & & & \\
[a : \Sigma] & \mathsf{iso} & [\{a\} : \Sigma] & [A : [B : \Sigma]] & \mathsf{iso} & [B : [A : \Sigma]] \\
[A : \Sigma] & \mathsf{iso} & [(A \cup \widehat{A}) : \Sigma] & [A : [B : \Sigma]] & \mathsf{iso}_{\mathrm{T}} & [(A \cup B) : \Sigma] \,.
\end{array}$$

Exercise 4.4.34. Show that generalised restriction distributes over all control flow operators, including the generalised ones.

4.4.13 Generalised Iterations

In this section we turn to the binary and unary versions of the iteration operator. Let us first see how the problems already identified in Sect. 3.3.1 are rendered in the Petri net framework.

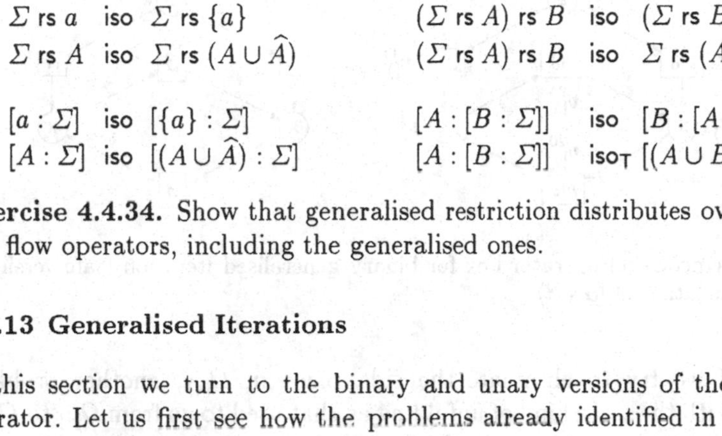

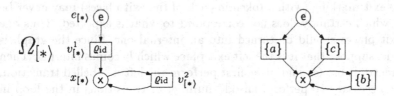

Fig. 4.42. Operator box for binary generalised iteration (unbounded version), and the translation of $[a * b\rangle \,\square\, a$

A possible translation for the $[E * F\rangle$ construct would be to use the operator box $\Omega_{[*\rangle}$ shown in Fig. 4.42. Apart from potential problems due to its side-loop (cf. discussion in Sect. 4.4.9; if fact, in such a case the problem is still worse since the nets generated by this operator box may be unbounded as shown at the end of Sect. 6.3.3), $\Omega_{[*\rangle}$ is not x-directed since there is an arc leaving its only exit place. Of course, one could ask at that point whether this is really important; so let us have a look at the reason why violating x-directedness (and, by symmetry, e-directedness) may lead to problems.

Let us consider again the expression $[a * b\rangle \,\square\, c$. Using $\Omega_{[*\rangle}$, it would be translated into the box shown in Fig. 4.42. It may now be noticed that, from the entry marking, a step sequence $\{\{c\}\}\{\{b\}\}$ is allowed, which corresponds to the SOS semantics already considered for this operator. But this is not what should be expected from a choice construct since the loop is entered 'from the end', after executing the right branch of the choice. However, if it is ascertained that such a loop does not occur, in particular, right at the end of an enclosing choice, then the problem disappears. A full discussion of this issue is deferred to Chaps. 7 and 8.

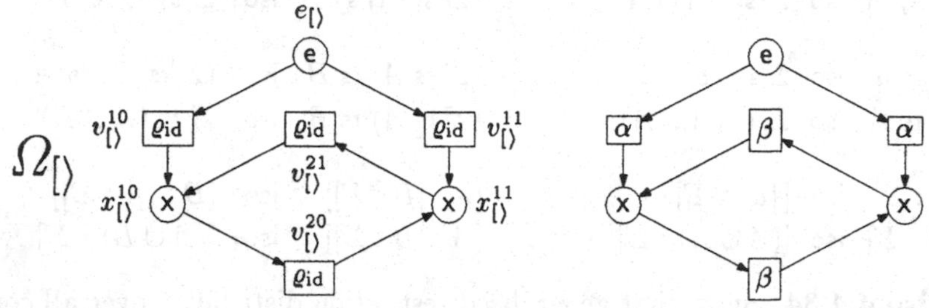

Fig. 4.43. (Incorrect) operator box for binary generalised iteration (safe version), and the translation of $[\alpha * \beta\rangle$

Now, if we try to eliminate the side place in $\Omega_{[*\rangle}$, another problem emerges. With the same kind of unfolding as that used to go from $\Omega_{[*]}$ to $\Omega_{[**]}$ (cf. Sect. 4.4.9), we would obtain the operator box shown in Fig. 4.43. After applying it to the expression $[\alpha * \beta\rangle$ we obtain the net shown in Fig. 4.43. But then, the exit marking (with a token in each of the exit places) may never be reached, which certainly does not correspond to what is expected. Thus one of the exit places should be turned into an internal one. Since the graph is symmetric, suppose that it is the left exit place which becomes internal. Then the symmetry is broken and, if we first perform the left α-labelled transition, it will be necessary to perform an odd number of transitions in the loop in order to reach the exit marking (an even if we first choose the right α-labelled transition). This is clearly unacceptable since we do not know in advance the number of the required executions of the looping part. The reader will easily verify that similar (but not identical) problems occur for the translation of the $\langle *]$-operator for whom the operator box $\Omega_{\langle *]}$ is shown in Fig. 4.44.

Exercise 4.4.35. Carry out the analysis of the operator box $\Omega_{\langle *]}$.

The last version of iteration, i.e., the unary operator $\langle E\rangle$, leads to an additional problem. Its direct translation would lead to the 'operator box' shown in Fig. 4.44, but its labelling is not legal, since it uses a place which is 'both' an entry and an exit place. It is possible to relax our labelling requirements in order to allow such mixed entry/exit places, but we shall refrain from doing it here, as it would require us to rewrite a sizable part of the theory (in particular, the definition of net refinement) and lead to serious

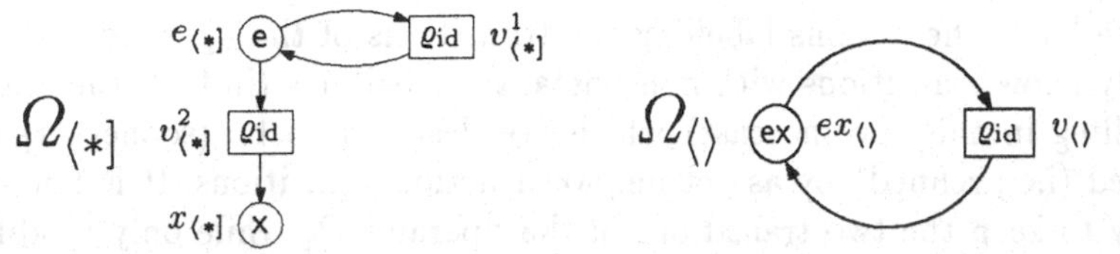

Fig. 4.44. Operator box for another binary generalised iteration, and (pseudo) operator box for the unary iteration

behavioural problems similar to those discussed above. However, 'combined' entry/exit places would make it possible to define a box which behaves as a neutral element for the sequential composition.

Exercise 4.4.36. Develop a framework allowing ex-labelled places.

The above discussion reassures us that the decision not to incorporate the generalised iteration operators into the standard theory was a justified one. However, we will later show that the first two generalised versions of iteration can be added to the standard framework at the price of placing syntactic restrictions on the allowed PBC expressions. More precisely:

- The ternary iteration operator (with explicit initialisation and termination) can be used liberally in any context.
- The two binary iteration operators can be used with care in restricted contexts. A discussion of this will be the province of Chap. 8 (in particular, Sect. 8.1.3).

4.5 Refined Operators

Since the general simultaneous refinement operation developed in this chapter allows us to construct new plain boxes, one may wonder whether it is also possible to construct in the same way new operator boxes. This idea has been exploited in a slightly different context in previous work (see [6, 30]). We shall now sketch how the results obtained there could be adapted to our present framework.

A first problem we have to face concerns the definition of the transition set, which determines the arity of a newly formed operator. Indeed, while it is clear how to apply a general relabelling to transitions labelled by actions, i.e., constant relabellings, this is by far less obvious if transitions are themselves labelled by general relabellings. Let us consider, for instance, an operator box expression $\Omega_{\|}$ sy a, or $\Omega_{\mathsf{sy}\,a}(\Omega_{\|})$ in a more refinement-oriented notation. The relabelling $\varrho_{\mathsf{sy}\,a}$ of the unique transition $v_{\mathsf{sy}\,a}$ of $\Omega_{\mathsf{sy}\,a}$ allows us to group transitions from the operand box when there is an adequate linkage graph corresponding to them, as explained in Sect. 4.4.6. But this assumes that we know the actions labelling the transitions of the operand, and here we only know transitions with nonconstant relabellings (in fact, the identity relabelling in this specific case) which are destined to be (sooner or) later replaced (i.e., refined) by as yet unknown action transitions. It is not satisfactory to keep the two transitions of the operand $\Omega_{\|}$ while only modifying their relabellings, as illustrated on the left of Fig. 4.45, since it will not be later possible to combine (i.e., synchronise) transitions arising from the refinement of the left transition of the new operator box with transitions arising from the refinement of the right one. But it is not satisfactory either to add one (or more) transitions anticipating such combinations, like the middle transition in the operator box on the right of Fig. 4.45, since this middle transition will be later refined by another operand box, which may not have anything to do with the operands refining the right and left transitions.

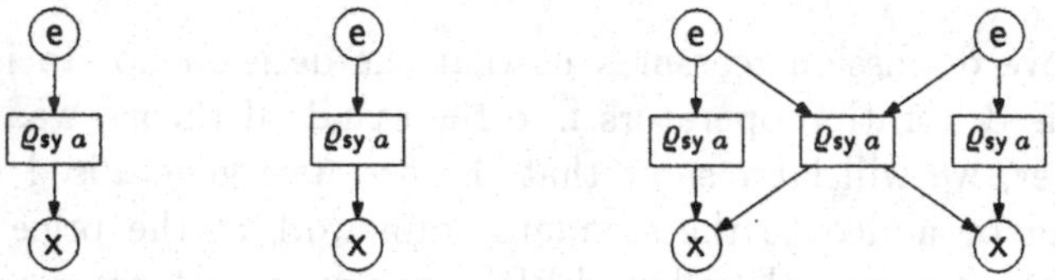

Fig. 4.45. Tentative operator boxes for the refinement $\Omega_{\mathsf{sy}\,a}(\Omega_{\|})$

A possible solution to this problem is to restrict relabellings in an operator box which is to be refined by other operator boxes, in order to avoid the need to combine several transitions which are unknown at the time of this refinement. This may be expressed by the condition

$$(\Gamma, \alpha) \in \lambda_\Omega(v) \;\Rightarrow\; |\Gamma| = 1\,,$$

which amounts to saying that each relabelling is equivalent to a binary relation on the set Lab of action labels; we shall call *relational* operator boxes whose relabellings fulfill this condition. In such a case, the new transitions in a refined operator $\Omega(\Omega_v \mid v \in T_\Omega)$ will be of the form $(v, \varrho)\triangleleft\{u\} \in \mathsf{T}^v_{\mathsf{new}}$ where $u \in T_{\Omega_v}$ and

$$\varrho \;=\; \Big\{(\Gamma, \alpha) \;\Big|\; \exists\beta : (\Gamma, \beta) \in \lambda_{\Omega_v}(u) \;\wedge\; (\{\beta\}, \alpha) \in \lambda_\Omega(v)\Big\}\,.$$

The result is a simple net, as required for operator boxes, provided (see the formulae (4.7) and (4.8) for the weight function in net refinement) that there are no transitions, v in Ω and u in Ω_v, such that v has a side place, and u is ex-skewed; in particular, this will always be the case if each Ω_u is ex-directed.

Another, less severe, problem is that we assumed that operator boxes have their places and transitions in $\mathsf{P}_{\mathsf{root}} \cup \mathsf{T}_{\mathsf{root}}$, but a box $\Omega(\Omega_v \mid v \in T_\Omega)$ will not satisfy this. A possible solution could be to relax the restriction on the type of name trees in operator boxes. But then it would be necessary to introduce a new rule for combining name trees. Such a procedure has been provided by the *tree expansion rule* in [24, 30], but it requires other restrictions to be imposed, which are fortunately always fulfilled in those cases where an operator box has arisen through a sequence of refinements from operator boxes with place and transition names in $\mathsf{P}_{\mathsf{root}} \cup \mathsf{T}_{\mathsf{root}}$. These restrictions are rather complex to formulate, and we shall refrain from presenting them here.

Another solution would be to rename places and transitions after the refinement $\Omega(\Omega_v \mid v \in T_\Omega)$ so as to remain within $\mathsf{P}_{\mathsf{root}}$ and $\mathsf{T}_{\mathsf{root}}$, respectively. The result would be somewhat arbitrary, not with respect to the structure and behaviour of the resulting operator box (which would be unaffected), but with respect to the naming of the nodes, which is essential to define an inclusion order used to solve recursive systems, as described in the next chapter. Fortunately, such a renaming may be delayed until the constructed operator box needs to be applied as an operator proper (not as an operand) in a refinement.

Exercise 4.5.1. Formalise the approach outlined above.

Exercise 4.5.2. Let Ω and each Ω_v, for $v \in T_\Omega$, be a relational operator box. Show that

$$\Omega\left(\Omega_v\left(\Omega_{v,u} \,\middle|\, u \in T_{\Omega_v}\right) \middle|\, v \in T_\Omega\right)$$

$$\text{iso } \Omega\left(\Omega_v \,\middle|\, v \in T_\Omega\right)\left(\Omega_{v,u} \,\middle|\, (v,\varrho) \lhd \{u\} \in T_{\Omega(\Omega_v \mid v \in T_\Omega)}\right)$$

provided that $\Omega_{v,u}$ is an operator box, for all $v \in T_\Omega$ and $u \in T_{\Omega_v}$.

Exercise 4.5.3. Let Ω be a relational operator box, and each Ω_v, for $v \in T_\Omega$, be an operator box. Show that

$$\Omega\left(\Omega_v\left(\Sigma_{v,u} \,\middle|\, u \in T_{\Omega_v}\right) \middle|\, v \in T_\Omega\right)$$

$$\text{iso } \Omega\left(\Omega_v \,\middle|\, v \in T_\Omega\right)\left(\Sigma_{v,u} \,\middle|\, (v,\varrho) \lhd \{u\} \in T_{\Omega(\Omega_v \mid v \in T_\Omega)}\right)$$

provided that $\Sigma_{v,u}$ is a plain box, for all $v \in T_\Omega$ and $u \in T_{\Omega_v}$.

These two properties correspond to what has been called the expansion law for successive refinements in [6]. They may also be interpreted as a form of associativity of net refinement. As a consequence, they may be used to obtain proofs for associativity properties in Excs. 4.4.1, 4.4.2, and 4.4.3.

4.6 Literature and Background

The set of new places resulting from a refinement is a generalisation of the Cartesian product of sets of places which is sometimes used to refine a single transition in a pure net by a whole refining net (see [46]). The setting we use here allows one to refine simultaneously all transitions, even infinitely many.

The notion of net refinement described in Sect. 4.3 is similar to that usually found in the literature (e.g., [46]). The tree view of refinement leads to a number of interesting and useful results, especially in a more general setting when only some of the transitions in Ω are refined, and the interested reader is referred to [25, 31] for a fuller exposition.

Other generalisations of the refinement operation could allow mixing of operator and plain boxes in operand lists, and consequently to allow simultaneously transformational and constant relabellings in boxes, so that the distinction between plain and operator boxes would disappear in favour of a more general notion. Moreover, transitions with transformational relabellings could be partitioned into families characterised by a separate *hierarchical label*, the same refining operand being used for all transitions of the refined operator with the same hierarchical label. This would lead to operations such as $\Omega(X_i.\Sigma_i \mid i \in I)$ or $\Omega(X_i.\Omega_i \mid i \in I)$, or more generally $Box(X_i.Box_i \mid i \in I)$, meaning that all transitions with the hierarchical label X_i in Box are refined by the operand Box_i, for i belonging to some (finite or infinite) indexing set I. This approach has been investigated, for static expressions and boxes, in some of the papers on which this book is based (see [6, 31]).

5. Adding Recursion

This chapter completes the treatment of Petri net semantics of PBC-like algebras, by developing a theory of associating nets with process variables defined by systems of recursive equations. We shall apply the standard methods of using fixpoints and limit construction to obtain solutions of such equations [18], but also show that if they are applied to Petri nets, a number of specific problems arise. In addressing these, the tree device for naming places and transitions in net refinement will prove to be crucial.

The chapter is structured as follows. In Sect. 5.1, we introduce an ordering relation on labelled nets which turns them into a domain in which the existence and construction of solutions are subsequently investigated. Section 5.2 is central to this chapter, as it develops a theory for solving arguably the most general kind of recursive equations in the domain of labelled nets. Section 5.3 then discusses two special instances of this theory. Examples are presented in Sect. 5.4, and Sect. 5.5 completes the development of the solvability of systems of equations.

5.1 Inclusion Order on Labelled Nets

To facilitate solving of recursive equations, we equip the set of labelled nets with an *ordering* relation, $\sqsubseteq$. For two labelled nets Σ and Θ, we denote $\Sigma \sqsubseteq \Theta$ if $S_\Sigma \subseteq S_\Theta$, $T_\Sigma \subseteq T_\Theta$, and the following hold (where $|$ denotes function restriction):

$$W_\Sigma = W_\Theta|_{(S_\Sigma \times T_\Sigma)\cup(T_\Sigma \times S_\Sigma)}$$

$$\lambda_\Sigma = \lambda_\Theta|_{S_\Sigma \cup T_\Sigma}$$

$$M_\Sigma = M_\Theta|_{S_\Sigma} ,$$

that is, if (as relations) $W_\Sigma \subseteq W_\Theta$, $\lambda_\Sigma \subseteq \lambda_\Theta$, and $M_\Sigma \subseteq M_\Theta$. The relation $\sqsubseteq$ is clearly a partial order, where a *partial order* is a pair (X, R) such that X is a set and $R \subseteq X \times X$ is a reflexive ($id_X \subseteq R$), transitive ($R \circ R \subseteq R$), and antisymmetric ($R \cap R^{-1} = id_X$) relation. Moreover, if $\Sigma \sqsubseteq \Theta$, then ${}^\circ\Sigma \subseteq {}^\circ\Theta$ and $\Sigma^\circ \subseteq \Theta^\circ$; and if $\Sigma \sqsubseteq \Theta$ and $S_\Theta \subseteq S_\Sigma$ and $T_\Theta \subseteq T_\Sigma$, then $\Sigma = \Theta$.

A $\sqsubseteq$-*chain* is an infinite sequence of labelled nets

$$\chi = \Sigma^0 \; \Sigma^1 \; \Sigma^2 \; \ldots$$

such that $\Sigma^k \sqsubseteq \Sigma^{k+1}$ for every $k \geq 0$. Each such chain has a unique least upper bound or *limit*, denoted by $\bigsqcup \chi$. We use the terminology that the Σ^k's are *approximations* of $\bigsqcup \chi$; more exactly, they are lower approximations since $\Sigma^k \sqsubseteq \bigsqcup \chi$ for every $k \geq 0$.

The places and transitions of $\bigsqcup \chi$ are respectively the union of the places and transitions of all labelled nets in the chain. For two elements x and y of $\bigsqcup \chi$ (one of them a place, the other a transition), the weight $W(x, y)$ is defined as the weight between x and y in the first Σ^k to which both of them belong; note that, by the definition of $\sqsubseteq$, the weight does not change in subsequent approximations. Finally, for a place or transition x in $\bigsqcup \chi$, the label $\lambda(x)$ is defined as the label of x in the first and hence in all Σ^k containing x, and for a place s in $\bigsqcup \chi$, the marking $M(s)$ is defined as the marking of s in the first and hence in all Σ^k containing s. In other words,

$$\bigsqcup \chi = \left(\bigcup_{k=0}^{\infty} S_{\Sigma^k} \, , \, \bigcup_{k=0}^{\infty} T_{\Sigma^k} \, , \, \bigcup_{k=0}^{\infty} W_{\Sigma^k} \, , \, \bigcup_{k=0}^{\infty} \lambda_{\Sigma^k} \, , \, \bigcup_{k=0}^{\infty} M_{\Sigma^k} \right) . \tag{5.1}$$

Let C be a property of labelled nets. We will say that C is *limit-preserved* if for every $\sqsubseteq$-chain χ of labelled nets satisfying C, it is the case that $\bigsqcup \chi$ satisfies C as well. As the next result shows, several important properties turn out to be limit-preserved.

Proposition 5.1.1. The following properties of labelled nets are limit-preserved.

(1) All markings reachable from the initial marking are safe. (safeness)
(2) There is at least one entry place and at least one exit place. (ex-restrictedness)
(3) There is no arc incoming to an entry place. (e-directedness)
(4) There is no arc outgoing from an exit place. (x-directedness)
(5) There is no marking reachable from the initial marking which simultaneously marks an entry place and an exit place. (ex-exclusiveness)
(6) Every transition has a nonempty pre-set and a nonempty post-set. (T-restrictedness)
(7) The initial marking is empty. (being an unmarked net)
(8) All transition labels are constant relabellings. (being a plain net)
(9) All markings reachable from the initial marking are clean. (cleanness)

Proof. Let $\chi = \Sigma^0 \, \Sigma^1 \, \Sigma^2 \, \ldots$ be a $\sqsubseteq$-chain and $\Theta = \bigsqcup \chi$ be its limit.

To show (1), suppose that all markings reachable from every M_{Σ^k} are safe. If Θ has a non-safe reachable marking, then there is a step sequence σ and a place s such that $M_\Theta \; [\sigma\rangle \; M$ in Θ and $M(s) > 1$. Let k be the first index such that s and all transitions of σ belong to Σ^k (such a k always exists since σ is a finite sequence of finite steps of transitions). From the inclusion property, $\Sigma^k \sqsubseteq \Sigma$, it follows that $M_\Theta|_{S_{\Sigma^k}} \; [\sigma\rangle \; M|_{S_{\Sigma^k}}$ in Σ^k. We also have that $M_\Theta|_{S_{\Sigma^k}} = M_{\Sigma^k}$ since every marking in the chain is a restriction of the marking of the limit, and $M|_{S_{\Sigma^k}}(s) = M(s) > 1$, which contradicts the assumption that all nets in the chain are safe.

The proofs of (5) and (9) are similar. Moreover, (2,3,4,6,7,8) follow directly from (5.1). □

Let $\mathcal{N}$ be a set of labelled nets containing the limits of all $\sqsubseteq$-chains made up of nets included in it (in particular, $\mathcal{N}$ can be the set of all labelled nets). A mapping $\psi : \mathcal{N} \to \mathcal{N}$ is $\sqsubseteq$-*monotonic* if, for all $\Sigma, \Theta \in \mathcal{N}$, $\Sigma \sqsubseteq \Theta$ implies $\psi(\Sigma) \sqsubseteq \psi(\Theta)$. Moreover, a $\sqsubseteq$-monotonic ψ is $\sqsubseteq$-*continuous* if for every $\sqsubseteq$-chain $\chi = \Sigma^0\ \Sigma^1\ \ldots$ of nets in $\mathcal{N}$, $\psi(\bigsqcup\chi) = \bigsqcup\psi(\chi)$, where

$$\psi(\chi) = \psi(\Sigma^0)\ \psi(\Sigma^1)\ \ldots .$$

$\bigsqcup\psi(\chi)$ is well defined as for a monotonic ψ, $\psi(\chi)$ is a $\sqsubseteq$-chain in $\mathcal{N}$.

Proposition 5.1.2. If $\psi \in \{\lfloor.\rfloor, \overline{(.)}, \underline{(.)}, (.)_{\mathrm{sr}}\}$, then ψ is $\sqsubseteq$-monotonic and $\sqsubseteq$-continuous mapping in the domain of labelled nets.

Proof. Immediate from (5.1) and the inclusion of entry and exit places for $\sqsubseteq$-chains. □

In this book, we shall only consider chains of unmarked nets as well as those marked with entry or exit markings.

5.2 Solving Recursive Equations

Let $\mathcal{X}$ be a nonempty, possibly infinite, set of *recursion variables*. We consider a system of *equations on boxes* of the following form (one equation per variable X in $\mathcal{X}$):

$$X \stackrel{\mathrm{df}}{=} \Omega_X(\boldsymbol{\Delta}_X) . \tag{5.2}$$

On the right-hand side, Ω_X is an unmarked operator box, and $\boldsymbol{\Delta}_X$ is an Ω_X-tuple of recursion variables and unmarked plain boxes, the latter being called in this chapter *rec-boxes*, $\mathsf{Box}^{\mathrm{rec}}$. That is, $\boldsymbol{\Delta}_X : T_{\Omega_X} \to \mathcal{X} \cup \mathsf{Box}^{\mathrm{rec}}$.

Now, given a mapping $\mathsf{sol} : \mathcal{X} \to \mathsf{Box}^{\mathrm{rec}}$ assigning a rec-box to every variable in $\mathcal{X}$, we will denote by $\boldsymbol{\Delta}_X[\mathsf{sol}]$ the Ω_X-tuple obtained from $\boldsymbol{\Delta}_X$ by replacing each variable Y by $\mathsf{sol}(Y)$. That is, $\boldsymbol{\Delta}_X[\mathsf{sol}] : T_{\Omega_X} \to \mathsf{Box}^{\mathrm{rec}}$ is such that for every $v \in T_{\Omega_X}$,

$$\boldsymbol{\Delta}_X[\mathsf{sol}](v) = \begin{cases} \mathsf{sol}(Y) & \text{if } \ \Delta_{X,v} = Y \in \mathcal{X} \\ \Delta_{X,v} & \text{if } \ \Delta_{X,v} \in \mathsf{Box}^{\mathrm{rec}} . \end{cases}$$

Then, a *solution* of the system of equations (5.2) is an assignment sol such that, for every variable $X \in \mathcal{X}$,

$$\mathsf{sol}(X) = \Omega_X(\boldsymbol{\Delta}_X[\mathsf{sol}]) .$$

Notice that since all nets in $\boldsymbol{\Delta}_X[\mathsf{sol}]$ are unmarked, the application of Ω_X to $\boldsymbol{\Delta}_X[\mathsf{sol}]$ is legal as the condition (4.3) stated in Sect. 4.3.1 is clearly satisfied.

For example, a solution of the recursive net equation $X \stackrel{\mathrm{df}}{=} \Omega_{\|}(\mathsf{N}_\alpha, X)$ will yield a box model for the hierarchical variable X with the defining equation $X \stackrel{\mathrm{df}}{=} \alpha \| X$, and a solution of $X \stackrel{\mathrm{df}}{=} \Omega_{;}(\mathsf{N}_\alpha, X)$ will do the same for $X \stackrel{\mathrm{df}}{=} \alpha; X$. The defining equation $X \stackrel{\mathrm{df}}{=} X$ will be treated in a slightly different way as we need a net operator on the right-hand side; we will model it by the net equation $X \stackrel{\mathrm{df}}{=} \Omega_{[id]}(X)$ (see Sects. 3.2.11 and 4.4.5). Since we only allow a single net operator on the right-hand side, more complex defining equations need to be decomposed into equivalent simple recursive systems, as explained in Sect. 3.1.

In order to simplify the presentation, until Sect. 5.5, we shall only consider a *single* recursive equation for a singleton set $\mathcal{X} = \{X\}$:

$$X \stackrel{\mathrm{df}}{=} \Omega(\boldsymbol{\Delta}) . \tag{5.3}$$

In this case, we identify an assignment $\mathsf{sol} : \{X\} \to \mathsf{Box}^{\mathrm{rec}}$ with the rec-box $\Sigma = \mathsf{sol}(X)$ and, accordingly, denote $\boldsymbol{\Delta}[\Sigma]$ rather than $\boldsymbol{\Delta}[\mathsf{sol}]$, calling Σ a *solution* of (5.3) whenever $\Sigma = \Omega(\boldsymbol{\Delta}[\Sigma])$. Without loss of generality, we shall also assume that for every rec-box Δ_v in $\boldsymbol{\Delta}$, it is the case that $T_{\Delta_v} \subseteq \mathsf{T}_{\mathsf{root}} \setminus T_\Omega$ and $S_{\Delta_v} \subseteq \mathsf{P}_{\mathsf{root}} \setminus S_\Omega$. (Recall that, by the definition of an operator box, $T_\Omega \subseteq \mathsf{T}_{\mathsf{root}}$ and $S_\Omega \subseteq \mathsf{P}_{\mathsf{root}}$.)

In general, the recursive equation (5.3) may have several different, even non-isomorphic, solutions; moreover, each such solution is uniquely identified by its sets of entry and exit places. A formal statement of this fact is preceded by an auxiliary lemma which shows that all solutions of (5.3) have the same set of transitions, and that the inclusion order $\sqsubseteq$ on the set of possible solutions is equivalent to the ordering resulting from the ordinary set inclusion on their entry and exit places.

Lemma 5.2.1. Let Σ and Θ be solutions of (5.3).

(1) The sets of transitions of Σ and Θ and their labellings are the same.
(2) The following are equivalent:
   (a) ${}^\circ\Sigma \subseteq {}^\circ\Theta$ and $\Sigma^\circ \subseteq \Theta^\circ$.
   (b) $\Sigma \sqsubseteq \Theta$.

Proof. Let us denote $\Sigma = (S_0, T_0, W_0, \lambda_0)$ and $\Theta = (S_1, T_1, W_1, \lambda_1)$.

(1) Suppose $T_0 \not\subseteq T_1$ and $t = (v, \alpha) \lhd R \in T_0$ is a smallest (with respect to the number of nodes) transition tree such that $t \notin T_1$ (we can assume that such a t exists since all transition trees are finite). Knowing that $\Sigma = \Omega(\boldsymbol{\Delta}[\Sigma])$, we proceed as follows.
If $\Delta_v \in \mathsf{Box}^{\mathrm{rec}}$, then R is a multiset of transitions of Δ_v and, since $\Delta[\Sigma]_v = \Delta_v = \Delta[\Theta]_v$ and $\Theta = \Omega(\boldsymbol{\Delta}[\Theta])$, we have $t \in T_1$ which produces a contradiction.

If $\Delta_v = X$, then all transitions in R are also in T_0. Now, if R is also a multiset of transition in T_1, then by $\Theta = \Omega(\mathbf{\Delta}[\Theta]))$ and $\Delta[\Theta]_v = \Theta$, $t \in T_1$ which contradicts our original assumption. Hence there is u in R such that $u \notin T_1$. Moreover, u is a proper subtree of t, which produces a contradiction with the minimality of t.
Hence $T_0 \subseteq T_1$. Similarly, one can show that $T_1 \subseteq T_0$, and so $T_0 = T_1$.

That the labellings of the transitions in Σ and Θ are the same follows from the definition of net refinement and $\Sigma = \Omega(\mathbf{\Delta}[\Sigma])$ and $\Theta = \Omega(\mathbf{\Delta}[\Theta])$, which together imply that each transition in Σ and Θ is of the form $(v,\alpha)\lhd R$ and has the label α.

(2) The (a) $\Leftarrow$ (b) implication follows directly from the definition of $\sqsubseteq$. Below we prove that (a) $\Rightarrow$ (b).

We first show that all places of Σ are also in Θ, and with the same labels. Since every place p of Σ is a tree of the form $t_1 \lhd \cdots \lhd t_k \lhd s \lhd \mathcal{S}$ (see Sect. 4.3.3), where $t_1,\ldots,t_k \in \mathsf{T_{root}}$ and $s \in \mathsf{P_{root}}$, we can proceed by induction on k. For $k=0$, we have $p = s \lhd \mathcal{S}$, where $s \in \mathsf{P_{root}}$. Thus, by the definition of net refinement and $\Sigma = \Omega(\mathbf{\Delta}[\Sigma])$, $p \in \mathsf{SP}^s_{\mathsf{new}}$ where $\mathsf{SP}^s_{\mathsf{new}}$ is as in the definition of net refinement. Since, from (a), we have ${}^\circ(\Delta[\Sigma]_v) \subseteq {}^\circ(\Delta[\Theta]_v)$ and $(\Delta[\Sigma]_v)^\circ \subseteq (\Delta[\Theta]_v)^\circ$ for every $v \in T_\Omega$, p is also a place in $\Omega(\mathbf{\Delta}[\Theta]) = \Theta$, with the same label $\lambda_\Omega(s)$. The induction step for $k>0$ is straightforward. If $p = t_1 \lhd \cdots \lhd t_k \lhd s \lhd \mathcal{S}$, then $p \in \mathsf{ST}^{t_1}_{\mathsf{new}}$ and the place $t_2 \lhd \cdots \lhd t_k \lhd s \lhd \mathcal{S}$ belongs either to some rec-box in $\mathbf{\Delta}$ or to Σ; hence, by $\Theta = \Omega(\mathbf{\Delta}[\Theta])$ and the definition of net refinement and the induction hypothesis, we obtain that p also belongs to Θ, with the same i-label.

Hence $S_0 \subseteq S_1$ and, clearly, $\lambda_0|_{S_0} = \lambda_1|_{S_1}$. Moreover, by (1), we have $T_0 = T_1$ and $\lambda_0|_{T_0} = \lambda_1|_{T_1}$. Thus $\lambda_0 \subseteq \lambda_1$ is also satisfied. The fact that $W_0 = W_1|_{(S_0\times T_0)\cup(T_0\times S_0)}$ immediately results from the fact that in a refinement the arc weights are determined by the structure of name trees, and by the same kind of argument on the size of the transition names as in the proof of part (1). □

Theorem 5.2.1. If Σ and Θ are solutions of (5.3) such that ${}^\circ\Sigma = {}^\circ\Theta$ and $\Sigma^\circ = \Theta^\circ$, then $\Sigma = \Theta$.

Proof. Follows from Lemma 5.2.1(2) and $\sqsubseteq$ being a partial order. □

5.2.1 Using Fixpoints to Solve Recursive Equations

From Prop. 4.3.1, it follows that the right-hand side of the recursive equation (5.3) induces a mapping $\mathbf{F} : \mathsf{Box}^{\mathsf{rec}} \to \mathsf{Box}^{\mathsf{rec}}$ defined, for every rec-box Σ, by

$$\mathbf{F}(\Sigma) = \Omega(\mathbf{\Delta}[\Sigma]) . \tag{5.4}$$

Hence solving equation (5.3) amounts to finding fixpoint(s) of $\mathbf{F}$, and we have already defined the inclusion order on labelled nets, $\sqsubseteq$, which will be suitable for the latter purpose. As we have seen in Sect. 5.1, it is possible to form

$\sqsubseteq$-chains and limits in the usual way. Moreover, every $\sqsubseteq$-chain of rec-boxes χ has a limit $\bigsqcup\chi$ and, by Prop. 5.1.1(2,6,7,8), this limit is also a rec-box. The usual treatment of fixpoint theory would now proceed by establishing the monotonicity and continuity of the mapping $\mathbf{F}$ defined by (5.4).

Proposition 5.2.1. $\mathbf{F}$ is $\sqsubseteq$-monotonic.

Proof. Follows from the definitions of net refinement and $\sqsubseteq$. □

A similar result does not hold for $\sqsubseteq$-continuity without making an additional assumption. We will say that (5.3) is *X-finitary* if, for every $s \in \Omega$,

$$\left|\{v \in {}^\bullet s \cup s^\bullet \mid \Delta_v = X\}\right| \; < \; \infty\,. \tag{5.5}$$

In other words, if each place in the operator box has 'access' to only finitely many instances of X.

Proposition 5.2.2. $\mathbf{F}$ is $\sqsubseteq$-continuous if and only if (5.3) is X-finitary.

Proof. ($\Longrightarrow$) Suppose that (5.3) is not X-finitary. Then there is $s \in S_\Omega$ such that, without loss of generality, the set $V = \{v \in {}^\bullet s \mid \Delta_v = X\}$ is infinite. In particular, this means that V contains a countably infinite set of transitions $U = \{v_0, v_1, \ldots\}$. We now define a $\sqsubseteq$-chain $\chi = \Sigma^0\ \Sigma^1\ \ldots$ of rec-boxes such that, for every $k \geq 0$,

$$\Sigma^k \;=\; \Big(\{e, x_0, \ldots, x_k\}, \emptyset, \emptyset, \{e\} \times \{\mathsf{e}\} \cup \{x_0, \ldots, x_k\} \times \{\mathsf{x}\}, \emptyset\Big),$$

where $e, x_0, x_1, \ldots$ are distinct places (i.e., χ is a chain of unmarked ex-boxes, see Sect. 4.4.7). For every transition $v \in s^\bullet$, let e_v be a place in ${}^\circ(\Delta_v)$ if $\Delta_v \in \mathsf{Box}^{\mathrm{rec}}$, and $e_v = e$ if $\Delta_v = X$. Moreover, for every transition $z \in Z = {}^\bullet s \setminus U$, let x_z be a place in $(\Delta_z)^\circ$ if $\Delta_z \in \mathsf{Box}^{\mathrm{rec}}$, and $x_z = x_0$ if $\Delta_z = X$. Then the following place

$$p \;=\; s \lhd \Big(\{v \lhd e_v\}_{v \in s^\bullet} + \{w \lhd x_z\}_{z \in Z} + \{v_i \lhd x_i\}_{i \in \mathbf{N}}\Big)$$

belongs to the rec-box $\mathbf{F}(\bigsqcup\chi)$, but it does not belong to any $\mathbf{F}(\Sigma^i)$ and thus neither to $\bigsqcup\mathbf{F}(\chi)$, proving that $\mathbf{F}$ is not $\sqsubseteq$-continuous.

($\Longleftarrow$) Suppose that $\chi = \Sigma^0\ \Sigma^1\ \ldots$ is a $\sqsubseteq$-chain of rec-boxes. We need to show that $\Theta = \mathbf{F}(\Sigma)$, where $\Sigma = \bigsqcup\chi$ and $\Theta = \bigsqcup\ \mathbf{F}(\chi)$. Since $\Sigma^i \sqsubseteq \Sigma$ and $\mathbf{F}$ is $\sqsubseteq$-monotonic, $\mathbf{F}(\Sigma^i) \sqsubseteq \mathbf{F}(\Sigma)$ for every $i \geq 0$, and thus $\Theta \sqsubseteq \mathbf{F}(\Sigma)$. As a consequence, to prove $\Theta = \mathbf{F}(\Sigma)$ it suffices to show that $T_{\mathbf{F}(\Sigma)} \subseteq T_\Theta$ and $S_{\mathbf{F}(\Sigma)} \subseteq S_\Theta$.

The set of transitions of $\mathbf{F}(\Sigma)$ is included in Θ. Indeed, if $(v, \alpha) \lhd R \in T_{\mathbf{F}(\Sigma)}$ and $\Delta_v = X$, then, since there are only finitely many transitions in R, there is $i \geq 0$ such that all transitions in R belong to Σ^i, which implies $(v, \alpha) \lhd R \in T_{\mathbf{F}(\Sigma^i)}$. Moreover, if Δ_v is a rec-box, then $(v, \alpha) \lhd R \in T_{\mathbf{F}(\Phi)}$ for any rec-box Φ. Hence $(v, \alpha) \lhd R \in T_{\mathbf{F}(\Sigma^0)}$.

To demonstrate that $S_{\mathbf{F}(\Sigma)} \subseteq S_\Theta$ we observe that, by (5.3) being X-finitary, to construct $p \in S_{\mathbf{F}(\Sigma)}$, we only need finitely many (possibly none) places which belong to Σ and therefore there is $i \geq 0$ such that all these places belong to Σ^i and hence $p \in S_{\mathbf{F}(\Sigma^i)} \subseteq S_\Theta$. □

Although the $\sqsubseteq$-continuity of $\mathbf{F}$ does not hold in the general case, there is a closely related and for our purposes sufficient, property of $\mathbf{F}$ obtained after strengthening the inclusion order. For two rec-boxes, $\Sigma \sqsubseteq \Theta$, we shall denote $\Sigma \tilde{\sqsubseteq} \Theta$ if, moreover, ${}^\circ\Sigma = {}^\circ\Theta$ and $\Sigma^\circ = \Theta^\circ$. We then define $\tilde{\sqsubseteq}$-chains, $\tilde{\sqsubseteq}$-chain preserved properties, $\tilde{\sqsubseteq}$-monotonicity, and $\tilde{\sqsubseteq}$-continuity in the same way as the corresponding notions for $\sqsubseteq$. Notice that since $\tilde{\sqsubseteq}$-chains are also $\sqsubseteq$-chains, Prop. 5.1.1 and other properties enjoyed by the latter hold for $\tilde{\sqsubseteq}$-chains as well.

Notice that right from the beginning we could have only considered the stronger inclusion order, $\tilde{\sqsubseteq}$, simplifying the presentation. However, in Sect. 5.3, we will need $\sqsubseteq$ rather than $\tilde{\sqsubseteq}$ to discuss solutions to (5.3) obtained as limits of finite approximations.

Theorem 5.2.2. $\mathbf{F}$ is $\tilde{\sqsubseteq}$-monotonic and $\tilde{\sqsubseteq}$-continuous.

Proof. Suppose that $\Sigma \tilde{\sqsubseteq} \Theta$. Then, by the $\sqsubseteq$-monotonicity of $\mathbf{F}$, $\mathbf{F}(\Sigma) \sqsubseteq \mathbf{F}(\Theta)$. As to the equality of the entry and exit places, we observe that from ${}^\circ\Sigma = {}^\circ\Theta$ and $\Sigma^\circ = \Theta^\circ$, it follows that for every $s \in S_\Omega$, $\mathsf{SP}^s_{\mathsf{new}}$ in $\mathbf{F}(\Sigma)$ is equal to $\mathsf{SP}^s_{\mathsf{new}}$ in $\mathbf{F}(\Theta)$, where each $\mathsf{SP}^s_{\mathsf{new}}$ is the auxiliary set of places used in the definition of net refinement in Sect. 4.3.4. Hence $\mathbf{F}$ is $\tilde{\sqsubseteq}$-monotonic, since

$$ {}^\circ\mathbf{F}(\Sigma) = \bigcup_{s \in {}^\circ\Omega} \mathsf{SP}^s_{\mathsf{new}} = {}^\circ\mathbf{F}(\Theta) \quad \text{and} \quad \mathbf{F}(\Sigma)^\circ = \bigcup_{s \in \Omega^\circ} \mathsf{SP}^s_{\mathsf{new}} = \mathbf{F}(\Theta)^\circ . $$

Suppose now that $\chi = \Sigma^0\ \Sigma^1\ \ldots$ is a $\tilde{\sqsubseteq}$-chain of rec-boxes. Moreover, let Σ and Θ be as in the second part of the proof of Prop. 5.2.2 whose careful inspection leads to a conclusion that it can be reused with one exception, namely we have to reprove that $S_{\mathbf{F}(\Sigma)} \subseteq S_\Theta$. This can be done in the following way.

By the equality of the sets of entry and exit places in the $\tilde{\sqsubseteq}$-chain, we have that ${}^\circ\Sigma = {}^\circ(\Sigma^0)$ and $\Sigma^\circ = (\Sigma^0)^\circ$. Hence, for every $s \in S_\Omega$, $\mathsf{SP}^s_{\mathsf{new}}$ in $\mathbf{F}(\Sigma)$ is equal to $\mathsf{SP}^s_{\mathsf{new}}$ in $\mathbf{F}(\Sigma^0)$. Moreover, if $v \lhd q \in \mathsf{ST}^v_{\mathsf{new}}$ in $\mathbf{F}(\Sigma)$ and $\Delta_v = X$, then there is $i \geq 0$ such that q belongs to Σ^i and hence $v \lhd q \in \mathsf{ST}^v_{\mathsf{new}}$ in $\mathbf{F}(\Sigma^i)$; and if $\Delta_v \in \mathsf{Box}^{\mathrm{rec}}$, then $v \lhd q \in \mathsf{ST}^v_{\mathsf{new}}$ in $\mathbf{F}(\Sigma^0)$. Thus we obtain $S_{\mathbf{F}(\Sigma)} \subseteq S_\Theta$. □

Since the recursive equation induces a $\tilde{\sqsubseteq}$-continuous mapping $\mathbf{F}$, we can attempt to solve it by constructing a $\tilde{\sqsubseteq}$-chain of rec-boxes formed through successive applications of $\mathbf{F}$ to some initial approximation Υ of a solution. For a rec-box Υ, we denote by $\mathbf{F}^*(\Upsilon)$ the sequence

$$ \mathbf{F}^0(\Upsilon)\ \mathbf{F}^1(\Upsilon)\ \mathbf{F}^2(\Upsilon)\ \ldots , $$

where $\mathbf{F}^0(\Upsilon) = \Upsilon$ and $\mathbf{F}^n(\Upsilon) = \mathbf{F}(\mathbf{F}^{n-1}(\Upsilon))$, for every $n \geq 1$. We will call Υ a $\tilde{\sqsubseteq}$-*seed* for the recursive equation (5.3) if $\Upsilon \tilde{\sqsubseteq} \mathbf{F}(\Upsilon)$; in such a case, by the $\tilde{\sqsubseteq}$-monotonicity of $\mathbf{F}$, we have that $\mathbf{F}^*(\Upsilon)$ is a $\tilde{\sqsubseteq}$-chain of rec-boxes. And such a chain always yields a solution.

Theorem 5.2.3. If Υ is a $\tilde{\sqsubseteq}$-seed, then $\bigsqcup \mathbf{F}^*(\Upsilon)$ is a solution of (5.3).

Proof. By Thm. 5.2.2,

$$\mathbf{F}(\bigsqcup \mathbf{F}^*(\Upsilon)) = \bigsqcup \ \mathbf{F}^1(\Upsilon)\ \mathbf{F}^2(\Upsilon)\ \ldots .$$

Moreover, by $\Upsilon \tilde{\sqsubseteq} \mathbf{F}(\Upsilon)$, we have

$$\bigsqcup \ \mathbf{F}^1(\Upsilon)\ \mathbf{F}^2(\Upsilon)\ \ldots = \bigsqcup \ \Upsilon\ \mathbf{F}^1(\Upsilon)\ \mathbf{F}^2(\Upsilon)\ \ldots = \bigsqcup \ \mathbf{F}^*(\Upsilon) .$$

Hence $\bigsqcup \mathbf{F}^*(\Upsilon)$ is a fixpoint of $\mathbf{F}$ and so a solution of (5.3). □

It is sufficient to consider ex-boxes (cf. Sect. 4.4.7) as the only seeds.

Theorem 5.2.4. Let Σ be a solution of (5.3) and

$$\Upsilon = \left({}^\circ\Sigma \cup \Sigma^\circ\ ,\ \emptyset\ ,\ \emptyset\ ,\ ({}^\circ\Sigma \times \{\mathsf{e}\}) \cup (\Sigma^\circ \times \{\mathsf{x}\})\ ,\ \emptyset\right) .$$

Then Υ is a $\tilde{\sqsubseteq}$-seed for (5.3) such that $\bigsqcup \mathbf{F}^*(\Upsilon) = \Sigma$.

Proof. We first observe that Υ is an unmarked ex-box and hence also a rec-box. Moreover, by $\Sigma = \mathbf{F}(\Sigma)$ and the definition of net refinement, it is the case that ${}^\circ\Upsilon = {}^\circ\Sigma = {}^\circ\mathbf{F}(\Upsilon)$ and $\Upsilon^\circ = \Sigma^\circ = \mathbf{F}(\Upsilon)^\circ$. Hence $\Upsilon \tilde{\sqsubseteq} \mathbf{F}(\Upsilon)$ and so Υ is a $\tilde{\sqsubseteq}$-seed. Furthermore,

$${}^\circ\left(\bigsqcup \mathbf{F}^*(\Upsilon)\right) = {}^\circ\Sigma \quad \text{and} \quad \left(\bigsqcup \mathbf{F}^*(\Upsilon)\right)^\circ = \Sigma^\circ .$$

Hence, by Thms. 5.2.1 and 5.2.3, $\bigsqcup \mathbf{F}^*(\Upsilon) = \Sigma$. □

What we still lack to complete the discussion of the recursive equation (5.3) is a theorem saying that it always has a solution. This result will be postponed until Sect. 5.2.4, after we have discussed in more detail seed boxes. First, however, we look at the form of place and transition trees in nets solving (5.3) and then present an example.

5.2.2 Places and Transitions in Net Solutions

A closer examination of the proof of Lemma 5.2.1(1) indicates that the uniqueness of the transition set in a solution of (5.3) is essentially due to the fact that we only allow *finite* transition trees. If we allowed infinite ones, then transition sets could be different for different solutions. The uniqueness result may also be derived using the following argument.

Let τ be a transition or place name in $\mathsf{T_{tree}} \cup \mathsf{P_{tree}}$ and ξ be one of its nodes. The *level* of ξ is the length of the path from the root of τ to ξ defined as the number of edges on that path; in particular, the level of the root is 0. For a finite τ, its *depth*, denoted by $\mathsf{depth}(\tau)$, is the maximal level of the nodes of τ.

Proposition 5.2.3. Let Υ and Υ' be $\tilde{\sqsubseteq}$-seeds for (5.3). Then, for every $n \geq 0$ and all $k, l \geq n$,

$$\{t \in T_{\mathbf{F}^k(\Upsilon)} \mid \mathsf{depth}(t) \leq n\} \;=\; \{t \in T_{\mathbf{F}^l(\Upsilon')} \mid \mathsf{depth}(t) \leq n\} .$$

Proof. For all $k, l \geq n \geq 0$, let

$$T_{nk} = \{t \in T_{\mathbf{F}^k(\Upsilon)} \mid \mathsf{depth}(t) \leq n\}$$
$$T'_{nl} = \{t \in T_{\mathbf{F}^l(\Upsilon')} \mid \mathsf{depth}(t) \leq n\} .$$

We proceed by induction on n. In the base case ($n = 0$), we have $T_{0k} = \emptyset = T'_{0l}$ which follows from the definition of net refinement (which never yields a transition of zero depth) and the following facts: $\mathbf{F}^0(\Upsilon) = \Upsilon \tilde{\sqsubseteq} \mathbf{F}(\Upsilon)$ and $\mathbf{F}^0(\Upsilon') = \Upsilon' \tilde{\sqsubseteq} \mathbf{F}(\Upsilon')$ and $\mathbf{F}^k(\Upsilon), \mathbf{F}^l(\Upsilon') \in \mathbf{F}(\mathsf{Box}^{\mathsf{rec}})$, for all $k, l \geq 1$.

Suppose that $T_{nk} = T'_{nl}$ for some $n \geq 0$ and all $k, l \geq n$. Let $k, l \geq n + 1$ and $u = (v, \alpha) \lhd R \in T_{(n+1)k}$. If $\Delta_v \neq X$, then $u \in T'_{(n+1)l}$, since $R \in \mathsf{mult}(T_{\Delta_v})$ and $\mathbf{F}^l(\Upsilon') \in \mathbf{F}(\mathsf{Box}^{\mathsf{rec}})$. If $\Delta_v = X$, then $u \in T'_{(n+1)l}$, since $R \in \mathsf{mult}(T_{n(k-1)})$ (and so, by the induction hypothesis, $R \in \mathsf{mult}(T'_{n(l-1)})$) and $\mathbf{F}^l(\Upsilon') = \mathbf{F}(\mathbf{F}^{l-1}(\Upsilon'))$. Hence $T_{(n+1)k} \subseteq T'_{(n+1)l}$. By symmetry, $T'_{(n+1)l} \subseteq T_{(n+1)k}$ and so $T_{(n+1)k} = T'_{(n+1)l}$. □

Corollary 5.2.1. Let Σ and Θ be solutions of (5.3). Then, for every $n \geq 0$,

$$\{t \in T_\Sigma \mid \mathsf{depth}(t) \leq n\} \;=\; \{t \in T_\Theta \mid \mathsf{depth}(t) \leq n\} .$$

Proof. Follows from Prop. 5.2.3 after setting $\Upsilon = \Sigma$ and $\Upsilon' = \Theta$. □

The last result did not rely on the fact that transition trees are finite; hence if we allowed infinite transition trees, then every solution of (5.3) would contain all (finite) transition trees generated from the ex-box appearing in Thm. 5.2.4. As a consequence, the solutions so constructed are *minimal* in terms of transition sets, and this minimality is sealed by the finiteness of transition trees.

It is not possible to obtain the same result for places because restricting place trees to finite ones could lead to empty nets, as illustrated later in Sect. 5.4.4, which is not allowed by the ex-restrictedness of boxes (whereas the transition sets may be empty, see again Sect. 5.4.4). Thus one cannot be certain that there will be a unique minimal solution in terms of places. But it is possible to obtain a result similar to Prop. 5.2.3, after allowing for the presence of infinite trees. Below, for every $n \geq 0$, the *n-prefix* of a (possibly infinite) $\tau \in \mathsf{P}_{\mathsf{tree}}$ is the tree $\mathsf{pref}_n(\tau)$ obtained by deleting all nodes whose level is greater than n; in particular, $\mathsf{pref}_0(\tau)$ consists just of the root of τ.

Proposition 5.2.4. Let Υ and Υ' be $\tilde{\sqsubseteq}$-seeds for the equation (5.3). Then $\mathsf{pref}_n(S_{\mathbf{F}^k(\Upsilon)}) = \mathsf{pref}_n(S_{\mathbf{F}^l(\Upsilon')})$, for every $n \geq 0$ and all $k, l \geq n + 2$.

Proof. Let $S_k = S_{\mathbf{F}^k(\Upsilon)}$ and $S'_l = S_{\mathbf{F}^l(\Upsilon')}$, for all $k, l \geq 0$. In general, each place tree has either the form $s \lhd \mathcal{S}$ or $t_1 \lhd \cdots \lhd t_k \lhd s \lhd \mathcal{S}$, where $t_1, \ldots, t_k \in \mathsf{T}_{\mathsf{root}}$ and $k \geq 1$ and $s \in \mathsf{P}_{\mathsf{root}}$. Moreover, no (nonempty) prefix of a place of one form is a prefix of a place of the other form. As a result, we can split the proof in two parts. Below, S^I denotes the set of trees of the first form, and S^{II} of the second form.

Part 1: For every $n \geq 0$ and all $k, l \geq \lfloor \frac{n}{2} \rfloor + 1$,

$$\mathsf{pref}_n(S_k \cap S^I) = \mathsf{pref}_n(S'_l \cap S^I) .$$

We proceed by induction on n. In the base case ($n \in \{0, 1\}$), we have

$$\mathsf{pref}_0(S_k \cap S^I) = S_\Omega = \mathsf{pref}_0(S'_l \cap S^I)$$
$$\mathsf{pref}_1(S_k \cap S^I) = \{s \lhd ({}^\bullet s + s^\bullet) \mid s \in S_\Omega\} = \mathsf{pref}_1(S'_l \cap S^I) ,$$

for all $k, l \geq 1$, which follows from the definition of net refinement (cf. (4.9) and (4.10) in the proof of Prop. 4.3.1) and $\mathbf{F}^k(\Upsilon), \mathbf{F}^l(\Upsilon') \in \mathbf{F}(\mathsf{Box}^{\mathsf{rec}})$.

Suppose that $\mathsf{pref}_{2n+1}(S_k \cap S^I) = \mathsf{pref}_{2n+1}(S'_l \cap S^I)$, for some $n \geq 0$ and all $k, l \geq n+1$. We will show that, for all $k, l \geq n+2$,

$$\mathsf{pref}_{2n+3}(S_k \cap S^I) \subseteq \mathsf{pref}_{2n+3}(S'_l \cap S^I) .$$

Let $k, l \geq n+2$ and $p = s \lhd \mathcal{S} \in S_k$. We have

$$p \;=\; s \lhd \begin{pmatrix} \{v \lhd p_v\}_{v \in {}^\bullet s \cap V} & + \{v \lhd q_v\}_{v \in {}^\bullet s \cap U} + \\ \{w \lhd r_w\}_{w \in s^\bullet \cap V} & + \{w \lhd z_w\}_{w \in s^\bullet \cap U} \end{pmatrix} ,$$

where $V = \{v \in T_\Omega \mid \Delta_v = X\}$ and $U = T_\Omega \backslash V$ and the following are satisfied: $p_v \in \mathbf{F}^{k-1}(\Upsilon)^\circ \subseteq S^I$, $r_w \in {}^\circ\mathbf{F}^{k-1}(\Upsilon) \subseteq S^I$, $q_v \in \Delta_v^\circ$, and $z_w \in {}^\circ\Delta_w$. Thus

$$\mathsf{pref}_{2n+3}(p) \;=\; s \lhd \begin{pmatrix} \{v \lhd \mathsf{pref}_{2n+1}(p_v)\}_{v \in {}^\bullet s \cap V} & + \{v \lhd q_v\}_{v \in {}^\bullet s \cap U} + \\ \{w \lhd \mathsf{pref}_{2n+1}(r_w)\}_{w \in s^\bullet \cap V} & + \{w \lhd z_w\}_{w \in s^\bullet \cap U} \end{pmatrix} .$$

By the induction hypothesis, for every $v \in {}^\bullet s \cap V$, there is $p'_v \in S'_{l-1} \cap S^I$ such that $\mathsf{pref}_{2n+1}(p_v) = \mathsf{pref}_{2n+1}(p'_v)$; similarly, for every $w \in s^\bullet \cap V$, there is $r'_w \in S'_{l-1} \cap S^I$ such that $\mathsf{pref}_{2n+1}(r_w) = \mathsf{pref}_{2n+1}(r'_w)$. Then the following place belongs to $S'_l \cap S^I$ (note that each p'_v is an exit place, and each r'_w is an exit place, by the definition of net refinement):

$$p' \;=\; s \lhd \begin{pmatrix} \{v \lhd p'_v\}_{v \in {}^\bullet s \cap V} & + \{v \lhd q_v\}_{v \in {}^\bullet s \cap U} + \\ \{w \lhd r'_w\}_{w \in s^\bullet \cap V} & + \{w \lhd z_w\}_{w \in s^\bullet \cap U} \end{pmatrix} .$$

And $\mathsf{pref}_{2n+3}(p') = \mathsf{pref}_{2n+3}(p)$. Hence $\mathsf{pref}_{2n+3}(S_k \cap S^I) \subseteq \mathsf{pref}_{2n+3}(S'_l \cap S^I)$ and, since the reverse inclusion holds by symmetry,

$$\mathsf{pref}_{2n+3}(S_k \cap S^I) = \mathsf{pref}_{2n+3}(S'_l \cap S^I) .$$

We have shown the result for all odd values of n. Clearly, the result holds then also for all even n.

Part 2: For every $n \geq 0$ and all $k, l \geq n+2$,

$$\mathsf{pref}_n(S_k \cap S^{II}) = \mathsf{pref}_n(S'_l \cap S^{II}) .$$

We first observe that if no net occurring in (5.3) has an internal place, then the same is true of $\Psi = \bigsqcup \mathbf{F}^*(\Upsilon)$ (as well as $\Psi = \bigsqcup \mathbf{F}^*(\Upsilon')$). Indeed, since Ψ is a fixpoint of $\mathbf{F}$, an internal place is (i) $s \lhd \mathcal{S}$ in S^I with $s \in \ddot{\Omega}$, but $\ddot{\Omega}$ is empty; or (ii) $t_1 \lhd \cdots \lhd t_k \lhd s$ in S^{II} with $s \in \ddot{\Delta}_v$, for some $v \in T_\Omega$, but $\ddot{\Delta}_v$ is empty too; or (iii) $t_1 \lhd \cdots \lhd t_k \lhd s \lhd \mathcal{S}$ in S^{II} with $s \in \ddot{\Psi} \cap S^I$, but we have already argued in (i) that $\ddot{\Psi} \cap S^I$ is empty (the last two facts result from a straightforward induction on the number of t_i's). Hence $S_k \cap S^{II} = \emptyset = S'_l \cap S^{II}$, for all $k, l \geq 0$.

Otherwise, we proceed by induction on n. In the base case ($n = 0$),

$$\mathsf{pref}_0(S_k \cap S^{II}) \;=\; T_\Omega \backslash \{v \mid \Delta_v \neq X \;\wedge\; \ddot{\Delta}_v = \emptyset\} \;=\; \mathsf{pref}_0(S'_l \cap S^{II}) ,$$

for all $k, l \geq 2$. Indeed, if $\Delta_v \neq X$ and $\ddot{\Delta}_v = \emptyset$, then there is no internal place of the kind $v \lhd q$ in any $\mathbf{F}(\Psi)$, whatever the rec-box Ψ; in particular, not in $\mathbf{F}^k(\Upsilon)$ nor in $\mathbf{F}^l(\Upsilon')$ for $k, l \geq 1$. If $\Delta_v \neq X$ and $q \in \ddot{\Delta}_v \neq \emptyset$, then the internal place $v \lhd q$ does occur in every $\mathbf{F}(\Psi)$, whatever the rec-box Ψ; in particular, in $\mathbf{F}^k(\Upsilon)$ and $\mathbf{F}^l(\Upsilon')$ for $k, l \geq 1$. If $\Delta_v = X$ and $s \in \ddot{\Omega}$, then there is some internal place $s \lhd \mathcal{S}$ in $\mathbf{F}(\Upsilon)$, an internal place $v \lhd s \lhd \mathcal{S}$ in $\mathbf{F}^2(\Upsilon)$, and an internal place $v \lhd v \lhd v \lhd \cdots \lhd v \lhd s \lhd \mathcal{S}$ in each $\mathbf{F}^k(\Upsilon)$ for $k \geq 2$, and similarly for $\mathbf{F}^l(\Upsilon')$, where $l \geq 2$. Finally, if $\Delta_v = X$, $\Delta_w \neq X$, and $q \in \ddot{\Delta}_w \neq \emptyset$ (for some w), then $w \lhd q$ does occur in every $\mathbf{F}(\Psi)$ and $v \lhd w \lhd q$ does occur in each $\mathbf{F}^k(\Upsilon)$ for $k \geq 2$; and similarly for $\mathbf{F}^l(\Upsilon')$, where $l \geq 2$. (Notice that

$$\ddot{\Upsilon} = \emptyset \implies \mathsf{pref}_0(S_1 \cap S^{II}) = \{v \mid \Delta_v \neq X \;\wedge\; \ddot{\Delta}_v \neq \emptyset\}$$

$$\ddot{\Upsilon}' \neq \emptyset \implies \mathsf{pref}_0(S'_1 \cap S^{II}) = \{v \mid \Delta_v \neq X \;\wedge\; \ddot{\Delta}_v \neq \emptyset\} \cup \{v \mid \Delta_v = X\} ,$$

so that it may happen that $\mathsf{pref}_0(S_1 \cap S^{II}) \neq \mathsf{pref}_0(S'_1 \cap S^{II})$.)

The inductive step results from Part 1, the induction hypothesis, and the fact that $\mathsf{pref}_n(S_k \cap S^{II})$ is equal to

$$\left\{ v \lhd q \,\middle|\, \left(\Delta_v \neq X \;\wedge\; q \in \ddot{\Delta}_v\right) \;\vee\; \left(\Delta_v = X \;\wedge\; q \in \mathsf{pref}_{n-1}(\ddot{\Psi})\right) \right\} ,$$

where $\Psi = \mathbf{F}^{k-1}(\Upsilon)$, for $n \geq 1$ and $k \geq n+2$ (and we can proceed similarly for $\mathsf{pref}_n(S'_l \cap S^{II})$). □

Corollary 5.2.2. Let Σ and Θ be solutions of (5.3). Then $\mathsf{pref}_n(S_\Sigma) = \mathsf{pref}_n(S_\Theta)$, for every $n \geq 0$.

Proof. Follows from Prop. 5.2.4 after setting $\Upsilon = \Sigma$ and $\Upsilon' = \Theta$. □

In other words, all solutions have the same finite prefixes of place trees and, by Prop. 5.2.4, the n-prefixes are all present in the $(n+2)^{\text{nd}}$ iteration of the limit construction, whatever the seed..

It is possible to derive new places and transitions of a solution of (5.3) as *suffixes* of known ones, through a repeated application of the next result.

Proposition 5.2.5. Let Σ be a solution of (5.3), and v be a transition in Ω such that $\Delta_v = X$.

(1) If $(v, \alpha) \lhd (\{u\} + R)$ is a transition in Σ, then u is a transition in Σ.
(2) If $v \lhd q$ is a place in Σ, then q is an internal place in Σ.
(3) If $s \lhd (\{v{\lhd}q\} + S)$ is a place in Σ, then q is an entry or exit place in Σ.

Proof. Follows from $\Sigma = \mathbf{F}(\Sigma)$ and the definition of net refinement. □

Conversely, by applying $\mathbf{F}$, or more generally $\mathbf{F}^n$, it is possible to construct new places and transitions from those already known by suitable prefixing. This may be useful to determine a minimal solution containing some specific place, as will be shown in Sect. 5.3.1 and applied in Sect. 5.4.3.

5.2.3 An Example of the Limit Construction

Let us consider the following instance of (5.3):

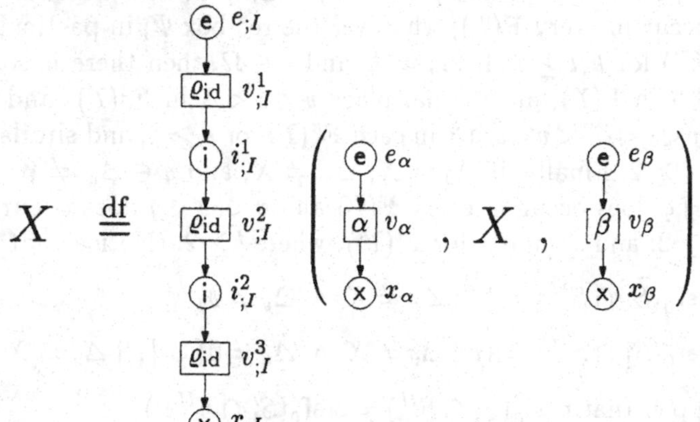

(5.6)

This equation is nothing but $X \stackrel{\mathrm{df}}{=} \Omega_{;I}(\mathsf{N}_\alpha, X, \mathsf{N}_\beta)$, where $\Omega_{;I}$ is the generalised sequence operator box for $I = \{1,2,3\}$, and N_α and N_β are the static boxes associated with two basic PBC expressions, α and β. In this way, (5.6) corresponds to the recursive equation $X \stackrel{\mathrm{df}}{=} ;_{\{1,2,3\}}(\alpha, X, \beta)$ or, less formally, to the equation $X \stackrel{\mathrm{df}}{=} a; (X; b)$ in example 10 of Sect. 3.1, assuming that $\alpha = \{a\}$ and $\beta = \{b\}$.

Let us examine how a solution of the net equation (5.6) might be derived using the limit construction. Suppose there exists already an approximation Υ given as a starting point, a hypothetical $\tilde{\sqsubseteq}$-seed. Inserting that approximation in the place occupied by X on the right-hand side of (5.6) yields a new box

$$\mathbf{F}(\Upsilon) = \Omega_{;I}(\mathsf{N}_\alpha, \Upsilon, \mathsf{N}_\beta)$$

which is a better approximation. We then construct another approximation

$$\mathbf{F}^2(\Upsilon) = \Omega_{;I}(\mathsf{N}_\alpha, \mathbf{F}(\Upsilon), \mathsf{N}_\beta)$$

by the same principle, and so on. We already know that if Υ is a $\tilde{\sqsubseteq}$-seed, then the limit of the sequence so constructed exists and solves (5.6). We now claim that the ex-box

Υ 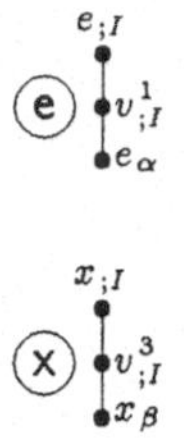

(5.7)

(with these particular place names) is a good seed for (5.6). Figure 5.1 shows the first three approximations resulting from this seed, except that we do not represent all name trees explicitly (the reader may easily guess the missing ones by analogy); in particular, $\mathbf{F}(\Upsilon)$ shows that Υ is indeed a good seed box, satisfying $\Upsilon \tilde{\sqsubseteq} \mathbf{F}(\Upsilon)$.

Figure 5.1 also shows $\Sigma = \bigsqcup \mathbf{F}^*(\Upsilon)$, which is the limit of an infinite series of boxes whose first three elements are shown in the upper part of Fig. 5.1. One may check that this box satisfies (5.6) in the strictest possible sense. To see this, replace the variable X on the right-hand side of the equation by the box Σ shown in Fig. 5.1, perform the net refinement $\Omega_{;I}(\mathsf{N}_\alpha, \Sigma, \mathsf{N}_\beta)$, and check that the result is not just isomorphic but even *identical* to Σ, i.e., it has the same places, the same transitions, the same connectivity, and the same labelling.

Thus, once an appropriate starting point Υ has been chosen, the approximation of a solution of a net equation is quite straightforward (we just keep refining successive approximations into the right-hand side of the equation and compute the limit) and, as we have proved, always leads to a solution. But it could be objected that the way in which the example equation has been solved was not fully satisfactory since the seed box was merely 'guessed' rather than derived. This will be remedied in the next section.

5.2.4 Deriving Seed Boxes

To find a $\tilde{\sqsubseteq}$-seed Υ for the recursive equation (5.3), we need to ensure that $\Upsilon \sqsubseteq \mathbf{F}(\Upsilon)$, ${}^\circ\Upsilon = {}^\circ\mathbf{F}(\Upsilon)$, and $\Upsilon^\circ = \mathbf{F}(\Upsilon)^\circ$ hold for Υ and the next approximation $\mathbf{F}(\Upsilon)$. One might ask why the rec-box (5.7) was a suitable starting point for the equation (5.6). First, it is a good idea to start always with an ex-box because such boxes are minimal with respect to the number of transitions and because, as a consequence of Thm. 5.2.4, all solutions can be so obtained. Moreover, one only needs to ensure that ${}^\circ\Upsilon = {}^\circ\mathbf{F}(\Upsilon)$ and $\Upsilon^\circ = \mathbf{F}(\Upsilon)^\circ$, since then $\Upsilon \sqsubseteq \mathbf{F}(\Upsilon)$ is automatically fulfilled. It is even better (if possible) to start with an ex-box containing only one entry and one exit place, because such boxes are minimal with respect to the ordering $\sqsubseteq$. We will come back to this point in Sect. 5.3.

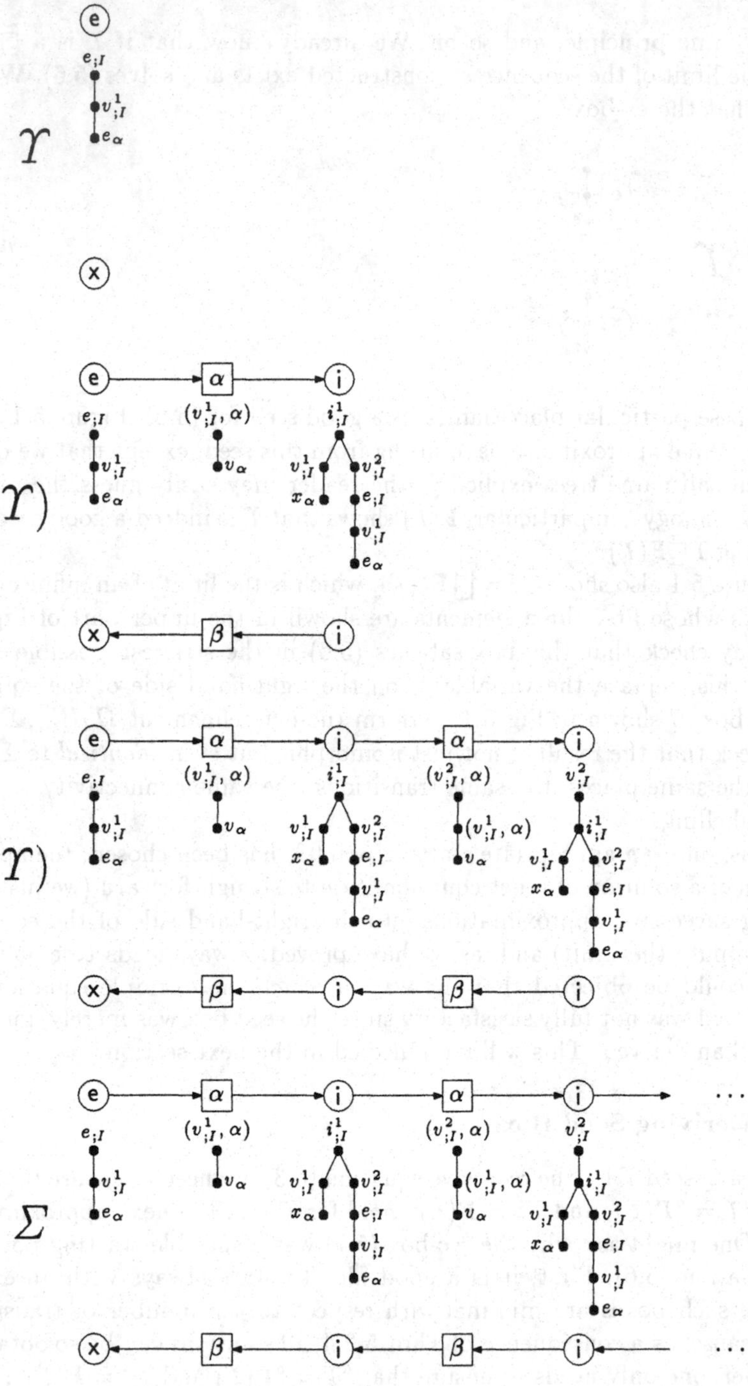

Fig. 5.1. Three approximations for equation (5.6) and the resulting solution

Let us now fix a hypothetical starting ex-box

$$\Upsilon = \Big(\mathcal{P} \cup \mathcal{R}\,,\, \emptyset\,,\, \emptyset\,,\, (\mathcal{P} \times \{\mathsf{e}\}) \cup (\mathcal{R} \times \{\mathsf{x}\})\,,\, \emptyset\Big) \tag{5.8}$$

and develop conditions for it to be a $\widetilde{\sqsubseteq}$-seed. First, due to ex-restrictedness, $\mathcal{P}$ and $\mathcal{R}$ should both be nonempty. They should also be disjoint, otherwise the labelling function would be ill-defined. Finally, with a view to obtain ${}^\circ\Upsilon = {}^\circ\mathbf{F}(\Upsilon)$ and $\Upsilon^\circ = \mathbf{F}(\Upsilon)^\circ$, some conditions need to be imposed on $\mathcal{P}$ and $\mathcal{R}$, essentially by performing the refinement at the entry and exit places. As both place sets are sets of trees, these conditions can be translated into equalities over sets of place trees defined in Sect. 4.3.3.

We define a mapping $\mathbf{f} : 2^{\mathsf{P_{tree}}} \times 2^{\mathsf{P_{tree}}} \to 2^{\mathsf{P_{tree}}} \times 2^{\mathsf{P_{tree}}}$ associated with $\mathbf{F}$, and through $\mathbf{F}$ with the recursive equation (5.3), so that for all $P, R \subseteq \mathsf{P_{tree}}$,

$$\mathbf{f}(P,R) = \Big(\bigcup_{s \in {}^\circ\Omega} \mathbf{f}_s(P,R) \;,\; \bigcup_{s \in \Omega^\circ} \mathbf{f}_s(P,R) \Big)\,, \tag{5.9}$$

where each $\mathbf{f}_s(P,R)$ comprises all place trees $s \lhd (\{v \lhd p_v\}_{v \in s^\bullet} + \{w \lhd r_w\}_{w \in {}^\bullet s})$ such that

$$p_v \in \begin{cases} P & \text{if } \Delta_v = X \\ {}^\circ\Delta_v & \text{if } \Delta_v \neq X \end{cases} \quad \text{and} \quad r_w \in \begin{cases} R & \text{if } \Delta_w = X \\ \Delta_w^\circ & \text{if } \Delta_w \neq X\,. \end{cases}$$

Notice that if s is an isolated place, then $\mathbf{f}_s(P,R) = \{s\}$, and that we have $\mathbf{f}_s(P,R) = \emptyset$ if and only if $P = \emptyset$ and $\Delta_v = X$ for some v, or $R = \emptyset$ and $\Delta_w = X$ for some w.

A pair of sets $(P,R) \in 2^{\mathsf{P_{tree}}} \times 2^{\mathsf{P_{tree}}}$ is a *fixpoint* of $\mathbf{f}$ if $\mathbf{f}(P,R) = (P,R)$. A fixpoint (P,R) is *positive* if $P \neq \emptyset \neq R$; moreover (P,R) is *min/pos* if it is minimal among all positive fixpoints. (That is, for every positive fixpoint (P',R') such that $P' \subseteq P$ and $R' \subseteq R$, it is the case that $(P',R') = (P,R)$.) It is a matter of simple check to verify that the function $\mathbf{f}$ can be used to calculate the entry and exit places of $\mathbf{F}(\Upsilon)$:

$$\mathbf{f}(\mathcal{P},\mathcal{R}) = ({}^\circ\mathbf{F}(\Upsilon)\,,\, \mathbf{F}(\Upsilon)^\circ)\,. \tag{5.10}$$

We then have

Lemma 5.2.2. The following are equivalent:

(1) Υ is a $\widetilde{\sqsubseteq}$-seed.

(2) $(\mathcal{P},\mathcal{R})$ is a positive fixpoint of $\mathbf{f}$.

Proof. If Υ is a $\widetilde{\sqsubseteq}$-seed, then $\emptyset \neq \mathcal{P} = {}^\circ\Upsilon = {}^\circ\mathbf{F}(\Upsilon)$ and $\emptyset \neq \mathcal{R} = \Upsilon^\circ = \mathbf{F}(\Upsilon)^\circ$. This and (5.10) means that $(\mathcal{P},\mathcal{R})$ is a positive fixpoint of $\mathbf{f}$. The reverse implication follows from (5.10), $\mathcal{P} \neq \emptyset \neq \mathcal{R}$, $\mathcal{P} \cap \mathcal{R} = \emptyset$ (which in turn follows from ${}^\circ\Omega \cap \Omega^\circ = \emptyset$) and the definition of net refinement. □

Thus we have reduced the problem of finding a $\tilde{\sqsubseteq}$-seed for (5.3) to that of finding a positive fixpoint of $\mathbf{f}$. Using the notation introduced in (5.9), the latter amounts to finding nonempty sets $\mathcal{P}$ and $\mathcal{R}$ solving the following *tree equations*:

$$\mathcal{P} = \bigcup_{s \in {}^{\circ}\Omega} \mathbf{f}_s(\mathcal{P}, \mathcal{R})$$

$$\mathcal{R} = \bigcup_{s \in \Omega^{\circ}} \mathbf{f}_s(\mathcal{P}, \mathcal{R}) .$$

Let us examine the function $\mathbf{f}$ and its fixpoints more closely. The first observation is that $(2^{\mathsf{P_{tree}}} \times 2^{\mathsf{P_{tree}}}, \subseteq)$ is a complete lattice with set union as the 'join' and intersection as the 'meet' operation, where all operations, including $\subseteq$, are carried out component-wise. Moreover, $\mathbf{f}$ is a monotonic mapping on this domain.

Lemma 5.2.3. $\mathbf{f}$ is monotonic.

Proof. Follows directly from the definition of $\mathbf{f}$. □

It therefore follows that Tarski's theorem ([18]) can be applied.

Theorem 5.2.5. The set of fixpoints of $\mathbf{f}$ is a complete lattice whose maximal element is positive.

Proof. That the set of fixpoints of $\mathbf{f}$ is a complete lattice follows from [18] and Lemma 5.2.3. To show the second part, it suffices to demonstrate that there are two trees in $\mathsf{P_{tree}}$, τ_{e} and τ_{x}, such that $(\{\tau_{\mathsf{e}}\}, \{\tau_{\mathsf{x}}\}) \subseteq \mathbf{f}(\{\tau_{\mathsf{e}}\}, \{\tau_{\mathsf{x}}\})$. The following construction achieves the desired effect.

For every $v \in T_\Omega$ such that $\Delta_v \neq X$, take any basic names $e_v \in {}^{\circ}\Delta_v$ and $x_v \in \Delta_v^{\circ}$ (this is possible since rec-boxes are ex-restricted and $S_{\Delta_v} \subseteq \mathsf{P_{root}}$). Moreover, for every $v \in T_\Omega$ such that $\Delta_v = X$, let $e_v = s$ and $x_v = r$, where $s \in {}^{\circ}\Omega$ and $r \in \Omega^{\circ}$ are arbitrary but fixed places (this is possible because Ω is ex-restricted). The nodes of τ_{e} and τ_{x} are all strings of the following form (together with their nonempty prefixes):

$$q_0\,(v_0, z_0)\,q_1\,(v_1, z_1)\,\ldots\,q_{k-1}\,(v_{k-1}, z_{k-1})\,q_k \qquad (k \geq 0)$$

such that $q_0, q_1, \ldots, q_{k-1} \in \{s, r\}$ and, for every $i < k$, one of the following holds: $v_i \in {}^{\bullet}q_i \wedge z_i = x \wedge q_{i+1} = x_{v_i}$, or $v_i \in q_i^{\bullet} \wedge z_i = e \wedge q_{i+1} = e_{v_i}$. Then the nodes of τ_{e} are all strings beginning with s, and of τ_{x} those beginning with r. There is an arc from node y to node y' if and only if $y' = ya$ for some symbol a, where we treat each q_i and each (v_i, z_i) as a symbol. The label of a node ending with q_i is q_i, and of a node ending with (v_i, z_i) is v_i.

It is easy to verify that τ_{e} and τ_{x} are valid place trees (notice that adding the z_i's was needed to cope with possible side places in Ω), and that

$$(\{\tau_{\mathsf{e}}\}, \{\tau_{\mathsf{x}}\}) \subseteq \mathbf{f}(\{\tau_{\mathsf{e}}\}, \{\tau_{\mathsf{x}}\})$$

since we defined place trees up to isomorphism. □

We will call the two trees constructed in the last proof *spines* associated with **f**; of course, for a given **f** there may be many different spines, depending on the choice of s, r, the e_v's, and the x_v's. The construction is illustrated in Fig. 5.2, where it is assumed that ${}^\bullet s = \{v_1, v_2\}$, $s^\bullet = \{v_1, v_3\}$, ${}^\bullet r = \{v_2\}$, $r^\bullet = \emptyset$, $\Delta_{v_1} = \Delta_{v_2} = X$, and $\Delta_{v_3} \neq X$. Note that

$$\tau_e = s \lhd \{v_1 \lhd \tau_e, v_1 \lhd \tau_x, v_2 \lhd \tau_x, v_3 \lhd e_{v_3}\}$$
$$\tau_x = r \lhd v_2 \lhd \tau_x \ .$$

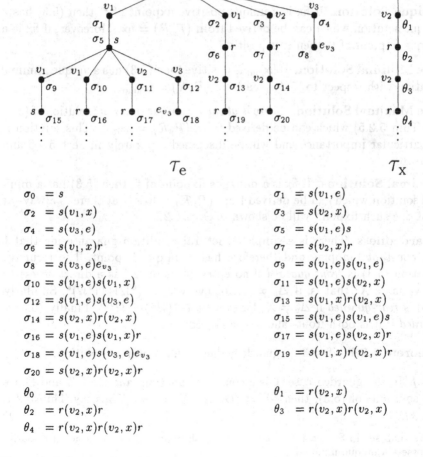

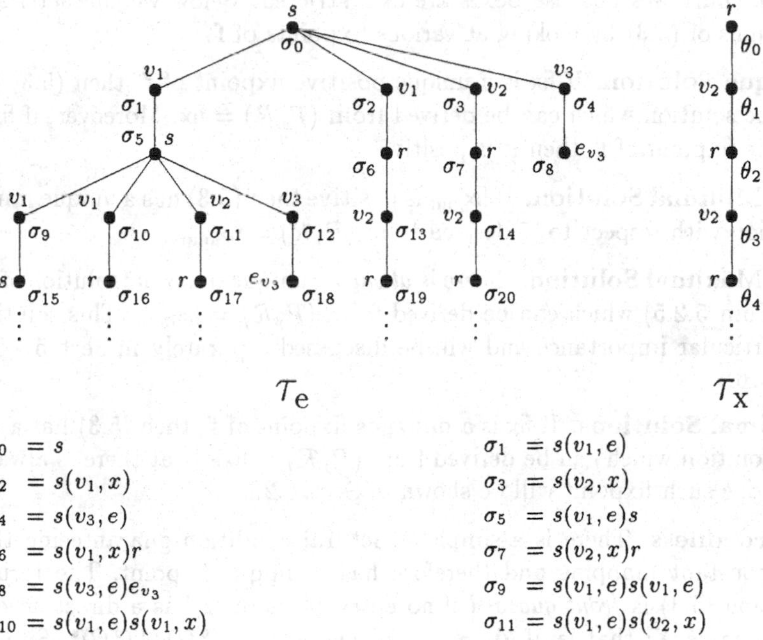

$\sigma_0 = s$ $\qquad \sigma_1 = s(v_1, e)$

$\sigma_2 = s(v_1, x)$ $\qquad \sigma_3 = s(v_2, x)$

$\sigma_4 = s(v_3, e)$ $\qquad \sigma_5 = s(v_1, e)s$

$\sigma_6 = s(v_1, x)r$ $\qquad \sigma_7 = s(v_2, x)r$

$\sigma_8 = s(v_3, e)e_{v_3}$ $\qquad \sigma_9 = s(v_1, e)s(v_1, e)$

$\sigma_{10} = s(v_1, e)s(v_1, x)$ $\qquad \sigma_{11} = s(v_1, e)s(v_2, x)$

$\sigma_{12} = s(v_1, e)s(v_3, e)$ $\qquad \sigma_{13} = s(v_1, x)r(v_2, x)$

$\sigma_{14} = s(v_2, x)r(v_2, x)$ $\qquad \sigma_{15} = s(v_1, e)s(v_1, e)s$

$\sigma_{16} = s(v_1, e)s(v_1, x)r$ $\qquad \sigma_{17} = s(v_1, e)s(v_2, x)r$

$\sigma_{18} = s(v_1, e)s(v_3, e)e_{v_3}$ $\qquad \sigma_{19} = s(v_1, x)r(v_2, x)r$

$\sigma_{20} = s(v_2, x)r(v_2, x)r$

$\theta_0 = r$ $\qquad \theta_1 = r(v_2, x)$

$\theta_2 = r(v_2, x)r$ $\qquad \theta_3 = r(v_2, x)r(v_2, x)$

$\theta_4 = r(v_2, x)r(v_2, x)r$

Fig. 5.2. Spine trees τ_e and τ_x

As a result of Thm. 5.2.5, **f** has a unique minimal fixpoint, $\mathsf{fix}_{\mathsf{min}}$, and a unique maximal positive fixpoint, $\mathsf{fix}_{\mathsf{max}}$. This leads us to the main result of this chapter.

Theorem 5.2.6. The recursive equation (5.3) has at least one solution. Moreover, the set of solutions has a unique maximal element.[1]

Proof. Follows from Lemmata 5.2.1(2) and 5.2.2 together with Thms. 5.2.3 and 5.2.5. □

One might be tempted to always take $\mathsf{fix}_{\mathsf{min}}$, and treat the corresponding solution as *the* solution of (5.3) since, by Lemma 5.2.1(2), such a solution would be the minimal solution in the sense of the inclusion order $\sqsubseteq$. Unfortunately, the minimal fixpoint may fail to be positive, making it unsuitable for our purposes because boxes are ex-restricted. Below we characterise the solutions of (5.3) by looking at various fixpoints of **f**.

Unique Solution. If fix is a unique positive fixpoint of **f**, then (5.3) has a unique solution which can be derived from $(\mathcal{P}, \mathcal{R}) = \mathsf{fix}$. Moreover, if fix is a unique fixpoint of **f**, then it is positive.

The Minimal Solution. If $\mathsf{fix}_{\mathsf{min}}$ is positive then (5.3) has a unique minimal solution with respect to $\sqsubseteq$ derived from $(\mathcal{P}, \mathcal{R}) = \mathsf{fix}_{\mathsf{min}}$.

The Maximal Solution. There is *always* a unique maximal solution of (5.3) (cf. Thm. 5.2.5) which can be derived from $(\mathcal{P}, \mathcal{R}) = \mathsf{fix}_{\mathsf{max}}$. This solution is of particular importance and will be discussed separately in Sect. 5.2.5 and Chap. 6.

Minimal Solutions. If fix is a min/pos fixpoint of **f**, then (5.3) has a minimal solution which can be derived from $(\mathcal{P}, \mathcal{R}) = \mathsf{fix}$. That there is always at least one such fixpoint will be shown in Sect. 5.2.6.

Guardedness. There is a simple structural condition guaranteeing that **f** is a constant mapping and therefore has a unique fixpoint. The recursive equation (5.3) is *front-guarded* if no entry place of Ω has a direct access to the variable X; that is, if $\Delta_v \neq X$, for every $v \in {}^\bullet({}^\circ\Omega) \cup ({}^\circ\Omega)^\bullet$. Similarly, (5.3) is *rear-guarded* if $\Delta_v \neq X$, for every $v \in {}^\bullet(\Omega^\circ) \cup (\Omega^\circ)^\bullet$. Finally, (5.3) is *guarded* if it is both front- and rear-guarded.

Theorem 5.2.7. If (5.3) is guarded, then it has a unique solution.

Proof. In the guarded case, **f** is a constant function, and so if Σ and Θ are two solutions of (5.3), then ${}^\circ\Sigma = {}^\circ\Theta$ and $\Sigma^\circ = \Theta^\circ$. Thus, by Thm. 5.2.1, $\Sigma = \Theta$. □

[1] We shall see in Sect. 5.4.3 that the set of solutions of (5.3) does not necessarily possess a unique minimal element.

The reverse implication does not hold, and we shall present a counterexample to that effect in Sect. 5.4.4.

As in the case of net refinement, various structural equivalences are respected by the solutions of recursive equations on boxes. Below we consider two instances of the recursive equation (5.3), $X \stackrel{\mathrm{df}}{=} \Omega(\boldsymbol{\Delta})$ and $X \stackrel{\mathrm{df}}{=} \Omega'(\boldsymbol{\Delta}')$.

Exercise 5.2.1. Let $\sim\in\{\ \mathsf{iso}\ ,\ \mathsf{iso_T}\ ,\ \mathsf{iso_S}\ ,\ \mathsf{iso_{ST}}\ \}$ and *isom* be an isomorphism from Ω to Ω' such that $\Delta_v = X = \Delta'_{isom(v)}$ or $\Delta_v \sim \Delta'_{isom(v)}$, for every $v \in T_\Omega$. Show that, for every solution Σ of the first equation, there is a solution Σ' of the second equation such that $\Sigma \sim \Sigma'$.

Exercise 5.2.2. Let Ω $\mathsf{iso_S}$ Ω' and *isom* be an isomorphism from Ω_0 to Ω'_0, where Ω_0 and Ω'_0 are obtained from respectively Ω and Ω' after replacing all places by their duplication equivalence classes. Moreover, let $\sim\in\{\ \mathsf{iso_S}\ ,\ \mathsf{iso_{ST}}\ \}$, and $\Delta_v = X = \Delta'_{isom(v)}$ or $\Delta_v \sim \Delta'_{isom(v)}$, for every $v \in T_\Omega$. Show that, for every solution Σ of the first equation, there is a solution Σ' of the second equation such that $\Sigma \sim \Sigma'$.

5.2.5 A Closed Form of the Maximal Solution

We have seen that there is always a unique maximal solution of the recursive equation (5.3) and that this solution, as well as any other solution, can be derived as the limit of successive approximations starting from an ex-box. In this section, we will present a direct construction of the maximal solution $\Sigma^{\max}$ of (5.3). Below

$$V = \{v \in T_\Omega \mid \Delta_v \neq X\} \quad \text{and} \quad U = \{v \in T_\Omega \mid \Delta_v = X\}\ .$$

Transitions. The set of transitions of $\Sigma^{\max}$ comprises all transition names with at least two nodes such that for every nonleaf node ξ, the *subtree* appended at ξ (i.e., ξ is the root of $\mathsf{subtr}(\xi)$), denoted by $\mathsf{subtr}(\xi)$, has one of the following two forms:

- $\mathsf{subtr}(\xi) = (v,\alpha) \lhd R$
  where $v \in V$ and $R \in \mathsf{mult}(T_{\Delta_v})$ and $(\lambda_{\Delta_v}(R),\alpha) \in \lambda_\Omega(v)$.
- $\mathsf{subtr}(\xi) = (v,\alpha) \lhd \{(v_1,\alpha_1) \lhd R_1, \ldots, (v_n,\alpha_n) \lhd R_n\}$
  where $v \in U$ and $(\{\alpha_1,\ldots,\alpha_n\},\alpha) \in \lambda_\Omega(v)$.

The label of a transition with (v,α)-labelled root is α. Note that the multisets of transition tree names R_i, for $i \in \{1,\ldots,n\}$, are constrained by the fact that (v_i,α_i) is itself the label of a nonleaf node, so that the above condition can be applied again.

Places. There are three types of places in $\Sigma^{\max}$. The first type comprises all finite and infinite place names τ such that the label of the root is a place r in Ω, and if a node ξ has the label $s \in S_\Omega$, then

$$\mathsf{subtr}(\xi) \;=\; s \lhd \Big(\{v \lhd p_v\}_{v\in{}^\bullet s} + \{w \lhd r_w\}_{w\in s^\bullet}\Big)$$

and the following are satisfied:

- if $v \in U$, then $p_v = x_v \lhd p'_v$ where $x_v \in \Omega^\circ$;
- if $w \in U$, then $r_w = e_w \lhd r'_w$ where $e_w \in {}^\circ\Omega$;
- if $v \in V$, then $p_v \in \Delta_v^\circ$;
- if $w \in V$, then $r_w \in {}^\circ\Delta_w$.

The label of τ is $\lambda_\Omega(r)$. Note that the place tree names p'_v and r'_v are constrained by the fact that x_v and e_w are nodes in operator boxes, so that the above condition can be applied again.

The second type of places, which are internal, comprises all place names

$$v_1 \lhd v_2 \lhd \cdots \lhd v_n \lhd \tau$$

such that $n \geq 1$, $v_1, \ldots, v_n \in U$, and τ is an internal place of the first type.

The third type of places, which are again internal, are all place names

$$v_1 \lhd v_2 \lhd \cdots \lhd v_n \lhd v \lhd i$$

such that $n \geq 0$, $v \in V$, $i \in \ddot{\Delta}_v$, and $v_1, \ldots, v_n \in U$.

Notice that in a place τ of the first type, only the root node can ever be labelled by an internal place (of Ω or some Δ_v).

Weight Function. To define the weight function of $\Sigma^{\max}$ we need an auxiliary notion of the *trails* of a place or transition name.

Let τ be a tree used as a place or transition name by $\Sigma^{\max}$, i.e., $\tau \in S_{\Sigma^{\max}} \cup T_{\Sigma^{\max}}$. To start with, for every node ξ of τ, we denote by $\mathsf{path}(\xi)$ the sequence of node labels along the path from the root of τ to ξ; we also say that ξ is a *place node* if the label of ξ (which is also the last element in $\mathsf{path}(\xi)$) belongs to $\mathsf{P}_{\mathsf{root}}$. Notice that all leaves (if any) of $\tau \in S_{\Sigma^{\max}}$ are place nodes.

For every sequence $\psi = (v_1, \alpha_1) \cdots (v_n, \alpha_n)v$ such that $\psi = \mathsf{path}(\xi)$ for some leaf node ξ of a tree $\tau \in T_{\Sigma^{\max}}$, we denote $\mathsf{trail}(\psi) = v_1 \cdots v_n v$. And, for every sequence

$$\phi = v_1 \ldots v_m s_0 v_{m+1} \ldots s_{n-1} v_{m+n} s_n$$

such that $\phi = \mathsf{path}(\xi)$ for some place node of a tree $\tau \in S_{\Sigma^{\max}}$, we denote

$$\mathsf{trail}(\phi) = v_1 \ldots v_m v_{m+1} \ldots v_{m+n} s_n \ .$$

Then, for every tree $\tau \in S_{\Sigma^{\max}} \cup T_{\Sigma^{\max}}$, the set of *trails* is the multiset

$$\mathsf{trails}(\tau) \;=\; \sum_{\xi \text{ is a leaf in } \tau} \Big\{\mathsf{trail}(\mathsf{path}(\xi))\Big\}.$$

The connectivity in $\Sigma^{\max}$ is determined by the trails of a place $p \in S_{\Sigma^{\max}}$ and transition $u \in T_{\Sigma^{\max}}$, as follows:

$$W_{\Sigma^{\max}}(p,u) = \sum_{\substack{\sigma, v, s, t \text{ such that} \\ \sigma v t \in \mathsf{trails}(u),\, \sigma v s \in \mathsf{trails}(p)}} \mathsf{trails}(u)(\sigma v t) \cdot \mathsf{trails}(p)(\sigma v s) \cdot W_{\Delta_v}(s,t)$$

$$W_{\Sigma^{\max}}(u,p) = \sum_{\substack{\sigma, v, s, t \text{ such that} \\ \sigma v t \in \mathsf{trails}(u),\, \sigma v s \in \mathsf{trails}(p)}} \mathsf{trails}(u)(\sigma v t) \cdot \mathsf{trails}(p)(\sigma v s) \cdot W_{\Delta_v}(t,s)\ .$$

Exercise 5.2.3. Show that $\Sigma^{\max}$ is the maximal solution of (5.3).

Exercise 5.2.4. Show that for $\tau \in S_{\Sigma^{\max}}$ it may happen that $\mathsf{trails}(\tau) = \emptyset$, if there is no leaf in τ; and that it may happen that $\mathsf{trails}(\tau)$ is a true multiset.

5.2.6 Minimal Solutions

Every min/pos fixpoint of $\mathbf{f}$ determines a minimal solution of (5.3). We shall now show that there is at least one such fixpoint, and so each recursive net equation has at least one minimal solution. We first make a general observation concerning the fixpoints of $\mathbf{f}$. Recall that in the domain $(2^{\mathsf{P}_{\mathsf{tree}}} \times 2^{\mathsf{P}_{\mathsf{tree}}}, \subseteq)$, the operations of union and intersection are carried out component-wise.

Proposition 5.2.6. Let fix^i be a fixpoint of $\mathbf{f}$, for every $i \in I \neq \emptyset$. Then

$$\mathsf{fix} = \bigcap_{i \in I} \mathsf{fix}^i$$

is also a fixpoint of $\mathbf{f}$.

Proof. Let us assume that $\mathsf{fix} = (P, R)$, $\mathbf{f}(\mathsf{fix}) = (P', R')$, and $\mathsf{fix}^i = (P^i, R^i)$, for every $i \in I$. That $\mathbf{f}(\mathsf{fix}) \subseteq \mathsf{fix}$ follows immediately from the monotonicity of $\mathbf{f}$ and $\mathbf{f}(\mathsf{fix}^i) \subseteq \mathsf{fix}^i$, for every $i \in I$. To show the reverse inclusion, we observe that from $\mathbf{f}(\mathsf{fix}^i) \supseteq \mathsf{fix}^i$, for every $i \in I$, it follows that each $p \in P$ (if $P \neq \emptyset$) has the form (see also Prop. 5.2.5(3))

$$p = s \lhd \left(\{v \lhd p_v\}_{v \in s^\bullet} + \{w \lhd r_w\}_{w \in {}^\bullet s} \right),$$

where $s \in {}^\circ\Omega$, $p_v \in \bigcap_{i \in I} P^i$ if $\Delta_v = X$; $r_w \in \bigcap_{i \in I} R^i$ if $\Delta_w = X$; $p_v \in {}^\circ\Delta_v$ if $\Delta_v \neq X$; and $r_w \in \Delta_w^\circ$ if $\Delta_w \neq X$. Thus $p \in P'$ and so $P \subseteq P'$. By a symmetric argument, $R \subseteq R'$. Hence $\mathbf{f}(\mathsf{fix}) \supseteq \mathsf{fix}$. □

There are two immediate consequences of the last result.

Corollary 5.2.3. If (P, R) and (P', R') are min/pos fixpoints of $\mathbf{f}$ such that $\emptyset \neq P \cap P' \notin \{P, P'\}$, then $R \cap R' = \emptyset$; similarly, if $\emptyset \neq R \cap R' \notin \{R, R'\}$, then $P \cap P' = \emptyset$.

Corollary 5.2.4. If $(P, R) \subseteq \mathsf{fix}_{\max}$, then there is a unique fixpoint of $\mathbf{f}$, later denoted by $\mathsf{sol}_{\min}(P, R)$, which is minimal among all those containing (P, R).

Proof. From Prop. 5.2.6 it follows that the intersection of all fixpoints of $\mathbf{f}$ containing P and R (and there is at least one such fixpoint, e.g., the maximal one) is a fixpoint of $\mathbf{f}$. □

Moreover, it follows directly from the definitions of $\mathsf{sol}_{\min}$ and $\mathsf{fix}_{\max}$ that $(P,R) \subseteq \mathsf{sol}_{\min}(P,R) \subseteq \mathsf{fix}_{\max}$, and that $\mathsf{sol}_{\min}$ is monotonic, i.e., $(P',R') \subseteq (P,R)$ implies $\mathsf{sol}_{\min}(P',R') \subseteq \mathsf{sol}_{\min}(P,R)$.

The result formulated as the last corollary is of an existential nature, but it is possible to give it a more constructive flavour. The *first-level suffix-set* of $(P,R) \subsetneq \mathsf{fix}_{\max}$ is the pair of sets $\mathsf{suf}(P,R) = (P',R')$ such that P' (respectively, R') comprises all places q such that there is $p \in P \cup R$ satisfying $p = s \lhd (\{v \lhd q\} + \mathcal{S})$ with $\Delta_v = X$ and the root of q is labelled by a place in ${}^\circ\Omega$ (respectively, in Ω°). Notice that, by the definition of $\mathbf{f}$, the root of each place in $\mathsf{fix}_{\max}$ is labelled by a place in ${}^\circ\Omega$ or Ω°, and that p is a place in $\mathbf{f}(\mathsf{fix}_{\max})$.

From $(P,R) \subseteq \mathsf{fix}_{\max} \subseteq \mathbf{f}(\mathsf{fix}_{\max})$, it follows that $\mathsf{suf}(P,R) \subseteq \mathsf{fix}_{\max}$. We thus can define the *suffix-set* of (P,R) as

$$\mathsf{Suf}(P,R) = \bigcup_{i=0}^{\infty} \mathsf{suf}_i(P,R) ,$$

where $\mathsf{suf}_0(P,R) = (P,R)$ and $\mathsf{suf}_i(P,R) = \mathsf{suf}(\mathsf{suf}_{i-1}(P,R))$, for every $i > 0$. Notice that in $\mathsf{Suf}(P,R) = (P',R')$, the set P' (respectively, R') comprises all subtrees q of the places in $P \cup R$ with the root q being labelled by a place in ${}^\circ\Omega$ (respectively, in Ω°).

Now, for a (possibly transfinite) ordinal β and $K, M \subseteq \mathsf{P}_{\mathsf{tree}}$, let us define

$$\mathbf{f}^\beta(K,M) = \begin{cases} (K,M) & \text{if } \beta = 0 \\ \mathbf{f}(\bigcup_{\gamma<\beta} \mathbf{f}^\gamma(K,M)) & \text{if } \beta > 0 . \end{cases}$$

Note that $\mathbf{f}^\beta$ is well defined as there is no infinite descending chain of ordinals.

Since $\mathbf{f}$ is monotonic, if $(K,M) \subseteq \mathbf{f}(K,M)$, then by transfinite induction, we obtain that $(K,M) \subseteq \mathbf{f}^\beta(K,M) \subseteq \mathbf{f}(\mathbf{f}^\beta(K,M))$, for every ordinal β. Hence from Hartog's theorem [73] which says that for any cardinal there is an ordinal which is strictly greater, we obtain[2] that if $(K,M) \subseteq \mathbf{f}(K,M)$ then there is an ordinal α such that, for every $\beta \geq \alpha$,

$$\mathbf{f}^\alpha(K,M) = \mathbf{f}(\mathbf{f}^\alpha(K,M)) = \mathbf{f}^\beta(K,M) .$$

We will denote such an α (for instance, the smallest one) by $\alpha(K,M)$, and then we obtain that $\mathbf{f}^{\alpha(K,M)}(K,M)$ is the smallest fixpoint of $\mathbf{f}$ containing (K,M).

Proposition 5.2.7. For every pair $(P,R) \subseteq \mathsf{fix}_{\max}$,

$$\mathsf{sol}_{\min}(P,R) = \mathbf{f}^{\alpha(\mathsf{Suf}(P,R))}(\mathsf{Suf}(P,R)) .$$

Proof. We first show that $\mathsf{Suf}(P,R) \subseteq \mathbf{f}(\mathsf{Suf}(P,R))$ and so $\mathsf{Suf}(P,R) \subseteq \mathsf{fix}_{\max}$. By the definitions of $\mathbf{f}$ and suf, we have $\mathsf{suf}_i(P,R) \subseteq \mathbf{f}(\mathsf{suf}_{i+1}(P,R))$, for every $i \geq 0$. Hence

[2] This is a general result of the fixpoint theory.

$$\begin{aligned}\mathsf{Suf}(P,R) &= \bigcup_{i=0}^{\infty} \mathsf{suf}_i(P,R) \subseteq \bigcup_{i=0}^{\infty} \mathbf{f}(\mathsf{suf}_{i+1}(P,R)) \\ &\subseteq \bigcup_{i=0}^{\infty} \mathbf{f}(\mathsf{suf}_i(P,R)) \subseteq_{(*)} \mathbf{f}\Big(\bigcup_{i=0}^{\infty} \mathsf{suf}_i(P,R)\Big) \\ &= \mathbf{f}(\mathsf{Suf}(P,R))\,,\end{aligned}$$

where (*) follows from the monotonicity of $\mathbf{f}$. Thus $\alpha(\mathsf{Suf}(P,R))$ is a well-defined ordinal and $\mathbf{f}^{\alpha(\mathsf{Suf}(P,R))}(\mathsf{Suf}(P,R))$ is the smallest fixpoint of $\mathbf{f}$ containing $\mathsf{Suf}(P,R)$. As a result, to complete the proof, it suffices to show that

$$(P,R) \subseteq \mathsf{Suf}(P,R) \subseteq \mathsf{sol}_{\min}(P,R).$$

The first inclusion is obvious. We prove the second one by induction, in the following way. Clearly, $\mathsf{suf}_0(P,R) = (P,R) \subseteq \mathsf{sol}_{\min}(P,R)$. Then, assuming $\mathsf{suf}_i(P,R) \subseteq \mathsf{sol}_{\min}(P,R)$, we obtain

$$\mathsf{suf}_{i+1}(P,R) = \mathsf{suf}(\mathsf{suf}_i(P,R)) \subseteq_{(**)} \mathsf{suf}(\mathsf{sol}_{\min}(P,R)) \subseteq_{(***)} \mathsf{sol}_{\min}(P,R).$$

Notice that (**) follows from the induction hypothesis and suf being monotonic, and (***) follows from $\mathsf{sol}_{\min}(P,R)$ being a fixpoint of $\mathbf{f}$, more precisely, from $\mathsf{sol}_{\min}(P,R) \subseteq \mathbf{f}(\mathsf{sol}_{\min}(P,R))$. □

Corollary 5.2.5. Let $(P,R) \subseteq \mathsf{fix}_{\max}$. Then, for every place p in $\mathsf{sol}_{\min}(P,R)$, on every infinite path (if any) from the root of p, there is a node ξ such that $\mathsf{subtr}(\xi)$ belongs to $\mathsf{Suf}(P,R)$.

Proof. Follows by transfinite induction from Prop. 5.2.7. An alternative argument is that if $\mathsf{sol}_{\min}(P,R)$ contained trees p not fulfilling the condition, then by dropping all of them we would get a strictly smaller pair of sets, which would still be a fixpoint of $\mathbf{f}$ containing P and R, contradicting the minimality of $\mathsf{sol}_{\min}(P,R)$. □

Let $(P,R) \subseteq \mathsf{fix}_{\max}$ be a pair of nonempty sets of places. Although $\mathsf{sol}_{\min}(P,R)$ is a positive fixpoint of $\mathbf{f}$ and minimal among those containing P and R, it may not be minimal among all positive fixpoints of $\mathbf{f}$ (even if P and R are singletons; see, for example, p and q in (5.13) for which $\mathsf{sol}_{\min}(\{p\},\{r\})$ is not a min/pos fixpoint). We therefore need to be careful when choosing P and R if we want to guarantee that $\mathsf{sol}_{\min}(P,R)$ is a min/pos fixpoint. Interestingly, the spine trees constructed in the proof of Thm. 5.2.5 will prove useful in looking for the right P and R. Below, we use an auxiliary mapping $g : {}^{\circ}\Omega \cup \Omega^{\circ} \to 2^{\{\mathrm{e},\mathrm{x}\}}$ such that $\mathrm{e} \in g(s)$ if and only if $\{v \in s^{\bullet} \mid \Delta_v = X\} \neq \emptyset$, and $\mathrm{x} \in g(s)$ if and only if $\{v \in {}^{\bullet}s \mid \Delta_v = X\} \neq \emptyset$. Intuitively, $g(s)$ reflects the non-guardedness of an entry or exit place s in the operator box Ω, and (5.3) is guarded if and only if $g({}^{\circ}\Omega \cup \Omega^{\circ}) = \{\emptyset\}$.

Theorem 5.2.8. $\mathbf{f}$ has a min/pos fixpoint.

Proof. In this proof we will make several references of the spine trees, τ_e and τ_x, defined as in the proof of Thm. 5.2.5 and illustrated in Fig. 5.2, for different pairs of places $s \in {}^\circ\Omega$ and $r \in \Omega^\circ$. As already observed, we have $(T_e, T_x) \subseteq \mathbf{f}(T_e, T_x) \subseteq \mathsf{fix}_{max}$, where here and later in this proof we use T_e and T_x to denote $\{\tau_e\}$ and $\{\tau_x\}$, respectively. Moreover, it is easy to see that $\mathsf{Suf}(T_e, T_x) = (T_e, T_x)$. Thus, by Cor. 5.2.5,

For every place p in $\mathsf{sol}_{min}(T_e, T_x)$, on every infinite path (if any) from the root of p, there is a node ξ such that $\mathsf{subtr}(\xi) \in \{\tau_e, \tau_x\}$. (*)

Suppose first that $\mathbf{f}(\emptyset, \emptyset) = (K, M) \neq (\emptyset, \emptyset)$. Without loss of generality, we assume that $K \neq \emptyset$. Then there is $s \in {}^\circ\Omega$ such that $g(s) = \emptyset$, and we consider two cases.

Case 1: There is $r \in \Omega^\circ$ such that $\mathrm{x} \notin g(r)$. Let τ_e and τ_x be defined for the current s and r. Then $(T_e, T_x) \subseteq \mathbf{f}^2(\emptyset, \emptyset)$ and so $\mathsf{fix}_{min} = \mathsf{sol}_{min}(\emptyset, \emptyset)$ is positive. Clearly, in such a case, it is the only min/pos fixpoint of $\mathbf{f}$, since $\mathsf{sol}_{min}(\emptyset, \emptyset)$ is always the minimal fixpoint of $\mathbf{f}$.

Case 2: For all $r \in \Omega^\circ$, $\mathrm{x} \in g(r)$. Take any $r \in \Omega^\circ$. Let τ_e and τ_x be defined for the current s and r. Suppose that $(P, R) \subseteq \mathsf{sol}_{min}(T_e, T_x)$ is a positive fixpoint. Clearly, $\tau_e \in P$. To show that $\tau_x \in R$, take any $q \in R$. By our assumption (and the discussion in Sect. 5.2.5 on the form of the place trees in the maximal - hence any - solution), there is an infinite path ϕ from the root of q such that no node of ϕ is labelled by a place in ${}^\circ\Omega$. Moreover, by (*), there is a node ξ on ϕ such that $\mathsf{subtr}(\xi) \in \{\tau_e, \tau_x\}$. Hence, since the root of τ_e belongs to ${}^\circ\Omega$, we obtain that τ_x belongs to $\mathsf{Suf}(\emptyset, \{q\})$. Hence $(T_e, T_x) \subseteq \mathsf{Suf}(P, R) \subseteq \mathsf{sol}_{min}(P, R) = (P, R)$. Thus, by sol_{min} being monotonic, $\mathsf{sol}_{min}(T_e, T_x) \subseteq (P, R)$, and so $\mathsf{sol}_{min}(T_e, T_x)$ is a min/pos fixpoint of $\mathbf{f}$.

Having shown the result for the case $\mathbf{f}(\emptyset, \emptyset) \neq (\emptyset, \emptyset)$, we now assume the opposite, i.e., $\mathbf{f}(\emptyset, \emptyset) = (\emptyset, \emptyset)$. This means that $\emptyset \notin g({}^\circ\Omega \cup \Omega^\circ)$ which in turn implies (again, by the discussion in Sect. 5.2.5) that

Every tree in fix_{max} has an infinite path from its root. (**)

Suppose that there are $s \in {}^\circ\Omega$ and $r \in \Omega^\circ$ such that $\mathrm{x} \in g(s)$ and $\mathrm{e} \in g(r)$. Let τ_e and τ_x be defined for the current s and r. Moreover, let $(P, R) \subseteq \mathsf{sol}_{min}(T_e, T_x)$ be a positive fixpoint of $\mathbf{f}$ and $p \in P$. By $\emptyset \notin g({}^\circ\Omega \cup \Omega^\circ)$ and (*) and (**), $\tau_e \in P'$ or $\tau_x \in R'$ for $(P', R') = \mathsf{Suf}(\{p\}, \emptyset)$. Moreover, for the current s and r it is the case that $\mathsf{Suf}(\emptyset, T_x) = \mathsf{Suf}(T_e, \emptyset) = (T_e, T_x)$. Hence $(T_e, T_x) \subseteq \mathsf{Suf}(P, R)$ and, as in Case 2, $\mathsf{sol}_{min}(T_e, T_x)$ is a min/pos fixpoint of $\mathbf{f}$.

If s and r as above cannot be found, we can assume without loss of generality that $g({}^\circ\Omega) = \{\{\mathrm{e}\}\}$. Then we choose any $s \in {}^\circ\Omega$ and consider two cases.

Case 3: There is $r \in \Omega^\circ$ such that $g(r) = \{\mathrm{e}\}$. Let τ_e and τ_x be defined for the current s and r. Notice that we have in this case $\mathsf{Suf}(T_e, \emptyset) = (T_e, \emptyset)$, $\mathsf{Suf}(\emptyset, T_x) = (T_e, T_x)$, and $(\emptyset, T_x) \subseteq \mathbf{f}(T_e, \emptyset) \subseteq \mathsf{sol}_{min}(T_e, \emptyset)$. It can then be seen that $\mathsf{sol}_{\min}(T_e, T_x) = \mathsf{sol}_{\min}(T_e, \emptyset)$ is a min/pos fixpoint of $\mathbf{f}$. Indeed, let $(P, R) \subseteq \mathsf{sol}_{\min}(T_e, T_x)$ be a positive fixpoint. By (*) and (**), $\tau_e \in P'$ or $\tau_x \in R'$ for $(P', R') = \mathsf{Suf}(\{p\}, \emptyset)$. Thus, by $\mathsf{Suf}(\emptyset, T_x) = (T_e, T_x)$, $\tau_e \in P'$. Hence $(T_e, \emptyset) \subseteq \mathsf{Suf}(P, R) \subseteq \mathsf{sol}_{\min}(P, R) = (P, R)$ and so, by $\mathsf{sol}_{\min}(T_e, T_x) = \mathsf{sol}_{\min}(T_e, \emptyset)$, we obtain $\mathsf{sol}_{\min}(T_e, T_x) = (P, R)$.

Case 4: For all $r \in \Omega^\circ$, $\mathrm{x} \in g(r)$. Take any $r \in \Omega^\circ$. Let τ_e and τ_x be defined for the current s and r. Notice that in this case $\mathsf{Suf}(T_e, \emptyset) = (T_e, \emptyset)$. Let $(P, R) \subseteq \mathsf{sol}_{\min}(T_e, T_x)$ be a positive fixpoint of $\mathbf{f}$. Take any $p \in P$. By (*) and (**) and the fact that no node in p is labelled by a place in Ω°, we obtain $(T_e, \emptyset) \subseteq \mathsf{Suf}(P, R)$. Similarly, $(\emptyset, T_x) \subseteq \mathsf{Suf}(P, R)$ follows from (*) and (**) and the fact that there is an infinite path in r such that no node along this path is labelled by a place in ${}^\circ\Omega$. Hence $(T_e, T_x) \subseteq \mathsf{Suf}(P, R)$ and $\mathsf{sol}_{\min}(T_e, T_x)$ is a min/pos fixpoint of $\mathbf{f}$. □

The first part of the above proof has identified a sufficient condition for the existence of a unique min/pos fixpoint of $\mathbf{f}$.

Corollary 5.2.6. Suppose that the operator box Ω in (5.3) is such that one of the following holds:

- $\emptyset \in g({}^\circ\Omega) \cap g(\Omega^\circ)$.
- $\emptyset \in g({}^\circ\Omega)$ and $\{\mathrm{e}\} \in g(\Omega^\circ)$.
- $\emptyset \in g(\Omega^\circ)$ and $\{\mathrm{x}\} \in g({}^\circ\Omega)$.

Then there is a unique min/pos fixpoint of $\mathbf{f}$. □

5.3 Finitary Equations and Finite Operator Boxes

We now look at the conclusions that can be drawn from assuming that (5.3) is an X-finitary equation (cf. Sect. 5.2.1), or that the operator box appearing in it is finite. The interest in the latter is motivated by the application of our theory to giving compositional net semantics to the standard PBC, where all the operators are finite.

5.3.1 Finitary Equation

For an X-finitary equation the auxiliary mapping $\mathbf{f}$ is not only monotonic but also continuous.

Proposition 5.3.1. If (5.3) is X-finitary, then $\mathbf{f}$ is continuous.

Proof. Let $P^0 \subseteq P^1 \subseteq \ldots$ be a chain in $(2^{\mathsf{P}_{\text{tree}}} \times 2^{\mathsf{P}_{\text{tree}}}, \subseteq)$ with the limit $P = \bigcup_{i=0}^{\infty} P^i$. The inclusion $\mathbf{f}(P) \supseteq \bigcup_{i=0}^{\infty} \mathbf{f}(P^i)$ follows from the monotonicity of $\mathbf{f}$ (see Lemma 5.2.3). As for the reverse inclusion, suppose that

$$p = s \lhd \left(\{v \lhd p_v\}_{v \in {}^\bullet s} + \{w \lhd r_w\}_{w \in s^\bullet} \right)$$

is a place such that, without loss of generality, $(\{p\}, \emptyset) \subseteq \mathbf{f}(P)$. Let $U = \{u \in {}^\bullet s \cup s^\bullet \mid \Delta_u = X\}$. Since U is finite (as (5.3) is X-finitary), from the definition of the chain, it follows that there is $i \geq 0$ such that for every $u \in U$, each p_u and r_u belongs to P^i. Hence $(\{p\}, \emptyset) \subseteq \mathbf{f}(P^i)$ and so $(\{p\}, \emptyset) \subseteq \bigcup_{i=0}^{\infty} \mathbf{f}(P^i)$. Thus $\mathbf{f}(P) \subseteq \bigcup_{i=0}^{\infty} \mathbf{f}(P^i)$. □

Continuity is useful in that we can calculate some fixpoints of $\mathbf{f}$ by taking

$$(\mathcal{P}, \mathcal{R}) = \bigcup_{i=0}^{\infty} \mathbf{f}^i(\{\tau_e\}, \{\tau_x\})$$

provided that τ_e and τ_x are place trees satisfying

$$(\{\tau_e\}, \{\tau_x\}) \subseteq \mathbf{f}(\{\tau_e\}, \{\tau_x\}) ,$$

without resorting to transfinite construction. And recall that the existence of such (spine) trees has been demonstrated in the proof of Thm. 5.2.5.

But the discussion of X-finitary equations does not end here. As we have shown in Prop. 5.2.2, if (5.3) is X-finitary, then the mapping $\mathbf{F}$ is $\sqsubseteq$-continuous rather than only $\tilde{\sqsubseteq}$-continuous. An immediate consequence of this fact is that the definition of a seed box can be changed, in the following way.

A rec-box Υ is a $\sqsubseteq$-*seed* for equation (5.3) if $\Upsilon \sqsubseteq \mathbf{F}(\Upsilon)$. What is interesting is that the results obtained previously for $\tilde{\sqsubseteq}$-seed boxes carry over to the modified setting.

Theorem 5.3.1. If Υ is a $\sqsubseteq$-seed for an X-finitary equation (5.3), then $\bigsqcup \mathbf{F}^*(\Upsilon)$ is a solution.

Proof. Similar to the proof of Thm. 5.2.3. □

The search for $\sqsubseteq$-seeds can be carried out similarly as in Sect. 5.2.4.

Lemma 5.3.1. Let Υ be an ex-box and $\mathcal{P} = {}^\circ\Upsilon$ and $\mathcal{R} = \Upsilon^\circ$. Then the following are equivalent:

(1) Υ is a $\sqsubseteq$-seed.
(2) The sets $\mathcal{P}$ and $\mathcal{R}$ are nonempty and $(\mathcal{P}, \mathcal{R}) \subseteq \mathbf{f}(\mathcal{P}, \mathcal{R})$.

Proof. Similar to the proof of Lemma 5.2.2. □

What is more, from the proof of Thm. 5.2.5, it is always possible to find a pair of *singleton* sets $\mathcal{P} = \{\tau_e\}$ and $\mathcal{R} = \{\tau_x\}$ satisfying $(\mathcal{P}, \mathcal{R}) \subseteq \mathbf{f}(\mathcal{P}, \mathcal{R})$. It therefore follows that in the X-finitary case, it is always possible to obtain a solution of (5.3) by starting from an ex-box with exactly two places. In terms of approximating a solution, we obtain

Theorem 5.3.2. If (5.3) is X-finitary, then it has a solution which can be approximated from a $\sqsubseteq$-seed with exactly two places.
Note: This does not mean that *any* solution can be so approximated.

Proof. Follows directly from Thm. 5.3.1 and Lemma 5.3.1, as well as the above remark. □

Since there is a solution which can be approximated by starting from a seed box with exactly one entry and one exit place, if all boxes occurring in the recursive equation are countable, then all rec-boxes in the chain derived from such a seed box will have countably many places, and the solution given by the limit will also have countably many places. However, the same reasoning cannot be made for transitions, due to the fact that there may be uncountably infinitely many labels α such that $(\lambda_{\Sigma_v}(R), \alpha)$ belongs to $\lambda_\Omega(v)$, and so each such α may give rise to a separate transition $(v, \alpha) \lhd R$. This observation motivates the next definition. A relabelling ϱ is *image-countable* if, for every finite multiset of labels Γ, $\{\alpha \mid (\Gamma, \alpha) \in \varrho\}$ is countable.

Theorem 5.3.3. Let (5.3) be an X-finitary recursive equation such that all boxes occurring in it are countable, and all relabellings of the operator box Ω are image-countable. Then the equation has a countable solution which can be approximated by a $\sqsubseteq$-chain of countable rec-boxes which begins with an ex-box comprising exactly two places. □

Another interesting feature of the X-finitary case is that sometimes it is not necessary to use transfinite constructions to obtain min/pos fixpoints of $\mathbf{f}$ containing some sets of places (see Sect. 5.2.6). Indeed, in the X-finitary case, from the continuity of the auxiliary function $\mathbf{f}$, it follows that $\alpha(\mathsf{Suf}(P, R)) \leq \omega$, where ω is the first transfinite ordinal.

Proposition 5.3.2. If (5.3) is X-finitary, then for every $(P, R) \subseteq \mathsf{fix}_{\mathsf{max}}$,

$$\mathsf{sol}_{\mathsf{min}}(P, R) = \bigcup_{i=0}^{\infty} \mathbf{f}^i(\mathsf{Suf}(P, R)) .$$

Proof. Follows from Props. 5.2.7 and 5.3.1. □

Corollary 5.3.1. If all boxes occurring in an X-finitary (5.3) are countable, then each min/pos fixpoint of $\mathbf{f}$ is countable.

Proof. Let (P, R) be a min/pos fixpoint of $\mathbf{f}$, and $p \in P$ and $r \in R$. Since the equation is X-finitary, for each countable $(K, M) \subseteq \mathsf{fix}_{\mathsf{max}}$, the first-level suffix $\mathsf{suf}(K, M)$ is countable too. Hence $\mathsf{Suf}(\{p\}, \{r\})$ is countable and, since all boxes occurring in (5.3) are countable and the equation is X-finitary, so is each $\mathbf{f}^i(\mathsf{Suf}(\{p\}, \{r\}))$. Thus, by Prop. 5.3.2, $\mathsf{sol}_{\mathsf{min}}(\{p\}, \{r\}) = (P, R)$ is countable. □

5.3.2 Finite Operator Box

We now assume that the operator box in (5.3) is finite (hence X-finitary), and that each box in the Ω-tuple $\mathbf{\Delta}$ is finite too. Our aim is to look at the conditions that need to be satisfied to guarantee that the equation has countable solutions which can be approximated by $\sqsubseteq$-chains of finite boxes.

To begin with, we can apply Thm. 5.3.2 to conclude that there is a solution which can be approximated by starting from a $\sqsubseteq$-seed with exactly two places and no transitions. Hence, since all nets occurring in (5.3) are finite, all rec-boxes in the chain derived from such a seed will have finitely many places. And the solution given by the limit will have at most countably many places. However, again, the same reasoning cannot be made for transitions. Indeed, since the definition of net refinement $\Omega(\boldsymbol{\Sigma})$ in Sect. 4.3.4 uses finite multisets R to construct new transitions of the form $(v, \alpha) \lhd R$, there is a possibility of constructing infinitely many transitions even if Ω and $\boldsymbol{\Sigma}$ only have finitely many transitions. As a matter of fact, there are two potential sources of infinity.

The first one is the possibility of there being infinitely many labels α such that $(\lambda_{\Sigma_v}(R), \alpha)$ belongs to $\lambda_\Omega(v)$, since each such α may give rise to a separate transition $(v, \alpha) \lhd R$. This motivates the next definition, strengthening the one already used in the previous section. A relabelling ϱ is *image-finitary* if, for every finite multiset of labels Γ, $\{\alpha \mid (\Gamma, \alpha) \in \varrho\}$ is finite. And it is worth noting in this context that all relabellings used to define operator boxes in the translation of PBC expressions are image-finitary. The second source of infinity is that we allow R to be a multiset of transitions rather than a set, as discussed in Sect. 4.3.5. This suggests a further strengthening of the condition imposed on relabellings. A relabelling ϱ is *image-finite* if, for every finite subset Lab' of Lab, $\{(\Gamma, \alpha) \in \varrho \mid \Gamma \in \mathsf{mult}(\mathsf{Lab}')\}$ is finite. For instance, the reduced version of $\varrho_{\mathsf{sy}'\,a}$, described in the last part of Sect. 4.4.6, is image-finite on account of Prop. 4.4.1, while $\varrho_{\mathsf{sy}\,a}$ is not.

Theorem 5.3.4. If all boxes occurring in (5.3) are finite, and all relabellings in Ω are image-finite, then the equation has a countable solution which can be approximated by a $\sqsubseteq$-chain of finite rec-boxes which begins with a $\sqsubseteq$-seed comprising exactly two places and no transitions.

Proof. Under the assumptions made about Ω and $\boldsymbol{\Delta}$, if Σ is a rec-box with finitely many transitions, then $\mathbf{F}(\Sigma)$ has finitely many transitions too. One then only needs to use Thm. 5.3.2. □

Taking the limit of a $\sqsubseteq$-chain of boxes will typically produce an infinite box even if the equation was based on finite boxes. However, we do not consider this to be a problem since the main purpose of this chapter is to provide theoretical foundations for solving recursive equations on boxes. In practical applications, initial fragments of the limit construction can be used to derive finite approximations of an infinite solution. We also expect that the results obtained here will provide a basis for high-level net solutions derived by folding infinite rec-boxes into finite high-level representations.

5.4 Further Examples

Let us first return to the example from Sect. 5.2.3. In this case recursion is guarded and the function **f** constant; hence it has a unique positive fixpoint

$$(\mathcal{P}\ ,\ \mathcal{R})\ =\ \Big(\ \{\ e_{;I} \lhd v^1_{;I} \lhd e_\alpha\ \}\ \ ,\ \ \{\ x_{;I} \lhd v^3_{;I} \lhd x_\beta\ \}\ \Big)$$

which gives the entry and exit places of the box Υ we have chosen earlier for this example in (5.7). And we can conclude that the net equation (5.6) has a unique solution, namely the box Σ shown in Fig. 5.1.

5.4.1 Unbounded Parallel Composition

The tree equation $(\mathcal{P},\mathcal{R}) = \mathbf{f}(\mathcal{P},\mathcal{R})$ may become more complicated for unguarded recursion. This is illustrated by the next example which is neither front- nor rear-guarded:

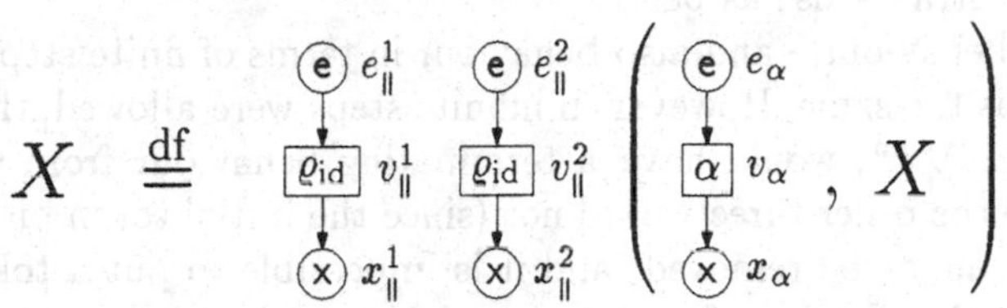

(5.11)

The above equation, $X \stackrel{\mathrm{df}}{=} \Omega_{\|}(\mathsf{N}_\alpha, X)$, corresponds to example 12 in Sect. 3.1, $X \stackrel{\mathrm{df}}{=} a\|X$, assuming that $\alpha = \{a\}$. We will now apply the theory developed earlier in this chapter in order to construct a $\tilde{\sqsubseteq}$-seed for (5.11). It is easily seen that we obtain the following instance of the tree equations used to derive fixpoints of the auxiliary function **f**:

$$\begin{aligned}\mathcal{P} &= \{\, e^1_{\|} \lhd v^1_{\|} \lhd e_\alpha \,\} \cup \{\, e^2_{\|} \lhd v^2_{\|} \lhd q \mid q \in \mathcal{P} \,\}\\ \mathcal{R} &= \{\, x^1_{\|} \lhd v^1_{\|} \lhd x_\alpha \,\} \cup \{\, x^2_{\|} \lhd v^2_{\|} \lhd q \mid q \in \mathcal{R} \,\}\,.\end{aligned} \tag{5.12}$$

And there is a one-to-one correspondence between positive fixpoints of **f** and the solutions of (5.11). But now we have more than one fixpoint; for example, the minimal fixpoint containing only finite trees, and the maximal fixpoint which contains infinite trees as well. More precisely, we have $\mathsf{fix}_{\mathsf{min}} = (P, R)$ and $\mathsf{fix}_{\mathsf{max}} = (P \cup \{p\}, R \cup \{r\})$, where P and R ar two sets of finite trees:

$$P = \left\{\begin{array}{l} e^1_{\|} \lhd v^1_{\|} \lhd e_\alpha\ \ ,\\ e^2_{\|} \lhd v^2_{\|} \lhd e^1_{\|} \lhd v^1_{\|} \lhd e_\alpha\ \ ,\\ e^2_{\|} \lhd v^2_{\|} \lhd e^2_{\|} \lhd v^2_{\|} \lhd e^1_{\|} \lhd v^1_{\|} \lhd e_\alpha\ \ ,\ \ \ldots \end{array}\right\}$$

$$R = \left\{\begin{array}{l} x^1_{\|} \lhd v^1_{\|} \lhd x_\alpha\ \ ,\\ x^2_{\|} \lhd v^2_{\|} \lhd x^1_{\|} \lhd v^1_{\|} \lhd x_\alpha\ \ ,\\ x^2_{\|} \lhd v^2_{\|} \lhd x^2_{\|} \lhd v^2_{\|} \lhd x^1_{\|} \lhd v^1_{\|} \lhd x_\alpha\ \ ,\ \ \ldots \end{array}\right\}$$

and p and r are two infinite trees:

$$
\begin{aligned}
p &= e^2_{\parallel} \lhd v^2_{\parallel} \lhd e^2_{\parallel} \lhd v^2_{\parallel} \lhd e^2_{\parallel} \lhd v^2_{\parallel} \cdots \\
r &= x^2_{\parallel} \lhd v^2_{\parallel} \lhd x^2_{\parallel} \lhd v^2_{\parallel} \lhd x^2_{\parallel} \lhd v^2_{\parallel} \cdots .
\end{aligned}
\tag{5.13}
$$

In all, there are four positive fixpoints which determine four starting boxes to begin the approximation of a solution of (5.11):

$$
\begin{array}{lll}
\Sigma^{\mathsf{min}} & \text{using} & \mathsf{fix}_{\mathsf{min}} \\
\Sigma^{\mathsf{max:min}} & \text{using} & (P \cup \{p\}, R) \\
\Sigma^{\mathsf{min:max}} & \text{using} & (P, R \cup \{r\}) \\
\Sigma^{\mathsf{max}} & \text{using} & \mathsf{fix}_{\mathsf{max}} .
\end{array}
$$

Depending on the choice, one of the four different solutions of (5.11) is obtained, as shown in Fig. 5.3 (as before, we do not represent all transition and place names). It is not hard to check that all four boxes are indeed solutions of (5.11) in the same strict sense as before.

The interleaving behaviour – and also behaviour in terms of finite steps – of the four solutions is the same. However, if infinite steps were allowed, then the minimal solution, Σ^{min}, would have a terminating behaviour from the initial marking while the other three would not (since the initial token on an isolated entry place cannot be removed, and it is impossible to put a token onto an isolated exit place). Thus, for instance, if X was the first part of a sequential composition, then the second part of that composition could be fully executed using the minimal solution but not using the remaining three. Even more puzzling is that if we allowed infinite steps, then the initially marked $\Sigma^{\mathsf{max:min}}$ would not be clean, and no longer safe if we added the **redo** transition. We will return to this problem in Sect. 6.4.

5.4.2 Rear-unguardedness

In the rest of this section, we discuss four examples, though sometimes not to the same degree of detail as before. First, consider

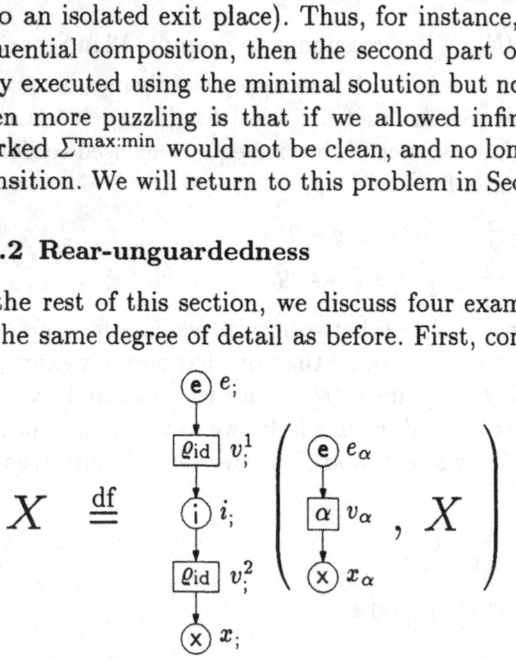

(5.14)

which is nothing but $X \stackrel{\mathrm{df}}{=} \Omega_;(\mathsf{N}_\alpha, X)$ and thus corresponds to example 13 in Sect. 3.1, $X \stackrel{\mathrm{df}}{=} a; X$, assuming that $\alpha = \{a\}$. This leads to the following tree equations:

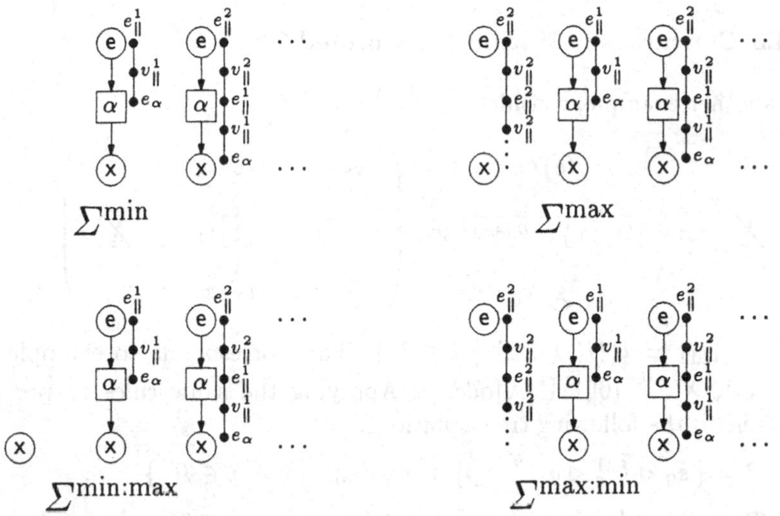

Fig. 5.3. The four solutions of (5.14)

$$\begin{aligned} \mathcal{P} &= \{\, e_{;} \lhd v_{;}^{1} \lhd e_{\alpha} \,\} \\ \mathcal{R} &= \{\, x_{;} \lhd v_{;}^{2} \lhd q \mid q \in \mathcal{R} \,\} . \end{aligned} \tag{5.15}$$

There are exactly two fixpoints for **f** in this case:

$$\begin{aligned} \mathsf{fix}_{\min} &= \left(\{ e_{;} \lhd v_{;}^{1} \lhd e_{\alpha} \} \,,\, \emptyset \right) \\ \mathsf{fix}_{\max} &= \left(\{ e_{;} \lhd v_{;}^{1} \lhd e_{a} \} \,,\, \{ x_{;} \lhd v_{;}^{2} \lhd x_{;} \lhd v_{;}^{2} \lhd x_{;} \lhd \cdots \} \right) . \end{aligned}$$

As we must start the approximation with a box, only the latter is allowed to determine the entry and exit places. When begun with the one-e-place one-x-place box defined by $\mathsf{fix}_{\max}$, the approximation leads to the only solution of (5.14) shown in Fig. 5.4.

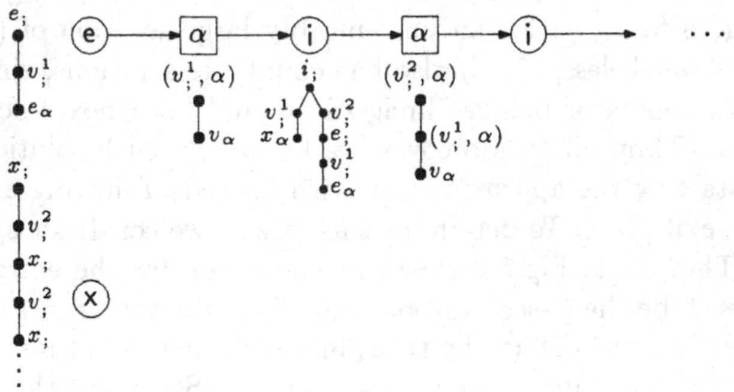

Fig. 5.4. The only solution of (5.14)

5.4.3 Concurrency Within Unbounded Choice

As another example, consider

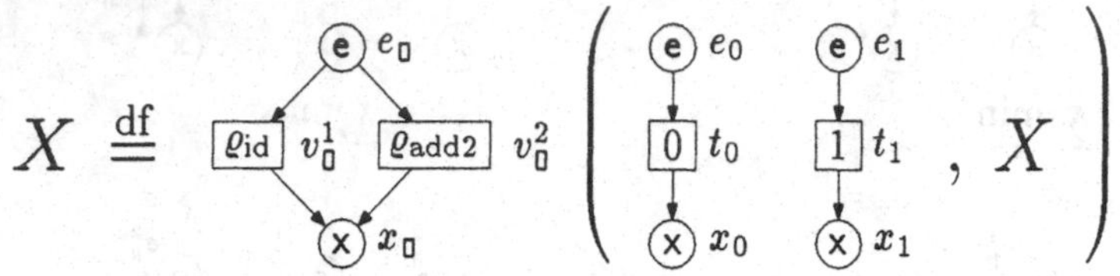

(5.16)

where $\varrho_{\mathsf{add2}} = \{(\{i\}, i+2) \mid i \in \mathbf{N}\}$. This corresponds to example 15 in Sect. 3.1, $X \stackrel{\mathrm{df}}{=} (0\|1)\,\square\,(X[add2])$. Applying the same rules as previously leads us to the following tree equations:

$$\begin{aligned} \mathcal{P} &= \{\, e_\square \lhd \{v_\square^1 \lhd q, v_\square^2 \lhd p\} \mid q \in \{e_0, e_1\} \wedge p \in \mathcal{P} \,\} \\ \mathcal{R} &= \{\, x_\square \lhd \{v_\square^1 \lhd q, v_\square^2 \lhd r\} \mid q \in \{x_0, x_1\} \wedge r \in \mathcal{R} \,\}\,. \end{aligned} \tag{5.17}$$

This case is interesting because, as it turns out, the maximal fixpoint of $\mathbf{f}$ has uncountably many trees. The generic trees for $\mathsf{fix}_{\mathsf{max}}$ are shown in Fig. 5.5 where $n_i \in \{e_0, e_1\}$ and $m_i \in \{x_0, x_1\}$, for every i, as explained in Sect. 5.2.5.

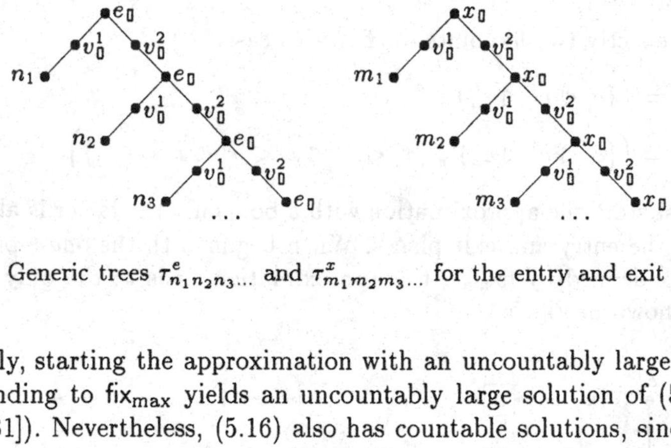

Fig. 5.5. Generic trees $\tau^e_{n_1 n_2 n_3 \ldots}$ and $\tau^x_{m_1 m_2 m_3 \ldots}$ for the entry and exit places

Clearly, starting the approximation with an uncountably large seed box corresponding to $\mathsf{fix}_{\mathsf{max}}$ yields an uncountably large solution of (5.16) (see also [6, 31]). Nevertheless, (5.16) also has countable solutions, since the relabellings in the operator box are image-finite and both boxes occurring in it are finite (see Thm. 5.3.4). Moreover, we know that such solutions can be obtained by starting the approximation with $\sqsubseteq$-seeds Υ having exactly one entry and one exit place. To determine such seeds, we construct spines as in the proof of Thm. 5.2.5; Fig. 5.6 shows two of them, for the entry and exit place of Υ. Let Υ be the $\sqsubseteq$-seed ex-box defined by the spines in Fig. 5.6, and Υ' be the $\sqsubseteq$-seed ex-box defined by the spines obtained from those in Fig. 5.6 by replacing each node label e_0 by e_1, and x_0 by x_1. Starting either with Υ or with Υ' yields a countable solution of (5.16). Due to Lemma 5.2.1(1), the two solutions have the same transitions and, according to the results presented in Chap. 7, also the same behaviour. But they are not only incomparable in the sense of $\sqsubseteq$ but also non-isomorphic: the first — but not the second — contains a place connected to all even-number-labelled but to no odd-number-labelled transitions, and the second — but not the first — contains a place connected to all odd-number-labelled but to no even-number-labelled transitions. This is despite the fact that at every step in the approximation, isomorphic nets are obtained (though with incompatible isomorphisms).

Exercise 5.4.1. Verify this and also show that, for $n \geq 0$,

$$(v^2_\square, 2 \cdot n) \lhd (v^2_\square, 2 \cdot n - 2) \lhd \cdots \lhd (v^2_\square, 2) \lhd (v^1_\square, 0) \lhd t_0$$
$$(v^2_\square, 2 \cdot n + 1) \lhd (v^2_\square, 2 \cdot n - 1) \lhd \cdots \lhd (v^2_\square, 3) \lhd (v^1_\square, 1) \lhd t_1$$

are two generic forms of transition trees in the solutions of (5.16).

Note also that the maximal solution cannot be obtained by starting with a countable seed (*a fortiori*, neither can it be obtained from a two-place seed).

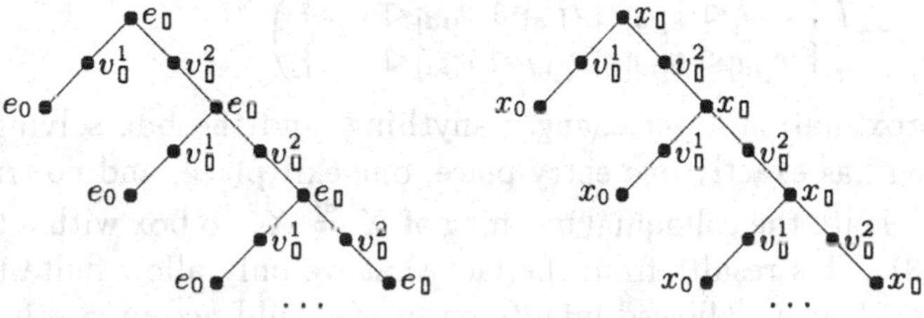

Fig. 5.6. Two spines for equation (5.17)

This example is of further interest since it has uncountably many different minimal solutions. For consider the trees of the maximal fixpoint, $\mathsf{fix}_{\max}$, shown in Fig. 5.5. It is easy to see that if we take any $P = \{\tau^e_{n_1 n_2 n_3 \ldots}\}$ and $R = \{\tau^x_{m_1 m_2 m_3 \ldots}\}$, then

$$\mathsf{sol}_{\min}(P, R) = \begin{pmatrix} \{ \tau^e_{l_1 \ldots l_k n_i n_{i+1} \ldots} & \mid i \geq 1 \wedge l_1, \ldots, l_k \in \{e_0, e_1\} \ \} \\ \{ \tau^x_{l_1 \ldots l_k m_i m_{i+1} \ldots} & \mid i \geq 1 \wedge l_1, \ldots, l_k \in \{x_0, x_1\} \} \end{pmatrix}$$

is a countable min/pos fixpoint of $\mathbf{f}$. Let Sol be the set of all such fixpoints.

Proposition 5.4.1. Sol is an uncountable set of min/pos fixpoints of $\mathbf{f}$.

Proof. We have

$$\mathsf{fix}_{\max} = \bigcup_{\mathsf{sol}_{\min}(P,R) \in \mathsf{Sol}} \mathsf{sol}_{\min}(P, R) .$$

Since $\mathsf{fix}_{\max}$ is uncountable and each $\mathsf{sol}_{\min}(P, R) \in \mathsf{Sol}$ countable, it follows that the set Sol is uncountable. $\square$

5.4.4 Extreme Unguardedness

As the next equation – which corresponds to example 16 in Sect. 3.1, $X \stackrel{\mathrm{df}}{=} X$ – consider the following:

$X \stackrel{\mathrm{df}}{=}$

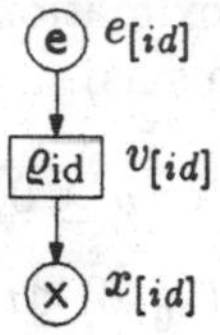

$\left(X \right)$. (5.18)

It leads to the following tree equations for the entry and exit places of the starting box:

$$\begin{aligned} \mathcal{P} &= \{ e_{[id]} \lhd v_{[id]} \lhd q \mid q \in \mathcal{P} \} \\ \mathcal{R} &= \{ x_{[id]} \lhd v_{[id]} \lhd q \mid q \in \mathcal{R} \} . \end{aligned} \tag{5.19}$$

Clearly, in such a case $\mathbf{f}$ has a unique positive fixpoint

$$\begin{pmatrix} \mathcal{P} \\ \mathcal{R} \end{pmatrix} = \begin{pmatrix} \{ e_{[id]} \lhd v_{[id]} \lhd e_{[id]} \lhd v_{[id]} \lhd \ldots \} \\ \{ x_{[id]} \lhd v_{[id]} \lhd x_{[id]} \lhd v_{[id]} \lhd \ldots \} \end{pmatrix} .$$

Hence approximation never changes anything, and the box solving the last net equation has exactly one entry place, one exit place, and no transitions. Note that despite the colloquial meaning of $X \stackrel{\mathrm{df}}{=} X$, no box with a transition solves (5.18). This results from the fact that we only allow finite transition names. Indeed, if we allowed infinite ones, we could accept a solution with transitions

$$(v_{[id]}, \alpha) \lhd (v_{[id]}, \alpha) \lhd (v_{[id]}, \alpha) \lhd \ldots$$

for any label α; the reader may verify that it would be faithfully reproduced by $\mathbf{F}$. Thus, solving recursive equation by boxes with the least behaviour is an in-built feature in our approach.

We feel that this gives an intuitive semantics to the 'recursive call' $X \stackrel{\mathrm{df}}{=} X$; in particular, we do not model procedure entry and procedure exit as separate (silent) actions. Thus, $X \stackrel{\mathrm{df}}{=} X$ essentially means that X is continually 'rewritten' by itself and effects no behaviour at all, be it silent or observable. This is quite in contrast to the view that $X \stackrel{\mathrm{df}}{=} X$ implies continual entry to a procedure and thus an infinite succession of silent moves; to model such behaviour, we would rather use $X \stackrel{\mathrm{df}}{=} \emptyset; X$ (or even $X \stackrel{\mathrm{df}}{=} \emptyset; X; \emptyset$, to model exit as well as entry, even though none of the transitions coming from the $\emptyset$ following X would be executable).

5.4.5 (Non)use of Empty Nets in the Limit Construction

Finally, consider the following equation:

$X \stackrel{\text{df}}{=}$ 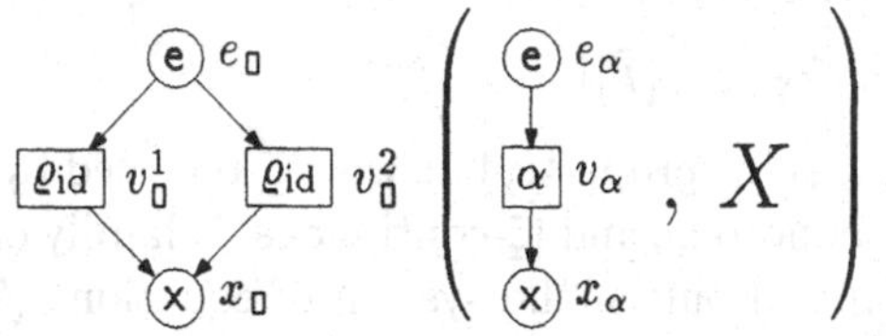

. (5.20)

That is, $X \stackrel{\text{df}}{=} \Omega_{\Box}(\mathsf{N}_\alpha, X)$, which corresponds to $X \stackrel{\text{df}}{=} a \,\Box\, X$, assuming that $\alpha = \{a\}$. Such an equation can be dealt with in a similar way as (5.17), and so we omit the details. The reason why we give this example is to show that the empty net is not suitable as a starting point of the approximation. For consider that the empty net were taken as a seed box. Refinement would work as usual, and we would obtain as a solution net containing infinitely many isolated α-labelled transitions. Such a solution would allow an arbitrarily large step of α-labelled transitions, which is both counterintuitive and inconsistent with the operational semantics of the choice construct given in Chap. 3. If the approximation is instead realised in the way we described, then we obtain as solution a choice between infinitely many α-labelled transitions, which is both intuitively correct and consistent with the operational semantics of $\Box$.

Exercise 5.4.2. Discuss examples 14 and 17 of Sect. 3.1.

5.5 Solving Systems of Recursive Equations

The results obtained for a single recursive equation are easily lifted to the general case, i.e., to the recursive system (5.2), using the standard techniques on Cartesian product spaces (for a finite set of equations) or on function spaces (for an infinite set of equations). Below we consider various $\mathcal{X}$*-families*[3] of boxes, sets, functions, etc. An $\mathcal{X}$-family will be denoted as $\check{\mathcal{F}}$, and its member corresponding to $X \in \mathcal{X}$ by $\mathcal{F}_X$. The inclusion orders $\sqsubseteq$ and $\widetilde{\sqsubseteq}$, and more generally any predicate $\sim$, will be extended to $\mathcal{X}$-families, by stipulating that $\check{\mathcal{F}} \sim \check{\mathcal{G}}$ if $\mathcal{F}_X \sim \mathcal{G}_X$, for every $X \in \mathcal{X}$. Similarly as for (5.3), we assume that all rec-boxes occurring in (5.2) have their transitions and places in, respectively,

$$\mathsf{T}_{\mathsf{root}} \setminus \bigcup_{X \in \mathcal{X}} T_{\Omega_X} \quad \text{and} \quad \mathsf{P}_{\mathsf{root}} \setminus \bigcup_{X \in \mathcal{X}} S_{\Omega_X} .$$

[3] Formally, an $\mathcal{X}$-family is a mapping from $\mathcal{X}$ to some set.

5.5.1 Approximations, Existence, and Uniqueness

A $\tilde{\sqsubseteq}$-seed for the system of equations (5.2) is an $\mathcal{X}$-family of rec-boxes $\check{\Upsilon}$ such that $\check{\Upsilon}\tilde{\sqsubseteq}\Theta$, where, for every $X \in \mathcal{X}$,

$$\Theta_X = \mathbf{F}_X(\check{\Upsilon}) = \Omega_X(\boldsymbol{\Delta}_X[\check{\Upsilon}]) .$$

A $\sqsubseteq$-seed satisfies the same property, but with $\tilde{\sqsubseteq}$ replaced by $\sqsubseteq$. As for a single equation, $\check{\mathbf{F}}$ is a $\sqsubseteq$-monotonic and $\tilde{\sqsubseteq}$-continuous $\mathcal{X}$-family of mappings.[4] And it is $\sqsubseteq$-continuous if and only if the system of equations (5.2) is $\mathcal{X}$-*finitary*, i.e., for every $X \in \mathcal{X}$ and every $s \in S_{\Omega_X}$, $\{v \in {}^\bullet s \cup s^\bullet \mid \Delta_{X,v} \in \mathcal{X}\}$ is a finite set.

If $\check{\Upsilon}$ is a $\tilde{\sqsubseteq}$-seed for (5.2) (or a $\sqsubseteq$-seed, if the system is $\mathcal{X}$-finitary), then $\bigsqcup \check{\mathbf{F}}^*(\check{\Upsilon})$ is a solution of this system of equations. For the same reasons as before, the transitions and finite prefixes of places of a solution are unique (for each $X \in \mathcal{X}$), and the solution generated from a $\tilde{\sqsubseteq}$-seed only depends on the entry and exit places of the latter. It is always possible to start from ex-boxes as seeds, and their entry/exit place sets have to satisfy the same kind of properties as for a single equation, but lifted to $\mathcal{X}$-families, $(\check{\mathcal{P}}, \check{\mathcal{R}}) = \check{\mathbf{f}}(\check{\mathcal{P}}, \check{\mathcal{R}})$ (or $(\check{\mathcal{P}}, \check{\mathcal{R}}) \subseteq \check{\mathbf{f}}(\check{\mathcal{P}}, \check{\mathcal{R}})$ for $\tilde{\sqsubseteq}$-seeds). Here, for all $X \in \mathcal{X}$, and all $\mathcal{X}$-families of place trees $\check{P}$ and $\check{R}$,

$$\mathbf{f}_X(\check{P}, \check{R}) = \left(\bigcup_{s \in {}^\circ\Omega_X} \mathbf{f}_{X,s}(\check{P}, \check{R}) \;,\; \bigcup_{s \in \Omega_X^\circ} \mathbf{f}_{X,s}(\check{P}, \check{R}) \right) ,$$

where $\mathbf{f}_{X,s}(\check{P}, \check{R})$ comprises all place trees $s \lhd (\{v \lhd p_v\}_{v \in s^\bullet} + \{w \lhd r_w\}_{w \in {}^\bullet s})$ such that

$$p_v \in \begin{cases} P_Y & \text{if } \Delta_{X,v} = Y \in \mathcal{X} \\ {}^\circ\Delta_{X,v} & \text{if } \Delta_{X,v} \notin \mathcal{X} \end{cases} \qquad r_w \in \begin{cases} R_Y & \text{if } \Delta_{X,w} = Y \in \mathcal{X} \\ \Delta_{X,w}^\circ & \text{if } \Delta_{X,w} \notin \mathcal{X} . \end{cases}$$

That there is always a unique positive maximal solution, and possibly many min/pos ones, arises from a generalisation of the construction of spine trees, τ_e and τ_x, in the proof of Thm. 5.2.5. We first build a labelled directed graph G in the following way.

Let us choose, for every $X \in \mathcal{X}$, an entry place $e_X \in {}^\circ\Omega_X$ and an exit place $x_X \in \Omega_X^\circ$, and for every $\Delta_{X,v} \notin \mathcal{X}$, an entry place $e_{X,v} \in {}^\circ\Delta_{X,v}$ and an exit place $x_{X,v} \in \Delta_{X,v}^\circ$. The set of nodes of G is given by

$$\begin{array}{ll} \{\varepsilon_X \mid X \in \mathcal{X}\} & \cup \\ \{\varepsilon_{X,v} \mid X \in \mathcal{X} \land v \in e_X^\bullet \cup x_X^\bullet\} & \cup \\ \{\varepsilon'_{X,v} \mid X \in \mathcal{X} \land v \in e_X^\bullet \cup x_X^\bullet \land \Delta_{X,v} \notin \mathcal{X}\} & \cup \\ \{\xi_X \mid X \in \mathcal{X}\} & \cup \\ \{\xi_{X,v} \mid X \in \mathcal{X} \land v \in {}^\bullet e_X \cup {}^\bullet x_X\} & \cup \\ \{\xi'_{X,v} \mid X \in \mathcal{X} \land v \in {}^\bullet e_X \cup {}^\bullet x_X \land \Delta_{X,v} \notin \mathcal{X}\} . & \end{array}$$

[4] Where the limit of an $\mathcal{X}$-family of $\sqsubseteq$-chains is defined component-wise (and is itself an $\mathcal{X}$-family).

Each node ε_X is labelled by e_X, each ξ_X by x_X, each $\varepsilon_{X,v}$ and $\xi_{X,v}$ by v, each $\varepsilon'_{X,v}$ by $e_{X,v}$ and each $\xi'_{X,v}$ by $x_{X,v}$. For every node ε_X there is an arc to $\varepsilon_{X,v}$ (for $v \in e_X^\bullet$) and to $\xi_{X,v}$ (for $v \in {}^\bullet e_X$). Similarly, for every node ξ_X there is an arc to $\varepsilon_{X,v}$ (for $v \in x_X^\bullet$) and to $\xi_{X,v}$ (for $v \in {}^\bullet x_X$). For every node $\varepsilon_{X,v}$ there is an arc to ε_Y if $\Delta_{X,v} = Y$ and to $\varepsilon'_{X,v}$ if $\Delta_{X,v} \notin \mathcal{X}$. Similarly, for every node $\xi_{X,v}$ there is an arc to ξ_Y if $\Delta_{X,v} = Y$ and to $\xi'_{X,v}$ if $\Delta_{X,v} \notin \mathcal{X}$. There are no arcs outgoing from the nodes $\varepsilon'_{X,v}$ and $\xi'_{X,v}$.

Figure 5.7(a) shows a possible graph G for the recursive system:

$$X \overset{\text{df}}{=} \Omega_{[*\rangle}(Y, X)$$
$$Y \overset{\text{df}}{=} \Omega_{\|}(\mathsf{N}_a, X) \,.$$

Here, we use the notation of Sect. 4.4 (in particular, Figs. 4.21 and 4.42).

Having defined the graph G, for every $X \in \mathcal{X}$, the spine tree $\tau_{e,X}$ (respectively, $\tau_{x,X}$) has as its nodes all finite directed paths ϕ in G starting from the node ε_X (respectively, ξ_X), with the label of the last visited node (of G) as its own label. Moreover, there is an arc from ϕ to ϕ' if and only if ϕ' is ϕ extended by exactly one arc and node θ (of G).

Exercise 5.5.1. Show that $(\{\breve{\tau}_{\mathsf{e}}\}, \{\breve{\tau}_{\mathsf{x}}\}) \subseteq \breve{\mathbf{f}}(\{\breve{\tau}_{\mathsf{e}}\}, \{\breve{\tau}_{\mathsf{x}}\})$.

Figure 5.7(b) shows the graph G which would give rise to the spine trees shown in Fig. 5.2.

Exercise 5.5.2. Show that for $\mathcal{X} = \{X\}$ the two different ways of defining spine trees (i.e., that in the proof of Thm. 5.2.5, and the one just described) are equivalent.

5.5.2 A Closed Form of the Maximal Solution

The closed form of the maximal solution given in Sect. 5.2.5 may be generalised for the recursive system of equations (5.2), by constructing its maximal solution, $\breve{\Sigma}^{\max}$. To simplify the presentation, but without loss of generality, we shall assume that the sets of transitions of the operator boxes $\breve{\Omega}$ in (5.2) are disjoint; thus, with each transition v in $\breve{\Omega}$ we can unambiguously associate $X(v) \in \mathcal{X}$ such that v is a transition in $\Omega_{X(v)}$. Moreover, for all $X, Y \in \mathcal{X}$,

$$V_X \quad = \{v \in T_{\Omega_X} \mid \Delta_{X,v} \notin \mathcal{X}\}$$
$$U_{X,Y} = \{v \in T_{\Omega_X} \mid \Delta_{X,v} = Y\} \,.$$

We now give the definition of $\Sigma_X^{\max}$, for $X \in \mathcal{X}$.

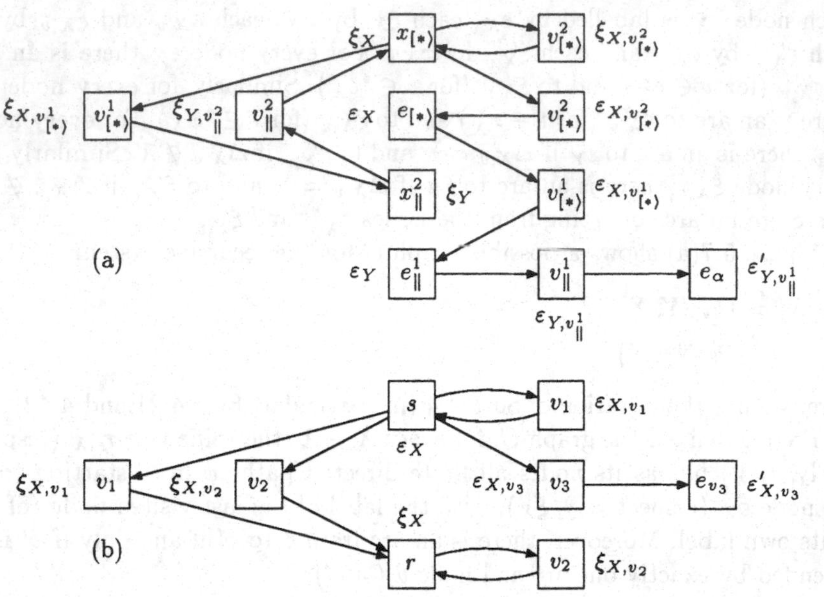

Fig. 5.7. Graphs generating spine trees

Transitions. The set of transitions of $\Sigma_X^{\max}$ comprises all transition names with at least two nodes such that the root is labelled by a pair in $T_{\Omega_X} \times \mathsf{Lab}$, and for every nonleaf node ξ, the subtree appended at ξ has one of the following two forms:

- $\mathsf{subtr}(\xi) = (v, \alpha) \lhd R$
  where $v \in V_Y$ and $R \in \mathsf{mult}(T_{\Delta_{Y,v}})$ and $(\lambda_{\Delta_{Y,v}}(R), \alpha) \in \lambda_{\Omega_Y}(v)$, for some $Y \in \mathcal{X}$.
- $\mathsf{subtr}(\xi) = (v, \alpha) \lhd \{(v_1, \alpha_1) \lhd R_1, \ldots, (v_n, \alpha_n) \lhd R_n\}$
  where $v \in U_{Y,Z}$ and $\{v_1, \ldots, v_n\} \subseteq T_{\Omega_Z}$ and $(\{\alpha_1, \ldots, \alpha_n\}, \alpha) \in \lambda_{\Omega_Y}(v)$, for some $Y, Z \in \mathcal{X}$.

The label of a transition with (v, α)-labelled root is α. Note that the multisets of transition tree names R_i, for $i \in \{1, \ldots, n\}$, are constrained by the fact that each (v_i, α_i) is itself the label of a nonleaf node, so that the above condition can be applied again.

Places. There are three types of places in $\Sigma_X^{\max}$. The first type comprises all finite and infinite place names τ such that the label of the root is a place r in Ω_X, and if a node ξ has the label $s \in S_{\Omega_Y}$, for some $Y \in \mathcal{X}$, then

$$\mathsf{subtr}(\xi) \;=\; s \lhd \Big(\{v \lhd p_v\}_{v \in {}^\bullet s} + \{w \lhd r_w\}_{w \in s^\bullet}\Big)$$

and the following are satisfied:

- if $v \in U_{Y,Z}$, then $p_v = x_v \lhd p'_v$ where $x_v \in \Omega_Z^\circ$;

- if $w \in U_{Y,Z}$, then $r_w = e_w \lhd r'_w$ where $e_w \in {}^\circ\Omega_Z$;
- if $v \in V_Y$, then $p_v \in \Delta^\circ_{Y,v}$;
- if $w \in V_Y$, then $r_w \in {}^\circ\Delta_{Y,w}$.

The label of τ is $\lambda_{\Omega_X}(r)$. Note that the place tree names p'_v and r'_v are constrained by the fact that x_v and e_w are nodes in operator boxes, so that the above condition can be applied again.

The second type of places, which are internal, are all place names

$$v_1 \lhd v_2 \lhd \cdots \lhd v_n \lhd \tau$$

such that $n \geq 1$, $X(v_1) = X$, $v_{j+1} \in U_{X(v_j),X(v_{j+1})}$ for $j = 1,\ldots,n-1$; $v_n \in U_{X(v_n),Z}$, and τ is an internal place of the first type in $\Sigma_Z^{\max}$.

The third type of places, which are again internal, comprises all place names

$$v_1 \lhd v_2 \lhd \cdots \lhd v_n \lhd i$$

such that $n \geq 1$, $X(v_1) = X$, $v_{j+1} \in U_{X(v_j),X(v_{j+1})}$ for $j = 1,\ldots,n-1$, $v_n \in V_{X(v_n)}$, and $i \in \ddot{\Delta}_{X(v_n),v_n}$.

Weight Function. The connectivity in $\Sigma_X^{\max}$ is directly driven by the structure of a place $p \in S_{\Sigma_X^{\max}}$ and transition $u \in T_{\Sigma_X^{\max}}$. Using trails of name trees[5] introduced in Sect. 5.2.5, we have the following:

$$W_{\Sigma_X^{\max}}(p,u) = \sum_{\substack{\sigma,v,s,t \text{ such that} \\ \sigma vt \in \mathsf{trails}(u),\, \sigma vs \in \mathsf{trails}(p)}} \mathsf{trails}(u)(\sigma vt) \cdot \mathsf{trails}(p)(\sigma vs) \cdot W_{\Delta_{X(v),v}}(s,t) \quad (5.21)$$

$$W_{\Sigma_X^{\max}}(u,p) = \sum_{\substack{\sigma,v,s,t \text{ such that} \\ \sigma vt \in \mathsf{trails}(u),\, \sigma vs \in \mathsf{trails}(p)}} \mathsf{trails}(u)(\sigma vt) \cdot \mathsf{trails}(p)(\sigma vs) \cdot W_{\Delta_{X(v),v}}(t,s)\,. \quad (5.22)$$

Exercise 5.5.3. Show that $\Sigma_X^{\max}$ is the maximal solution of (5.2).

5.5.3 Guarded Systems

The notion of guardedness introduced for a single equation in Sect. 5.2.4 may be lifted to a system of recursive equations. The system (5.2) is

- *front-guarded* if $\Delta_{X,v} \notin \mathcal{X}$, for all $X \in \mathcal{X}$ and $v \in {}^\bullet({}^\circ\Omega_X) \cup ({}^\circ\Omega_X)^\bullet$;
- *rear-guarded* if $\Delta_{X,v} \notin \mathcal{X}$, for all $X \in \mathcal{X}$ and $v \in {}^\bullet(\Omega_X^\circ) \cup (\Omega_X^\circ)^\bullet$;
- *guarded* if it is both front- and rear-guarded.

Proposition 5.5.1. If (5.2) is front-guarded, then, in every solution, all entry places (treated as trees) have the depth of at most two. The same is true of the exit places, if the system is rear-guarded. And, each place tree has a finite depth, if the system is guarded.

---

[5] Note that trails have been defined purely on the basis of the shape of name trees and are thus still applicable.

Proof. Follows immediately from the form of place trees of the maximal solution. □

The last result may be further extended in the following way. Let us first construct a graph whose nodes are ε_X and ξ_X, for all $X \in \mathcal{X}$. There are the following arcs:

- $\varepsilon_X \to \varepsilon_Y$ if $U_{X,Y} \cap ({}^\circ\Omega_X)^\bullet \neq \emptyset$;
- $\varepsilon_X \to \xi_Y$ if $U_{X,Y} \cap {}^\bullet({}^\circ\Omega_X) \neq \emptyset$;
- $\xi_X \to \varepsilon_Y$ if $U_{X,Y} \cap (\Omega_X^\circ)^\bullet \neq \emptyset$;
- $\xi_X \to \xi_Y$ if $U_{X,Y} \cap {}^\bullet(\Omega_X^\circ) \neq \emptyset$;

where, as in Sect. 5.5.2, $U_{X,Y} = \{v \in T_{\Omega_X} \mid \Delta_{X,v} = Y\}$. Then (5.2) is *n-front-guarded* if all directed paths in this graph from the nodes ε_X have the length of at most n (in terms of nodes), so that front-guardedness amounts to 1-front-guardedness. Similarly, (5.2) is *n-rear-guarded* if all directed paths in this graph from the nodes ξ_X have the length of at most n, so that rear-guardedness amounts to 1-rear-guardedness. Finally, (5.2) is *n-guarded* if it is both n-front-guarded and n-rear-guarded.

Proposition 5.5.2. If (5.2) is n-front-guarded, then in every solution, all entry places have the depth of at most $2 \cdot n$. The same is true of the exit places, if the system is n-rear-guarded. And, each place tree has a finite depth, if the system is n-guarded.

Proof. Again, this follows immediately from the form of the place trees of the maximal solution. □

As a consequence, if the system of recursive equations (5.2) is n-guarded and $\mathcal{X}$-finitary, all relabellings appearing in it are image-countable, and all boxes are countable, then all its solutions are countable (i.e., each box which belongs to an $\mathcal{X}$-family of rec-boxes solving the system of equations is countable).

Exercise 5.5.4. Show this.

5.6 Literature and Background

For an introductory text on ordinals, the reader is referred to [98].

A treatment of (unbounded) recursion in terms of high-level nets can be found in [53]. A different way of treating recursion in nets is described in [64].

The literature is divided over the treatment of unboundedly recursive calls such as $X \stackrel{\mathrm{df}}{=} X$. For instance, [2] agrees with our interpretation; other authors [16] advocate, e.g., that $X \stackrel{\mathrm{df}}{=} X$ generates an infinite sequence of silent steps (modelling continuous entry to the procedure).

6. S-invariants

In Chap. 4 we stated that our main objects of interest in the whole class of labelled nets were static and dynamic (plain and operator) boxes. Their definitions have a strong behavioural flavour due to being based on the safeness and cleanness properties, in addition to more structural conditions related to the graph of a net and its labelling. As a result, checking that a specific marked or unmarked box is dynamic or static may require a considerable amount of work, especially if the net is infinite. But, as we shall see, structural analysis techniques developed for Petri nets can often be of help. In particular, a variant of S-invariant analysis lends itself to a characterisation of the property of being a static or dynamic box, as well as other behavioural properties of boxes.

The theory we shall develop in this chapter is general, in fact much more general than really needed for our specific aim. We shall adopt this general setup for the following reasons. First, the overhead required to achieve this generality is relatively low when compared with the amount of work needed to obtain a more restricted framework. Second, the general setup allows a better understanding of the phenomena underlying S-invariant analysis of compositionally defined nets. And, last but not least, it is likely that in the future development of the theory, one would need some parts of this general setup.

We shall first define S-invariants and see how they are related to safeness, boundedness, absence of concurrency and auto-concurrency, cleanness, and isomorphism-based net equivalences. Then we shall consider the synthesis problem for net refinement, i.e., we shall investigate the construction of an S-invariant of a refined net from S-invariants of the operator and operand boxes. It will then be possible to answer one of the questions left pending in Chap. 4, i.e., when it is possible to guarantee that the result of net refinement is a static or dynamic box. Afterwards, we consider similar issues for nets obtained as maximal solutions of systems of recursive expressions. Finally, we shall introduce a partial order semantics of Petri nets, and investigate infinite occurrence nets of boxes defined through net refinement and recursive systems of net equations.

6.1 S-invariants, S-components, and S-aggregates

In this chapter we will use the set of non-negative real numbers extended by infinity, and operations thereon. The set is defined as $\mathsf{Re} = \Re^+ \cup \{\infty\}$, where $\Re^+$ denotes the set of non-negative reals. We define two binary operations on Re, viz. addition (denoted by $+$) and multiplication (denoted by $\cdot$). Both are extensions of the corresponding operations on reals, satisfying

$$\infty + r = r + \infty = \infty \cdot r_0 = r_0 \cdot \infty = \infty \,,$$

for all $r \in \mathsf{Re}$ and $r_0 \in \mathsf{Re}\setminus\{0\}$, and $0\cdot\infty = \infty\cdot 0 = 0$. The standard arithmetic orderings, $<$ and $\leq$, are lifted to Re by setting $r < \infty$ and $r' \leq \infty$, for all $r \in \Re^+$ and $r' \in \mathsf{Re}$. The following properties of the standard multiplication and addition carry over to Re: commutativity and associativity of both operators, and distributivity of multiplication over addition.

Generalised Sums. Let Z be a set and $\theta : Z \to \mathsf{Re}$ be a mapping. A *generalised sum*

$$\sum_{z \in Z} \theta(z)$$

is the element of Re defined by

$$\sum_{z\in Z} \theta(z) \;=\; \begin{cases} \infty & \text{if } \infty \in \theta(Z) \text{ or } Z' \text{ is uncountable} \\ \displaystyle\sum_{z\in Z'} \theta(z) & \text{otherwise ,} \end{cases}$$

where $Z' = Z \setminus \theta^{-1}(0) = \{z \in Z \mid \theta(z) > 0\}$ and

$$\sum_{z\in Z'} \theta(z) \;=\; \begin{cases} \displaystyle\sum_{i=1}^{n} \theta(z_i) & \text{if } \; Z' = \{z_1,\ldots,z_n\} \;\; (n \geq 0) \\ \displaystyle\sum_{i=1}^{\infty} \theta(z_i) & \text{if } \; Z' = \{z_1, z_2, \ldots\} \,. \end{cases}$$

In both cases, the series are defined in the standard way. If $n = 0$, then $\sum_{i=1}^{n} \theta(z_i) = 0$, and $\sum_{z\in Z'} \theta(z)$ is well defined for an infinite Z' since all $\theta(z_i)$'s are non-negative reals, hence the value of $\sum_{i=1}^{\infty} \theta(z_i)$ does not depend on the particular order in which the elements of Z' were enumerated (the series is absolutely converging or diverging).

Exercise 6.1.1. Show that if Z is the disjoint union of a family of sets $\{Z_y\}_{y\in Y}$, and $\theta : Z \to \mathsf{Re}$, then

$$\sum_{z\in Z} \theta(z) \;=\; \sum_{y\in Y} \sum_{z\in Z_y} \theta(z) \,, \tag{6.1}$$

where we follow the usual conventions for denoting nested applications of generalised sums, i.e., the right-hand side of (6.1) is

$$\sum_{y \in Y} \theta'(y) \quad \text{with} \quad \theta' : Y \to \mathsf{Re} \quad \text{defined by} \quad \theta'(y) = \sum_{z \in Z_y} \theta(z) \,.$$

Generalised sums enjoy several properties which we will find useful in the arithmetic manipulations throughout the rest of this chapter:

$$\sum_{z \in Z} \Big(\theta(z) + \theta'(z)\Big) = \sum_{z \in Z} \theta(z) + \sum_{z \in Z} \theta'(z) \tag{6.2}$$

$$\sum_{z \in Z} r \cdot \theta(z) = r \cdot \sum_{z \in Z} \theta(z) \tag{6.3}$$

$$\sum_{z \in Z \cup Z'} \theta(z) = \sum_{z \in Z} \theta(z) + \sum_{z \in Z'} \theta(z) \quad (Z \cap Z' = \emptyset) \tag{6.4}$$

$$\sum_{z \in Z} \sum_{z' \in Z'} \theta(z) \cdot \theta'(z') = \left(\sum_{z \in Z} \theta(z)\right) \cdot \left(\sum_{z' \in Z'} \theta'(z')\right) . \tag{6.5}$$

Exercise 6.1.2. Show that the above equalities are satisfied.

Generalised Products. Let Z be a set and $\theta : Z \to \mathsf{Re}$ be a mapping. A *generalised product*

$$\prod_{z \in Z} \theta(z)$$

is defined if $0 \in \theta(Z)$, or if $Z' = Z \setminus \theta^{-1}(\{0,1\})$ is a countable set and then also $\theta^{-1}((0,1))$ or $\theta^{-1}((1,\infty])$ is a finite set. The value of the generalised product is defined by

$$\prod_{z \in Z} \theta(z) \;=\; \begin{cases} 0 & \text{if } 0 \in \theta(Z) \\ \displaystyle\prod_{z \in Z'} \theta(z) & \text{otherwise} \,, \end{cases}$$

where

$$\prod_{z \in Z'} \theta(z) \;=\; \begin{cases} \displaystyle\prod_{i=1}^{n} \theta(z_i) & \text{if} \;\; Z' = \{z_1, \ldots, z_n\} \;\; (n \geq 0) \\ \displaystyle\prod_{i=1}^{\infty} \theta(z_i) & \text{if} \;\; Z' = \{z_1, z_2, \ldots\} \,. \end{cases}$$

In both cases, the products are defined in the standard way. If $n = 0$, then $\prod_{i=1}^{n} \theta(z_i) = 1$, and $\prod_{z \in Z'} \theta(z)$ is well defined for an infinite Z' since all the $\theta(z_i)$'s are then positive reals and almost all (that is, all but finitely many) $\theta(z_i)$'s are on the same side of 1, hence the value of $\prod_{i=1}^{\infty} \theta(z_i)$ does not depend on the particular order in which the elements of Z' were enumerated (the product is absolutely converging or diverging).

6.1.1 S-invariants

In the literature (see [20, 72, 88, 90]), an S-invariant is defined as a weighting of the places of a (finite) Petri net such that the global weights of net markings are preserved by the occurrence rule, and hence by possible evolutions of the net. Here we allow infinite nets, so care is needed in order to ensure meaningful operations on weights of possibly infinite, even uncountable, sets of places. This was the reason why we explicitly introduced the set Re and the operation of generalised sum.

An *S-invariant* of a labelled net Σ is a mapping $\theta : S_\Sigma \cup T_\Sigma \to \Re^+$ such that, for every transition $t \in T_\Sigma$,

$$\sum_{s\in S_\Sigma} W_\Sigma(s,t)\cdot\theta(s) \;=\; \theta(t) \;=\; \sum_{s\in S_\Sigma} W_\Sigma(t,s)\cdot\theta(s)\,. \tag{6.6}$$

A node z of Σ is *covered* by θ if $\theta(z) > 0$. Given a class of S-invariants $\mathcal{J}$, Σ is said to be *S-covered* (*T-covered*) by $\mathcal{J}$, if for each place (transition) of Σ there is an S-invariant in $\mathcal{J}$ covering it. The mapping θ satisfying $\theta(S_\Sigma \cup T_\Sigma) = \{0\}$ is a *trivial* S-invariant. Figure 6.1 shows three different nontrivial S-invariants of the labelled net $\Sigma_{\mathbf{bLs}}$ (this box has already been shown in Fig. 4.37 as the result of a compositional translation of a PBC expression); we shall refer to them as the 'bold', 'large', and 'small' S-invariants of $\Sigma_{\mathbf{bLs}}$.

Exercise 6.1.3. Show that if a labelled net is S-covered by a class of S-invariants $\mathcal{J}$, then it is also T-covered by $\mathcal{J}$. The converse, however, does not hold in general (see example in Fig. 6.2).

An S-invariant θ may be extended to real and complex markings of Σ in the following way:

$$\theta(M) = \sum_{s\in S_\Sigma} M(s)\cdot\theta(s)$$

$$\theta(M,Q) = \sum_{s\in S_\Sigma} M(s)\cdot\theta(s) \;+\; \sum_{t\in T_\Sigma} Q(t)\cdot\theta(t)\,.$$

Note that, unlike $\theta(s)$ and $\theta(t)$, the value of $\theta(M)$ and $\theta(M,Q)$ can be equal to ∞. In this chapter, we will consider labelled nets Σ with both real and complex markings; in particular, M_Σ may be such a marking. We will call $\theta(M_\Sigma)$ the *weight* of θ.

The definition of an S-invariant means that the number of tokens, weighted by θ, consumed by any transition, is finite and the same as that produced by it. This invariance property is the cornerstone of the invariant analysis in Petri net theory.

Proposition 6.1.1. Let θ be an S-invariant of a labelled net Σ. If $\mathcal{M}$ is a real or complex marking of Σ and $\mathcal{M}'$ is reachable from $\mathcal{M}$, then $\theta(\mathcal{M}) = \theta(\mathcal{M}')$.

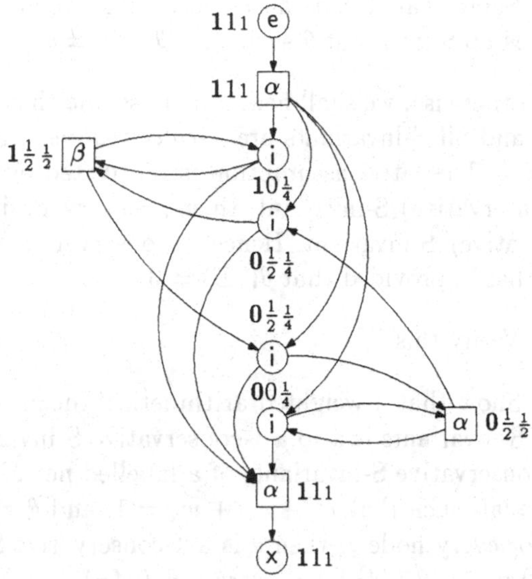

Fig. 6.1. Three S-invariants of Σ_{bLs}

Conservative and 1-conservative S-invariants. An S-invariant θ of an ex-restricted labelled net Σ will be called *conservative* if $\theta({}^\circ\Sigma) = \theta(\Sigma^\circ) < \infty$. The interest in such a condition derives from the observation that every conservative S-invariant θ is an S-invariant of the net Σ_{sr} with $\theta(\mathsf{redo}) = \theta(\mathsf{skip}) = \theta({}^\circ\Sigma) = \theta(\Sigma^\circ)$; and conversely, if θ is an S-invariant of Σ_{sr}, then it yields a conservative S-invariant when restricted to Σ.

Exercise 6.1.4. Verify this.

Exercise 6.1.5. Show that the sum of finitely many (conservative) S-invariants is also a (conservative) S-invariant; and that the same is true for infinite sets of S-invariants provided that we do not introduce diverging series. That is, show that if $\{\theta_h\}_{h\in\mathcal{H}}$ is an indexed set of S-invariants and

$$\theta(z) = \sum_{h\in\mathcal{H}} \theta_h(z) < \infty\,,$$

for every node z, then θ is an S-invariant. Moreover, show that θ is also conservative, if each θ_h is conservative and

$$\sum_{h\in\mathcal{H}} \theta_h({}^\circ\Sigma) < \infty\,.$$

We further observe that if ${}^\circ\Sigma$ and Σ° have the same weight (not necessarily finite), then it may be possible for Σ to evolve from the entry marking to the exit marking; though it is not guaranteed that this evolution is really possible.

Exercise 6.1.6. Show that if Σ is an ex-restricted labelled net and $\Sigma^\circ \in [{}^\circ\Sigma\rangle$, then there is no S-invariant θ such that $\theta({}^\circ\Sigma) \neq \theta(\Sigma^\circ)$.

Unless stated otherwise, we shall henceforth assume that all labelled nets are ex-restricted and all S-invariants are *1-conservative*, which means that $\theta({}^\circ\Sigma) = \theta(\Sigma^\circ) = 1$. The latter assumption is motivated by the observation that if θ is a (conservative) S-invariant, then for every positive real r, $r \cdot \theta$ is also a (conservative) S-invariant. Hence a conservative S-invariant may always be 'normalised', provided that $\theta({}^\circ\Sigma) \neq 0$.

Exercise 6.1.7. Verify this.

Exercise 6.1.8. Show that a weighted arithmetical mean of a finite family of 1-conservative S-invariants is also a 1-conservative S-invariant. That is, if $\theta_1, \ldots, \theta_n$ are 1-conservative S-invariants of a labelled net Σ and $w_1, \ldots, w_n$ are non-negative reals such that $w_1 + \cdots + w_n = 1$, and $\theta(z) = w_1 \cdot \theta_1(z) + \cdots + w_n \cdot \theta_n(z)$ for every node z, then θ is a 1-conservative S-invariant with the weight $\theta(M_\Sigma) = w_1 \cdot \theta_1(M_\Sigma) + \cdots + w_n \cdot \theta_n(M_\Sigma)$.

The three S-invariants indicated in Fig. 6.1 are all 1-conservative. Trivial S-invariants are conservative but not 1-conservative. T-covering by 1-conservative S-invariants does not guarantee that an S-covering by 1-conservative S-invariants exists, as shown in Fig. 6.2, where the only 1-conservative S-invariant is indicated.

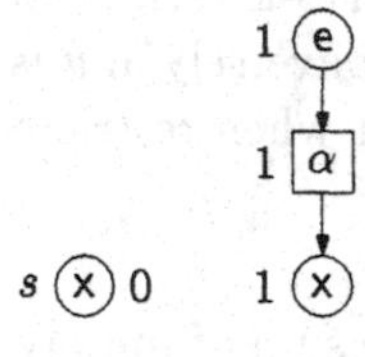

Fig. 6.2. A net T-covered but not S-covered by 1-conservative S-invariants

S-invariants and Marking Reachability. Let us consider a labelled net Σ with a 1-conservative S-invariant θ and a nonempty initial real marking M_0. If M_0 is reachable from ${}^\circ\Sigma$ or Σ°, or if ${}^\circ\Sigma$ or Σ° is reachable from M_0, then we know that $\theta(M_0) = 1$. Otherwise, M_0 may in principle have any weight; however, the case $\theta(M_0) = 1$ is especially interesting. Indeed, if we delete some of the transitions and/or unmarked isolated internal places, then the restriction of θ to the remaining places and transitions is still an S-invariant of the resulting net, with the same weight, and the same weighting of the entry and exit markings, hence the same conservation properties. For example, the marking in $\mathsf{box}_{\mathrm{PBC}}((a; \overline{b})\ \mathsf{rs}\ a)$ is unreachable from the entry marking, but it arises from $\mathsf{box}_{\mathrm{PBC}}(a; \overline{b})$, where the corresponding marking is reachable, after dropping the $\{a\}$-labelled transition. In general, if $\theta(M_0) = 1$, then it may be the case that the net could be obtained from another net, say Θ, where M_0 is reachable from ${}^\circ\Theta$ and Θ° is reachable from M_0, but from which some transitions have been removed (for instance, through restriction) to yield Σ; this can never be the case if $\theta(M_0) \neq 1$.

Indeed, if $\theta(M_0) = 1$, one can imagine an instance of the situation just described with a single extra transition, t_{new}, whose suppression led from Θ to Σ, such that ${}^\bullet t_{\mathrm{new}} = {}^\circ\Theta = {}^\circ\Sigma$, all the arcs incoming to t_{new} are unitary, and for every place s in Θ (and so in Σ), $W_\Theta(t_{\mathrm{new}}, s) = M_0(s)$. Then M_0 is reachable from the entry marking in Θ, and θ extended with $\theta(t) = 1$ is a 1-conservative S-invariant of weight 1 in Θ.

This explains why, in general, for a labelled net Σ with a family $\{\theta_h\}_{h\in\mathcal{H}}$ of 1-conservative S-invariants and a nonempty real initial marking M_0, we will require that $\theta_h(M_0) = 1$, for every $h \in \mathcal{H}$.

Sometimes we will also consider the case when the weight of the initial marking is an arbitrary natural number, $\theta(M_0) \in \mathbf{N}$. This includes the unmarked case, $\theta(M_0) = 0$, the default case, $\theta(M_0) = 1$, as well as other cases, like that of the initial marking $n \cdot {}^\circ\Sigma$, i.e., when n tokens are placed in each entry place, and none elsewhere. The case $\theta(M_0) = n$ may be seen as one where M_0 may be reached from $n \cdot {}^\circ\Sigma$. Notice that if θ is a 1-conservative S-invariant of Σ of weight n, then it is also an S-invariant of weight 0 for the net $\lfloor\Sigma\rfloor$, of weight 1 for $\overline{\Sigma}$ and $\underline{\Sigma}$, and of weight m for Σ with the marking $m \cdot {}^\circ\Sigma$ or $m \cdot \Sigma^\circ$.

Exercise 6.1.9. Verify this.

S-invariants and Net Behaviour. Allowing only S-invariants which have values in non-negative real numbers results in the monotonicity of marking weights.

Proposition 6.1.2. Let θ be an S-invariant of a labelled net Σ. If $\mathcal{M}$ and $\mathcal{M}'$ are two complex markings of Σ such that $\mathcal{M} \subseteq \mathcal{M}'$, then $\theta(\mathcal{M}) \leq \theta(\mathcal{M}')$. *Note:* $\mathcal{M} = (M, Q) \subseteq \mathcal{M}' = (M', Q')$ if $M \subseteq M'$ and $Q \subseteq Q'$.

Despite being straightforward, such a property has direct implications on the boundedness, safeness, cleanness, and concurrency of nets. When combined with the existence of coverings by S-invariants, it yields a number of behavioural characteristics.

Proposition 6.1.3. Let Σ be a labelled net, and θ be an S-invariant of weight n.

(1) If a place s is covered by θ, then for all real and complex markings reachable from M_Σ,
   - s never holds more than $\left\lfloor \frac{n}{\theta(s)} \right\rfloor$ tokens.

   Note: Hence s is structurally bounded [83].

- If $\theta(s) > n$, then s is never marked and no transition connected to it is ever enabled.
- If $n \geq \theta(s) > \frac{n}{2}$, then s is safe and no transition connected to it is auto-concurrent.

(2) If a transition t is covered by θ, then for all real and complex markings reachable from M_Σ,
- t may occur at most $\left\lfloor \frac{n}{\theta(t)} \right\rfloor$ times concurrently with itself.
  Note: This amounts to saying that if (M, Q) is a complex marking reachable from M_Σ, then $Q(t) \leq \left\lfloor \frac{n}{\theta(t)} \right\rfloor$. Such a property can be interpreted as a form of structural boundedness.
- If $\theta(t) > n$, then t can never be enabled. And, more generally, no step of transitions U can ever be enabled if $\sum_{u \in U} U(u) \cdot \theta(u) > n$.
- If $n \geq \theta(t) > \frac{n}{2}$, then t is never auto-concurrent.

Proof. Follows immediately from Props. 6.1.1 and 6.1.2. □

Proposition 6.1.4. Let Σ be a labelled net.

(1) If for every place $s \in S_\Sigma$ there is an S-invariant θ of weight n such that $n \geq \theta(s) > \frac{n}{2}$, then Σ is safe and exhibits no auto-concurrency.
(2) If for every transition $t \in T_\Sigma$ there is an S-invariant θ of weight n such that $n \geq \theta(t) > \frac{n}{2}$, then Σ exhibits no auto-concurrency.
(3) If Σ is S-covered by 1-conservative S-invariants of weight 1, then Σ is clean.

Proof. The first two parts follow immediately from Prop. 6.1.3. The third part follows from the observation that if $s \in S_\Sigma$ is covered by a 1-conservative S-invariant θ, then $\theta(\Sigma^\circ + \{s\}) > 1$, so that no marking containing $\Sigma^\circ + \{s\}$ is reachable, and similarly for ${}^\circ\Sigma + \{s\}$. □

The last two propositions will typically be used for $n = 1$.

S-invariants and Structural Net Equivalences. If Σ and Θ are isomorphic labelled nets, then they have up to isomorphism the same S-invariants. If two transitions t and t' in a net duplicate each other (more generally, if they have the same connectivity, as their labels are not important here), then for every S-invariant θ, it follows that $\theta(t) = \theta(t')$. Hence, if two unmarked nets are iso_T-equivalent, then they have the same S-invariants, up to isomorphism and transition duplication. The same is true for labelled nets marked with real markings. The case of iso_S-equivalence is more complicated. For consider a set $S' \subseteq S$ of duplicate places and an S-invariant θ. It is then possible to freely redistribute the weights inside S' without destroying the S-invariance and 1-conservativeness properties, provided that the total weight of S' and the weight of the initial marking on S' remain unchanged. Hence two iso_S-equivalent nets will have the same S-invariants up to isomorphism and redistribution of weights inside each duplication equivalence class. The case of the $\mathsf{iso}_{\mathsf{ST}}$-equivalence combines the discussion of both $\mathsf{iso}_{\mathsf{T}}$-equivalence and $\mathsf{iso}_{\mathsf{S}}$-equivalence. Finally, if a transition t in a labelled net is equivalent in terms of connectivity to a step U, i.e., for all $s \in S$,

$$W(s,t) = \sum_{u \in U} U(u) \cdot W(s,u) \quad \text{and} \quad W(t,s) = \sum_{u \in U} U(u) \cdot W(u,s) ,$$

then for every S-invariant θ, we have that

$$\theta(t) = \sum_{u \in U} U(u) \cdot \theta(u) .$$

Hence, if two labelled nets with real markings are step-equivalent, then they have the same S-invariants, up to isomorphism, transition duplication, and transition grouping.

6.1.2 S-components

An *S-component* of a labelled net Σ is a 1-conservative S-invariant θ such that $\theta(S_\Sigma \cup T_\Sigma) \subseteq \{0,1\}$, i.e., it is a 'binary' 1-conservative S-invariant. For instance, the bold S-invariant in Fig. 6.1 is an S-component, but the other two are not. The terminology 'S-component' is due to the observation that if we select all the nodes covered by θ and retain the arc weights between them, as well as their labels, then the result is a subnet of Σ generated by its places (i.e., the places have the same pre- and post-sets in the subnet as in Σ) and each transition has exactly one predecessor and one successor in the subnet, connected to it by unitary arcs. Moreover, since θ is assumed to be 1-conservative, the result has exactly one entry place and one exit place, and it is a box whenever Σ was a box. And, if Σ is e-directed, x-directed or ex-directed, so is the result. Such a subnet is called a *component* of Σ.

Exercise 6.1.10. Verify these properties.

As an immediate consequence of Props. 6.1.1–6.1.4, we obtain

Proposition 6.1.5. Let Σ be a labelled net, and θ be an S-component of weight 1 (thus, in particular, M_Σ can be $^\circ\Sigma$). Then, for all markings reachable from M_Σ, the following hold.

(1) If a place is covered by θ, then it holds at most one token, and no transition connected to it is auto-concurrent; moreover, if two places are covered by θ, then they are never marked simultaneously.

(2) If a transition is covered by θ, then it is never auto-concurrent; moreover, if two transitions are covered by θ, then they are never enabled concurrently.

Proposition 6.1.6. Let Σ be a labelled net.

(1) If Σ is S-covered by S-components of weight 1, then it is safe, clean, and exhibits no auto-concurrency.

(2) If Σ is T-covered by S-components of weight 1, then it exhibits no auto-concurrency.

Proposition 6.1.7. Let Σ be a box.

(1) If Σ is unmarked and S-covered by S-components, then it is a static box.
(2) If Σ is marked by a real marking and S-covered by S-components of weight 1, then it is a dynamic box.

The interest in the coverability by S-components is motivated by the above results and other properties developed in the rest of this chapter. Notice that all operator and plain boxes introduced in Sect. 4.4 for the translation of PBC expressions into boxes are S-covered by S-components.

6.1.3 S-aggregates

S-components and coverability by S-components are not preserved by the step-equivalence discussed in Sect. 4.4.6, nor by operations like PBC synchronisation since in both cases the nets to be compared may differ by transitions with arc weights greater than 1, which is incompatible with the coverability by S-components. This problem may be solved by introducing another special kind of S-invariants.

An *S-aggregate* θ of a labelled net Σ is a 1-conservative S-invariant with integer weights, $\theta(S_\Sigma \cup T_\Sigma) \subseteq \mathbf{N}$. In particular, every S-component is an S-aggregate. Notice that exactly one entry (exit) place of Σ has the unit weight, all remaining entry (exit) places having the zero weight. Moreover, if the marking of Σ has the weight 1, then exactly one marked place or transition (if the marking is complex) has the weight 1 (and it is marked/engaged once), all the remaining ones having the zero weight.

Exercise 6.1.11. Verify this.

The connection between structural equivalences and coverability by S-invariants carries over to S-aggregates. As an immediate consequence of Props. 6.1.1–6.1.4, we obtain the following.

Proposition 6.1.8. Let Σ be a labelled net, and θ be an S-aggregate of weight 1. Then, for all markings reachable from M_Σ, the following hold.

(1) If a place is covered by θ, then it holds at most one token, and no transition connected to it is auto-concurrent; moreover, if two places are covered by θ, then they are never marked simultaneously.
(2) If a transition is covered by θ, then it is never auto-concurrent; moreover, if two transitions are covered by θ, then they are never enabled concurrently.

Proposition 6.1.9. Let Σ be a labelled net.

(1) If Σ is S-covered by S-aggregates of weight 1, then it is safe, clean, and exhibits no auto-concurrency.
(2) If Σ is T-covered by S-aggregates of weight 1, then it exhibits no auto-concurrency.

Proposition 6.1.10. Let Σ be a box.

(1) If Σ is unmarked and S-covered by S-aggregates, then it is a static box.
(2) If Σ is marked by a real marking and S-covered by S-aggregates of weight 1, then it is a dynamic box.

6.2 The Synthesis Problem for Net Refinement

The Petri net semantics of PBC, including recursion, is based on net refinement. A natural question may now be raised, namely, is it possible to construct S-invariants from those of the operator and operand boxes? This question, as well as the consequences of some of the answers, concern us in this section.

Given an instance of net refinement, $\Omega(\boldsymbol{\Sigma})$, our aim will not be to construct all S-invariants of the refined net $\Omega(\boldsymbol{\Sigma})$, but to do so for a family of S-invariants rich enough to allow reasoning about some relevant behavioural properties. To see why it does not seem reasonable to aim at constructing all nontrivial S-invariants, we consider the boxes in Fig. 6.3.

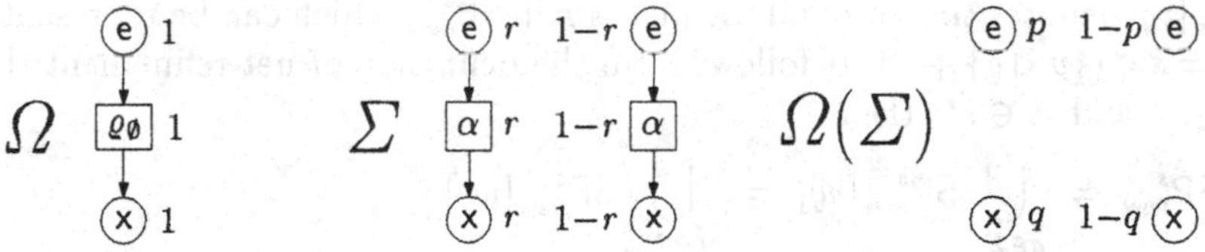

Fig. 6.3. S-invariant explosion phenomenon (where $\varrho_\emptyset = \emptyset$)

The operator box, Ω, has a single 1-conservative S-invariant, and each 1-conservative S-invariant of Σ can be characterised by a real number r ($0 \leq r \leq 1$) and the weight distribution shown in Fig. 6.3. Now, in the refined net $\Omega(\Sigma)$, each 1-conservative S-invariant is characterised by two real numbers, p and q ($0 \leq p, q \leq 1$), and the weight distribution shown in Fig. 6.3. Hence, unlike in Σ, there is no link between the distribution of weights for the exit and entry places. But, if the relabelling of Ω was the identity relabelling, ϱ_{id}, then the refinement would be isomorphic to Σ and thus have the same S-invariants. This situation would not be compatible with a hypothetical purely structural construction of all S-invariants of a refined net based solely on the connectivity in the nets involved in the refinement, without taking into account the (possibly complex) properties of transition labellings. Having said that, it is possible to devise a general scheme for constructing sufficiently rich sets of S-invariants for nets constructed through net refinement, which will be described next.

Assumptions. Until the end of Sect. 6.2, we shall consider a fixed instance of net refinement $\Omega(\boldsymbol{\Sigma})$, relying extensively on the notation used in Sect. 4.3, and some S-invariants of the boxes involved in it. In particular, we shall assume the following.

- Ω is an operator box with a complex marking $\mathcal{M} = (M, Q)$ (note that M and Q may be empty).
- Each Σ_v in $\boldsymbol{\Sigma}$ is a plain box with a real marking which is nonempty if and only if $v \in Q$.
- θ is an S-invariant of Ω; we denote respectively by T_θ and S_θ the transitions and places of Ω covered by θ.
- θ_v is a 1-conservative S-invariant of Σ_v, for every $v \in T_\theta$.

Exercise 6.2.1. Show that ${}^\bullet S_\theta = S_\theta^\bullet = T_\theta$.

For every non-isolated place s in Ω, we introduce the following notation. Let $v \in {}^\bullet s$ and $q \in \Sigma_v^\circ$, or symmetrically, $v \in s^\bullet$ and $q \in {}^\circ\Sigma_v$. Then

$$\mathsf{SP}^s_{\mathsf{new}}[vq] \;=\; \left\{ p \in \mathsf{SP}^s_{\mathsf{new}} \;\middle|\; v \lhd q \in \mathsf{subtrees}(p) \right\},$$

where $\mathsf{subtrees}(p)$ is the set of all subtrees of p directly appended to the root (formally, $\mathsf{subtrees}(p) = \{v \lhd q \mid p = s \lhd (\{v \lhd q\} + \mathcal{S})\}$). In other words, $\mathsf{SP}^s_{\mathsf{new}}[vq]$ denotes the set of all the places p in $\mathsf{SP}^s_{\mathsf{new}}$ which can be represented as $p = s \lhd (\{v \lhd q\} + \mathcal{S})$. It follows from the definition of net refinement that if $v \in {}^\bullet s$ and $w \in s^\bullet$, then

$$\mathsf{SP}^s_{\mathsf{new}} \;=\; \bigcup_{q \in \Sigma_v^\circ} \mathsf{SP}^s_{\mathsf{new}}[vq] \;=\; \bigcup_{q \in {}^\circ\Sigma_w} \mathsf{SP}^s_{\mathsf{new}}[wq] \tag{6.7}$$

and that both unions are in fact disjoint unions.

Distribution Functions. We aim at constructing an S-invariant of $\Omega(\boldsymbol{\Sigma})$, taking as the starting point the S-invariants defined for Ω and the Σ_v's (for $v \in T_\theta$). In general, there may be a variety of ways in which one could combine them into a valid S-invariant of the refined net $\Omega(\boldsymbol{\Sigma})$. Here, we shall do it parametrically, by devising a scheme based on distribution functions.

A *distribution function* is a mapping $\partial_s : \mathsf{SP}^s_{\mathsf{new}} \to \Re^+$ defined for a place s in S_θ. If s is an isolated place, then $\mathsf{SP}^s_{\mathsf{new}} = \{s\}$ and we set $\partial_s(s) = 1$. If s is not isolated, then we require that, for all $v \in {}^\bullet s$ and $q \in \Sigma_v^\circ$ as well as for all $v \in s^\bullet$ and $q \in {}^\circ\Sigma_v$,

$$\theta_v(q) \;=\; \sum_{p \in \mathsf{SP}^s_{\mathsf{new}}[vq]} \partial_s(p)\,. \tag{6.8}$$

The above condition is well formed as θ_v exists due to ${}^\bullet S_\theta = S_\theta^\bullet = T_\theta$ (see Exc. 6.2.1). We shall later see how to construct distribution functions for non-isolated places. The name of ∂_s derives from it distributing the weight of 1 over all the places in $\mathsf{SP}^s_{\mathsf{new}}$.

Lemma 6.2.1. For every place s in S_θ,

$$\sum_{p \in \mathsf{SP}^s_{\mathsf{new}}} \partial_s(p) = 1 .$$

Proof. If s is an isolated place, then the property clearly holds; otherwise, we may assume without loss of generality that there is $v \in {}^\bullet s$, and then

$$\sum_{p \in \mathsf{SP}^s_{\mathsf{new}}} \partial_s(p) =_{(6.1,6.7)} \sum_{q \in \Sigma_v^\circ} \sum_{p \in \mathsf{SP}^s_{\mathsf{new}}[vq]} \partial_s(p)$$

$$=_{(6.8)} \sum_{q \in \Sigma_v^\circ} \theta_v(q) = \theta_v(\Sigma_v^\circ) = 1 ,$$

where the last equality follows from θ_v being 1-conservative. □

6.2.1 Composing S-invariants

Given a set of distribution functions $\partial = \{\partial_s\}_{s \in S_\theta}$, we define a mapping $\theta_\partial : S_{\Omega(\boldsymbol{\Sigma})} \cup T_{\Omega(\boldsymbol{\Sigma})} \to \Re^+$, the intended S-invariant of $\Omega(\boldsymbol{\Sigma})$, in such a way that for every place or transition z in $\Omega(\boldsymbol{\Sigma})$,

$$\theta_\partial(z) = \begin{cases} \theta(v) \cdot \sum_{t \in R} R(t) \cdot \theta_v(t) & \text{if } z = (v,\alpha) \lhd R \in \mathsf{T}^v_{\mathsf{new}} \text{ and } v \in T_\theta , \\ \theta(v) \cdot \theta_v(i) & \text{if } z = v \lhd i \in \mathsf{ST}^v_{\mathsf{new}} \text{ and } v \in T_\theta , \\ \theta(s) \cdot \partial_s(z) & \text{if } z \in \mathsf{SP}^s_{\mathsf{new}} \text{ and } s \in S_\theta , \\ 0 & \text{otherwise} . \end{cases} \tag{6.9}$$

That is, θ_∂ is a composition of an S-invariant of the operator box with S-invariants of the operand boxes, parameterised by a set of distribution functions.

Theorem 6.2.1. θ_∂ is an S-invariant of $\Omega(\boldsymbol{\Sigma})$ which is 1-conservative if the same holds for θ. Moreover, if the weight of each θ_v is $Q(v)$, then the weight of θ_∂ is that of θ.

Proof. Let $\Sigma = \Omega(\boldsymbol{\Sigma})$ and S_{is} (S_{nis}) be the set of all isolated (respectively, non-isolated) places in S_θ. We first show that (6.6) holds, for every transition $u = (v,\alpha) \lhd R \in \mathsf{T}^v_{\mathsf{new}}$. If $v \notin T_\theta$, then $\theta_\partial(u) = 0$ and (6.6) holds by $({}^\bullet v \cup v^\bullet) \cap S_\theta = \emptyset$, the definition of net refinement, and the fact that $\theta_\partial(z) = 0$, for every $z \in \mathsf{ST}^v_{\mathsf{new}} \cup \mathsf{SP}^{({}^\bullet v \cup v^\bullet)}_{\mathsf{new}}$. If $v \in T_\theta$, then by the definition of net refinement and

(6.1)—(6.3) and (6.7)—(6.9), we can proceed as follows (let us also recall that operator nets are simple):

$$\begin{aligned}
&\sum_{p\in \mathsf{ST}^{v}_{\mathsf{new}}} \theta_{\partial}(p)\cdot W_{\Sigma}(p,u)\\
&= \sum_{i\in \ddot{\Sigma}_v} \theta(v)\cdot\theta_v(i)\cdot\sum_{t\in R} R(t)\cdot W_{\Sigma_v}(i,t)\\
&= \theta(v)\cdot\sum_{t\in R} R(t)\cdot\sum_{i\in \ddot{\Sigma}_v}\theta_v(i)\cdot W_{\Sigma_v}(i,t)
\end{aligned}$$

as well as

$$\begin{aligned}
&\sum_{s\in {}^\bullet v\cup v^\bullet}\ \sum_{p\in \mathsf{SP}^{s}_{\mathsf{new}}} \theta_{\partial}(p)\cdot W_{\Sigma}(p,u)\\
&= \sum_{s\in {}^\bullet v\setminus v^\bullet}\ \sum_{q\in {}^\circ\Sigma_v}\ \sum_{p\in \mathsf{SP}^{s}_{\mathsf{new}}[vq]} \theta(s)\cdot\partial_s(p)\cdot\sum_{t\in R} R(t)\cdot W_{\Sigma_v}(q,t)\\
&\quad+\sum_{s\in v^\bullet\setminus {}^\bullet v}\ \sum_{q\in \Sigma_v^\circ}\ \sum_{p\in \mathsf{SP}^{s}_{\mathsf{new}}[vq]} \theta(s)\cdot\partial_s(p)\cdot\sum_{t\in R} R(t)\cdot W_{\Sigma_v}(q,t)\\
&\quad+\sum_{s\in v^\bullet\cap {}^\bullet v}\ \sum_{q\in {}^\circ\Sigma_v}\ \sum_{q'\in \Sigma_v^\circ}\ \sum_{p\in \mathsf{SP}^{s}_{\mathsf{new}}[vq]\cap \mathsf{SP}^{s}_{\mathsf{new}}[vq']} \theta(s)\cdot\partial_s(p)\\
&\qquad\qquad\cdot\sum_{t\in R} R(t)\cdot\left(W_{\Sigma_v}(q,t)+W_{\Sigma_v}(q',t)\right)\\
&= \sum_{t\in R} R(t)\cdot\Bigg(\sum_{s\in {}^\bullet v\setminus v^\bullet}\theta(s)\cdot\sum_{q\in {}^\circ\Sigma_v} W_{\Sigma_v}(q,t)\cdot\sum_{p\in \mathsf{SP}^{s}_{\mathsf{new}}[vq]}\partial_s(p)\\
&\qquad+\sum_{s\in v^\bullet\setminus {}^\bullet v}\theta(s)\cdot\sum_{q\in \Sigma_v^\circ} W_{\Sigma_v}(q,t)\cdot\sum_{p\in \mathsf{SP}^{s}_{\mathsf{new}}[vq]}\partial_s(p)\\
&\qquad+\sum_{s\in v^\bullet\cap {}^\bullet v}\theta(s)\cdot\sum_{q\in {}^\circ\Sigma_v}\ \sum_{q'\in \Sigma_v^\circ} W_{\Sigma_v}(q,t)\\
&\qquad\qquad\cdot\sum_{p\in \mathsf{SP}^{s}_{\mathsf{new}}[vq]\cap \mathsf{SP}^{s}_{\mathsf{new}}[vq']}\partial_s(p)\\
&\qquad+\sum_{s\in v^\bullet\cap {}^\bullet v}\theta(s)\cdot\sum_{q\in {}^\circ\Sigma_v}\ \sum_{q'\in \Sigma_v^\circ} W_{\Sigma_v}(q',t)\\
&\qquad\qquad\cdot\sum_{p\in \mathsf{SP}^{s}_{\mathsf{new}}[vq]\cap \mathsf{SP}^{s}_{\mathsf{new}}[vq']}\partial_s(p)\Bigg)\\
&= \sum_{t\in R} R(t)\cdot\Bigg(\sum_{s\in {}^\bullet v\setminus v^\bullet}\theta(s)\cdot\sum_{q\in {}^\circ\Sigma_v} W_{\Sigma_v}(q,t)\cdot\sum_{p\in \mathsf{SP}^{s}_{\mathsf{new}}[vq]}\partial_s(p)\\
&\qquad+\sum_{s\in v^\bullet\setminus {}^\bullet v}\theta(s)\cdot\sum_{q\in \Sigma_v^\circ} W_{\Sigma_v}(q,t)\cdot\sum_{p\in \mathsf{SP}^{s}_{\mathsf{new}}[vq]}\partial_s(p)
\end{aligned}$$

$$+ \sum_{s \in v^\bullet \cap {}^\bullet v} \theta(s) \cdot \sum_{q \in {}^\circ\Sigma_v} W_{\Sigma_v}(q,t) \cdot \sum_{p \in \mathsf{SP}^s_{\mathsf{new}}[vq]} \partial_s(p)$$

$$\left. + \sum_{s \in v^\bullet \cap {}^\bullet v} \theta(s) \cdot \sum_{q' \in \Sigma_v^\circ} W_{\Sigma_v}(q',t) \cdot \sum_{p \in \mathsf{SP}^s_{\mathsf{new}}[vq']} \partial_s(p) \right)$$

$$= \sum_{t \in R} R(t) \cdot \left(\sum_{s \in {}^\bullet v} \theta(s) \cdot \sum_{q \in {}^\circ\Sigma_v} W_{\Sigma_v}(q,t) \cdot \sum_{p \in \mathsf{SP}^s_{\mathsf{new}}[vq]} \partial_s(p) \right.$$

$$\left. + \sum_{s \in v^\bullet} \theta(s) \cdot \sum_{q \in \Sigma_v^\circ} W_{\Sigma_v}(q,t) \cdot \sum_{p \in \mathsf{SP}^s_{\mathsf{new}}[vq]} \partial_s(p) \right)$$

$$= \sum_{t \in R} R(t) \cdot \left(\left(\sum_{s \in {}^\bullet v} \theta(s) \right) \cdot \left(\sum_{q \in {}^\circ\Sigma_v} W_{\Sigma_v}(q,t) \cdot \theta_v(q) \right) \right.$$

$$\left. + \left(\sum_{s \in v^\bullet} \theta(s) \right) \cdot \left(\sum_{q \in \Sigma_v^\circ} W_{\Sigma_v}(q,t) \cdot \theta_v(q) \right) \right)$$

$$= \theta(v) \cdot \sum_{t \in R} R(t) \cdot \sum_{q \in {}^\circ\Sigma_v \cup \Sigma_v^\circ} W_{\Sigma_v}(q,t) \cdot \theta_v(q) .$$

The two calculations can now be combined:

$$\sum_{p \in S_\Sigma} \theta_\partial(p) \cdot W_\Sigma(p,u)$$

$$= \sum_{p \in \mathsf{ST}^v_{\mathsf{new}}} \theta_\partial(p) \cdot W_\Sigma(p,u) + \sum_{s \in {}^\bullet v \cup v^\bullet} \sum_{p \in \mathsf{SP}^s_{\mathsf{new}}} \theta_\partial(p) \cdot W_\Sigma(p,u)$$

$$= \theta(v) \cdot \sum_{t \in R} R(t) \cdot \sum_{i \in \ddot{\Sigma}_v} \theta_v(i) \cdot W_{\Sigma_v}(i,t)$$

$$+ \theta(v) \cdot \sum_{t \in R} R(t) \cdot \sum_{q \in {}^\circ\Sigma_v \cup \Sigma_v^\circ} \theta_v(q) \cdot W_{\Sigma_v}(q,t)$$

$$= \theta(v) \cdot \sum_{t \in R} R(t) \cdot \sum_{q \in S_{\Sigma_v}} \theta_v(q) \cdot W_{\Sigma_v}(q,t)$$

$$= \theta(v) \cdot \sum_{t \in R} R(t) \cdot \theta_v(t) \;=\; \theta_\partial(v) .$$

Showing that $\sum_{p \in S_\Sigma} \theta_\partial(p) \cdot W_\Sigma(u,p) = \theta_\partial(v)$ holds is similar. Thus θ_∂ is an S-invariant. Moreover, if θ is 1-conservative, then

$$\theta_\partial({}^\circ\Sigma)$$

$$= \sum_{p \in {}^\circ\Sigma} \theta_\partial(p)$$

$$= \sum_{s \in {}^\circ\Omega \cap S_{\mathrm{is}}} \theta(s) + \sum_{s \in {}^\circ\Omega \cap S_{\mathrm{nis}}} \sum_{p \in \mathsf{SP}^s_{\mathsf{new}}} \theta(s) \cdot \partial_s(p)$$

$$= \sum_{s\in{}^\circ\Omega\cap S_{\mathrm{is}}} \theta(s) + \sum_{s\in{}^\circ\Omega\cap S_{\mathrm{nis}}} \theta(s)\cdot \sum_{p\in \mathsf{SP}^s_{\mathsf{new}}} \partial_s(p)$$

$$= \sum_{s\in{}^\circ\Omega\cap S_{\mathrm{is}}} \theta(s) + \sum_{s\in{}^\circ\Omega\cap S_{\mathrm{nis}}} \theta(s) \qquad \text{(by Lemma 6.2.1)}$$

$$= \theta({}^\circ\Omega) = 1$$

and, similarly, $\theta_\partial(\Sigma^\circ) = 1$. Hence θ_∂ is 1-conservative. The last part of the theorem is proved thus.

$$\begin{aligned}
&\theta_\partial(M_\Sigma)\\
&= \sum_{p\in S_\Sigma} \theta_\partial(p)\cdot M_\Sigma(p)\\
&= \sum_{s\in S_{\mathrm{nis}}}\sum_{p\in \mathsf{SP}^s_{\mathsf{new}}} \theta(s)\cdot\partial_s(p)\cdot M_\Sigma(p)\\
&\quad + \sum_{v\in T_\theta}\sum_{p\in \mathsf{ST}^v_{\mathsf{new}}} \theta_\partial(p)\cdot M_\Sigma(p) + \sum_{s\in S_{\mathrm{is}}} \theta(s)\cdot M_\Sigma(s)\\
&= \sum_{s\in S_{\mathrm{nis}}}\sum_{p\in \mathsf{SP}^s_{\mathsf{new}}} \theta(s)\cdot\partial_s(p)\cdot\left(M(s) + \sum_{v\lhd q\in \mathsf{subtrees}(p)} M_{\Sigma_v}(q)\right)\\
&\quad + \sum_{v\in T_\theta}\sum_{i\in \ddot{\Sigma}_v} \theta(v)\cdot\theta_v(i)\cdot M_{\Sigma_v}(i) + \sum_{s\in S_{\mathrm{is}}} \theta(s)\cdot M(s)\\
&= \sum_{s\in S_{\mathrm{nis}}}\sum_{p\in \mathsf{SP}^s_{\mathsf{new}}} \theta(s)\cdot\partial_s(p)\cdot M(s)\\
&\quad + \sum_{s\in S_{\mathrm{nis}}}\sum_{v\in s^\bullet}\sum_{q\in{}^\circ\Sigma_v}\sum_{p\in \mathsf{SP}^s_{\mathsf{new}}[vq]} \theta(s)\cdot\partial_s(p)\cdot M_{\Sigma_v}(q)\\
&\quad + \sum_{s\in S_{\mathrm{nis}}}\sum_{v\in {}^\bullet s}\sum_{q\in\Sigma_v^\circ}\sum_{p\in \mathsf{SP}^s_{\mathsf{new}}[vq]} \theta(s)\cdot\partial_s(p)\cdot M_{\Sigma_v}(q)\\
&\quad + \sum_{v\in T_\theta}\sum_{i\in \ddot{\Sigma}_v} \theta(v)\cdot\theta_v(i)\cdot M_{\Sigma_v}(i) + \sum_{s\in S_{\mathrm{is}}} \theta(s)\cdot M(s)\\
&= \sum_{s\in S_{\mathrm{nis}}} \theta(s)\cdot M(s)\cdot \sum_{p\in \mathsf{SP}^s_{\mathsf{new}}} \partial_s(p)\\
&\quad + \sum_{s\in S_{\mathrm{nis}}}\sum_{v\in s^\bullet}\sum_{q\in{}^\circ\Sigma_v} \theta(s)\cdot M_{\Sigma_v}(q)\cdot \sum_{p\in \mathsf{SP}^s_{\mathsf{new}}[vq]} \partial_s(p)\\
&\quad + \sum_{s\in S_{\mathrm{nis}}}\sum_{v\in {}^\bullet s}\sum_{q\in\Sigma_v^\circ} \theta(s)\cdot M_{\Sigma_v}(q)\cdot \sum_{p\in \mathsf{SP}^s_{\mathsf{new}}[vq]} \partial_s(p)\\
&\quad + \sum_{v\in T_\theta}\sum_{i\in \ddot{\Sigma}_v} \theta(v)\cdot\theta_v(i)\cdot M_{\Sigma_v}(i) + \sum_{s\in S_{\mathrm{is}}} \theta(s)\cdot M(s)\\
&= \sum_{s\in S_{\mathrm{nis}}} \theta(s)\cdot M(s) \qquad \text{(by Lemma 6.2.1)}
\end{aligned}$$

$$+\sum_{s\in S_{\text{nis}}}\sum_{v\in s^\bullet}\sum_{q\in {}^\circ\Sigma_v}\theta(s)\cdot\theta_v(q)\cdot M_{\Sigma_v}(q) \qquad \text{(by (6.8))}$$

$$+\sum_{s\in S_{\text{nis}}}\sum_{v\in {}^\bullet s}\sum_{q\in\Sigma_v^\circ}\theta(s)\cdot\theta_v(q)\cdot M_{\Sigma_v}(q) \qquad \text{(by (6.8))}$$

$$+\sum_{v\in T_\theta}\sum_{i\in\ddot{\Sigma}_v}\theta(v)\cdot\theta_v(i)\cdot M_{\Sigma_v}(i)+\sum_{s\in S_{\text{is}}}\theta(s)\cdot M(s)$$

$$=\sum_{s\in S_{\text{nis}}}\sum_{v\in s^\bullet}\sum_{q\in {}^\circ\Sigma_v}\theta(s)\cdot\theta_v(q)\cdot M_{\Sigma_v}(q)$$

$$+\sum_{s\in S_{\text{nis}}}\sum_{v\in {}^\bullet s}\sum_{q\in\Sigma_v^\circ}\theta(s)\cdot\theta_v(q)\cdot M_{\Sigma_v}(q)$$

$$+\sum_{v\in T_\theta}\sum_{i\in\ddot{\Sigma}_v}\theta(v)\cdot\theta_v(i)\cdot M_{\Sigma_v}(i)+\sum_{s\in S_\Omega}\theta(s)\cdot M(s)$$

$$=\sum_{v\in T_\theta}\left(\sum_{q\in {}^\circ\Sigma_v}\sum_{s\in {}^\bullet v}\theta(s)\cdot\theta_v(q)\cdot M_{\Sigma_v}(q)+\sum_{q\in\Sigma_v^\circ}\sum_{s\in v^\bullet}\theta(s)\cdot\theta_v(q)\cdot M_{\Sigma_v}(q)\right)$$

$$+\sum_{v\in T_\theta}\sum_{i\in\ddot{\Sigma}_v}\theta(v)\cdot\theta_v(i)\cdot M_{\Sigma_v}(i)+\sum_{s\in S_\Omega}\theta(s)\cdot M(s)$$

$$=\sum_{v\in T_\theta}\left(\sum_{q\in {}^\circ\Sigma_v}\theta_v(q)\cdot M_{\Sigma_v}(q)\cdot\sum_{s\in {}^\bullet v}\theta(s)+\sum_{q\in\Sigma_v^\circ}\theta_v(q)\cdot M_{\Sigma_v}(q)\cdot\sum_{s\in v^\bullet}\theta(s)\right)$$

$$+\sum_{v\in T_\theta}\sum_{i\in\ddot{\Sigma}_v}\theta(v)\cdot\theta_v(i)\cdot M_{\Sigma_v}(i)+\sum_{s\in S_\Omega}\theta(s)\cdot M(s)$$

$$=\sum_{v\in T_\theta}\theta(v)\cdot\left(\sum_{q\in {}^\circ\Sigma_v}\theta_v(q)\cdot M_{\Sigma_v}(q)+\sum_{q\in\Sigma_v^\circ}\theta_v(q)\cdot M_{\Sigma_v}(q)\right) \qquad (*)$$

$$+\sum_{v\in T_\theta}\sum_{i\in\ddot{\Sigma}_v}\theta(v)\cdot\theta_v(i)\cdot M_{\Sigma_v}(i)+\sum_{s\in S_\Omega}\theta(s)\cdot M(s)$$

$$=\sum_{s\in S_\Omega}\theta(s)\cdot M(s)+\sum_{v\in T_\theta}\theta(v)\cdot\sum_{q\in S_{\Sigma_v}}\theta_v(q)\cdot M_{\Sigma_v}(q)$$

$$=\sum_{s\in S_\Omega}\theta(s)\cdot M(s)+\sum_{v\in T_\theta}\theta(v)\cdot Q(v)$$

$$=\sum_{s\in S_\Omega}\theta(s)\cdot M(s)+\sum_{v\in T_\Omega}\theta(v)\cdot Q(v)$$

$$=\theta(M,Q)\ .$$

Note that (*) holds since Ω is simple. □

The method of obtaining an S-invariant of the refined net $\Omega(\boldsymbol{\Sigma})$ is not yet fully constructive since we said nothing about how to find distribution functions. In fact, it may even happen that such functions do not exist, due to the choice of S-invariants.

For consider the nets in Fig. 6.4 and the refinement $\Omega(\boldsymbol{\Sigma})$, where for every transition v_i of Ω, $\Sigma_{v_i} = \Sigma$. Let θ be the all-1 constant S-invariant of Ω, and, for every v_i, let θ_{v_i} be the 1-conservative S-invariant of Σ such that $\theta_{v_i}(t) = 1 = \theta_{v_i}(x_0)$, $\theta_{v_i}(e_j) = \frac{1}{i}$ for every $j < i$, and $\theta_{v_i}(e_j) = 0$ otherwise (i.e., the first i entry places have the weight $\frac{1}{i}$, and the remaining ones have the zero weight). Then no distribution function ∂_e can be found for the place e. For suppose that ∂_e is such a function. Then, by Lemma 6.2.1, there is $p \in \mathsf{SP}^e_{\mathsf{new}}$ such that $\partial_e(p) > 0$. Let $i \geq 1$ be such that $\frac{1}{i} < \partial_e(p)$ and $p = e \lhd (\{v_i \lhd q\} + \mathcal{S})$. Then we obtain a contradiction with (6.8) since

$$\theta_{v_i}(q) \leq \frac{1}{i} < \partial_e(p) \leq \sum_{r \in \mathsf{SP}^e_{\mathsf{new}}[v_i q]} \partial_e(r) .$$

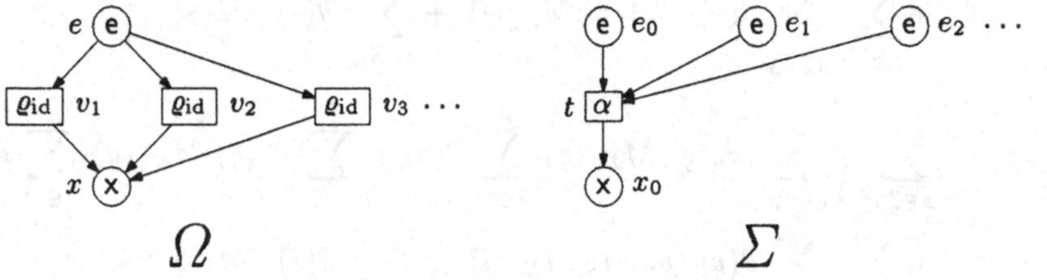

Fig. 6.4. Distribution function may not exist

Nevertheless, in many cases, it is not only possible to find at least one distribution function, but it can even be done in many different ways. It should also be stressed that choosing a distribution function can be done separately and independently for each place $s \in S_\theta$.

6.2.2 Multiplicative Distribution Functions

A scheme for constructing distribution functions can be based on the following definition. For a place s in S_θ, let $\partial_s^{\mathrm{mult}} : \mathsf{SP}^s_{\mathsf{new}} \to \Re^+$ be the mapping such that, for every p,

$$\partial_s^{\mathrm{mult}}(p) = \prod_{v \lhd q \in \mathsf{subtrees}(p)} \theta_v(q) \tag{6.10}$$

provided that the generalised product always exists and is finite (which is not in general guaranteed). But even if $\partial_s^{\mathrm{mult}}$ always returns real numbers, it may fail to be a distribution function, as illustrated in Fig. 6.5.

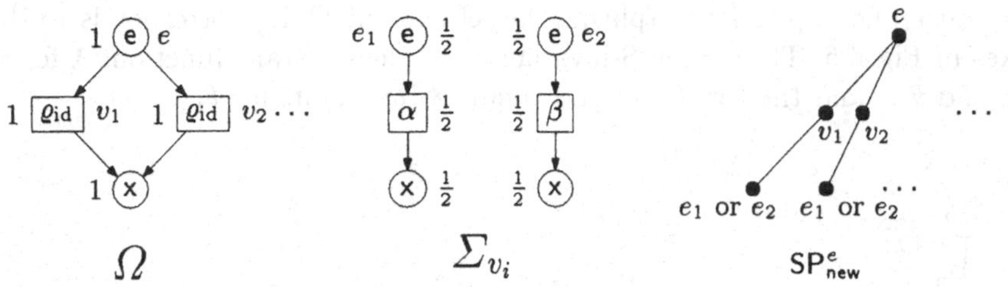

Fig. 6.5. (6.8) is not fulfilled since $\partial_e^{\text{mult}}(p) = \frac{1}{2} \cdot \frac{1}{2} \cdot \ldots = 0$, for all $p \in \mathsf{SP}^e_{\text{new}}$

In all those cases that ∂_s^{mult} is a distribution function, we will call it *the multiplicative* distribution function for s. We shall now discuss some important cases for which multiplicative distribution functions do exist.

Degree-finiteness. A place s in Ω is *degree-finite* if ${}^\bullet s \cup s^\bullet$ is a finite set. For such a place, (6.10) always yields a distribution function.

Proposition 6.2.1. If a place s in S_θ is degree-finite, then ∂_s^{mult} is a distribution function.

Proof. Let $s^\bullet = \{v_1, \ldots, v_k\}$ and ${}^\bullet s = \{w_1, \ldots, w_m\}$. Suppose that $k, m \geq 1$. Then, for every $q \in {}^\circ\Sigma_{v_1}$,

$$\begin{aligned}
&\sum_{p \in \mathsf{SP}^s_{\text{new}}[v_1 q]} \partial_s^{\text{mult}}(p) \\
&= \sum_{e_2 \in {}^\circ\Sigma_{v_2}} \cdots \sum_{e_k \in {}^\circ\Sigma_{v_k}} \sum_{x_1 \in \Sigma^\circ_{w_1}} \cdots \sum_{x_m \in \Sigma^\circ_{w_m}} \\
&\qquad\qquad \theta_{v_1}(q) \cdot \theta_{v_2}(e_2) \cdot \ldots \cdot \theta_{v_k}(e_k) \cdot \theta_{w_1}(x_1) \cdot \ldots \cdot \theta_{w_m}(x_m) \\
&= \theta_{v_1}(q) \cdot \left(\sum_{e_2 \in {}^\circ\Sigma_{v_2}} \theta_{v_2}(e_2) \right) \cdot \ldots \cdot \left(\sum_{e_k \in {}^\circ\Sigma_{v_k}} \theta_{v_k}(e_k) \right) \\
&\qquad \cdot \left(\sum_{x_1 \in \Sigma^\circ_{w_1}} \theta_{w_1}(x_1) \right) \cdot \ldots \cdot \left(\sum_{x_m \in \Sigma^\circ_{w_m}} \theta_{w_m}(x_m) \right) \\
&= \theta_{v_1}(q) \,.
\end{aligned}$$

The last equality holds since S-invariants for the Σ_v's ($v \in T_\theta$) are 1-conservative. For $v_2, \ldots, v_k, w_1, \ldots, w_m$, we proceed in a similar way. If $k \geq 1$ and $m = 0$, or $k = 0$ and $m \geq 1$, the proof is similar; and the case $k = 0 = m$ is trivial since $\prod_{v \lhd q \in \emptyset} \theta_v(q) = 1$. □

The bold and small S-invariants in Fig. 6.1 can both be derived through multiplicative distribution functions. We may observe that the net there is isomorphic to that in Fig. 4.37, obtained as the result of the refinement with the operator $\Omega_{[*]}$ (see Fig. 4.36) and operands $\mathsf{box}_{\mathsf{PBC}}(\alpha)$, $\mathsf{box}_{\mathsf{PBC}}(\beta \| \alpha)$, and

$\mathsf{box}_{\mathsf{PBC}}(\alpha)$, i.e., up to isomorphism, the refinement $\Omega(\boldsymbol{\Sigma})$ corresponds to the boxes in Fig. 6.6. The chosen S-invariants are the constant functions **1** for θ, θ_{v_1}, and θ_{v_3}, and the two (bold and small) S-invariants for θ_{v_2}.

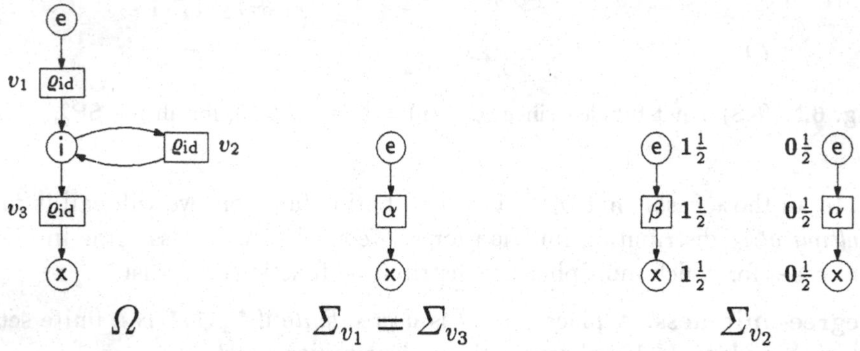

Fig. 6.6. Refinement and S-invariants for the net in Fig. 6.1

The interest in multiplicative distribution functions stems from the fact that all the operator boxes introduced in Chap. 4 for the translation of PBC have degree-finite places, with the sole exception of the infinite generalised choice. An interesting consequence of the last result is a characterisation of S-coverings and T-coverings of $\Omega(\boldsymbol{\Sigma})$.

Proposition 6.2.2. Suppose that all places in Ω are degree-finite. If Ω and $\boldsymbol{\Sigma}$ are all T-covered (S-covered), then so is $\Omega(\boldsymbol{\Sigma})$.

Proof. For T-covering, the result follows from Prop. 6.2.1, Thm. 6.2.1, and the fact that for each $u = (v, \alpha) \lhd R$ in $\Omega(\boldsymbol{\Sigma})$, one can choose θ such that v is covered by θ, and θ_v such that at least one transition in R is covered by θ_v. For S-covering we proceed as follows. If $p = v \lhd i \in \mathsf{ST}^v_{\mathsf{new}}$, one can choose θ such that v is covered by θ, and θ_v such that i is covered by θ_v. For an isolated place s in Ω and $p = s \in \mathsf{SP}^s_{\mathsf{new}}$, one simply needs to choose θ such that s is covered by θ. For a non-isolated place s in Ω and $p = s \lhd (\{v \lhd e_v\}_{v \in s^\bullet} + \{w \lhd x_w\}_{w \in {}^\bullet s}) \in \mathsf{SP}^s_{\mathsf{new}}$, one can choose θ such that s is covered by θ and

- For every $v \in {}^\bullet s \setminus s^\bullet$, one can choose θ_v such that x_v is covered by θ_v.
- For every $v \in s^\bullet \setminus {}^\bullet s$, one can choose θ_v such that e_v is covered by θ_v.
- For every $v \in s^\bullet \cap {}^\bullet s$, one can choose θ_v such that both e_v and x_v are covered by θ_v.

Notice that the latter is always possible, by taking the average of an S-invariant covering e_v and an S-invariant covering x_v. □

6.2.3 Ex-binary S-invariants

We now partially lift the degree-finiteness condition, which is not too restrictive for the translation of the basic PBC expressions but may lead to difficulties when considering more general operators. A 1-conservative S-invariant θ of a labelled net Σ is *ex-binary* if $\theta({}^\circ\Sigma \cup \Sigma^\circ) \subseteq \{0,1\}$. In such a case there is exactly one entry place and one exit place with weight 1, all remaining entry and exit places having the zero weight. Notice that S-components and S-aggregates are always ex-binary. Crucially, for ex-binary S-invariants, multiplicative distribution functions may be used in full generality.

Proposition 6.2.3. If s is a non-isolated place in S_θ and, for every transition $v \in {}^\bullet s \cup s^\bullet$, θ_v is an ex-binary 1-conservative S-invariant, then $\partial_s^{\mathsf{mult}}$ is a distribution function for s.

Proof. It suffices to observe that in $\mathsf{SP}^s_{\mathsf{new}}$ there is exactly one place $p = s \triangleleft (\{v \triangleleft e_v\}_{v \in s^\bullet} + \{w \triangleleft x_w\}_{w \in {}^\bullet s})$ for which $\partial_s^{\mathsf{mult}}$ returns a nonzero value. Moreover, we have $\partial_s^{\mathsf{mult}}(p) = 1$ and $\theta_v^{-1}(1) \cap {}^\circ\Sigma_v = \{e_v\}$, for every $v \in s^\bullet$, and $\theta_w^{-1}(1) \cap \Sigma_w^\circ = \{x_w\}$, for every $w \in {}^\bullet s$. □

We then combine Props. 6.2.1 and 6.2.3 into a single result.

Theorem 6.2.2. If s is a non-isolated place in S_θ and almost all of the S-invariants θ_v, for $v \in {}^\bullet s \cup s^\bullet$, are ex-binary, then $\partial_s^{\mathsf{mult}}$ is a distribution function for s.

Proof. Similar to the proof of Prop. 6.2.1, since the products which need to be considered are finite, and the problem indicated in Fig. 6.5 cannot occur. □

An important consequence of the existence of multiplicative distribution functions is that T-covering by S-aggregates is preserved by net refinement.

Proposition 6.2.4. Let Ω be T-covered by S-aggregates of weight n and, for every $v \in T_\Omega$, Σ_v be T-covered by S-aggregates of weight $Q(v)$. Then $\Omega(\boldsymbol{\Sigma})$ is T-covered by S-aggregates of weight n.

Proof. Follows from Thm. 6.2.1, Prop. 6.2.3 and the facts that S-aggregates are ex-binary; multiplicative distribution functions always lead to S-aggregates if we start from S-aggregates; and a transition $(v, \alpha) \triangleleft R$ in $\Omega(\boldsymbol{\Sigma})$ may be covered by choosing an S-aggregate θ such that v is covered by θ, and then an S-aggregate θ_v such that at least one transition in R is covered by θ_v. □

A similar property for S-coverings does not always hold. For example, all the nets in Fig. 6.6 are S-covered by S-components while the result of the refinement, shown in Fig. 6.1, is not S-covered by S-components, nor by S-aggregates, since it is not safe from the entry marking (cf. Prop. 6.1.10). A closer look at the reasons why this phenomenon occurs reveals that it is due to the fact that if there is a side-loop, $\{s, v\}$ in Ω, and in Σ_v there is an entry place e_v and exit place x_v which are not covered by the same S-aggregate, then the construction fails to cover places $s \lhd (\{v \lhd e_v\} + \{v \lhd x_v\} + \mathcal{S})$. (Notice that we cannot use the average of two S-aggregates, as in the proof of Prop. 6.2.2, since this could lead to non-integer weights.) As a result, for S-coverings, we only have a weaker counterpart of Prop. 6.2.4.

Proposition 6.2.5. Let Ω be S-covered by S-aggregates of weight n and, for every $v \in T_\Omega$, Σ_v be S-covered by S-aggregates of weight $Q(v)$. Moreover, let us assume that if $v \in T_\Omega$ and ${}^\bullet v \cap v^\bullet \neq \emptyset$, then for all $e \in {}^\circ\Sigma_v$ and $x \in \Sigma_v^\circ$, there is an S-aggregate of weight $Q(v)$ covering both e and x. Then $\Omega(\boldsymbol{\Sigma})$ is S-covered by S-aggregates of weight n.

Proof. Follows from Prop. 6.2.3, Thm. 6.2.1, and the last remark. □

Corollary 6.2.1. Let Ω be S-covered by S-aggregates of weight n, and each Σ_v be S-covered by S-aggregates of weight $Q(v)$. Moreover, let S_{pure} be the set of all places s in Ω such that ${}^\bullet s \cap s^\bullet = \emptyset$.

(1) The following nodes of $\Omega(\boldsymbol{\Sigma})$ are covered by S-aggregates of weight n:

$$T_{\Omega(\boldsymbol{\Sigma})} \cup \bigcup_{v \in T_\Omega} \mathsf{ST}^v_{\mathsf{new}} \cup \bigcup_{s \in S_{\text{pure}}} \mathsf{SP}^s_{\mathsf{new}} .$$

(2) For $s \in S_\Omega \setminus S_{\text{pure}}$, the places in $\mathsf{SP}^s_{\mathsf{new}}$ are also covered by S-aggregates of weight n, if for every $v \in {}^\bullet s \cap s^\bullet$ and all $e \in {}^\circ\Sigma_v$ and $x \in \Sigma_v^\circ$, there is an S-aggregate of weight $Q(v)$ covering both e and x.
(3) If ${}^\circ\Omega \subseteq S_{\text{pure}}$ (or $\Omega^\circ \subseteq S_{\text{pure}}$), then the entry (respectively, exit) places of $\Omega(\boldsymbol{\Sigma})$ are covered by S-aggregates of weight n.
(4) If Ω is e-directed (or x-directed), then the entry (respectively, exit) places of $\Omega(\boldsymbol{\Sigma})$ are covered by S-aggregates of weight n.

Next, directly from Prop. 6.2.5, with $n = 0$ and $n = 1$, and Prop. 6.1.10, we obtain the following.

Corollary 6.2.2. The following hold.

(1) Let Ω and the nets in $\boldsymbol{\Sigma}$ be unmarked and S-covered by S-aggregates. Moreover, let for every $v \in T_\Omega$ with a side place, all pairs of entry/exit places of Σ_v be covered by S-aggregates. Then $\Omega(\boldsymbol{\Sigma})$ is a static box, S-covered by S-aggregates.
(2) Let Ω and the Σ_v's (for $v \in Q$) be marked and S-covered by S-aggregates of weight 1, and the Σ_v's (for $v \in T_\Omega \setminus Q$) be unmarked and S-covered by S-aggregates. Moreover, let for every $v \in T_\Omega$ with a side place, all pairs of entry/exit places of Σ_v be covered by S-aggregates (of weight 1 if $v \in Q$). Then $\Omega(\boldsymbol{\Sigma})$ is a dynamic box, S-covered by S-aggregates of weight 1.

Corollary 6.2.3. Let Ω be a pure operator box.

(1) Let Ω and $\boldsymbol{\Sigma}$ be unmarked and S-covered by S-aggregates. Then $\Omega(\boldsymbol{\Sigma})$ is a static box, S-covered by S-aggregates.
(2) Let Ω and the Σ_v's (for $v \in Q$) be marked and S-covered by S-aggregates of weight 1, and the Σ_v's (for $v \in T_\Omega \setminus Q$) be unmarked and S-covered by S-aggregates. Then $\Omega(\boldsymbol{\Sigma})$ is a dynamic box, S-covered by S-aggregates of weight 1.

Exercise 6.2.2. Determine sufficient conditions guaranteeing that all pairs of entry/exit places in $\Omega(\boldsymbol{\Sigma})$ are covered by S-aggregates.

The above discussion has important consequences for net algebras such as that one used in Sect. 4.4 to model PBC expressions.

Theorem 6.2.3. Let $\mathcal{O}$ be a set of pure operator boxes, and $\mathcal{P}$ be a set of plain boxes.

(1) If the boxes in $\mathcal{O}$ and $\mathcal{P}$ are unmarked and S-covered by S-aggregates, then all the boxes obtained through successive refinements from $\mathcal{O}$ and $\mathcal{P}$ are static plain boxes S-covered by S-aggregates.
(2) If the boxes in $\mathcal{O}$ and $\mathcal{P}$ are marked and S-covered by S-aggregates of weight 1, then all the boxes obtained through successive refinements from $\mathcal{O}$ and $\mathcal{P} \cup \lfloor \mathcal{P} \rfloor$ are dynamic plain boxes S-covered by S-aggregates of weight 1.

Proof. Follows from Cor. 6.2.3 and Prop. 6.1.10. □

This explains why we preferred to use $\Omega_{[**]}$ (see Fig. 4.38) instead of $\Omega_{[*]}$ (see Fig. 4.36) for the Petri net translation of the PBC iteration operation when we did not want to constrain the operands, and why we were allowed to state it was a safe version. Corollary 6.2.2 also explains why we emphasized the need to avoid using the parallel operator directly inside the core of a loop modelled through $\Omega_{[*]}$, $\Omega_{\langle *]}$, or $\Omega_{[*\rangle}$, explicitly or through recursion variables.

6.2.4 Rational Groupings

The large S-invariant in Fig. 6.1 cannot be obtained using multiplicative distribution functions from any 1-conservative S-invariants of the nets shown in Fig. 6.6. For the only 1-conservative S-invariants for Ω, Σ_{v_1} and Σ_{v_3}, have the constant weights 1, and each 1-conservative S-invariant for Σ_{v_2} is characterised by a real number r ($0 \le r \le 1$) such that r is the weight for the nodes on the left, and $1{-}r$ is the weight for the nodes on the right. Using the formula characterising multiplicative distribution functions, these would lead to the following weights for the places of the net in Fig. 6.1 (taken from top to bottom): 1, r^2, $r(1{-}r)$, $r(1{-}r)$, $(1{-}r)^2$, and 1. Thus no r could possibly yield 1, 0, $\frac{1}{2}$, $\frac{1}{2}$, 0, and 1.

But the multiplicative scheme is not the only way to construct distribution functions. Referring again to the assumptions made at the beginning of Sect. 6.2, let us rephrase the condition for being a distribution function for a place s in S_θ. Suppose that $s^\bullet = \{\ldots, v_i, \ldots\}$ and ${}^\bullet s = \{\ldots, w_j, \ldots\}$. Defining a distribution function ∂_s for s amounts to constructing the following table.

$$\begin{array}{llll}
\text{Weight} \ldots & v_i \ldots & w_j \ldots & \\
\partial_1 & \ldots e_{i,1} \ldots & x_{j,1} \ldots & \text{corresp. to } s \lhd \{\ldots, v_i \lhd e_{i,1}, \ldots, w_j \lhd x_{j,1}, \ldots\} \\
\partial_2 & \ldots e_{i,2} \ldots & x_{j,2} \ldots & \text{corresp. to } s \lhd \{\ldots, v_i \lhd e_{i,2}, \ldots, w_j \lhd x_{j,2}, \ldots\} \\
\vdots & \vdots & \vdots & \vdots \\
\underbrace{\qquad}_{\partial_s} & & & \underbrace{\qquad\qquad\qquad\qquad\qquad\qquad}_{\mathsf{SP}^s_{\mathsf{new}} \setminus \partial_s^{-1}(0)} ,
\end{array}$$

where each ∂_k is a positive real number; each $e_{i,k}$ belongs to ${}^\circ\Sigma_{v_i} \setminus \theta_{v_i}^{-1}(0)$; each $x_{j,k}$ belongs to $\Sigma^\circ_{w_j} \setminus \theta_{w_j}^{-1}(0)$; each place in ${}^\circ\Sigma_{v_i} \setminus \theta_{v_i}^{-1}(0)$ and $\Sigma^\circ_{w_j} \setminus \theta_{w_j}^{-1}(0)$ must appear at least once in the column corresponding to v_i and w_j, respectively; and the following hold, for all i, j, k:

$$\theta_{v_i}(e_{i,k}) = \sum_{\substack{l \text{ such that} \\ e_{i,l} = e_{i,k}}} \partial_l \quad \text{and} \quad \theta_{w_j}(x_{j,k}) = \sum_{\substack{l \text{ such that} \\ x_{j,l} = x_{j,k}}} \partial_l .$$

The table may be constructed if, for example, the distribution of nonzero weights in the entry and exit places of the nets contributing to $\mathsf{SP}^s_{\mathsf{new}}$ is the same. More precisely, assume that there is a finite or infinite set $K = \{1, 2, \ldots\} \subseteq \mathbf{N}$ and a set $\{\partial_k \mid k \in K\}$ of positive reals such that, for every $v_i \in s^\bullet$, the entry places covered by θ_{v_i} can be enumerated as $\{e_{i,k}\}_{k \in K}$ and $\theta_{v_i}(e_{i,k}) = \partial_k$, for every $k \in K$; and, similarly, for every $w_j \in {}^\bullet s$, the exit places covered by θ_{w_j} can be enumerated as $\{x_{j,k}\}_{k \in K}$ and $\theta_{w_j}(x_{j,k}) = \partial_k$, for every $k \in K$. If we use these ∂_k's, $e_{i,k}$'s, and $x_{j,k}$'s to construct the table, then each $e_{i,k}$ or $x_{j,k}$ occurs exactly once in the corresponding column, and the resulting table yields a distribution function for s.

Exercise 6.2.3. Verify this.

A similar effect may be achieved even if the distributions of weights are not identical for all the v_i's and w_j's. In particular, if all weights for the places in $(\bigcup_{v \in s^\bullet} {}^\circ\Sigma_v) \cup (\bigcup_{w \in {}^\bullet s} \Sigma^\circ_w)$ belong to $\mathbf{N}/n$, i.e., are rational numbers with the same base n, we can take $\partial_1 = \cdots = \partial_n = \frac{1}{n}$, and choose $\{e_{i,k}\}_{k=1,\ldots,n}$ and $\{x_{j,k}\}_{k=1,\ldots,n}$ where we allow repetitions. Then it is possible to construct a table of the following form where we allow repetitions of rows.

$$\begin{array}{llll}
\text{Weight} \ldots & v_i \ \ldots & w_j \ \ldots & \\
\frac{1}{n} & \ldots e_{i,1} \ldots & x_{j,1} \ldots & \text{corresp. to } s \lhd \{\ldots, v_i \lhd e_{i,1}, \ldots, w_j \lhd x_{j,1}, \ldots\} \\
\frac{1}{n} & \ldots e_{i,2} \ldots & x_{j,2} \ldots & \text{corresp. to } s \lhd \{\ldots, v_i \lhd e_{i,2}, \ldots, w_j \lhd x_{j,2}, \ldots\} \\
\vdots & \quad \vdots & \vdots & \qquad\qquad\qquad \vdots \\
\frac{1}{n} & \ldots e_{i,n} \ldots & x_{j,n} \ldots & \text{corresp. to } s \lhd \{\ldots, v_i \lhd e_{i,n}, \ldots, w_j \lhd x_{j,n}, \ldots\} \\
\underbrace{\qquad}_{\partial_s} & & & \underbrace{\qquad\qquad\qquad\qquad\qquad\qquad}_{\mathrm{SP}^s_{new} \setminus \partial_s^{-1}(0)}
\end{array}$$

If some $e_{i,h}$ or $x_{j,h}$ has the weight $\frac{m}{n}$ in the net it originates from, it will occur exactly m times in the corresponding column; moreover, if a row occurs m times in the table, the corresponding place will have the weight $\frac{m}{n}$ in ∂_s. Such a construction is always possible, and generally in many different ways since we may reorder independently each column. As an example, let us consider the following weights (with the base $n = 2$):

$$\begin{array}{lll}
v_1 & v_2 & w_1 \\
e_{1,1} : \frac{1}{2} & e_{2,1} : \frac{2}{2} & x_{3,1} : \frac{1}{2} \\
e_{1,2} : \frac{1}{2} & & x_{3,2} : \frac{1}{2}
\end{array}$$

These give rise, for example, to the following two different tables.

$$\begin{array}{lllll}
\text{Weight} & v_1 & v_2 & w_1 & \\
\frac{1}{2} & e_{1,1} & e_{2,1} & x_{3,2} & \text{corresp. to } s \lhd \{v_1 \lhd e_{1,1}, v_2 \lhd e_{2,1}, w_1 \lhd x_{3,2}\} \\
\frac{1}{2} & e_{1,2} & e_{2,1} & x_{3,1} & \text{corresp. to } s \lhd \{v_1 \lhd e_{1,2}, v_2 \lhd e_{2,1}, w_1 \lhd x_{3,1}\}
\end{array}$$

$$\begin{array}{lllll}
\text{weight} & v_1 & v_2 & w_1 & \\
\frac{1}{2} & e_{1,2} & e_{2,1} & x_{3,2} & \text{corresp. to } s \lhd \{v_1 \lhd e_{1,2}, v_2 \lhd e_{2,1}, w_1 \lhd x_{3,2}\} \\
\frac{1}{2} & e_{1,1} & e_{2,1} & x_{3,1} & \text{corresp. to } s \lhd \{v_1 \lhd e_{1,1}, v_2 \lhd e_{2,1}, w_1 \lhd x_{3,1}\} \ .
\end{array}$$

Exercise 6.2.4. Formalise the above procedure and verify its correctness.

Semi-aggregates. The last example has interesting implications. Let us call a 1-conservative S-invariant θ of a box Σ a *semi-aggregate* if $\theta(S_\Sigma \cup T_\Sigma) \in \mathbf{N}/2$. Then a box with the empty marking is *semi-static* if it is T-covered by S-aggregates, its entry and exit places are covered by S-aggregates, and its internal places are covered by semi-aggregates. Similarly, a box with a nonempty marking is *semi-dynamic* if it is T-covered by S-aggregates of weight 1, its entry and exit places are covered by S-aggregates of weight 1, and its internal places are covered by semi-aggregates of weight 1.

It follows from Props. 6.1.3 and 6.1.8 that a semi-dynamic box, as well as a semi-static box when started from ${}^\circ\Sigma$, is clean, exhibits no auto-concurrency, its entry and exit places are safe, and its internal places are 2-bounded although not necessarily safe. In a semi-static or semi-dynamic box, every pair of entry/exit places is also covered by semi-aggregates of weight 1; one simply can form the average between an S-aggregate covering the entry place and an S-aggregate covering the exit place. Finally, a static box S-covered by S-aggregates is clearly semi-static as well; and a dynamic box S-covered by S-aggregates of weight 1 is clearly semi-dynamic and, after deleting all tokens, semi-static.

Proposition 6.2.6. Let Ω be such that no entry or exit place is involved in a side-loop (for instance, because Ω is ex-directed).

(1) Let Ω be an unmarked operator box S-covered by S-aggregates and, for every $v \in T_\Omega$, Σ_v be a semi-static plain box. Then $\Omega(\boldsymbol{\Sigma})$ is a semi-static plain box.
(2) Let Ω be a marked operator box S-covered by S-aggregates of weight 1 and, for every $v \in T_\Omega$, Σ_v be a semi-dynamic plain box if $v \in Q$ and a semi-static plain box otherwise. Then $\Omega(\boldsymbol{\Sigma})$ is a semi-dynamic plain box.

Proof. Follows from Props. 6.2.3 and 6.2.4, Cor. 6.2.1(3), the remark that every pair of entry/exit places of each Σ_v is simultaneously covered by a semi-aggregate, and the last table construction (for $n = 2$). □

One can then generalise Thm. 6.2.3.

Theorem 6.2.4. Let $\mathcal{O}$ be a set of operator boxes with no entry or exit place being involved in a side-loop, and $\mathcal{P}$ be a set of plain boxes (in particular, we may assume that the operator boxes in $\mathcal{O}$ are ex-directed).

(1) If the boxes in $\mathcal{O}$ are unmarked and S-covered by S-aggregates, and the boxes in $\mathcal{P}$ are semi-static, then all the boxes obtained through successive refinements from $\mathcal{O}$ and $\mathcal{P}$ are semi-static plain boxes.
(2) If the boxes in $\mathcal{O}$ are marked and S-covered by S-aggregates of weight 1, and the boxes in $\mathcal{P}$ are semi-dynamic, then all boxes obtained through successive refinements from $\mathcal{O}$ and $\mathcal{P} \cup \lfloor \mathcal{P} \rfloor$, are semi-dynamic plain boxes.

This delineates the consequences of allowing non-pure but ex-directed operator boxes, like $\Omega_{[*]}$, and it may be argued that these are not as dramatic as one could have feared. Indeed, after noticing in the example in Fig. 4.37, that the result can be 2-bounded, one was right to ask if the situation could become worse, i.e., if other examples could lead to true 3-boundedness, 4-boundedness, etc., or even to unboundedness. We now discovered, perhaps surprisingly, that in the context of ex-directed boxes, 2-boundedness is the worst case, and even though we may lose safeness, the resulting nets do not exhibit auto-concurrency.

The situation is less attractive if we allow operator boxes with side-loops involving entry or exit places, like $\Omega_{(*]}$ or $\Omega_{[*)}$. Indeed, in such a case, the result of a single refinement is not always semi-static or semi-dynamic (Prop. 6.2.6 does not apply), as demonstrated by the counterexample $\Omega_{(*]}(\Sigma_{v_2}, \Sigma_{v_3})$, where Σ_{v_2} and Σ_{v_3} are taken from Fig. 6.6. This net, which may be seen as modelling the extended PBC expression $\langle(\beta\|\alpha) * \alpha]$, is shown in Fig. 6.7, and it is not difficult to see that its entry places are not all covered by S-aggregates since two of them are only 2-bounded from the entry marking.

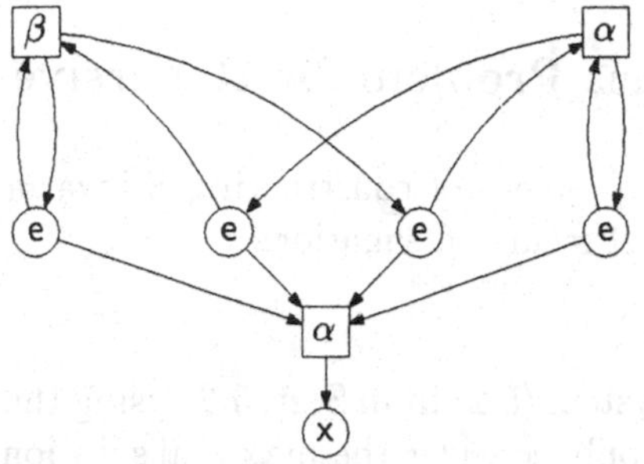

Fig. 6.7. The box $\Omega_{\langle *]}(\Sigma_{v_2}, \Sigma_{v_3})$

Still worse, with one more level of refinements, we may get a net which is only 3-bounded from its entry marking, as exhibited in Fig. 6.8 with the net modelling the extended PBC expression $\langle\ (\langle(\langle(\alpha\|\beta) * \gamma]\ \|\ \delta) * \epsilon\ ]$. However, Prop. 6.2.4 is applicable, and the nets in Figs. 6.8 and 6.7 exhibit no auto-concurrency from their entry markings.

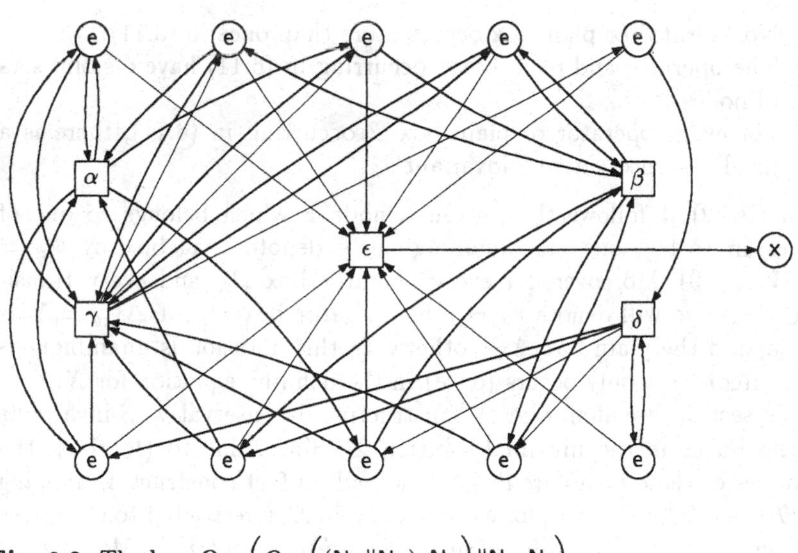

Fig. 6.8. The box $\Omega_{\langle *]}\left(\Omega_{\langle *]}\left((\mathsf{N}_\alpha\|\mathsf{N}_\beta), \mathsf{N}_\gamma\right)\|\mathsf{N}_\delta, \mathsf{N}_\epsilon\right)$

Exercise 6.2.5. Determine the name trees of the places and transitions in Fig. 6.8, and find the bound of each place (from the entry marking).

Exercise 6.2.6. For every $n \geq 1$, design a net built using $\Omega_{(*]}$ which is n-bounded from its entry marking, but not $(n-1)$-bounded.

6.3 The Synthesis Problem for Recursive Systems

We now turn to the problem of constructing S-invariants for the maximal solution of a system of recursive equations:

$$X \stackrel{\text{df}}{=} \Omega_X(\boldsymbol{\Delta}_X) \tag{6.11}$$

for $X \in \mathcal{X}$, i.e., the system (5.2) from Sect. 5.2. Using the notation introduced in Sect. 5.5, we shall only consider the maximal solution $\breve{\Sigma}^{\max}$ which will be denoted by $\breve{\Sigma}$; we will also denote by $S_{\breve{\Sigma}}$ and $T_{\breve{\Sigma}}$ the places and transitions in the boxes of $\breve{\Sigma}$, respectively.

As noticed in Sect. 5.4.1, nonmaximal solutions of recursive systems may not be clean if infinite steps are allowed, which is related to there being no covering by S-invariants (in Fig. 5.3, for example, the isolated place in $\Sigma^{\max:\min}$ is not covered by any 1-conservative S-invariant). Hence we shall concentrate on the maximal solution whose closed form was presented in Sect. 5.5.2.

Assumptions. Throughout this section we shall make some assumptions which will ease the presentation without losing its generality.

(Rec1) No operator or plain box occurs more than once in (6.11).
(Rec2) The operator and plain boxes occurring in (6.11) have disjoint sets of nodes.
(Rec3) For every operator or plain box Ψ occurring in (6.11), there is a fixed[1] 1-conservative S-invariant θ_Ψ.

From (Rec2) it follows that, given a node z which belongs to one of the boxes in (6.11), one can unambiguously denote that box by $\Psi_z = (S_z, T_z, W_z, \lambda_z, \emptyset)$. Moreover, for every operator box Ω_X and every transition $v \in T_{\Omega_X}$, we will denote by Θ_v the operator box Ω_Y, if $\Delta_{X,v} = Y$ is a variable, and the plain box $\Delta_{X,v}$ otherwise; this notation is unambiguous since, by (Rec1-2), v only occurs (once) in the defining equation for X.

In this section, we shall aim at constructing 1-conservative S-invariants for all the boxes in the maximal solution $\breve{\Sigma}$. Since, due to (Rec1-2), the sets of nodes of these boxes are disjoint, we will in fact construct a mapping $\breve{\theta} : S_{\breve{\Sigma}} \cup T_{\breve{\Sigma}} \to \Re^+$ such that, for each box Σ_X in $\breve{\Sigma}$, $\breve{\theta}$ restricted to the nodes of Σ_X is an S-invariant. We will call such a $\breve{\theta}$ an *S-invariant of* $\breve{\Sigma}$. Moreover, if

[1] In Sect. 6.3.3, we shall see how we can be more flexible without redeveloping the whole theory.

the resulting S-invariants are all ex-binary or are all S-aggregates, then $\breve{\theta}$ will also be called an *ex-binary* S-invariant or an *S-aggregate* of $\breve{\Sigma}$, respectively. By (Rec2), we can use the notation $\theta(z)$ instead of $\theta_{\Psi_z}(z)$, for every node z in the boxes occurring in (6.11). This extends to finite sequences of nodes $\sigma = z_1 \dots z_n$ by $\theta(\sigma) = \theta(z_1) \cdot \dots \cdot \theta(z_n)$.

6.3.1 Name Trees of Nets in the Maximal Solution

Dealing with the maximal solution of (6.11) can be made easier by a direct analysis of trees used as the names of places and transitions. We have already seen in Sects. 5.2.5 and 5.5.2 how the notion of a trail can be used to calculate the arc weight functions for maximal solutions of system(s) of recursive equations. But the applicability of trail-related notions goes further than that. We now provide a systematic presentation of some basic properties of such notions. All the results in this section follow from the definition of the closed form of the maximal solution of a recursive system in Sect. 5.5.2; we shall simply look closer at the set of finite and infinite paths occurring in the name trees.

For every place $p \in S_{\breve{\Sigma}}$, we will denote by $\mathsf{paths}(p)$ the multiset of *finite paths* of p:

$$\sum_{\xi \text{ is a place node in } p} \{\mathsf{path}(\xi)\}$$

and by $\mathsf{paths}^\infty(p)$ the multiset of *infinite paths* of p:

$$\sum_{\substack{\omega \text{ is an infinite path} \\ \text{from the root of } p}} \{\mathsf{path}(\omega)\}$$

where $\mathsf{path}(\omega)$ is the sequence of node labels along the infinite path ω. We will also denote by Paths and Paths^∞, respectively, the sets of all finite paths and infinite paths of all the places in $S_{\breve{\Sigma}}$. Note that if s is an isolated place in an operator box Ω_X in (6.11), then it is also an isolated place in the net Σ_X of $\breve{\Sigma}$ and $\mathsf{paths}(s) = \mathsf{trails}(s) = \{s\}$.

In what follows, s (possibly indexed) will be a place in a box of (6.11), and v (also possibly indexed) a transition.

Proposition 6.3.1. Let $p \in S_{\breve{\Sigma}}$ and

$$\phi = v_1 \dots v_m s_0 v_{m+1} \dots s_{n-1} v_{m+n} s_n \in \mathsf{paths}(p),$$

where $v_1, \dots, v_{m+n}$ are transition names and $s_0, \dots, s_n$ are place names.

(1) $\mathsf{trail}(\phi) = \sigma s_n$ with $\sigma = v_1 v_2 \dots v_{m+n}$.
(2) For every $0 \le i \le n$, s_i is a place in $\Theta_{v_{m+i}}$, and $\Psi_{s_n} = \Theta_{v_{m+n}}$ is a plain box if and only if ϕ leads to a leaf in p.
(3) For every $1 \le i \le n$, $v_{m+i} \in {}^\bullet s_{i-1} \cup s_{i-1}^\bullet$.

(4) p is an internal place if and only if s_0 is an internal place; moreover, if $m \geq 1$, then s_0 is an internal place (talking about s_0 'being an internal place' is allowed due to (Rec2)).
(5) If $n \geq 1$, then $s_1, \ldots, s_n$ are all entry or exit places.
(6) If the operator boxes in (6.11) are all ex-directed, then $s_1, \ldots, s_n$ are either all entry places or all exit places.
(7) If s_0 is an entry place, then so are $s_1, \ldots, s_n$, provided that operator boxes in (6.11) are all e-directed.
(8) If s_0 is an exit place, then so are $s_1, \ldots, s_n$, provided that all the operator boxes in (6.11) are all x-directed.
(9) $\mathsf{paths}(p)$ and $\mathsf{paths}^\infty(p)$ are sets.
(10) $\mathsf{trails}(p)$ is a set, provided that the operator boxes in (6.11) are all pure or all ex-directed.
(11) If $u \in T_{\check{\Sigma}}$ and $v_1 \cdots v_n v \in \mathsf{trails}(u)$, then $\Psi_v = \Theta_{v_n}$ is a plain box.

Two paths, ϕ and ϕ', in $\mathsf{Paths} \cup \mathsf{Paths}^\infty$ will be called *agreeable* if they are equal, or if they can be decomposed as $\phi = \eta s t \psi$ and $\phi' = \eta s t' \psi'$, where t and t' are distinct transitions and s is a place. One then easily obtains the following.

Proposition 6.3.2. Let $p \in S_{\check{\Sigma}}$ and the operator boxes in (6.11) be all pure. Then $\mathsf{paths}(p) \cup \mathsf{paths}^\infty(p)$ is a set of mutually agreeable paths. Conversely, if $Z \subseteq \mathsf{Paths} \cup \mathsf{Paths}^\infty$ is a set of mutually agreeable paths, then there is $p \in S_{\check{\Sigma}}$ such that $Z \subseteq \mathsf{paths}(p) \cup \mathsf{paths}^\infty(p)$. □

Paths leading to leaf nodes provide a characterisation of the pre- and post-sets of places. For $p \in S_{\check{\Sigma}}$ and $\delta \in \{\mathsf{e}, \mathsf{i}, \mathsf{x}\}$, we denote by $\mathsf{paths}^\delta(p)$ the set of all paths $\phi s \in \mathsf{paths}(p)$ leading to a leaf node such that $\lambda_{\Psi_s}(s) = \delta$. One can see that the following holds.

Proposition 6.3.3. Let p and r be places in the same net Σ_X of $\check{\Sigma}$.

(1) $\mathsf{paths}^\mathsf{i}(p) \neq \emptyset$ if and only if p is an internal place such that $\mathsf{paths}^\mathsf{i}(p) = \mathsf{paths}(p) = \mathsf{trails}(p)$. Moreover, a nonempty $\mathsf{paths}^\mathsf{i}(p)$ is a singleton set.
(2) Let the plain boxes occurring in (6.11) be all e-directed (or all x-directed) and $\mathsf{paths}^\mathsf{i}(p) = \emptyset = \mathsf{paths}^\mathsf{x}(p)$ (respectively, $\mathsf{paths}^\mathsf{i}(p) = \emptyset = \mathsf{paths}^\mathsf{e}(p)$). Then ${}^\bullet p = \emptyset$ (respectively, $p^\bullet = \emptyset$).
(3) Let the plain boxes occurring in (6.11) be all e-directed (or all x-directed) and $\mathsf{paths}^\mathsf{i}(p) = \mathsf{paths}^\mathsf{i}(r) = \emptyset$ and $\mathsf{trail}(\mathsf{paths}^\mathsf{x}(p)) = \mathsf{trail}(\mathsf{paths}^\mathsf{x}(r))$ (respectively, $\mathsf{trail}(\mathsf{paths}^\mathsf{e}(p)) = \mathsf{trail}(\mathsf{paths}^\mathsf{e}(r))$). Then ${}^\bullet p = {}^\bullet r$ (respectively, $p^\bullet = r^\bullet$).

6.3.2 Composing S-invariants for Recursive Boxes

For every transition u in the boxes $\check{\Sigma}$, we define

$$\check{\theta}(u) = \sum_{\sigma t \in \mathsf{trails}(u)} \theta(\sigma t) \cdot \mathsf{trails}(u)(\sigma t) \tag{6.12}$$

which is always a real number as u is a finite tree. This gives the weights for $T_{\check{\Sigma}}$. We do not give a similar recipe for the places in the nets $\check{\Sigma}$, since there may be several different ways to define their weight. Rather, we shall assume that $\check{\theta}$ is defined in such a way that, for every $\sigma s \in \mathsf{trail}(\mathsf{Paths})$,

$$\sum_{\substack{p \in S_{\check{\Sigma}} \text{ such that} \\ \sigma s \in \mathsf{trail}(\mathsf{paths}(p))}} \check{\theta}(p) \cdot \mathsf{trail}(\mathsf{paths}(p))(\sigma s) = \theta(\sigma s) . \tag{6.13}$$

Note that $\mathsf{trail}(\mathsf{paths}(p))(\sigma s)$ is the number of paths $\phi \in \mathsf{paths}(p)$ such that $\mathsf{trail}(\phi) = \sigma s$.

Theorem 6.3.1. A mapping $\check{\theta}$ satisfying (6.12) and (6.13) is a 1-conservative S-invariant of $\check{\Sigma}$.

Proof. Let $X \in \mathcal{X}$ and θ_X be $\check{\theta}$ restricted to the nodes of $\Sigma = \Sigma_X$. To show (6.6), suppose that u is a transition in Σ. We then proceed as follows.

$$\begin{aligned}
&\sum_{p \in S_\Sigma} W_\Sigma(p,u) \cdot \theta_X(p) \\
&= \sum_{p \in S_\Sigma} \check{\theta}(p) \cdot \sum_{\substack{\sigma, s, t \text{ such that} \\ \sigma t \in \mathsf{trails}(u),\, \sigma s \in \mathsf{trails}(p)}} \mathsf{trails}(u)(\sigma t) \cdot \mathsf{trails}(p)(\sigma s) \cdot W_t(s,t) && \text{(by (5.21))} \\
&= \sum_{p \in S_\Sigma} \sum_{\substack{\sigma, s, t \text{ such that} \\ \sigma t \in \mathsf{trails}(u),\, \sigma s \in \mathsf{trails}(p)}} \mathsf{trails}(u)(\sigma t) \cdot W_t(s,t) \cdot \check{\theta}(p) \cdot \mathsf{trails}(p)(\sigma s) \\
&= \sum_{\sigma t \in \mathsf{trails}(u)} \sum_{s \in S_t} \sum_{\substack{p \in S_\Sigma \text{ such that} \\ \sigma s \in \mathsf{trails}(p)}} \mathsf{trails}(u)(\sigma t) \cdot W_t(s,t) \cdot \check{\theta}(p) \cdot \mathsf{trails}(p)(\sigma s) \\
& && \text{(by (6.1))} \\
&= \sum_{\sigma t \in \mathsf{trails}(u)} \mathsf{trails}(u)(\sigma t) \cdot \sum_{s \in S_t} W_t(s,t) \cdot \sum_{\substack{p \in S_{\check{\Sigma}} \text{ such that} \\ \sigma s \in \mathsf{trails}(p)}} \check{\theta}(p) \cdot \mathsf{trails}(p)(\sigma s) && (*) \\
&= \sum_{\sigma t \in \mathsf{trails}(u)} \mathsf{trails}(u)(\sigma t) \cdot \sum_{s \in S_t} W_t(s,t) \cdot \theta(\sigma s) && (**) \\
&= \sum_{\sigma t \in \mathsf{trails}(u)} \mathsf{trails}(u)(\sigma t) \cdot \theta(\sigma) \cdot \sum_{s \in S_t} W_t(s,t) \cdot \theta(s) && \text{(by } \theta(\sigma s) = \theta(\sigma) \cdot \theta(s)) \\
&= \sum_{\sigma t \in \mathsf{trails}(u)} \mathsf{trails}(u)(\sigma t) \cdot \theta(\sigma) \cdot \theta(t) && (\theta_{\Psi_t} \text{ is an S-invariant}) \\
&= \theta_X(u). && \text{(by } \theta(\sigma t) = \theta(\sigma) \cdot \theta(t) \text{ and (6.12))}
\end{aligned}$$

Note that (*) holds since, by (Rec1-2), if p is a place such that $\sigma t \in \mathsf{trails}(u)$ and $\sigma s \in \mathsf{trails}(p)$, for some σ, then $p \in S_\Sigma$. Moreover, (**) holds by (6.13) and since $\sigma t \in \mathsf{trails}(u)$ implies that σ is nonempty and Ψ_t is a plain box, and so $\mathsf{trail}(\mathsf{paths}(p))(\sigma s) = \mathsf{trails}(p)(\sigma s)$.

Thus θ_X is an S-invariant of Σ, since we can show

$$\sum_{p \in S_\Sigma} W_\Sigma(u,p) \cdot \theta_X(p) = \theta_X(u)$$

in a similar way. Moreover, θ_X is 1-conservative since

$$\begin{aligned}\sum_{p \in {}^\circ\Sigma} \theta_X(p) &= \sum_{e \in {}^\circ\Omega_X} \sum_{\substack{p \in S_\Sigma \text{ such that} \\ e \in \mathsf{trail}(\mathsf{paths}(p))}} \breve{\theta}(p) \\ &=_{(*)} \sum_{e \in {}^\circ\Omega_X} \sum_{\substack{p \in \breve{\Sigma} \text{ such that} \\ e \in \mathsf{trail}(\mathsf{paths}(p))}} \breve{\theta}(p) \\ &=_{(6.13)} \sum_{e \in {}^\circ\Omega_X} \theta(e) =_{(**)} 1\end{aligned}$$

and, similarly, for the exit places. Note that (*) holds by (Rec1-2) and (**) holds by (Rec3). Moreover, (6.13) can be applied since $\mathsf{trails}(\mathsf{paths}(p))(e) = 1$ for p satisfying $e \in \mathsf{trail}(\mathsf{paths}(p))$. □

The way in which $\breve{\theta}$ has been specified is similar to that in which S-invariants were treated in the previous section. Again, the characterisation we provided is not fully constructive, the problem being here more complicated than before, due to the fact that we may have place trees which are not only infinitely wide but also infinitely deep. We shall see, however, that in some relevant cases the characterisation given in Thm. 6.3.1 may be fruitfully exploited. Notice also that once we obtained some S-invariants of $\breve{\Sigma}$, new ones may be constructed by forming (possibly weighted) averages of finite sets of those already known, and by applying techniques developed in Sect. 6.2 for net refinement to the equation $\breve{\Sigma} = \breve{\mathbf{F}}(\breve{\Sigma})$, since each $\mathbf{F}_X$ is defined in terms of net refinement; such a procedure may be applied iteratively.

The Node-finitary Case. The recursive system (6.11) is *node-finitary* if all name trees of places of its maximal solution $\breve{\Sigma}$ are finite (recall that transition trees are always finite). Thus, by taking into account the form of place trees in Sect. 5.5.2, (6.11) is node-finitary if the places in all its operator boxes are degree-finite, and there are no infinite paths in the graph defined in Sect. 5.5.3; the latter always holds if the system is n-guarded for some n. For a node-finitary system, it is possible to devise a generalisation of the multiplicative distribution functions.

Let $p = v_1 \lhd \cdots \lhd v_k \lhd s \lhd S$ be a place in $S_{\breve{\Sigma}}$ where $k \geq 0$. Then

$$\breve{\theta}(p) = \theta(v_1 \ldots v_k) \cdot \prod_{\xi \in \Xi} \theta(s_\xi) , \tag{6.14}$$

where Ξ is the set of all place nodes ξ in p and s_ξ is the label of ξ, which is a place in an operator or plain box in (6.11). The above formula always yields a real number since p is a finite tree. And, crucially, we can show that (6.13) is satisfied.

Proposition 6.3.4. For every $\sigma s \in \mathsf{trail}(\mathsf{Paths})$,

$$\sum_{\substack{p \text{ such that} \\ \sigma s \in \mathsf{trail}(\mathsf{paths}(p))}} \breve{\theta}(p) \cdot \mathsf{trail}(\mathsf{paths}(p))(\sigma s) \;=\; \theta(\sigma s) \,.$$

Proof. We shall in fact slightly extend the property to be proven, by allowing $\sigma s = s$ where s is a place in a plain box occurring in (6.11), and setting $\mathsf{paths}(s) = \{s\}$ and $\breve{\theta}(s) = \theta(s)$ in such a case; thus $\{p \mid s \in \mathsf{trail}(\mathsf{paths}(p))\} = \{p \mid s \in \mathsf{paths}(p)\} = \{s\}$.

We first show that for all $\phi = v_1 \ldots v_m s_0 v_{m+1} \ldots s_{n-1} v_{m+n} s_n \in \mathsf{Paths}$,

$$\sum_{\substack{p \text{ such that} \\ \phi \in \mathsf{paths}(p)}} \breve{\theta}(p) \;=\; \theta(\lceil \phi \rceil) \,, \tag{6.15}$$

where $\lceil \phi \rceil = v_1 \ldots v_m s_0 \ldots s_{n-1} s_n$. Note that we again allow $\phi = s$ where s is a place in a plain box occurring in (6.11). To show (6.15), we proceed by induction on the length of ϕ.

Base case for the proof of (6.15): We have $\phi = s$, where s is a place in one of the operator or plain boxes of the system. Let us introduce a relation $\rightsquigarrow$ on places of the boxes in (6.11) such that $s' \rightsquigarrow s''$ if s'' occurs in the tree of some place $p \in S_{\Sigma}$ whose root is labelled s'. It may be checked from the form of the maximal solution that $\rightsquigarrow$ is transitive and reflexive; it is also antisymmetric, otherwise the system would not be node-finitary (there would be an infinite place name tree with a path labelled $\ldots s' \ldots s'' \ldots s' \ldots s'' \ldots$ in the maximal solution of the recursive system). Hence $\rightsquigarrow$ is a partial order. For the same reason, it has no infinitely descending chains (i.e., infinite sequences of distinct elements such that $\ldots \rightsquigarrow s_2 \rightsquigarrow s_1 \rightsquigarrow s_0$) so that $\rightsquigarrow$ may be used to carry out induction-based proofs, as shown below.

In the base step, s is either an isolated place in one of the operator boxes or a place in a plain box of (6.11). In either case,

$$\sum_{\substack{p \text{ such that} \\ s \in \mathsf{paths}(p)}} \breve{\theta}(p) \;=\; \breve{\theta}(s) \;=\; \theta(s) \;=\; \theta(\lceil s \rceil) \,.$$

In the induction step, s is a non-isolated place in one of the operator boxes for whose predecessors in $\rightsquigarrow$ (6.15) does hold. Let ${}^\bullet s = \{v_1, \ldots, v_m\}$ and $s^\bullet = \{w_1, \ldots, w_n\}$. (Note that ${}^\bullet s \cup s^\bullet$ is finite since otherwise (6.11) would not be node-finitary.) Then,

$$\sum_{\substack{p \text{ such that} \\ s \in \mathsf{paths}(p)}} \breve{\theta}(p)$$

$$= \sum_{i=1}^{m} \sum_{s_i \in \Theta^{\circ}_{v_i}} \sum_{\substack{p_i \text{ such that} \\ s_i \in \mathsf{paths}(p_i)}} \sum_{j=1}^{n} \sum_{r_j \in {}^{\circ}\Theta_{w_j}} \sum_{\substack{q_j \text{ such that} \\ r_j \in \mathsf{paths}(q_j)}} \breve{\theta}\left(s \lhd \{v_1 \lhd p_1, \ldots, v_m \lhd p_m, w_1 \lhd q_1, \ldots, w_n \lhd q_n\}\right)$$

$$= \sum_{i=1}^{m} \sum_{s_i \in \Theta^{\circ}_{v_i}} \sum_{\substack{p_i \text{ such that} \\ s_i \in \mathsf{paths}(p_i)}} \sum_{j=1}^{n} \sum_{r_j \in {}^{\circ}\Theta_{w_j}} \sum_{\substack{q_j \text{ such that} \\ r_j \in \mathsf{paths}(q_j)}} \theta(s) \cdot \breve{\theta}(p_1) \cdot \ldots \cdot \breve{\theta}(p_m) \cdot \breve{\theta}(q_1) \cdot \ldots \cdot \breve{\theta}(q_n) \qquad (*)$$

$$= \theta(s) \cdot \left(\prod_{i=1}^{m} \sum_{s_i \in \Theta^{\circ}_{v_i}} \sum_{\substack{p_i \text{ such that} \\ s_i \in \mathsf{paths}(p_i)}} \breve{\theta}(p_i) \right) \qquad (**)$$

$$\cdot \left(\prod_{j=1}^{n} \sum_{r_j \in {}^{\circ}\Theta_{w_j}} \sum_{\substack{q_j \text{ such that} \\ r_j \in \mathsf{paths}(q_j)}} \breve{\theta}(q_j) \right)$$

$$= \theta(s) \cdot \left(\prod_{i=1}^{m} \sum_{s_i \in \Theta^{\circ}_{v_i}} \theta(s_i) \right) \cdot \left(\prod_{j=1}^{n} \sum_{r_j \in {}^{\circ}\Theta_{w_j}} \theta(r_j) \right) \qquad (\text{by } \rightsquigarrow\text{-induction})$$

$$= \theta(s) \cdot \left(\prod_{i=1}^{m} 1 \right) \cdot \left(\prod_{j=1}^{n} 1 \right) = \theta(s) . \qquad (\text{by (Rec3)})$$

Note that (*) and (**) hold by (6.14) and (6.5), respectively.

Inductive step for the proof of (6.15): We consider two cases.

Case 1: $\phi = v\phi'$. Then ϕ' is nonempty and $\{p \mid \phi \in \mathsf{paths}(p)\} = \{v \lhd p' \mid \phi' \in \mathsf{paths}(p')\}$, and we obtain the following.

$$\sum_{\substack{p \text{ such that} \\ \phi \in \mathsf{paths}(p)}} \breve{\theta}(p)$$

$$= \sum_{\substack{p' \text{ such that} \\ \phi' \in \mathsf{paths}(p')}} \theta(v) \cdot \breve{\theta}(p') \qquad (\text{by (6.14)})$$

$$= \theta(v) \cdot \sum_{\substack{p' \text{ such that} \\ \phi' \in \mathsf{paths}(p')}} \breve{\theta}(p')$$

$$= \theta(v) \cdot \theta(\lceil \phi' \rceil) \qquad (\text{by the induction hypothesis})$$

$$= \theta(\lceil \phi \rceil) . \qquad (\text{by } \lceil \phi \rceil = v \lceil \phi' \rceil)$$

Case 2: $\phi = sv\phi'$. Then ϕ' is nonempty and starts either with a place in ${}^\circ\Theta_v$ or in Θ_v°. Suppose that the first case holds (the other one is similar) so that $v \in s^\bullet$. Let $s^\bullet = \{w_1, \ldots, w_n\} \cup \{v\}$ and ${}^\bullet s = \{v_1, \ldots, v_m\}$. (Again, note that ${}^\bullet s \cup s^\bullet$ is finite since otherwise (6.11) would not be finitary.) Then the following holds, similar to the last part of the proof of the base case for (6.15).

$$\begin{aligned}
&\sum_{\substack{p \text{ such that} \\ \phi \in \mathsf{paths}(p)}} \breve{\theta}(p) \\
&= \sum_{\substack{p' \text{ such that} \\ \phi' \in \mathsf{paths}(p')}} \sum_{i=1}^{m} \sum_{s_i \in \Theta_{v_i}^\circ} \sum_{\substack{p_i \text{ such that} \\ s_i \in \mathsf{paths}(p_i)}} \sum_{j=1}^{n} \sum_{r_j \in {}^\circ\Theta_{w_j}} \sum_{\substack{q_j \text{ such that} \\ r_j \in \mathsf{paths}(q_j)}} \\
&\qquad \breve{\theta}(s \lhd \{v \lhd p', v_1 \lhd p_1, \ldots, v_m \lhd p_m, w_1 \lhd q_1, \ldots, w_n \lhd q_n\}) \\
&= \sum_{\substack{p' \text{ such that} \\ \phi' \in \mathsf{paths}(p')}} \sum_{i=1}^{m} \sum_{s_i \in \Theta_{v_i}^\circ} \sum_{\substack{p_i \text{ such that} \\ s_i \in \mathsf{paths}(p_i)}} \sum_{j=1}^{n} \sum_{r_j \in {}^\circ\Theta_{w_j}} \sum_{\substack{q_j \text{ such that} \\ r_j \in \mathsf{paths}(q_j)}} \\
&\qquad \theta(s) \cdot \breve{\theta}(p') \cdot \breve{\theta}(p_1) \cdots \breve{\theta}(p_m) \cdot \breve{\theta}(q_1) \cdots \breve{\theta}(q_n) \\
&= \theta(s) \cdot \left(\sum_{\substack{p' \text{ such that} \\ \phi' \in \mathsf{paths}(p')}} \breve{\theta}(p') \right) \cdot \left(\prod_{i=1}^{m} \sum_{s_i \in \Theta_{v_i}^\circ} \sum_{\substack{p_i \text{ such that} \\ s_i \in \mathsf{paths}(p_i)}} \breve{\theta}(p_i) \right) \\
&\qquad \cdot \left(\prod_{j=1}^{n} \sum_{r_j \in {}^\circ\Theta_{w_j}} \sum_{\substack{q_j \text{ such that} \\ r_j \in \mathsf{paths}(q_j)}} \breve{\theta}(q_j) \right) \\
&= \theta(s) \cdot \theta(\lceil \phi' \rceil) \cdot \left(\prod_{i=1}^{m} \sum_{s_i \in \Theta_{v_i}^\circ} \theta(s_i) \right) \cdot \left(\prod_{j=1}^{n} \sum_{r_j \in {}^\circ\Theta_{w_j}} \theta(r_j) \right) && (*) \\
&= \theta(s) \cdot \theta(\lceil \phi' \rceil) \cdot \left(\prod_{i=1}^{m} 1 \right) \cdot \left(\prod_{j=1}^{n} 1 \right) && \text{(by (Rec3))} \\
&= \theta(\lceil \phi \rceil) && (\text{by } \lceil \phi \rceil = s\lceil \phi' \rceil)
\end{aligned}$$

where (*) follows from the base case and induction hypothesis. This completes the proof of (6.15).

The next auxiliary result is that, for every $\sigma s \in \mathsf{trail}(\mathsf{Paths})$,

$$\sum_{\phi \in \mathsf{Paths}(\sigma s)} \theta(\lceil \phi \rceil) = \theta(\sigma s) , \tag{6.16}$$

where $\mathsf{Paths}(\sigma s) = \{\phi \in \mathsf{Paths} \mid \mathsf{trail}(\phi) = \sigma s\}$. Indeed, if s is an internal place, then $\mathsf{Paths}(\sigma s) = \{\sigma s\}$ and the result holds. Suppose that s is an entry place (if s is an exit place, the proof is similar). We proceed by induction on the length of σ. In the base case, $\mathsf{Paths}(s) = \{s\}$ and the result clearly holds. In the inductive step, we assume $\sigma = \sigma' v$. Then $\mathsf{Paths}(\sigma s) = \{\phi' v s \mid \phi' \in \mathsf{Paths}(\sigma' s') \wedge s' \in {}^{\bullet}v\}$, since if s (which is a non-internal place) was the first and only place, we would have $\sigma = \varepsilon$. As a result, we obtain

$$
\begin{aligned}
& \sum_{\phi \in \mathsf{Paths}(\sigma s)} \theta(\lceil \phi \rceil) \\
&= \sum_{s' \in {}^{\bullet}v} \sum_{\phi' \in \mathsf{Paths}(\sigma' s')} \theta(\lceil \phi' v s \rceil) \\
&= \sum_{s' \in {}^{\bullet}v} \sum_{\phi' \in \mathsf{Paths}(\sigma' s')} \theta(\lceil \phi' \rceil) \cdot \theta(s) && \text{(by } \lceil \phi' v s \rceil = \lceil \phi' \rceil \cdot s\text{)} \\
&= \theta(s) \cdot \sum_{s' \in {}^{\bullet}v} \sum_{\phi' \in \mathsf{Paths}(\sigma' s')} \theta(\lceil \phi' \rceil) \\
&= \theta(s) \cdot \sum_{s' \in {}^{\bullet}v} \theta(\sigma' s') && \text{(by the induction hypothesis)} \\
&= \theta(s) \cdot \theta(\sigma') \cdot \sum_{s' \in {}^{\bullet}v} \theta(s') \\
&= \theta(s) \cdot \theta(\sigma') \cdot \sum_{s' \in {}^{\bullet}v} \theta(s') \cdot W_v(s', v) && \text{(operator boxes are simple)} \\
&= \theta(s) \cdot \theta(\sigma') \cdot \theta(v) && (\theta_{\Psi_v} \text{ is an S-invariant}) \\
&= \theta(\sigma s) \, .
\end{aligned}
$$

Hence we have demonstrated that (6.16) also holds. We then obtain, for every $\sigma s \in \mathsf{trail}(\mathsf{Paths})$,

$$
\begin{aligned}
& \sum_{\substack{p \text{ such that} \\ \sigma s \in \mathsf{trail}(\mathsf{paths}(p))}} \breve{\theta}(p) \cdot \mathsf{trail}(\mathsf{paths}(p))(\sigma s) \\
&= \sum_{\phi \in \mathsf{Paths}(\sigma s)} \sum_{\substack{p \text{ such that} \\ \phi \in \mathsf{paths}(p)}} \breve{\theta}(p) && (*) \\
&= \sum_{\phi \in \mathsf{Paths}(\sigma s)} \theta(\lceil \phi \rceil) && \text{(by (6.15))} \\
&= \theta(\sigma s) \, . && \text{(by (6.16))}
\end{aligned}
$$

Note that (*) holds since the same p may occur many times, with different ϕ's satisfying $\mathsf{trail}(\phi) = \sigma s$; moreover, $\mathsf{paths}(p)$ is a set, by Prop. 6.3.1(9). Hence (6.13) is satisfied. □

From the last proposition and Thm. 6.3.1 we obtain that for a node-finitary recursive system (6.11), the multiplicative scheme given by (6.12) and (6.14) yields a 1-conservative S-invariant for each box in the maximal solution of (6.11).

Theorem 6.3.2. If the recursive system (6.11) is node-finitary, then $\breve{\theta}$ given by (6.12) and (6.14) is a 1-conservative S-invariant of $\breve{\Sigma}$.

The Ex-binary Case. The multiplicative scheme for constructing S-invariants works also for the ex-binary case.

Proposition 6.3.5. Let all S-invariants in (Rec3) be ex-binary 1-conservative S-invariants (S-aggregates), and $\breve{\theta}$ be given by (6.12) and (6.14). Then $\breve{\theta}$ is an ex-binary 1-conservative S-invariant (respectively, S-aggregate) of $\breve{\Sigma}$.

Proof. Let $p = v_1 \lhd \cdots \lhd v_m \lhd s \lhd \mathcal{S}$ be a place in $\breve{\Sigma}$. Then $\breve{\theta}(p) = \theta(v_1 \ldots v_m) \cdot \theta(s)$ if there is no node in $\mathcal{S}$ labelled by a place r such that $\theta(r) = 0$ (note Prop. 6.3.1(5) and the assumption that the S-invariants in (Rec3) are ex-binary); and $\breve{\theta}(p) = 0$ otherwise. Thus $\breve{\theta}$ is a real-valued function.

The proof then follows that of Prop. 6.3.4. We observe that (6.16) holds as its proof does not rely on (6.11) being node-finitary. Moreover, (6.15) can be shown to hold since for every path $\phi = v_1 \ldots v_m s_0 v_{m+1} \cdots s_{n-1} v_{m+n} s_n$ such that $\theta(\lceil\phi\rceil) \neq 0$, there is exactly one p such that $\phi \in \mathsf{paths}(p)$ and $\breve{\theta}(p) \neq 0$. And for such p, $\breve{\theta}(p) = \theta(\lceil\phi\rceil) = \theta(v_1 \ldots v_m) \cdot \theta(s_0)$ due to Prop. 6.3.1(5) and S-invariants being ex-binary. This observation can also be used to show that $\breve{\theta}$ is an ex-binary 1-conservative S-invariant (respectively, an S-aggregate). □

Exercise 6.3.1. Work out the full details of the last proof.

6.3.3 Coverability Results

The results we have obtained lead to

Proposition 6.3.6. Let (6.11) be node-finitary and the boxes occurring in it be all S-covered (T-covered) by 1-conservative S-invariants. Then all the boxes in $\breve{\Sigma}$ are S-covered (respectively, T-covered) by 1-conservative S-invariants.

Proof. Similar to that of Prop. 6.2.2, using Thm. 6.3.2 as well as (6.12) and (6.14). Note that since we are not interested here in integer-valued S-invariants, we can construct arbitrary weighted averages of finite sets of S-invariants, which allows one to cover all nodes occurring in a finite place or transition tree of $\breve{\Sigma}$. □

When constructing S-invariants for the system of recursive equations (6.11), it was assumed in (Rec3) that each operator and operand box comes with a fixed S-invariant. However, to obtain coverability results for specific places and transitions in the maximal solution, it may be advantageous to use different S-invariants for the nodes of the same box. There is a simple solution to this problem.

It is always possible to replace (6.11) by another system of recursive equations, equivalent in the sense of generating isomorphic box solutions, generally after introducing fresh variables in such a way that, for every variable X, no variable is used more than once if we look forward from X in the dependency graph of the system of equations. For example, if $\mathcal{X} = \{X\}$ where $X \stackrel{\mathrm{df}}{=} \Omega(X, \Delta, X)$, then we can replace this single-equation system by an infinite system such that $\mathcal{X}' = \{X\} \cup \{X_\sigma \mid \sigma \in \{0,1\}^+\}$ and

$$X \stackrel{\mathrm{df}}{=} \Omega(X_0, \Delta, X_1) \quad \text{and} \quad X_\sigma \stackrel{\mathrm{df}}{=} \Omega_\sigma(X_{\sigma 0}, \Delta_\sigma, X_{\sigma 1})$$

for all $\sigma \in \{0,1\}^+$, where Ω_σ and Δ_σ are disjoint isomorphic copies of Ω and Δ, respectively. Formally, we proceed as follows.

The *dependency graph* for the recursive system (6.11) is a directed graph $\mathcal{DG}$ where $\{w_X \mid X \in \mathcal{X}\} \cup \{w_{X,v} \mid X \in \mathcal{X} \ \wedge \ v \in T_{\Omega_X} \ \wedge \ \Delta_{X,v} \in \mathcal{X}\}$ is the set of nodes, for each w_X there is an arc to every node $w_{X,v}$, and for every $w_{X,v}$ there is an arc to w_Y where $Y = \Delta_{X,v}$. The dependence graph $\mathcal{DG}$ is *unfolded* if there are no two different nonempty paths originating at the same node and ending at the same node (in particular, this means that the graph is acyclic).

Exercise 6.3.2. Show that there is always an unfolded recursive system over a set of variables $\mathcal{X}' \supseteq \mathcal{X}$ which uses only operator and operand boxes isomorphic to those present in (6.11), such that the maximal solution for the former, after restricting to the variables in $\mathcal{X}$, is isomorphic to the maximal solution for the latter.
Hint: Generalise the construction used in the last example.

Thus, without loss of generality, in the discussion that follows we can assume that (6.11) is unfolded, in addition to the assumptions (Rec1-3) we have already made. We can then use the results from the previous section to obtain stronger coverability results for the nets in the maximal solutions of recursive net equations.

Proposition 6.3.7.

(1) Let the boxes in (6.11) be all T-covered by ex-binary 1-conservative S-invariants. Then all the boxes in $\check{\Sigma}$ are T-covered by ex-binary 1-conservative S-invariants.
(2) Let the boxes in (6.11) be all S-covered by ex-binary 1-conservative S-invariants and, moreover, let the operator boxes be all pure. Then all the boxes in $\check{\Sigma}$ are S-covered by ex-binary 1-conservative S-invariants.
(3) Let the boxes in (6.11) be all S-covered by ex-binary 1-conservative S-invariants, and p be a place in $S_{\check{\Sigma}}$. Moreover, if $\phi v s \in \mathsf{paths}(p)$ and $\phi' v s' \in \mathsf{paths}(p)$ are such that $s, s' \in {}^\circ\Psi_v \cup \Psi_v^\circ$, then there is an ex-binary

1-conservative S-invariant which covers both s and s'. Then p is covered by an ex-binary 1-conservative S-invariant.

(4) Let the boxes in (6.11) be all S-covered by ex-binary 1-conservative S-invariants, and p be an entry (exit) place in $S_{\breve{\Sigma}}$. Moreover, let the operator boxes be all e-directed (respectively, all x-directed). Then p is covered by an ex-binary 1-conservative S-invariant.

Note: The same properties hold if we consistently replace S-invariant(s) by S-aggregate(s) in their formulation.

Proof. (1) Let u be a transition in $\breve{\Sigma}$ and $v_1 \dots v_n \in \mathsf{trails}(u)$. Since (6.11) is unfolded and (Rec1) holds, $\Psi_{v_i} \neq \Psi_{v_j}$, for all $i \neq j$. Thus we can choose, for every $i \leq n$, $\theta_{\Psi_{v_i}}$ to be an ex-binary S-invariant (respectively, S-aggregate) covering v_i. Hence $\breve{\theta}$ constructed as in Prop. 6.3.5 is an ex-binary S-invariant (respectively, S-aggregate) covering u (cf. (6.12)).

(2) Follows from Prop. 6.3.5 and the observation that, since (6.11) is unfolded and (Rec1) holds and the operator boxes are pure, if $p = v_1 \lhd \cdots \lhd v_m \lhd s \lhd \mathcal{S} \in S_{\breve{\Sigma}}$, then $\Psi_{v_1}, \dots, \Psi_{v_m}$ and Ψ_{s_ξ} (for $\xi \in \Xi$, where s_ξ and Ξ are as in the formulation of (6.14)) are all different boxes; hence the corresponding ex-binary S-invariants (respectively, S-aggregates) may be chosen independently.

(3) We first notice that, if $\phi ve, \phi' ve' \in \mathsf{paths}(p)$ and $e, e' \in {}^\circ\Psi_v$, then $e = e'$, since otherwise it would not be possible to simultaneously cover e and e' by an ex-binary 1-conservative S-invariant or S-aggregate (and, similarly, for paths ending with exit places); as a consequence, if $\phi vs\, phi' vs', \phi'' vs'' \in \mathsf{paths}(p)$, then either $s = s''$ or $s' = s''$. Hence, since (6.11) is unfolded and (Rec1) holds, if $p = v_1 \lhd \cdots \lhd v_m \lhd s \lhd \mathcal{S} \in S_{\breve{\Sigma}}$, then $\Psi_{v_1}, \dots, \Psi_{v_m}$, and any Ψ_{s_ξ} for $\xi \in \Xi$ (where s_ξ and Ξ are as in the formulation of (6.14)) are all different boxes. Moreover, it follows from our assumptions that if $\Psi_{s_\xi} = \Psi_{s'_\xi}$, then it is possible to choose an ex-binary 1-conservative S-invariant or S-aggregate covering simultaneously s and s'. The property then results from Prop. 6.3.5 and the possibility to choose independently the invariants of the different boxes.

Note that we cannot guarantee the S-coverability in full generality since a side-loop may require that we choose an S-invariant covering simultaneously some entry and some exit place.

(4) If p is an entry place, then the condition from (3) holds immediately by Prop. 6.3.1(7) and the fact that (6.11) is unfolded. The case of an exit place is symmetrical. □

Theorem 6.3.3. Let $\mathcal{O}$ be a set of pure unmarked operator boxes and $\mathcal{P}$ be a set of unmarked plain boxes, all S-covered by S-aggregates. Then all the boxes obtained through successive refinements and maximal solutions of recursive systems from $\mathcal{O}$ and $\mathcal{P}$ are static plain boxes S-covered by S-aggregates.

Proof. Follows from Thm. 6.2.3(1) and Props. 6.1.10(1) and 6.3.7(2). □

Some places p may have different paths of arbitrary length, with the same transition sequence. We have already observed this for paths with three nodes, when discussing refinement with a side-loop in Sect. 6.2.3. And, in the extreme case, an infinite path $sv_1e_1v_2e_2\dots$, where each e_i is an entry place, may coexist with an infinite path $sv_1x_1v_2x_2\dots$, where each x_i is an exit place (if the operator boxes are not ex-directed, it could even happen that some e_i's are exit places, and some x_i's are entry places, from $i = 2$). For consider the following recursive system:

$$X \stackrel{\mathrm{df}}{=} \Omega_{[*]}(\mathsf{N}_a, Y, \mathsf{N}_a)$$
$$Y \stackrel{\mathrm{df}}{=} \Omega(\mathsf{N}_a, Y, \mathsf{N}_a)\ ,$$

where Ω is shown in Fig. 6.9. Then there is an internal place p in Σ_X such that the following two paths belong to $\mathsf{paths}^{\infty}(p)$:

$$i_{[*]} \lhd v^2_{[*]} \lhd e_1 \lhd v \lhd e_1 \lhd v \lhd e_1 \lhd v \lhd e_1 \dots$$

$$i_{[*]} \lhd v^2_{[*]} \lhd x_2 \lhd v \lhd x_2 \lhd v \lhd x_2 \lhd v \lhd x_2 \dots\ .$$

We then observe that although Ω is S-covered by S-components (and so are $\Omega_{[*]}$ and N_a), the pair e_1 and x_2 may not be covered by the same S-aggregate (the only S-aggregates of Ω assign either $e_1, v_1, x_1 \mapsto 1$ and $i_1 \mapsto n$ and $v, i_2 \mapsto n+1$ and $x_2, v_2, e_2 \mapsto 0$; or $e_2, v_2, x_2 \mapsto 1$ and $i_2 \mapsto n$ and $v, i_1 \mapsto n+1$ and $x_1, v_1, e_1 \mapsto 0$).

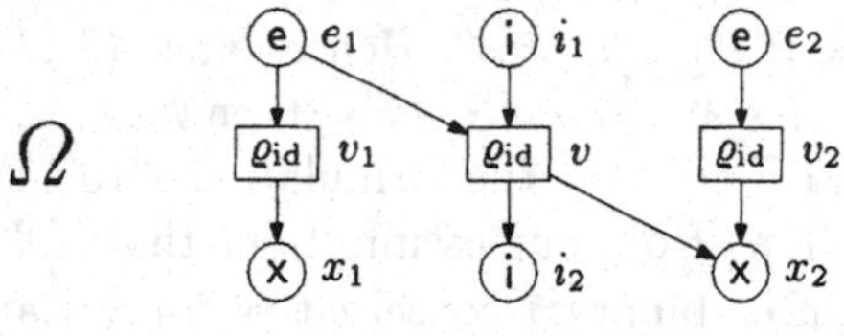

Fig. 6.9. An operator box, S-covered by S-components, without the entry/exit pair covering property

The last coverability result was formulated for pure operator boxes. A similar result may also be obtained for non-pure operators, after making restrictions allowing us to use Prop. 6.3.7(3).

Lemma 6.3.1. Let the boxes in (6.11) be all S-covered by ex-binary 1-conservative S-invariants, and the operator boxes be all ex-directed. Moreover, let p be a place in $\check{\Sigma}$ and $v_1 \dots v_n s \in \mathsf{paths}(p)$. Then, for every choice of ex-binary 1-conservative S-invariants $\theta_{\Psi_{v_i}}$ (for $i \le n$) and θ_{Ψ_s}, it is possible to construct a 1-conservative S-invariant $\check{\theta}$ for $\check{\Sigma}$ such that

$$\check{\theta}(p) = \theta(v_1 \dots v_n) \cdot \theta(s) \quad \text{or} \quad \check{\theta}(p) = \frac{1}{2} \cdot \theta(v_1 \dots v_n) \cdot \theta(s)\ .$$

Proof. Recall that (6.11) is assumed to be unfolded. The basic idea is to notice that for every $n \geq 0$, $\breve{\Sigma} = \breve{\mathbf{F}}^n(\breve{\Sigma})$, so that we may apply the construction techniques we developed for net refinement in Sect. 6.2, knowing already how to cover the entry and exit places in $\breve{\Sigma}$ using Prop. 6.3.7(4).
Let $p = v_1 \lhd \cdots \lhd v_n \lhd s \lhd \mathcal{S}$.

If Ψ_s is a plain box, then $n \geq 1$, s is an internal place, $p = v_1 \lhd \cdots \lhd v_n \lhd s$, and by applying n times Prop. 6.2.3 and Thm. 6.2.1 we see that, for $i = n, \ldots, 1$, it is possible to cover the place $v_i \lhd v_{i+1} \lhd \cdots \lhd v_n \lhd s$, by an S-aggregate with the weight $\theta(v_i v_{i+1} \ldots v_n s)$, the last one giving $\breve{\theta}(p) = \theta(v_1 \ldots v_n) \cdot \theta(s)$.

If $\Psi_s = \Omega_Y$ for some $Y \in \mathcal{X}$, we may first use the technique described in Sect. 6.2.4 (the **N**/2 case) in order to construct a covering of the place $s \lhd \mathcal{S}$, from the relationship $\Sigma_Y = \mathbf{F}_Y(\breve{\Sigma})$, and the coverings of the entry and exit places of $\breve{\Sigma}$ by ex-binary 1-conservative S-invariants (see Prop. 6.3.7(4)). This leads to a cover of that place with the weight $\theta(s)$ or $\frac{1}{2} \cdot \theta(s)$. Then, by applying n times Prop. 6.2.3 and Thm. 6.2.1, as above, we obtain that $\breve{\theta}(p) = \theta(v_1 \ldots v_n) \cdot \theta(s)$ or $\breve{\theta}(p) = \frac{1}{2} \cdot \theta(v_1 \ldots v_n) \cdot \theta(s)$. □

Finally, we obtain two additional coverability results.

Proposition 6.3.8.

(1) Let the boxes in (6.11) be all S-covered by ex-binary 1-conservative S-invariants, and the operator boxes be all ex-directed. Then all the boxes in $\breve{\Sigma}$ are S-covered by 1-conservative S-invariants.
(2) Let the operator boxes in (6.11) be all ex-directed and S-covered by S-aggregates, and the plain boxes be all semi-static. Then all the boxes in $\breve{\Sigma}$ are semi-static.

Proof. The first part follows from Lemma 6.3.1. In the second part, the covering of the entry and exit places follows from Prop. 6.3.7(4), and the covering of the internal places by semi-aggregates from Lemma 6.3.1. □

Theorem 6.3.4. Let $\mathcal{O}$ be a set of unmarked ex-directed operator boxes S-covered by S-aggregates, and $\mathcal{P}$ be a set of semi-static plain boxes. Then all the boxes obtained through successive refinements and maximal solutions of recursive systems from $\mathcal{O}$ and $\mathcal{P}$ are semi-static plain boxes.

Proof. Follows from Thm. 6.2.4 and Props. 6.3.8(2). □

We may now state that all boxes modelling extended PBC expressions given by the syntax in Sect. 3.4 are static ex-directed boxes if we use the safe version for the iteration operator, and semi-static if we use a simpler, but only 2-bounded, version; hence, Thms. 6.3.3 and 6.3.4 directly apply to the PBC theory. We may further observe that our results on recursive systems encompass those obtained for net refinements, since a net refinement may always be transformed into a recursion. Thus, in order to study specifications involving both refinement and recursion, we may restrict ourselves to studying those involving only recursion. We may also observe that, like in Sect. 6.2.4, the ex-directedness property in the above results could be replaced by a condition that no entry or exit place in an operator box is involved in a side-loop.

Unbounded Solutions. The situation deteriorates dramatically if we allow operator boxes with side-loops involving entry or exit places, since the above coverability results do not apply and, indeed, it is then possible to generate (auto-concurrency-free[2]) unbounded solutions of a system of recursive equations. Let us consider, for example, the system

$$\begin{aligned} X &\stackrel{\mathrm{df}}{=} \Omega_{(*]}(Y, \mathsf{N}_\beta) \\ Y &\stackrel{\mathrm{df}}{=} X \| \mathsf{N}_\alpha \,. \end{aligned} \tag{6.17}$$

From the form of the maximal solution described in Sect. 5.5.2 and the nets defined in Figs. 4.21 and 4.44, it follows that in the maximal solution for X there are no internal places and there are a single exit place,

$$\xi = x_{(*]} \lhd v^2_{(*]} \lhd x_\beta \,,$$

and two infinite families of entry places. The places of the first family are denoted by ε_σ, where σ is a finite nonempty string of α's and β's, and are defined by

$$\varepsilon_\alpha = e_{(*]} \lhd \left\{ \begin{array}{l} v^2_{(*]} \lhd e_\beta \,,\; v^1_{(*]} \lhd e^2_{\|} \lhd v^2_{\|} \lhd e_\alpha \,, \\ v^1_{(*]} \lhd x^2_{\|} \lhd v^2_{\|} \lhd x_\alpha \end{array} \right\}$$

$$\varepsilon_\beta = e_{(*]} \lhd \left\{ \begin{array}{l} v^2_{(*]} \lhd e_\beta \,,\; v^1_{(*]} \lhd e^2_{\|} \lhd v^2_{\|} \lhd e_\alpha \,, \\ v^1_{(*]} \lhd x^1_{\|} \lhd v^1_{\|} \lhd x_{(*]} \lhd v^2_{(*]} \lhd x_\beta \end{array} \right\}$$

$$\varepsilon_{\alpha\sigma} = e_{(*]} \lhd \left\{ \begin{array}{l} v^2_{(*]} \lhd e_\beta \,,\; v^1_{(*]} \lhd e^1_{\|} \lhd v^1_{\|} \lhd \varepsilon_\sigma \,, \\ v^1_{(*]} \lhd x^2_{\|} \lhd v^2_{\|} \lhd x_\alpha \end{array} \right\}$$

$$\varepsilon_{\beta\sigma} = e_{(*]} \lhd \left\{ \begin{array}{l} v^2_{(*]} \lhd e_\beta \,,\; v^1_{(*]} \lhd e^1_{\|} \lhd v^1_{\|} \lhd \varepsilon_\sigma \,, \\ v^1_{(*]} \lhd x^1_{\|} \lhd v^1_{\|} \lhd x_{(*]} \lhd v^2_{(*]} \lhd x_\beta \end{array} \right\} .$$

The second family of entry places, denoted by ε_ω, where ω is an infinite string of α's and β's, may be described by the following formulae.

$$\varepsilon_{\alpha\omega} = e_{(*]} \lhd \left\{ \begin{array}{l} v^2_{(*]} \lhd e_\beta \,,\; v^1_{(*]} \lhd e^1_{\|} \lhd v^1_{\|} \lhd \varepsilon_\omega \,, \\ v^1_{(*]} \lhd x^2_{\|} \lhd v^2_{\|} \lhd x_\alpha \end{array} \right\}$$

$$\varepsilon_{\beta\omega} = e_{(*]} \lhd \left\{ \begin{array}{l} v^2_{(*]} \lhd e_\beta \,,\; v^1_{(*]} \lhd e^1_{\|} \lhd v^1_{\|} \lhd \varepsilon_\omega \,, \\ v^1_{(*]} \lhd x^1_{\|} \lhd v^1_{\|} \lhd x_{(*]} \lhd v^2_{(*]} \lhd x_\beta \end{array} \right\} .$$

[2] By Props. 6.1.6(2) and 6.3.7(1) which imply that there is no auto-concurrency.

Finally, there are two families of transitions, defined as

$$\begin{aligned}
\tau_0^\alpha &= (v^1_{(*]},\alpha) \lhd (v^2_{\|},\alpha) \lhd v_\alpha \\
\tau_0^\beta &= (v^2_{(*]},\beta) \lhd v_\beta \\
\tau_{n+1}^\alpha &= (v^1_{(*]},\alpha) \lhd (v^1_{\|},\alpha) \lhd \tau_n^\alpha \\
\tau_{n+1}^\beta &= (v^1_{(*]},\beta) \lhd (v^1_{\|},\beta) \lhd \tau_n^\beta .
\end{aligned}$$

For example, the places $\varepsilon_{\alpha\beta\alpha}$ and $\varepsilon_{\alpha\alpha\alpha\ldots}$ are shown in Fig. 6.10. Note that the second family of entry places is uncountable.

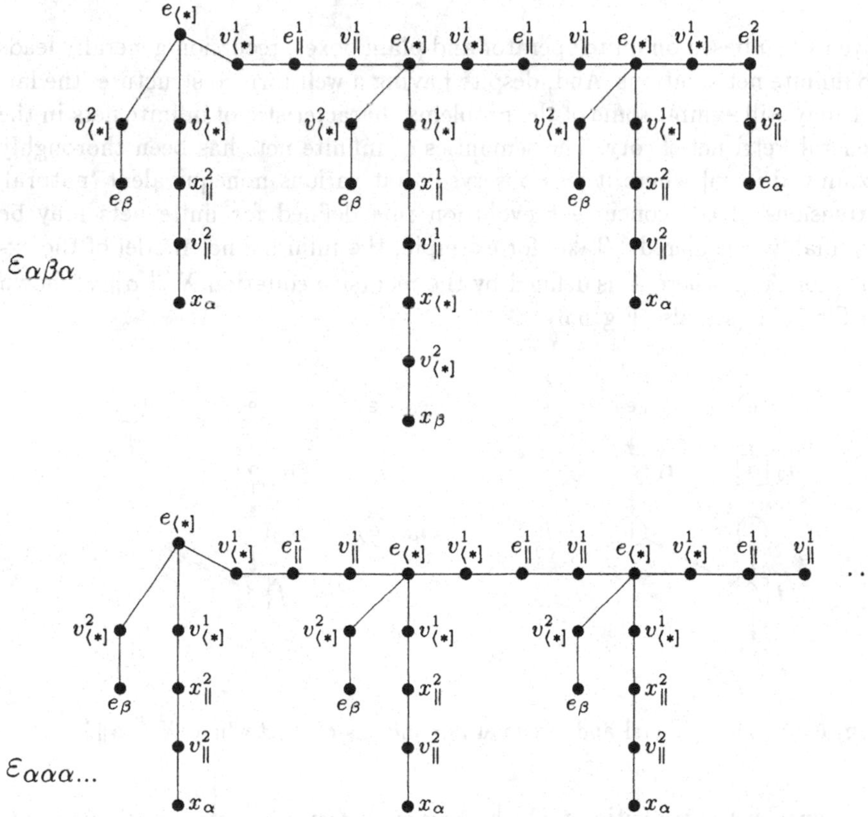

Fig. 6.10. Two entry places of the maximal solution of (6.17)

The resulting net is simple since N_α and N_β are simple, and $|\mathsf{trails}(\tau_n^\alpha)| = |\mathsf{trails}(\tau_n^\beta)| = 1$, for every $n \geq 0$. To show that it is not bounded, we first observe that, for every $n \geq 0$,

$$^\bullet(\tau_n^\alpha) = \{\varepsilon_{\sigma\alpha} \mid n = |\sigma|\} \quad \text{and} \quad \varepsilon_{\alpha\alpha\alpha\ldots} \in (\tau_n^\alpha)^\bullet .$$

Hence, for every $n \geq 0$, the transitions $t_0^\alpha, t_1^\alpha, \ldots, t_n^\alpha$ may occur (in sequence or concurrently) from the entry marking and lead to a marking M such that $M(\varepsilon_{\alpha\alpha\alpha\ldots}) = n+2$ which means that $\varepsilon_{\alpha\alpha\alpha\ldots}$ is not bounded.

Exercise 6.3.3. Find the connectivity between all transitions and places, and then determine the boundedness of places other than $\varepsilon_{\alpha\alpha\alpha\ldots}$.
Hint: Try to relate the boundedness of a place ε_σ or ε_ω with the number of the α's in σ or ω, respectively.

6.4 Finite Precedence Properties

Even when based on finite operator and plain boxes, recursion generally leads to infinite net solutions. And, despite having a well-formed structure, the latter may still exhibit some of the problems characteristic of infinite nets in the general Petri net theory. The semantics of infinite nets has been thoroughly examined in [5] where it was observed that various nonequivalent 'natural' extensions of the concurrent evolution rule defined for finite nets may be justifiably considered.[3] Take, for example, the minimal net model of the expression $X;\beta$ where X is defined by the recursive equation $X \stackrel{\text{df}}{=} \alpha \| X$, shown in Fig. 6.11 (see also Fig. 5.3).

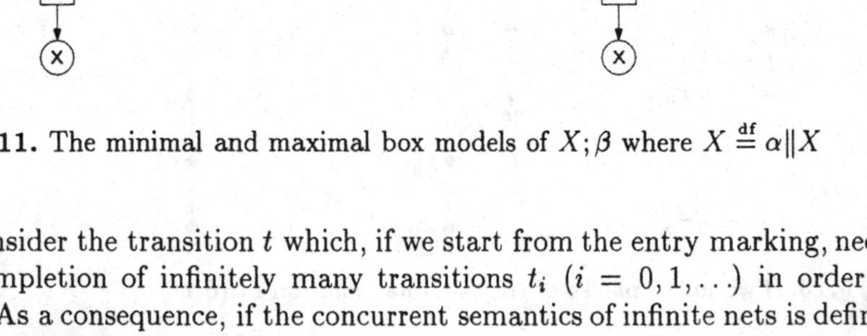

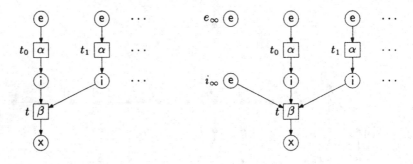

Fig. 6.11. The minimal and maximal box models of $X;\beta$ where $X \stackrel{\text{df}}{=} \alpha \| X$

Consider the transition t which, if we start from the entry marking, needs the completion of infinitely many transitions t_i $(i = 0, 1, \ldots)$ in order to occur. As a consequence, if the concurrent semantics of infinite nets is defined through finite steps, as in Sect. 4.2.3, t will never be enabled. But, as pointed out in [5], it is possible to allow infinite steps, by executing simultaneously infinitely many concurrent transitions. This could lead to evolutions which

[3] The problem is not limited to the extension to infinite nets as other extensions of the finite net model may also allow different nonequivalent concurrent evolution rules; for example, place capacities [21, 22] and inhibitor arcs [61].

are neither equivalent to any interleaving sequence of transitions (unless one is prepared to allow transfinite execution sequences), nor can be obtained as 'limits' of some infinite evolutions. For instance, if the infinite step rule was adopted, t would be allowed to occur after the infinite step $\{t_0, t_1, t_2, \ldots\}$. Thus the range of possible evolutions depends crucially on the choice of the basic occurrence rule.

By contrast, the picture changes if we consider the maximal box model for the same PBC expression, shown in Fig. 6.11 (see also Fig. 5.3). Here, it is the places e_∞ and i_∞ which are of our interest. Apart from arising naturally in the search for a unique solution of the recursive equation $X \stackrel{\text{df}}{=} \Omega_\|(\mathsf{N}_\alpha, X)$, they enjoy the following property: with e_∞ and i_∞ present, even after concurrently executing all t_i's in a single step, t will still be missing the token in i_∞. Indeed, we shall show in the next section that this is an instance of a general property of the maximal solutions of recursive systems. As a result, possible evolutions will not fundamentally rely on allowing or disallowing infinite steps, and infinite steps may always be regarded as limits of successive finite steps or even individual transition occurrences. The interleaving semantics will not lose any relevant evolutions, and the previous results based on S-invariant analysis may be applied in full generality, i.e., also for the semantics based on infinite steps.

Notice that the above problem does not arise from a lack of covering by S-aggregates, as was the case for the non-cleanness property of nonminimal/nonmaximal solutions for a recursive equation in Sect. 6.3, since the minimal solution as well as the maximal one are covered by S-components. Hence we shall conduct an additional investigation which will be based on some of the previous results on S-invariant coverings.

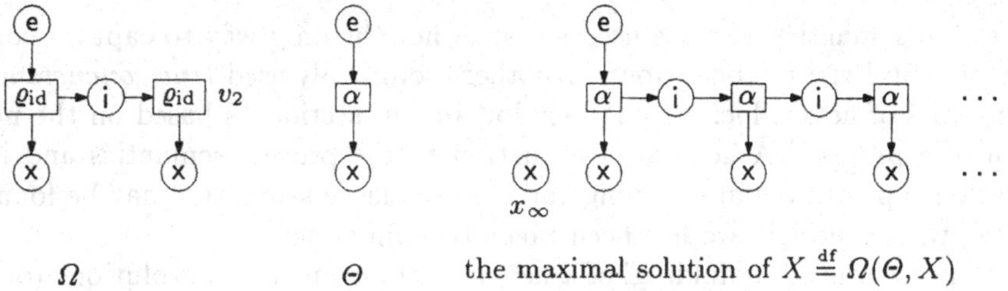

Fig. 6.12. Terminated evolution and finite precedence

Another observation of a similar nature concerns terminated evolutions, i.e., those after which all exit places are marked. Consider, for instance, the nets in Fig. 6.12. The recursive equation $X \stackrel{\text{df}}{=} \Omega(\Theta, X)$ has two solutions, the minimal and maximal one, $\Sigma^{\min}$ and $\Sigma^{\max}$. The latter is shown in Fig. 6.12, and $\Sigma^{\min}$ is simply $\Sigma^{\max}$ after deleting the isolated exit place x_∞. Both solutions are covered by S-components, which is also true of the operator box

Ω and the operand box Θ. Let us first consider the minimal solution, $\Sigma^{\min}$. It has a unique infinite evolution which may not be reduced to a finite succession of steps as the net is essentially sequential. The corresponding 'limit' marking puts one token in each of the exit places, so that the exit marking may be considered as reachable at a limit after infinitely many sequential transition occurrences. This has worrying consequences. Indeed, suppose that a box Σ is obtained by arranging two component boxes $\Sigma_1 = \Sigma^{\min}$ and Σ_2 in sequence. One would expect that valid finite or infinite evolutions of Σ (from its entry marking) may be obtained by taking an evolution of Σ_1, or by putting in sequence a complete evolution of Σ_1 and an evolution of Σ_2. But if a complete evolution of Σ_1 may involve an infinite sequence of actions, the result will not always be a valid evolution of Σ.

Again, the situation is radically different for the maximal solution, $\Sigma^{\max}$. With the isolated exit place x_∞, while the limit marking corresponding to the above infinite sequential evolution only contains exit places, the evolution is not complete since there will never be any token in the isolated place x_∞. Again, we are going to show that this is a general property, i.e., that the complete evolutions of a maximal solution are always finite and may not correspond to a limit case or, to put it the other way round, infinite evolutions may never be complete.

The properties we have just discussed further justify an interest in the maximal solutions of recursive systems despite the fact that they may lead to uncountably infinite nets; not only are they the only way to define the solutions uniquely in all generality, but they also retain important behavioural properties when infinite (sequential or concurrent) evolutions are allowed.

6.4.1 Process Semantics

The step semantics we have used so far is not the only way to capture concurrency in Petri net behaviours. Another, commonly used *true concurrency* semantics of nets, which we shall exploit in this section, is based on the notion of a *process*.[4] A general introduction to the process semantics and its relationship with the interleaving and step sequence semantics may be found in [5], where special care has been taken of infinite nets.

A process is an 'unfolding' of a net, modelling one of its evolutions (or a family of 'equivalent' evolutions with respect to causality and concurrency) through the absorption and production of tokens from the initial (real) marking. This is illustrated in Fig. 6.13, where the process π corresponds, for example, to the step sequences $\{v_1\}\{v_2, v_4\}\{v_2\}\{v_3\}$ and $\{v_1\}\{v_2\}\{v_2, v_4\}\{v_3\}$. In general, processes of a labelled net are labelled occurrence nets satisfying some additional criteria.

[4] The word *process* here is a technical term taken from the Petri net theory; such a process has nothing to do with process algebras and process expressions considered in the previous chapters.

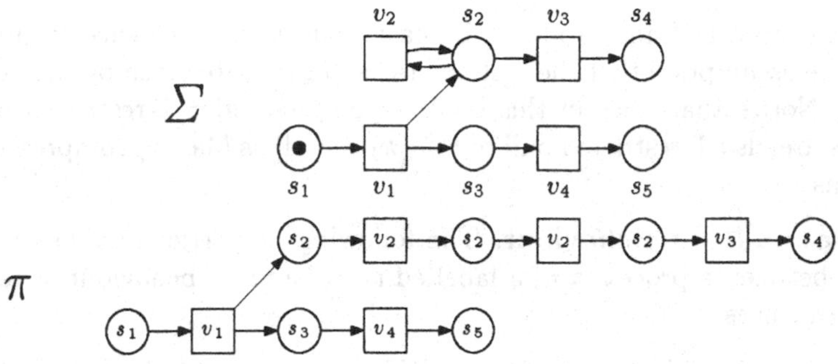

Fig. 6.13. A net and its processes

Occurrence Nets. An *occurrence net* is a quadruple $\pi = (B, E, F, l)$ where B is the set of conditions (places), E is the set of events (transitions), F is the flow function (a weight function for arcs), and l is a labelling function for $B \cup E$. It is assumed that the arc weights are binary, $F((B \times E) \cup (E \times B)) \subseteq \{0, 1\}$ (we shall treat F as if it were a relation on the nodes of π, i.e., we identify it with $F^{-1}(1)$); for every condition b in B, $|{}^\bullet b| \leq 1 \geq |b^\bullet|$; and when considered as a directed graph, π is acyclic. The latter means that it induces a partial order $\mathsf{poset}(\pi)$ on the set of its nodes:

$$\mathsf{poset}(\pi) = (B \cup E\, , \preceq_\pi) = (B \cup E\, , F^*)\,.$$

The set of the *minimal* and *maximal* elements[5] in $\mathsf{poset}(\pi)$ will be denoted by $\mathsf{Min}(\pi)$ and $\mathsf{Max}(\pi)$, respectively. Two distinct nodes of $\mathsf{poset}(\pi)$ are *concurrent* if they are not ordered. A *B-cut* is a maximal (w.r.t. set inclusion) set of mutually concurrent conditions. The *predecessors* of a node $x \in B \cup E$ are all the nodes y strictly preceding it in $\mathsf{poset}(\pi)$, i.e., $y \prec_\pi x$.

Processes. For a finite occurrence net to be a valid process of a finite labelled net, some additional conditions must be fulfilled by the former in order to faithfully represent an evolution of the latter.

A finite occurrence net $\pi = (B, E, F, l)$ is a *process* of a finite labelled net $\Sigma = (S, T, W, \lambda, M_0)$ if the following hold.

(SP) The conditions and events of π are respectively labelled by places and transitions of Σ, $l(B) \subseteq S$, and $l(E) \subseteq T$. (sort preservation)

(LC) For every $e \in E$ and $s \in S$, $W(s, l(e)) = |l^{-1}(s) \cap {}^\bullet e|$ and $W(l(e), s) = |l^{-1}(s) \cap e^\bullet|$. (local coherence)

(IC) For every $s \in S$, $M_0(s) = |l^{-1}(s) \cap \mathsf{Min}(\pi)|$. (initial coherence)

The conditions B of π correspond to tokens occurring at some time in the places of Σ during evolution(s) corresponding to π, and the events E of π correspond to transition occurrences. (LC) means that the correct number

[5] An element $x \in B \cup E$ is *minimal* (*maximal*) if there is no other $y \in B \cup E$ such that $y \preceq_\pi x$ (respectively, $x \preceq_\pi y$).

of tokens is absorbed and produced by each transition occurrence, and (IC) reflects the assumption that the system starts in the state given by the initial marking. Notice that since in this book we only consider T-restricted nets, processes are also T-restricted and so $\mathsf{Min}(\pi)$ as well as $\mathsf{Max}(\pi)$ comprise only conditions.

Processes and Net Behaviour. The following properties establish a formal link between a process π of a labelled net Σ and its behaviour in terms of step sequences:

(Lin) Every linearisation δ of a partition of the events of π, compatible with the associated partial order,[6] leads through its labels to a valid finite step sequence $\sigma(\delta)$ occurring from the initial marking.[7] Conversely, for every valid finite step sequence σ occurring from the initial marking, there is a process from which this sequence can be obtained in the way we have just described. (linearisation)

(FC) If M is the marking reached after the execution of the step sequence σ in (Lin), then for every $s \in S$,
$M(s) = |l^{-1}(s) \cap \mathsf{Max}(\pi)|$. (final coherence)

(Cut) For every B-cut μ of the process, the restriction of the process to μ and the predecessors of its conditions is a valid process; in particular, there is a reachable marking M_μ corresponding to μ such that $M_\mu(s) = |l^{-1}(s) \cap \mu|$, for every $s \in S$. Moreover, the part of the process comprising μ and all the nodes after μ is also a process, but for the net started from the marking M_μ. (cuts consistency)

(Con) If some conditions are mutually concurrent, then there is a reachable marking comprising their labels (and respecting their multiplicity), and if some events are mutually concurrent, then there is a reachable marking allowing the step corresponding to their labels (with their label multiplicity). (concurrency)

Infinite Processes of Finite Labelled Nets. To model infinite evolutions of a finite net Σ, another condition has to be added to the definition of a process π:

(FFA) Every node has finitely many predecessors. (first finite accessibility)

This avoids a counterintuitive situation where tokens 'come' from an infinite past. For example, (SP), (LC), and (IC) are all fulfilled by the occurrence net π in Fig. 6.14, but (FFA) does not hold. And, indeed, π does not correspond to any valid evolution of Σ.

The link between the extended processes and (possibly infinite) net evolutions is similar as before: (Lin) holds after assuming that δ (and consequently σ) can be infinite and all the sets appearing in it are finite; (Cut) holds but

---

[6] That is, $\delta = E_1 E_2 \ldots E_k$ where $E = E_1 \uplus E_2 \uplus \ldots \uplus E_k$ and, for all distinct $e \in E_i$ and $f \in E_j$, $e \preceq_\pi f$ implies $i < j$.

[7] That is, $\sigma(\delta) = l(E_1)l(E_2)\ldots l(E_k)$ and $M_0\,[\sigma(\delta)\rangle$.

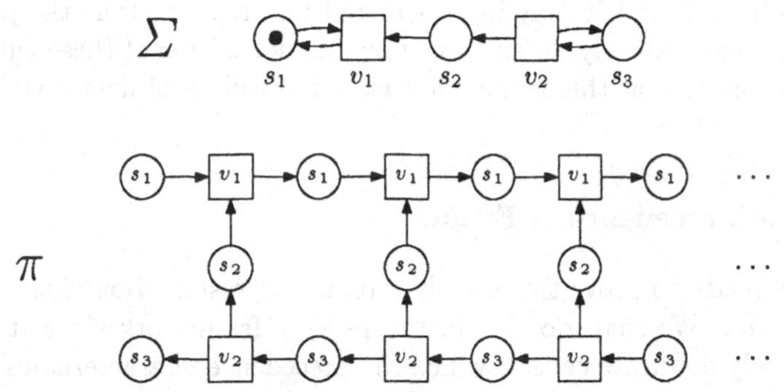

Fig. 6.14. An occurrence net which is not a process

M_μ exists only if μ is finite; (Con) holds without any changes for finite sets of conditions and events; and (FC) still holds for finite σ and π.

Processes of Infinite Labelled Nets with Finite Steps. To deal with infinite labelled nets, (FFA) has to be modified. There are at least two ways of proceeding, depending on whether we intend to allow infinite steps or not. If not, we may require that

(SFA) Every node has finitely many predecessor events. (second finite accessibility)

We ignore predecessor conditions since, in an infinite net, a single transition may insert/absorb tokens into/from infinitely many places. The properties of the new class of processes are similar to those stated above.

Processes of Infinite Labelled Nets with Infinite Steps. If we want to allow infinite steps, the finite accessibility property can be formulated as follows.

(TFA) For every node there is an upper bound on the length of the chains of nodes leading to it.[8] (third finite accessibility)

The link between processes satisfying (TFA) and possibly infinite evolutions of the net is again similar as before with one exception, namely, in (Lin) δ can contain infinite sets, and σ infinite steps. The *level* of a node in a process satisfying (TFA) is the maximal length (in terms of arcs) of a directed chain coming from a minimal condition; it may be seen that all the nodes at the same level are concurrent.

(TFA) implies that every node is accessible from $\mathsf{Min}(\pi)$ and that every event may be accessed through a finite sequence of possibly infinite steps. In general, (SFA) and (TFA) lead to different sets of processes, the latter

[8] A chain of nodes *leading to* (or *descending from*) x is a finite or infinite sequence of distinct nodes $x_1, x_2, \ldots$ such that $\ldots \preceq_\pi x_2 \preceq_\pi x_1 \preceq_\pi x$.

being more liberal, and it may be considered unfortunate that the possible evolutions of a system rely on an (arbitrary) choice of one of these rules. We shall see, however, that this is not the case for the class of nets investigated here.

6.4.2 Finite Precedence of Events

We are now ready to prove the already announced result about finite precedence of events. We shall do this in the process framework since it allows one to precisely delineate causality, i.e., the precedence characteristics of any event (transition occurrence). We shall adopt the most liberal definition of finite accessibility, (TFA), since one of our aims is to show that even being so liberal is of no consequence here. We shall only handle safe nets and assume that all the nets we start from are finite (so that they themselves respect the finite precedence property), ex-directed, and pure, since this is the case for the standard modelling of PBC expressions.

Theorem 6.4.1. Let Σ be a plain box obtained, through refinements and maximal solutions of recursive systems, from a set of finite pure unmarked ex-directed operator boxes and a set of finite unmarked ex-directed plain boxes, all S-covered by S-aggregates. Moreover, let π be a process of $\overline{\Sigma}$. Then every event of π has finitely many predecessor events.

Proof. Following the discussion at the beginning of Sect. 6.3.3, we may assume that $\Sigma = \Sigma_X$, for some $X \in \mathcal{X}$, is part of the maximal solution $\breve{\Sigma}$ of a recursive system (6.11) which is unfolded and satisfies (Rec1-2). We shall proceed by contradiction, retaining the notation and terminology used in Sect. 6.3.

Suppose that $\pi = (B, E, F, l)$ does not satisfy the theorem. Since each node of π has bounded descending chains, we can conduct inductive proofs based on the partial order $po = (E, F^*|_{E\times E})$. Hence there is an event $e \in E$ such that e has an infinite predecessor set E' in po, and each predecessor of e has only finitely many predecessors. Consider the poset $po' = (E', F^*|_{E'\times E'})$. Since the chains leading to e have bounded length in π (by (TFA)), E' can be partitioned into finitely many sets $E_1, \ldots, E_k$ ($k \geq 1$) such that for every $e' \in E_i$, the length of the longest path (in terms of arcs) from e' to e in po is i. Thus each E_i is a set of mutually concurrent events and $E_1 \times \{e\} \subseteq F \circ F$. Moreover, from the choice of e it follows that $E'' = E_1$ is infinite (otherwise, one of the events in E'' would have infinitely many predecessors).

From the finiteness of the transition tree $l(e)$ labelling the event e and the assumed finiteness of the boxes in (6.11), it follows that $\mathsf{trails}(l(e))$ is finite, and there are only finitely many $\phi \in \mathsf{Paths}$ for which there is $\sigma t \in \mathsf{trails}(l(e))$ and s such that $\mathsf{trail}(\phi) = \sigma s$. This, together with the infinite size of E'' and (5.21, 5.22) implies that there are infinitely many distinct events $e_1, e_2, \ldots \in E''$ and distinct conditions $b_1, b_2, \ldots \in B$ satisfying the following: (i) for every $i \geq 1$, $(e_i, b_i) \in F$ and $(b_i, e) \in F$; and (ii) there are $\sigma t \in \mathsf{trails}(l(e))$ and $s \in S_{\Psi_t}$ ($\Psi_t = \Psi_s$ being a plain box, see Prop. 6.3.1(11)) and $\phi \in \bigcap_{i=1}^{\infty} \mathsf{paths}(l(b_i))$ such that $\mathsf{trail}(\phi) = \sigma s \in \bigcap_{i=1}^{\infty} \mathsf{trails}(l(b_i))$ and $W_{\Psi_t}(s,t) > 0$ (in other words, each relationship $(b_i, e) \in F$ is 'established' by the path ϕ and trail σt). Notice that s is an entry place. Indeed, it cannot be an exit place since $t \in s^\bullet$ and Ψ_t is ex-directed. Moreover, it cannot be an internal place since then, by Prop. 6.3.1(6), $\mathsf{paths}(l(b_i)) = \{\sigma s\}$ for every $i \geq 1$, implying $l(b_1) = l(b_2) = \cdots$ which contradicts $\overline{\Sigma}$ being safe[9] and $b_1, b_2, \ldots$ being concurrent (since $e_1, e_2, \ldots$ are concurrent and $^\bullet b_i = \{e_i\}$ for all $i \geq 1$).

Let $\phi = v_1 \ldots v_m r_0 v_{m+1} r_1 \ldots v_{m+n} r_n$ (where $r_n = s$ and $r_1, \ldots, r_n$ are entry places, see Prop. 6.3.1(6)). Then $n > 0$ (otherwise $r_0 = s$ and $\mathsf{paths}(l(b_i)) = \{\sigma s\}$ which can be refuted in the same way as the fact that s cannot be an internal place), and so Ψ_{r_0} is an operator box. Moreover, r_0 is an internal place, which follows from Prop. 6.3.1(4) if $m \geq 1$, and from $^\bullet l(b_i) \neq \emptyset \neq l(b_i)^\bullet$ if $m = 0$.

We now turn to the relationships $(e_i, b_i) \in F$. For every $i \geq 1$, there are $\phi_i \in \mathsf{paths}(l(b_i))$ and $\sigma_i t_i \in \mathsf{trails}(l(e_i))$ such that $\mathsf{trail}(\phi_i) = \sigma_i s_i$, for some $s_i \in t_i^\bullet$. Such s_i and t_i belong to a plain box, so that we have: $s_i \neq r_0$;

$$\phi_i = v_1 \ldots v_m r_0 w_{m+1}^i r_1^i \ldots v_{m+n_i}^i s_i$$

with $n_i > 0$; s_i together with all r_j^i's are exit places (see Prop. 6.3.1(6)); and $v_{m+1} \neq w_{m+1}^i$ (operator nets are pure).

Suppose now that $i \neq j$ and ϕ_i is a prefix of ϕ_j. Then, since s_i and s_j are both places in plain boxes, $\phi_i = \phi_j$. But then $l(b_i) \in l(e_j)^\bullet$ and $l(b_j) \in l(e_i)^\bullet$, contradicting the fact that $\overline{\Sigma}$ is safe and e_i and e_j are concurrent, since we already have $l(b_i) \in l(e_i)^\bullet$ and $l(b_j) \in l(e_j)^\bullet$.

Thus $\Phi = \{\phi_1, \phi_2, \ldots\}$ is an infinite set such that no path in Φ is a prefix of another path. Let $\mathcal{T}$ be the tree generating the set Φ (notice that this tree has nothing to do with place or transition trees). From what we have shown it follows that $\mathcal{T}$ has infinitely many nodes. Moreover, by the assumed finiteness of the boxes in (6.11) and (Rec2), $\mathcal{T}$ is finitely branching. As a result, by König's Lemma, $\mathcal{T}$ contains an infinite sequence $\phi^\infty = v_1 \ldots v_m r_0 w_{m+1} \ldots$ starting at its root. Moreover, from the closed form of the maximal solution of a recursive system, it follows that $\phi^\infty \in \mathsf{Paths}^\infty$.

Since ϕ^∞ and ϕ are clearly agreeable (from $v_{m+1} \neq w_{m+1}$), by Prop. 6.3.2, there is a place p in Σ such that $\phi^\infty \in \mathsf{paths}^\infty(p)$ and $\phi \in \mathsf{paths}(p)$. Note also that p is an internal place since r_0 was an internal place (see Prop. 6.3.1(4)).

[9] By Thm. 6.3.3 and Prop. 6.1.10. Strictly speaking, the results we obtained in the previous sections about safeness, cleanness, etc, are only valid for finite reachability; nothing is said for cases needing occurrences of infinite steps, even if some extensions are possible. However, here all the nodes preceding e are finitely reachable and the problem disappears. Similar comment applies to the argument made in the proof of Thm. 6.4.2.

Since $\phi \in \mathsf{paths}(p)$, we have $p \in {}^\bullet l(e)$, and so there is $b \in B$ such that $l(b) = p$ and $(b, e) \in F$. Since p is an internal place, and π is a process of Σ, there must be $f \in E$ such that $(f, b) \in F$; hence there are $\psi \in \mathsf{paths}(p)$ and $\sigma' w \in \mathsf{trails}(l(f))$ such that $\mathsf{trail}(\psi) = \sigma' q'$ and $w \in {}^\bullet q'$. Moreover, since $\phi^\infty, \psi \in \mathsf{paths}(p) \cup \mathsf{paths}^\infty(p)$, from Prop. 6.3.2, ϕ^∞ and ψ are agreeable (recall that the operator boxes are pure).

Let $\eta = v_1 \dots v_m r_0 \eta'$ be the maximal common prefix of ψ and ϕ^∞, and $\widetilde{E}$ be the set of all $e_i \in E'' \setminus \{f\}$ such that the maximal common prefix of ϕ_i and ϕ^∞ is longer than η. Clearly, $\widetilde{E}$ is infinite.

Consider now $e_i \in \widetilde{E}$. To start with, we observe that ϕ_i and ψ are agreeable (as ϕ^∞ and ψ are; see also the definition of $\widetilde{E}$), so the set $Z = \{\phi_i, \psi\} \cup \mathsf{paths}^{\mathsf{e}}(p)$ is agreeable too (cf. Prop. 6.3.1(6)). Thus, by Prop. 6.3.2, there is $p_i \in S_\Sigma$ such that $Z \subseteq \mathsf{paths}(p_i)$. Moreover, from $\phi_i, \psi \in \mathsf{paths}(p_i)$, it follows that $p_i \in l(f)^\bullet \cap l(e_i)^\bullet$. Hence there are $c_i, d_i \in B$ such that $l(c_i) = l(d_i) = p_i$, $(f, c_i) \in F$, and $(e_i, d_i) \in F$. Clearly, c_i and d_i cannot be concurrent, so that either $(c_i, d_i) \in F^+$, or $(d_i, c_i) \in F^+$. The former implies $(f, e_i) \in F^+$, and the latter $(e_i, f) \in F^+$. We then observe that the latter can only hold for finitely many e_i's since $f \in E'$. Hence we may choose e_i so that $(c_i, d_i) \in F^+$. Thus there is $g \in E$ such that $(c_i, g) \in F$ and $(g, e_i) \in F^*$. Moreover, by Prop. 6.3.3(3), $l(g) \in p_i^\bullet = p^\bullet$ (note that $\mathsf{paths}^{\mathsf{e}}(p) = \mathsf{paths}^{\mathsf{e}}(p_i)$ follows from the fact that for every $z \in r_0^\bullet$, there is $\phi_z \in \mathsf{paths}^{\mathsf{e}}(p)$ such that $v_1 \dots v_m r_0 z$ is a prefix of ϕ_z). Hence there is $c \in B$ such that $l(c) = p$ and $(c, g) \in F$. But this means that b and c are two concurrent conditions (otherwise π would not be acyclic) labelled by $l(p)$, contradicting the safeness of Σ. (See also Fig. 6.15.) □

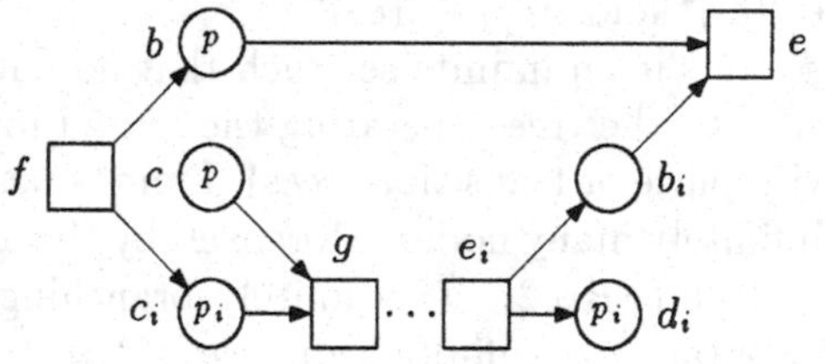

Fig. 6.15. The final configuration in the proof of Thm. 6.4.1

Thus in the context of the nets considered in this section, the two finite accessibility constraints, (SFA) and (TFA), are equivalent. Moreover, allowing infinite steps does not fundamentally change the semantics with respect to that based on finite steps. What may happen though is that some configurations of tokens are finitely reachable instead of being reachable 'at the limit'.

6.4.3 Finiteness of Complete Processes

Complete processes, i.e., those which lead from the entry to exit marking of a box, may be treated in a way similar to that used in the proof of Thm. 6.4.1, with some modifications, since we may not assert that infinitely many events producing exit conditions (i.e., conditions labelled by the exit places of the box) are at the same level. But exit conditions are trivially concurrent since all the boxes considered here are ex-directed, hence we obtain the following.

Theorem 6.4.2. Let Σ be a plain box obtained, through refinements and maximal solutions of recursive systems, from a set of finite pure unmarked ex-directed operator boxes and a set of finite unmarked ex-directed plain boxes, all S-covered by S-aggregates. Moreover, let π be a process of $\overline{\Sigma}$ which contain conditions corresponding to all the exit places of Σ. Then π has finitely many events.

Proof. Following the discussion at the beginning of Sect. 6.3.3, we may assume that $\Sigma = \Sigma_X$, for some $X \in \mathcal{X}$, is part of the maximal solution $\breve{\Sigma}$ of a recursive system (6.11) which is unfolded and satisfies (Rec1-2). We shall proceed by contradiction, retaining the notation and terminology used in Sect. 6.3.

Suppose that $\pi = (B, E, F, l)$ is a process of $\overline{\Sigma}$ which does not satisfy the theorem, i.e., $\Sigma^\circ \subseteq l(B)$ and E is infinite. Let

$$\widetilde{E} = \{e \in E \mid \exists b \in B \,:\, (e, b) \in F \wedge l(b) \in \Sigma^\circ\}\,.$$

We first show that

$$\widetilde{E} \text{ is infinite} \qquad (*)$$
$$\forall e, f \in \widetilde{E} \,:\, e \neq f \;\Rightarrow\; l(e)^\bullet \cap l(f)^\bullet = \emptyset\,. \qquad (**)$$

Indeed, if $\widetilde{E}$ is finite, then by Thm. 6.4.1 and E being infinite, there is $e \in E \backslash \widetilde{E}$ which is not a predecessor of any event in $\widetilde{E}$. Thus there is a subprocess of π, denoted by $\pi' = (B', \{e\} \cup \widetilde{E} \cup E', F', l')$, where E' is the set of all the predecessors of events in $\{e\} \cup \widetilde{E}$ and

$$B' = {}^\bullet(\{e\} \cup \widetilde{E} \cup E') \cup (\{e\} \cup \widetilde{E} \cup E')^\bullet \cup \{b \in B \mid {}^\bullet b = \emptyset\}\,.$$

By Thm. 6.4.1, π' has finitely many events. Thus its set of maximal conditions, $\mathsf{Max}(\pi')$, determines a marking $l(\mathsf{Max}(\pi'))$ which is finitely reachable from $\overline{\Sigma}$. Such a marking must be clean, by Thm. 6.3.3 and Prop. 6.1.10. But this contradicts $\Sigma^\circ + l(e^\bullet) \subseteq l(\mathsf{Max}(\pi'))$, and so (*) holds. To show (**) we assume that $s \in l(e)^\bullet \cap l(f)^\bullet$ and proceed similarly, after taking the subprocess $\pi'' = (B'', \{e, f\} \cup E'', F'', l'')$ of π where E'' is the set of all the predecessors of e and f. We then obtain a contradiction between the safeness of $\overline{\Sigma}$ and $\{s, s\} \subseteq l(\mathsf{Max}(\pi''))$. Thus (**) holds too.

By (*) and (**), there are infinitely many distinct events $e_1, e_2, \ldots \in \widetilde{E}$ and conditions $b_1, b_2, \ldots \in l^{-1}(\Sigma^\circ)$ such that $(e_i, b_i) \in F$, for every $i \geq 1$.

Thus, for every $i \geq 1$, there are $\sigma_i t_i \in \mathsf{trails}(l(e_i))$ and $\phi_i \in \mathsf{paths}(l(b_i))$ such that $\mathsf{trail}(\phi_i) = \sigma_i s_i$ and $s_i \in t_i^\bullet$. Moreover, since Ω_X is finite, we may choose the e_i's and b_i's in such a way that all the paths ϕ_i begin with the same exit place in Ω_X°. Thus the set $\Phi = \{\phi_1, \phi_2, \dots\}$ is generated by some tree $\mathcal{T}$; moreover, such a tree is infinite since $\phi_i \neq \phi_j$ for $i \neq j$ due to (**). We also observe that $\mathcal{T}$ is finitely branching, by the assumed finiteness of boxes in (6.11) and (Rec2). Hence, by König's Lemma, $\mathcal{T}$ contains an infinite path ϕ^∞ starting from its root. Since $\phi^\infty \in \mathsf{Paths}^\infty$, it follows from Prop. 6.3.2 that there is an exit place $\widetilde{x}$ in Σ such that $\phi^\infty \in \mathsf{paths}^\infty(\widetilde{x})$. Hence there are $\widetilde{e} \in \widetilde{E}$ and $\widetilde{b} \in l^{-1}(\widetilde{x})$ satisfying $(\widetilde{e}, \widetilde{b}) \in F$. Thus there are $\sigma t \in \mathsf{trails}(l(\widetilde{e}))$ and $\phi \in \mathsf{paths}(\widetilde{x})$ such that $\mathsf{trail}(\phi) = \sigma s$ and $s \in t^\bullet$. Suppose that ϕ has k elements. From the definition of $\mathcal{T}$ it follows that there is $l > 1$ such that $e_l \neq \widetilde{e}$ and ϕ_l shares with ϕ^∞ a prefix of length at least $k+1$. Now, since ϕ and ϕ^∞ are agreeable (as they are paths in $\widetilde{x}$), it follows that ϕ and ϕ_l are agreeable as well. Thus there is an exit place x in Σ such that $\phi, \phi_l \in \mathsf{paths}(x)$. But this means that $x \in l(\widetilde{e})^\bullet \cap l(e_l)^\bullet$, producing a contradiction with (**). □

We have obtained two basic finiteness properties of infinite process semantics. We have also demonstrated the power of the S-invariant analysis — part of the general Petri net theory — which proved to be fruitful in our specific framework. The above results rely heavily on the assumption that the operator boxes are all pure and that the boxes are all ex-directed, as is the case in the standard PBC framework. We conjecture that the introduction of operators like $\Omega_{[*]}$, $\Omega_{(*]}$ and $\Omega_{[*)}$ does not invalidate the present results, but in such a case a more general theory needs to be developed.

6.5 Literature and Background

The notion of an S-invariant as defined in this chapter is a restricted version of the more general notion where the values associated with places and transitions are allowed to be negative. S-invariants with negative values were investigated in [31]; however, this complicates the theory when dealing with infinite nets, without improving behavioural results obtained in this chapter.

The construction of S-invariants for boxes was first developed for finite refinements in [25]. The extension to arbitrary refinements was presented in [27], and recursion was taken into account in [31]. Several other results may be found in [26]. The finite precedence properties were proved in [28]. Other applications of the S-invariant analysis to the modelling of PBC may be found in [29].

7. The Box Algebra

In this chapter, we shall introduce a third framework for compositionality in Petri net theory, referred to in Sect. 4.1 as the SOS-compositionality. In Chap. 3, we defined a process algebra, called PBC, and its concurrent operational semantics; then, in Chap. 4, we introduced a denotational semantics for PBC in terms of two classes of labelled Petri nets, called static and dynamic boxes, and extended the resulting theory to handle recursion, in Chap. 5. However, we still have to show that these two kinds of semantics are equivalent. Due to the generality of the developed framework and, in particular, the way we treated refinements and solutions of recursive systems in the box domain, it will be possible to (i) define a general scheme to construct an algebra of boxes; then (ii) a process algebra for which the denotational semantics is given homomorphically by the box algebra; and (iii) a concurrent operational semantics equivalent to the denotational one. In Chap. 8, we will show that the PBC model is an instance of this general scheme, and so are other process algebras.

Our first aim will be to delineate a class of operator boxes which yield plain boxes with step sequence semantics obeying SOS operational rules. Section 7.1 contains the results of our investigation; towards the end of its first part we introduce the *SOS-operator boxes* which possess such a property. This is demonstrated in Sect. 7.2 where we show that SOS-operator boxes give rise to compositional nets for which a variant of the SOS semantics can be both defined and proved to be fully consistent with the standard net semantics. We shall take advantage of such a strong result and proceed, in Sect. 7.3, with the development of an algebra of process expressions whose operational semantics is derived from that satisfied by the SOS-operator boxes. In the resulting generic model — called *the box algebra* — we introduce both a version of the standard structured operational semantics and a denotational semantics in terms of plain boxes, in a similar — but much more general — way as it has been done for PBC.

7.1 SOS-operator Boxes

The discussion in this section is deliberately informal in order to better convey an intuition and motivation behind the definition of an SOS-operator box;

in particular, the notion of a factorisable operator box. Although we do not formulate nor prove any formal results, the reader should have little or no difficulty in extracting these from the text that follows.

Our aim is to find a class of operator boxes, of which operator boxes considered in Sect. 4.4 are special cases, that could be applied to static and dynamic boxes — the base plain boxes considered in the preceding chapters — in such a way that a counterpart of the process algebra SOS-rule could also be formulated in the domain of nets. We will define operator boxes satisfying this property, called SOS-operator boxes, by imposing two conditions on a generic SOS-operator box

$$\Omega = \Big(S, T, W, \lambda, \emptyset\Big).$$

To simplify the argument, but without losing its generality as we shall later see, we assume that only the identity relabelling ϱ_{id} is used to annotate the transitions in Ω. Another point arises from the observation that here a concrete net operator Ω aims at modelling a corresponding syntactic operator op_Ω on process expressions, to be used in prefix, postfix, infix, or functional mode, much in the same way as the net operator $\Omega_;$ was used to obtain the denotational semantics of a net composition $\Sigma; \Sigma'$ or process expression $E; G$. But, whereas Ω might in principle be marked, we have no means of specifying explicitly the marking of a symbol representing a net operator, as it is a mere lexical construct or sign. From the static or dynamic nature of the various operands it is, however, possible to associate a marking to an expression, but such a marking will be purely imaginary: $(\emptyset, Q)$ where Q is the set of all the transitions of Ω corresponding to a dynamic operand. (We already anticipate that our framework will only generate safe boxes, by assuming that Q is a set rather than a multiset of transitions: see the paragraph after Requirement 2.) Hence Ω and, indeed, all operator boxes considered in this chapter will have only imaginary markings which will simply be left implicit.

The two conditions defining an SOS-operator box, (OpNet1) and (OpNet2) in the following, will be derived by applying Ω to suitably chosen tuples of static boxes $\boldsymbol{\Theta}$, and then considering some properties that the composition $\Omega(\boldsymbol{\Theta})$ ought to satisfy. Our first requirement is as follows.

> Requirement 1
> Operations on static boxes should not lead outside the semantic domain of static boxes.

To analyse its consequences, consider an Ω-tuple of very simple[1] static boxes $\boldsymbol{\Theta}$ defined thus, for every $v \in T$:

---

[1] We implicitly assume that one should always be able to use these boxes, as well as those used later in this section to motivate (OpNet2), as refining nets.

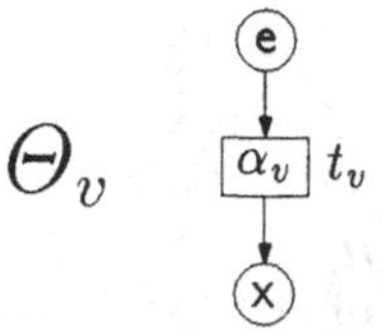

where $\alpha_v \neq \alpha_w$ for $v \neq w$. An easy observation is that, due to the simple form of the refining nets, $\Omega(\boldsymbol{\Theta})$ and Ω are isomorphic up to transition labelling. In other words, we can think of the net $\Omega(\boldsymbol{\Theta})$ as though it were Ω with each transition v being labelled by α_v. And so the first requirement implies that $\Omega(\boldsymbol{\Theta})$, and thus Ω itself, should be a static box,[2] i.e., all markings reachable from[3] ${}^\circ\Omega$ or Ω° — the entry or exit marking of Ω — should be safe and clean:

(OpNet1) Ω is a static box.

In the light of the discussion presented in the previous chapters, such a condition should not be too surprising. What is worth noting, however, is that it has been derived using a very general argument without, for instance, relying on the covering of Ω by S-invariants. But we still have to prove that, together with the second condition and some simple restrictions on the tuples of refining boxes, (OpNet1) will be sufficient to guarantee the existence of the operational semantics of composite boxes.

The second requirement is directly motivated by our wish to obtain an SOS-rule for compositionally defined nets:

> Requirement 2
> The net obtained as a result of an evolution of a composite net should itself be a composition of nets related to the original ones.

We interpret this as saying that no matter how $\overline{\Omega(\boldsymbol{\Theta})}$ (or $\underline{\Omega(\boldsymbol{\Theta})}$) evolves, the resulting net Σ should be derivable as a composition $\Sigma = \Omega(\boldsymbol{\Theta}')$ where the nets underlying $\boldsymbol{\Theta}$ and $\boldsymbol{\Theta}'$ are the same, i.e., $\lfloor \boldsymbol{\Theta}' \rfloor = \lfloor \boldsymbol{\Theta} \rfloor$. It turns out that this leads to a perhaps unexpected property of the complex markings reachable from the entry or exit marking of Ω. For suppose that Ω is an operator box satisfying (OpNet1) and (M, Q) is an arbitrary complex marking reachable from the entry marking ${}^\circ\Omega$ (the case where (M, Q) is reachable from the exit marking is symmetric). Then, from the safeness assumption made about Ω, it follows that M and Q are sets and $M \cap ({}^\bullet Q \cup Q^\bullet) = \emptyset$. Consider now another Ω-tuple $\boldsymbol{\Theta}''$ of simple static boxes obtained from the previously used $\boldsymbol{\Theta}$ by replacing, for every $w \in Q$, Θ_w by Θ''_w defined as follows.

---

[2] Strictly speaking, static boxes are plain. Here we extend the (purely marking-based) notion of being static to operator boxes as well.

[3] Notice that, if we consider Ω_{sr}, the reachability from ${}^\circ\Omega$ is equivalent to the reachability from Ω°, and to the reachability from ${}^\circ\Omega$ or Ω° in Ω.

Θ''_w

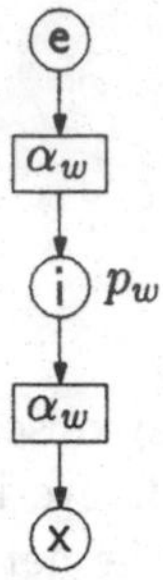

$\Omega(\Theta'')$ can be thought of as Ω with every transition $v \in T\backslash Q$ being labelled by α_v, and every transition $w \in Q$ being split into the 'beginning of w' and 'end of w', both transitions being labelled by α_w and connected by a single place p_w.

It is easy to see that $\overline{\Omega(\Theta'')}$ is a safe net which can evolve into the net Σ whose (safe and real) marking is $M \cup \{p_w \mid w \in Q\}$. Since we should be able to represent Σ as $\Omega(\Theta')$ where $\lfloor\Theta'\rfloor = \lfloor\Theta''\rfloor$, it follows from the definition of net refinement and safeness of Σ, that for every $v \in T\backslash Q$, Θ'_v is either Θ_v or $\overline{\Theta_v}$ or $\underline{\Theta_v}$, and for every $w \in Q$, Θ'_w is Θ''_w with exactly one token inside the place p_w. And, crucially, M is the disjoint union of the pre-sets of transitions v such that $\Theta'_v = \overline{\Theta_v}$ and of the post-sets of transitions v such that $\Theta'_v = \underline{\Theta_v}$. In other words, the 'real' part of the complex marking (M, Q) can be factored onto pre- and post-sets of some transitions in T.

Following the above observation, we shall say that a pair (μ_e, μ_x) of sets of transitions of Ω is a *factorisation* of a set of places $M \subseteq S$ if

$$M \;=\; \biguplus_{v\in\mu_e} {}^\bullet v \;\uplus\; \biguplus_{v\in\mu_x} v^\bullet \;.$$

Moreover, a quadruple of sets of transitions of Ω, $\mu = (\mu_e, \mu_d, \mu_x, \mu_s)$ will be called a *factorisation* of a complex safe marking (M, Q) of Ω if $\mu_d = Q$ and $\mu_s = T\backslash(\mu_e \cup \mu_d \cup \mu_x)$ and (μ_e, μ_x) is a factorisation of M. Then Ω will be called *factorisable* if for every safe complex marking (M, Q) reachable from the entry or exit marking of Ω, there is at least one factorisation. And a second condition to be imposed on an SOS-operator box is that

(OpNet2) Ω is factorisable.

It may easily be checked that all standard PBC operator boxes introduced in Sect. 4.4 — including those modelling the two binary iteration constructs — are factorisable. However, not all static operator boxes are factorisable, as illustrated in Fig. 7.1. For in the marking obtained (from the entry marking) through the occurrences of v_1 and v_3, the left token can (must, in fact) be factored out using v_3 and the right token can (must) be factored out using v_2, but the middle token is orphaned. Notice, however, that this nonfactorisable static operator box is S-covered by S-components, and thus most of the results obtained in Chap. 6 can be applied.

Factorisability of an operator is a behavioural property since it is necessary to check whether every reachable marking can be factored. Fortunately,

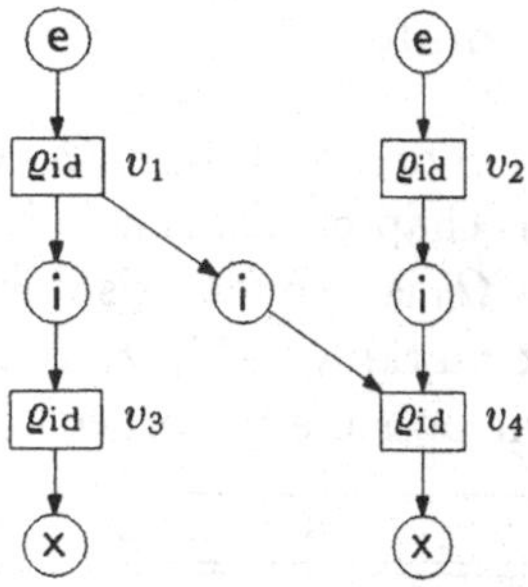

Fig. 7.1. A nonfactorisable static box

most operators used in practice are small, or have a straightforward symmetric structure, so that this will not normally be a real problem. Checking whether an operator box is a static box also involves checking the safeness of every marking reachable from the entry or exit marking, which may sometimes be done statically by establishing a covering by S-components (see Prop. 6.1.6); it is at present not known whether factorisability can be checked in a purely structural manner. Notice that it is possible to extend the notion of factorisation to markings and operators which are not safe, by using multisets and multiset sums instead of sets and disjoint set unions, but we shall not need this extra generality here.

To summarise, in this chapter we shall only consider operators fulfilling the conditions (OpNet1) and (OpNet2), i.e., simple static (hence safe and clean) factorisable operator boxes, called *SOS-operator boxes.* For an SOS-operator box Ω, we will denote by fact_Ω the set comprising all factorisations of all complex markings reachable from the entry or exit marking of Ω, as well as the only factorisation $(\emptyset, \emptyset, \emptyset, T_\Omega)$ of the empty marking of Ω. The domain of application of SOS-operator boxes will be defined after establishing some basic properties of factorisations.

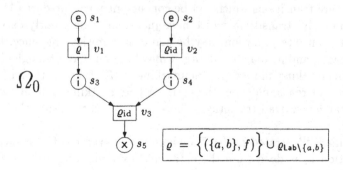

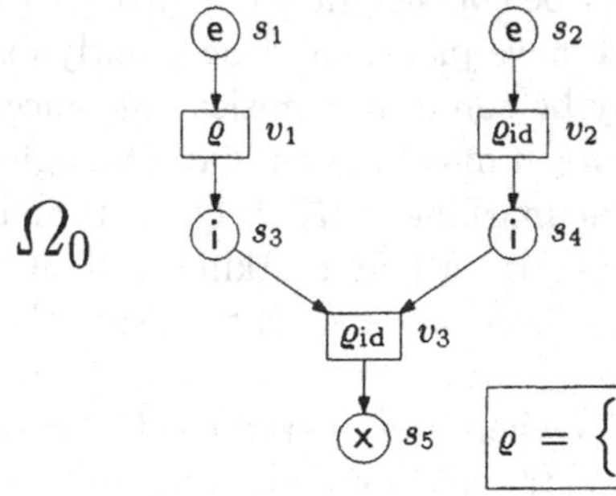

$$\varrho = \Big\{(\{a,b\},f)\Big\} \cup \varrho_{\mathsf{Lab}\setminus\{a,b\}}$$

Fig. 7.2. An SOS-operator box for the running example

7.1.1 A Running Example

Figure 7.2 shows an SOS-operator box Ω_0 which will serve as a running example throughout this chapter. Ω_0 is a finite, simple, pure, safe, and clean box. A justification that Ω_0 is also factorisable is provided by the table below which lists all complex markings (M, Q) reachable from the entry or exit marking of Ω_0, and their possible factorisations $(\mu_e, \mu_d, \mu_x, \mu_s)$:

M	Q	μ_e	μ_d	μ_x	μ_s
$\{s_1, s_2\}$	$\emptyset$	$\{v_1, v_2\}$	$\emptyset$	$\emptyset$	$\{v_3\}$
$\{s_1\}$	$\{v_2\}$	$\{v_1\}$	$\{v_2\}$	$\emptyset$	$\{v_3\}$
$\{s_2\}$	$\{v_1\}$	$\{v_2\}$	$\{v_1\}$	$\emptyset$	$\{v_3\}$
$\{s_1, s_4\}$	$\emptyset$	$\{v_1\}$	$\emptyset$	$\{v_2\}$	$\{v_3\}$
$\{s_2, s_3\}$	$\emptyset$	$\{v_2\}$	$\emptyset$	$\{v_1\}$	$\{v_3\}$
$\emptyset$	$\{v_1, v_2\}$	$\emptyset$	$\{v_1, v_2\}$	$\emptyset$	$\{v_3\}$
$\{s_3\}$	$\{v_2\}$	$\emptyset$	$\{v_2\}$	$\{v_1\}$	$\{v_3\}$
$\{s_4\}$	$\{v_1\}$	$\emptyset$	$\{v_1\}$	$\{v_2\}$	$\{v_3\}$
$\{s_3, s_4\}$	$\emptyset$	$\emptyset$	$\emptyset$	$\{v_1, v_2\}$	$\{v_3\}$
		$\{v_3\}$	$\emptyset$	$\emptyset$	$\{v_1, v_2\}$
$\emptyset$	$\{v_3\}$	$\emptyset$	$\{v_3\}$	$\emptyset$	$\{v_1, v_2\}$
$\{s_5\}$	$\emptyset$	$\emptyset$	$\emptyset$	$\{v_3\}$	$\{v_1, v_2\}$

(7.1)

Thus $\mathbf{fact}_{\Omega_0}$ comprises thirteen different factorisations as the only factorisation $(\emptyset, \emptyset, \emptyset, \{v_1, v_2, v_3\})$ of the empty marking $(M, Q) = (\emptyset, \emptyset)$ also belongs to $\mathbf{fact}_{\Omega_0}$. Notice, and this is crucial as we shall see, that a marking may have more than one factorisation, like $(M, Q) = (\{s_3, s_4\}, \emptyset)$ above.

7.1.2 Properties of Factorisations

A factorisation of a reachable marking (M, Q) is essentially a way of representing its real part, M, as the disjoint union of the pre-sets of a set of transitions, μ_e, and the post-sets of another set of transitions, μ_x. Intuitively, μ_e are transitions which can be concurrently executed at (M, Q), and μ_x are (possibly) transitions which have just been concurrently executed. However, such an interpretation may be somewhat misleading since, besides the fact that μ_e and μ_x may be infinite, it may happen that although (M, Q) is a reachable marking and $v \in \mu_x$, the markings $(M \backslash v^\bullet, Q \cup \{v\})$ and $((M \backslash v^\bullet) \cup {}^\bullet v, Q)$ are not reachable from the entry or exit marking of Ω; it may even happen that v is a dead transition, whose sole role is to ensure the factorisability of Ω.

This is illustrated in Fig. 7.3 where the operator is factorisable but the transition v_2 is dead from the entry and exit markings. Yet if we dropped this dead (hence supposedly useless) transition, the operator would become non-factorisable since after the occurrences of v_1 and v_3, leading to the marking $(\{i, x\}, \emptyset)$, the internal place i could no longer be factored out. Notice that,

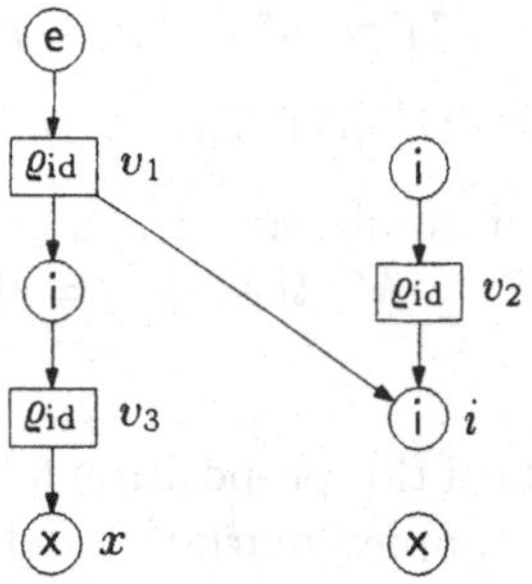

Fig. 7.3. An SOS-operator box with a dead transition

contrary to what we have said about μ_x, if $v \in \mu_e$, then both $(M \setminus {}^\bullet v, Q \cup \{v\})$ and $((M \setminus {}^\bullet v) \cup v^\bullet, Q)$ are reachable from the entry or exit marking of Ω.

In what follows, we will call a transition v of an SOS-operator box Ω *reversible* if, for every complex marking (M, Q) reachable from ${}^\circ\Omega$ or Ω°, $v^\bullet \subseteq M$ implies that $(M \setminus v^\bullet, Q \cup \{v\})$ is also a marking reachable from ${}^\circ\Omega$ or Ω°. In Fig. 7.3, both v_1 and v_3 are reversible, while v_2 is not. In Fig. 7.2, all three transitions are reversible. Notice also that all transitions in the operator nets used for the translation of PBC in Chap. 4 are reversible.

Throughout the rest of this chapter we will need several basic properties of factorisations which, for ease of reference, are gathered in the proposition below.

Proposition 7.1.1. Let Ω be an SOS-operator box, $\mu \in \mathsf{fact}_\Omega$ be a factorisation of a complex marking (M, Q) reachable from ${}^\circ\Omega$ or Ω°, and T_{rev} be the set of all reversible transitions of Ω.

(1) μ_e, μ_d, μ_x, and μ_s are disjoint sets of transitions, and μ_d is finite.

(2) For every finite $U \subseteq T_\Omega$ and every finite $Z \subseteq T_{\mathrm{rev}}$ such that

$$\biguplus_{v \in U} {}^\bullet v \;\uplus\; \biguplus_{v \in Z} v^\bullet \subseteq M\,,$$

it is the case that $(M \setminus ({}^\bullet U \cup Z^\bullet), Q \uplus U \uplus Z)$ is a marking reachable from ${}^\circ\Omega$ or Ω°. Moreover, both $(M \cup {}^\bullet Q, \emptyset)$ and $(M \cup Q^\bullet, \emptyset)$ are markings reachable from ${}^\circ\Omega$ or Ω°.

(3) For every $V \subseteq \mu_d$, every finite $U \subseteq \mu_e$, and every finite $Z \subseteq \mu_x \cap T_{\mathrm{rev}}$, the following are factorisations in fact_Ω:

$$\begin{array}{l}
(\ \mu_e \setminus U\ ,\ \mu_d \cup U\ ,\ \mu_x\ ,\ \mu_s\ )\\
(\ \mu_e \cup V\ ,\ \mu_d \setminus V\ ,\ \mu_x\ ,\ \mu_s\ )\\
(\ \mu_e\ ,\ \mu_d \setminus V\ ,\ \mu_x \cup V\ ,\ \mu_s\ )\\
(\ \mu_e\ ,\ \mu_d \cup Z\ ,\ \mu_x \setminus Z\ ,\ \mu_s\ )\,.
\end{array}$$

(4) The transitions of $\mu_e \cup \mu_d \cup (\mu_x \cap T_{\mathrm{rev}})$ are mutually independent, and the transitions in μ_x are such that

$$\emptyset = \mu_x^\bullet \cap ({}^\bullet\mu_e \cup \mu_e^\bullet) = \mu_x^\bullet \cap ({}^\bullet\mu_d \cup \mu_d^\bullet)$$
$$= (\mu_x \setminus T_{rev})^\bullet \cap ({}^\bullet(\mu_x \cap T_{rev}) \cup (\mu_x \cap T_{rev})^\bullet)$$

and $v^\bullet \cap w^\bullet = \emptyset$, for all distinct $v, w \in \mu_x$.

(5) If ${}^\circ\Omega \subseteq M$ or $\Omega^\circ \subseteq M$, then $\mu_d = \emptyset$ and $M = {}^\circ\Omega$ or $M = \Omega^\circ$, respectively.

Proof. The various parts of this proposition follow directly from the definition of a factorisation of a complex marking, and the safeness and cleanness of Ω. □

7.1.3 The Domain of Application of an SOS-operator Box

We first extend the notion of a factorisation to tuples of boxes. Let Ω be an SOS-operator box and $\boldsymbol{\Sigma}$ be an Ω-tuple of static and dynamic boxes. The *factorisation* of $\boldsymbol{\Sigma}$ is the quadruple $\mu = (\mu_e, \mu_d, \mu_x, \mu_s)$ such that, for $\delta \in \{e, d, x, s\}$,

$$\mu_\delta = \begin{cases} \{v \mid \Sigma_v \in \mathsf{Box}^\delta\} & \text{if } \delta \in \{e, x, s\} \\ \{v \mid \Sigma_v \in \mathsf{Box}^d \setminus (\mathsf{Box}^e \cup \mathsf{Box}^x)\} & \text{if } \delta = d\,. \end{cases}$$

The *domain of application* of Ω, denoted by dom_Ω, is then defined as the set comprising every Ω-tuple of static and dynamic boxes $\boldsymbol{\Sigma}$ whose factorisation belongs to fact_Ω, and such that, for every $v \in T_\Omega$:

(Dom1) If ${}^\bullet v \cap v^\bullet \neq \emptyset$, then Σ_v is ex-exclusive.
(Dom2) If v is not reversible, then Σ_v is x-directed.

It follows directly from Prop. 7.1.1(4) that for every $\boldsymbol{\Sigma}$ in the domain of Ω, $\Omega(\boldsymbol{\Sigma})$ is a legal net refinement (later, we will also prove Thm. 7.1.1 which asserts that $\Omega(\boldsymbol{\Sigma})$ has a marking which is not only well defined but even safe).

The motivations behind the introduction of (Dom1) have already been presented in the earlier parts of this book (see, for example, the discussion in Sect. 4.4.9); it is meant to prevent the non-safeness of refinements in which a box is refined into a transition with a side place. The other condition, (Dom2), is related to the possible presence of dead transitions in an operator box. Referring again to the SOS-operator box in Fig. 7.3, we may observe that if we were allowed to refine the transition v_2 with a non-x-directed box Σ_{v_2}, then some behaviours originating from Σ_{v_2} could be possible from $\Omega(\boldsymbol{\Sigma})$, while no behaviour beginning at the entry nor exit marking of the operator box can possibly involve v_2. It is therefore justifiable to insist that Σ_{v_2} be x-directed in this case (we could have required that no transition be enabled at $\Sigma_{v_2}^\circ$ after applying the relabelling ϱ_{v_2}, but x-directedness is a simple structural sufficient condition for this).

Taking the example of the SOS-operator box in Fig. 7.2 and the factorisations (7.1), we obtain (since Ω_0 is pure and all its transitions are reversible) that the following holds, where $\mathsf{Box}^{\tilde{d}}$ denotes $\mathsf{Box}^{d} \backslash (\mathsf{Box}^{e} \cup \mathsf{Box}^{x})$:

$$\begin{aligned}
\mathsf{dom}_{\Omega_0} = {} & \mathsf{Box}^{e} \times \mathsf{Box}^{e} \times \mathsf{Box}^{s} \;\cup\; \mathsf{Box}^{e} \times \mathsf{Box}^{\tilde{d}} \times \mathsf{Box}^{s} \;\cup \\
& \mathsf{Box}^{\tilde{d}} \times \mathsf{Box}^{e} \times \mathsf{Box}^{s} \;\cup\; \mathsf{Box}^{e} \times \mathsf{Box}^{x} \times \mathsf{Box}^{s} \;\cup \\
& \mathsf{Box}^{x} \times \mathsf{Box}^{e} \times \mathsf{Box}^{s} \;\cup\; \mathsf{Box}^{\tilde{d}} \times \mathsf{Box}^{\tilde{d}} \times \mathsf{Box}^{s} \;\cup \\
& \mathsf{Box}^{x} \times \mathsf{Box}^{\tilde{d}} \times \mathsf{Box}^{s} \;\cup\; \mathsf{Box}^{\tilde{d}} \times \mathsf{Box}^{x} \times \mathsf{Box}^{s} \;\cup \\
& \mathsf{Box}^{x} \times \mathsf{Box}^{x} \times \mathsf{Box}^{s} \;\cup\; \mathsf{Box}^{s} \times \mathsf{Box}^{s} \times \mathsf{Box}^{e} \;\cup \\
& \mathsf{Box}^{s} \times \mathsf{Box}^{s} \times \mathsf{Box}^{\tilde{d}} \;\cup\; \mathsf{Box}^{s} \times \mathsf{Box}^{s} \times \mathsf{Box}^{x} \;\cup \\
& \mathsf{Box}^{s} \times \mathsf{Box}^{s} \times \mathsf{Box}^{s} \\
= {} & \mathsf{Box}^{d} \times \mathsf{Box}^{d} \times \mathsf{Box}^{s} \;\cup\; \mathsf{Box}^{s} \times \mathsf{Box}^{s} \times \mathsf{Box}^{d} \;\cup \\
& \mathsf{Box}^{s} \times \mathsf{Box}^{s} \times \mathsf{Box}^{s}\,.
\end{aligned}$$

Figure 7.4 shows an Ω_0-tuple of boxes $\boldsymbol{\Sigma} = (\Sigma_{v_1}, \Sigma_{v_2}, \Sigma_{v_3})$ with the factorisation, $(\{v_1\}, \{v_2\}, \emptyset, \{v_3\})$, which belongs to fact_{Ω_0}. Thus,

$$\boldsymbol{\Sigma} \in \mathsf{Box}^{e} \times \mathsf{Box}^{\tilde{d}} \times \mathsf{Box}^{s}$$

is a tuple in the domain of application of the SOS-operator box Ω_0. The result of performing the refinement $\Omega_0(\boldsymbol{\Sigma})$ is also shown in Fig. 7.4.

In order not to make the running example too complicated, we did not choose an infinite operator with side-loops and non-reversible transitions, nor plain boxes which are not ex-directed (but Σ_1 is not ex-exclusive). However, in the discussion and proofs below, we shall carefully consider all these possibilities.

7.1.4 Static Properties of Refinements

Throughout this chapter we will make frequent references to various notions used in the definition of net refinement, $\Omega(\boldsymbol{\Sigma})$, in Sect. 4.3. In particular, we will use the notations $\mathsf{T}^{v}_{\mathsf{new}}$, $\mathsf{ST}^{v}_{\mathsf{new}}$, and $\mathsf{SP}^{s}_{\mathsf{new}}$, the latter also lifted to sets of places R, through the formula

$$\mathsf{SP}^{R}_{\mathsf{new}} = \bigcup_{s \in R} \mathsf{SP}^{s}_{\mathsf{new}}\,,$$

and the notations $\mathsf{trees}(u)$ and $\mathsf{trees}^{v}(p)$, where u and p are respectively a transition and a place in $\Omega(\boldsymbol{\Sigma})$, and v is a transition in Ω. Recall that $\mathsf{trees}^{v}(p)$ can contain at most two places: one entry and one exit place of Σ_v; also, though $\mathsf{trees}(u)$ is formally a multiset, we will sometimes treat it as if it were a set when writing, e.g., ${}^{\bullet}\mathsf{trees}(u)$. For the example in Fig. 7.4, we have

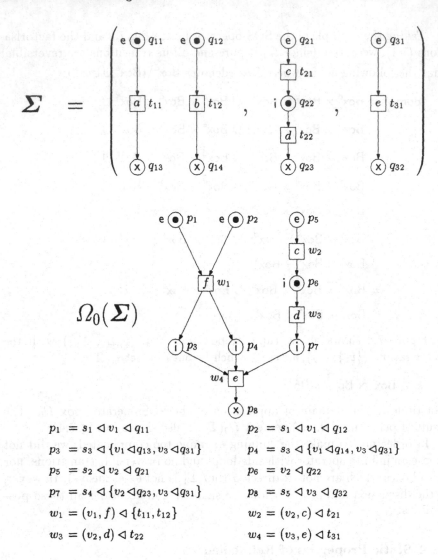

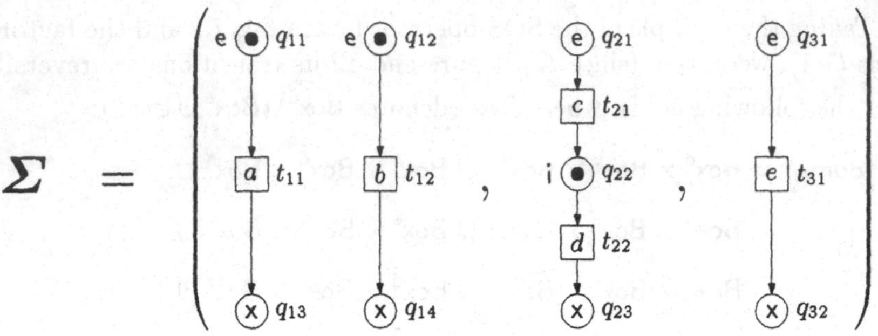

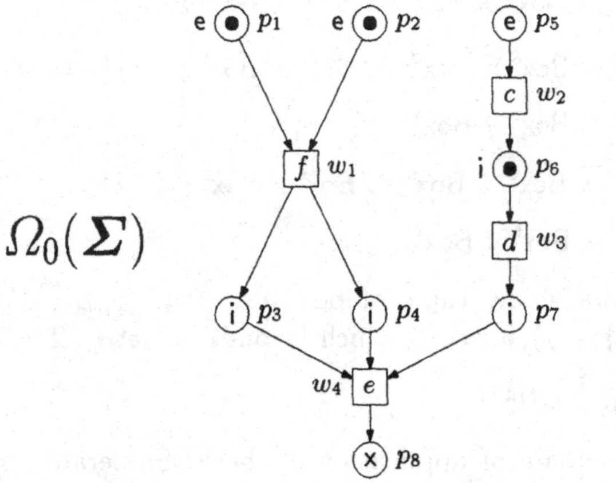

$$p_1 = s_1 \lhd v_1 \lhd q_{11} \qquad p_2 = s_1 \lhd v_1 \lhd q_{12}$$

$$p_3 = s_3 \lhd \{v_1 \lhd q_{13}, v_3 \lhd q_{31}\} \qquad p_4 = s_3 \lhd \{v_1 \lhd q_{14}, v_3 \lhd q_{31}\}$$

$$p_5 = s_2 \lhd v_2 \lhd q_{21} \qquad p_6 = v_2 \lhd q_{22}$$

$$p_7 = s_4 \lhd \{v_2 \lhd q_{23}, v_3 \lhd q_{31}\} \qquad p_8 = s_5 \lhd v_3 \lhd q_{32}$$

$$w_1 = (v_1, f) \lhd \{t_{11}, t_{12}\} \qquad w_2 = (v_2, c) \lhd t_{21}$$

$$w_3 = (v_2, d) \lhd t_{22} \qquad w_4 = (v_3, e) \lhd t_{31}$$

Fig. 7.4. A tuple of boxes $\boldsymbol{\Sigma} = (\Sigma_{v_1}, \Sigma_{v_2}, \Sigma_{v_3})$ for the running example, and the result of an application of the SOS-operator box Ω_0

$$\begin{array}{lll}
\mathsf{trees}^{v_2}(p_7) = \{q_{23}\} & \mathsf{trees}(w_1) = \{t_{11}, t_{12}\} & \mathsf{trees}^{v_2}(p_3) = \emptyset \\
\mathsf{trees}^{v_3}(p_7) = \{q_{31}\} & \mathsf{trees}(w_2) = \{t_{21}\} & \mathsf{trees}^{v_1}(p_7) = \emptyset \\
\mathsf{trees}^{v_2}(p_6) = \{q_{22}\} & \mathsf{trees}(w_4) = \{t_{31}\} & \mathsf{trees}^{v_3}(p_6) = \emptyset\,.
\end{array}$$

Using these notations, we can characterise several useful structural properties of the box $\Omega(\boldsymbol{\Sigma})$. Below, Ω is an SOS-operator box, v and w are distinct transitions in Ω, and $\boldsymbol{\Sigma}$ is a tuple in its domain of application. The first three propositions follow directly from the definition of net refinement and the ex-restrictedness of boxes (see also the proof of Prop. 4.3.1).

Proposition 7.1.2. Let $z \in \ddot{\Sigma}_v$, $q \in {}^{\circ}\Sigma_v$, and $r \in \Sigma_v^{\circ}$.

(1) There is $p \in \mathsf{ST}^v_{\mathsf{new}}$ such that $\mathsf{trees}^v(p) = \{z\}$.
(2) For every $s \in {}^{\bullet}v \backslash v^{\bullet}$ there is $p \in \mathsf{SP}^s_{\mathsf{new}}$ such that $\mathsf{trees}^v(p) = \{q\}$.
(3) For every $s \in v^{\bullet} \backslash {}^{\bullet}v$ there is $p \in \mathsf{SP}^s_{\mathsf{new}}$ such that $\mathsf{trees}^v(p) = \{r\}$.
(4) For every $s \in {}^{\bullet}v \cap v^{\bullet}$ there is $p \in \mathsf{SP}^s_{\mathsf{new}}$ such that $\mathsf{trees}^v(p) = \{q, r\}$. □

Proposition 7.1.3. Suppose that one of the following holds:

$$\begin{array}{llll} q \in {}^{\circ}\Sigma_v & \wedge\ r \in {}^{\circ}\Sigma_w & \wedge\ s \in {}^{\bullet}v \cap {}^{\bullet}w \\ q \in \Sigma_v^{\circ} & \wedge\ r \in {}^{\circ}\Sigma_w & \wedge\ s \in v^{\bullet} \cap {}^{\bullet}w \\ q \in \Sigma_v^{\circ} & \wedge\ r \in \Sigma_w^{\circ} & \wedge\ s \in v^{\bullet} \cap w^{\bullet} \\ q \in {}^{\circ}\Sigma_v & \wedge\ r \in \Sigma_w^{\circ} & \wedge\ s \in {}^{\bullet}v \cap w^{\bullet}\,. \end{array}$$

Then there is $p \in \mathsf{SP}^s_{\mathsf{new}}$ such that $q \in \mathsf{trees}^v(p)$ and $r \in \mathsf{trees}^w(p)$. □

Proposition 7.1.4. Let p be a place in $\Omega(\boldsymbol{\Sigma})$, and $t \in \mathsf{T}^v_{\mathsf{new}}$. Then

$$\begin{array}{lcl} p \in {}^{\bullet}t & \Longleftrightarrow & \mathsf{trees}^v(p) \cap {}^{\bullet}\mathsf{trees}(t) \neq \emptyset \\ p \in t^{\bullet} & \Longleftrightarrow & \mathsf{trees}^v(p) \cap \mathsf{trees}(t)^{\bullet} \neq \emptyset\,. \end{array}$$

Moreover, if $p \in \mathsf{SP}^s_{\mathsf{new}}$, then

$$\begin{array}{lcl} p \in {}^{\bullet}t & \Longleftrightarrow & (\mathsf{trees}^v(p) \cap {}^{\bullet}\mathsf{trees}(t) \cap \Sigma_v^{\circ} \neq \emptyset\ \wedge s \in v^{\bullet}) \quad \vee \\ & & (\mathsf{trees}^v(p) \cap {}^{\bullet}\mathsf{trees}(t) \cap {}^{\circ}\Sigma_v \neq \emptyset \wedge s \in {}^{\bullet}v) \\ p \in t^{\bullet} & \Longleftrightarrow & (\mathsf{trees}^v(p) \cap \mathsf{trees}(t)^{\bullet} \cap \Sigma_v^{\circ} \neq \emptyset\ \wedge s \in v^{\bullet}) \quad \vee \\ & & (\mathsf{trees}^v(p) \cap \mathsf{trees}(t)^{\bullet} \cap {}^{\circ}\Sigma_v \neq \emptyset \wedge s \in {}^{\bullet}v)\,. \end{array}$$

□

Applying the above proposition to the example in Fig. 7.4, we obtain that $p_3 \in {}^{\bullet}w_4$ since $\mathsf{trees}^{v_3}(p_3) = \{q_{31}\}$ and ${}^{\bullet}\mathsf{trees}(w_4) = {}^{\bullet}\{t_{31}\} = \{q_{31}\}$.

In the results that now follow, we use four predicates, In^{e}_t, In^{x}_t, Out^{e}_t, and Out^{x}_t, where t is a transition in $\mathsf{T}^v_{\mathsf{new}}$, defined by

$$\begin{array}{ll} In^{\mathsf{e}}_t\ = {}^{\bullet}\mathsf{trees}(t) \cap {}^{\circ}\Sigma_v \neq \emptyset & In^{\mathsf{x}}_t\ = {}^{\bullet}\mathsf{trees}(t) \cap \Sigma_v^{\circ} \neq \emptyset \\ Out^{\mathsf{e}}_t = \mathsf{trees}(t)^{\bullet} \cap {}^{\circ}\Sigma_v \neq \emptyset & Out^{\mathsf{x}}_t = \mathsf{trees}(t)^{\bullet} \cap \Sigma_v^{\circ} \neq \emptyset\,, \end{array}$$

expressing that one of the 'constituents' of t produces (Out) or absorbs (In) a token in an entry (e) or exit (x) place of Σ_v.

Proposition 7.1.5. Let $t, u \in \mathsf{T}^v_{\mathsf{new}}$. Then

$$\begin{array}{lcl} {}^{\bullet}t \cap {}^{\bullet}u \neq \emptyset & \Longleftrightarrow & {}^{\bullet}\mathsf{trees}(t) \cap {}^{\bullet}\mathsf{trees}(u) \neq \emptyset\ \vee \\ & & ({}^{\bullet}v \cap v^{\bullet} \neq \emptyset\ \wedge\ (In^{\mathsf{e}}_t \wedge In^{\mathsf{x}}_u\ \vee\ In^{\mathsf{x}}_t \wedge In^{\mathsf{e}}_u)) \\ {}^{\bullet}t \cap u^{\bullet} \neq \emptyset & \Longleftrightarrow & {}^{\bullet}\mathsf{trees}(t) \cap \mathsf{trees}(u)^{\bullet} \neq \emptyset\ \vee \\ & & ({}^{\bullet}v \cap v^{\bullet} \neq \emptyset\ \wedge\ (In^{\mathsf{e}}_t \wedge Out^{\mathsf{x}}_u\ \vee\ In^{\mathsf{x}}_t \wedge Out^{\mathsf{e}}_u)) \\ t^{\bullet} \cap {}^{\bullet}u \neq \emptyset & \Longleftrightarrow & \mathsf{trees}(t)^{\bullet} \cap {}^{\bullet}\mathsf{trees}(u) \neq \emptyset\ \vee \\ & & ({}^{\bullet}v \cap v^{\bullet} \neq \emptyset\ \wedge\ (Out^{\mathsf{e}}_t \wedge In^{\mathsf{x}}_u\ \vee\ Out^{\mathsf{x}}_t \wedge In^{\mathsf{e}}_u)) \\ t^{\bullet} \cap u^{\bullet} \neq \emptyset & \Longleftrightarrow & \mathsf{trees}(t)^{\bullet} \cap \mathsf{trees}(u)^{\bullet} \neq \emptyset\ \vee \\ & & ({}^{\bullet}v \cap v^{\bullet} \neq \emptyset\ \wedge\ (Out^{\mathsf{e}}_t \wedge Out^{\mathsf{x}}_u\ \vee\ Out^{\mathsf{x}}_t \wedge Out^{\mathsf{e}}_u))\,. \end{array}$$

Proof. We only prove the first equivalence, as the other three can be shown in basically the same way. Similar remark applies to the proofs of the next two propositions.

Suppose that $p \in {}^\bullet t \cap {}^\bullet u$. Then, by Prop. 7.1.4, there are $q \in \mathsf{trees}^v(p) \cap {}^\bullet\mathsf{trees}(t)$ and $r \in \mathsf{trees}^v(p) \cap {}^\bullet\mathsf{trees}(u)$. If $q = r$, then ${}^\bullet\mathsf{trees}(t) \cap {}^\bullet\mathsf{trees}(u) \neq \emptyset$. Otherwise, $\mathsf{trees}^v(p) = \{q, r\}$ and, without loss of generality, $q \in {}^\circ\Sigma_v$ and $r \in \Sigma_v^\circ$ (and so $In_t^{\mathsf{e}} \wedge In_u^{\mathsf{x}}$). Moreover, $|\mathsf{trees}^v(p)| > 1$ implies ${}^\bullet v \cap v^\bullet \neq \emptyset$. Hence the ($\Longrightarrow$) implication holds. The reverse implication ($\Longleftarrow$) follows from Props. 7.1.2(4) and 7.1.4. □

For the example in Fig. 7.4, we obtain $w_2^\bullet \cap {}^\bullet w_3 \neq \emptyset$ since

$$\{t_{21}\}^\bullet \cap {}^\bullet\{t_{22}\} = \{q_{22}\} \cap \{q_{22}\} \neq \emptyset .$$

Proposition 7.1.6. Let $t \in \mathsf{T}^v_{\mathsf{new}}$ and $u \in \mathsf{T}^w_{\mathsf{new}}$. Then

$$\begin{aligned}
{}^\bullet t \cap {}^\bullet u \neq \emptyset \iff\; & (In_t^{\mathsf{e}} \wedge In_u^{\mathsf{e}} \wedge {}^\bullet v \cap {}^\bullet w \neq \emptyset) \vee \\
& (In_t^{\mathsf{x}} \wedge In_u^{\mathsf{e}} \wedge v^\bullet \cap {}^\bullet w \neq \emptyset) \vee \\
& (In_t^{\mathsf{e}} \wedge In_u^{\mathsf{x}} \wedge {}^\bullet v \cap w^\bullet \neq \emptyset) \vee \\
& (In_t^{\mathsf{x}} \wedge In_u^{\mathsf{x}} \wedge v^\bullet \cap w^\bullet \neq \emptyset) \\
{}^\bullet t \cap u^\bullet \neq \emptyset \iff\; & (In_t^{\mathsf{e}} \wedge Out_u^{\mathsf{e}} \wedge {}^\bullet v \cap {}^\bullet w \neq \emptyset) \vee \\
& (In_t^{\mathsf{x}} \wedge Out_u^{\mathsf{e}} \wedge v^\bullet \cap {}^\bullet w \neq \emptyset) \vee \\
& (In_t^{\mathsf{e}} \wedge Out_u^{\mathsf{x}} \wedge {}^\bullet v \cap w^\bullet \neq \emptyset) \vee \\
& (In_t^{\mathsf{x}} \wedge Out_u^{\mathsf{x}} \wedge v^\bullet \cap w^\bullet \neq \emptyset) \\
t^\bullet \cap {}^\bullet u \neq \emptyset \iff\; & (Out_t^{\mathsf{e}} \wedge In_u^{\mathsf{e}} \wedge {}^\bullet v \cap {}^\bullet w \neq \emptyset) \vee \\
& (Out_t^{\mathsf{x}} \wedge In_u^{\mathsf{e}} \wedge v^\bullet \cap {}^\bullet w \neq \emptyset) \vee \\
& (Out_t^{\mathsf{e}} \wedge In_u^{\mathsf{x}} \wedge {}^\bullet v \cap w^\bullet \neq \emptyset) \vee \\
& (Out_t^{\mathsf{x}} \wedge In_u^{\mathsf{x}} \wedge v^\bullet \cap w^\bullet \neq \emptyset) \\
t^\bullet \cap u^\bullet \neq \emptyset \iff\; & (Out_t^{\mathsf{e}} \wedge Out_u^{\mathsf{e}} \wedge {}^\bullet v \cap {}^\bullet w \neq \emptyset) \vee \\
& (Out_t^{\mathsf{x}} \wedge Out_u^{\mathsf{e}} \wedge v^\bullet \cap {}^\bullet w \neq \emptyset) \vee \\
& (Out_t^{\mathsf{e}} \wedge Out_u^{\mathsf{x}} \wedge {}^\bullet v \cap w^\bullet \neq \emptyset) \vee \\
& (Out_t^{\mathsf{x}} \wedge Out_u^{\mathsf{x}} \wedge v^\bullet \cap w^\bullet \neq \emptyset) .
\end{aligned}$$

Proof. Suppose that $p \in {}^\bullet t \cap {}^\bullet u$. Then, by the definition of net refinement, there is $s \in ({}^\bullet v \cup v^\bullet) \cap ({}^\bullet w \cup w^\bullet)$ such that $p \in \mathsf{SP}^s_{\mathsf{new}}$. Hence the ($\Longrightarrow$) implication holds by Prop. 7.1.4. The reverse implication ($\Longleftarrow$) follows directly from Props. 7.1.3 and 7.1.4. □

For the example in Fig. 7.4, we obtain $w_1^\bullet \cap {}^\bullet w_4 \neq \emptyset$ since

$$\begin{aligned}
\{t_{11}, t_{12}\}^\bullet \cap \Sigma_{v_1}^\circ &= \{q_{13}, q_{14}\} \cap \{q_{13}, q_{14}\} \neq \emptyset && \text{(i.e., } Out^{\mathsf{x}}_{w_1}) \\
{}^\bullet\{t_{31}\} \cap {}^\circ\Sigma_{v_3} &= \{q_{31}\} \cap \{q_{31}\} \neq \emptyset && \text{(i.e., } In^{\mathsf{e}}_{w_4}) \\
v_1^\bullet \cap {}^\bullet v_3 &= \{s_3\} \cap \{s_3, s_4\} \neq \emptyset .
\end{aligned}$$

Proposition 7.1.7. Let $t \in \mathsf{T}^v_{\mathsf{new}}$. Then

$$\begin{aligned}
{}^\bullet t \cap {}^\circ\Omega(\boldsymbol{\Sigma}) \neq \emptyset &\iff (In^{\mathrm{e}}_t \wedge {}^\bullet v \cap {}^\circ\Omega \neq \emptyset) \vee (In^{\mathrm{x}}_t \wedge v^\bullet \cap {}^\circ\Omega \neq \emptyset)\\
t^\bullet \cap {}^\circ\Omega(\boldsymbol{\Sigma}) \neq \emptyset &\iff (Out^{\mathrm{e}}_t \wedge {}^\bullet v \cap {}^\circ\Omega \neq \emptyset) \vee (Out^{\mathrm{x}}_t \wedge v^\bullet \cap {}^\circ\Omega \neq \emptyset)\\
t^\bullet \cap \Omega(\boldsymbol{\Sigma})^\circ \neq \emptyset &\iff (Out^{\mathrm{e}}_t \wedge {}^\bullet v \cap \Omega^\circ \neq \emptyset) \vee (Out^{\mathrm{x}}_t \wedge v^\bullet \cap \Omega^\circ \neq \emptyset)\\
{}^\bullet t \cap \Omega(\boldsymbol{\Sigma})^\circ \neq \emptyset &\iff (In^{\mathrm{e}}_t \wedge {}^\bullet v \cap \Omega^\circ \neq \emptyset) \vee (In^{\mathrm{x}}_t \wedge v^\bullet \cap \Omega^\circ \neq \emptyset)\,.
\end{aligned}$$

Proof. Suppose that $p \in {}^\bullet t \cap {}^\circ\Omega(\boldsymbol{\Sigma})$. Then, by the definition of net refinement, there is $s \in {}^\circ\Omega \cap ({}^\bullet v \cup v^\bullet)$ such that $p \in \mathsf{SP}^s_{\mathsf{new}}$. Hence the ($\Longrightarrow$) implication holds by Prop. 7.1.4. The reverse implication ($\Longleftarrow$) follows directly from Props. 7.1.2 and 7.1.4. □

Thus, for the example in Fig. 7.4, we obtain ${}^\bullet w_2 \cap {}^\circ\Omega_0(\boldsymbol{\Sigma}) \neq \emptyset$ since

$$\begin{aligned}
{}^\bullet\{t_{21}\} \cap {}^\circ\Sigma_{v_2} &= \{q_{21}\} \cap \{q_{21}\} \neq \emptyset \qquad (\text{i.e., } In^{\mathrm{e}}_{w_2})\\
{}^\bullet v_2 \cap {}^\circ\Omega_0 &= \{s_2\} \cap \{s_1, s_2\} \neq \emptyset\,.
\end{aligned}$$

7.1.5 Markings of Nets

It is yet far from clear that the definition of an SOS-operator box is really what we were looking for. This will be accomplished in the rest of the present chapter. In this section, we will show that the marking of any net obtained by applying an SOS-operator box to a tuple of boxes in its domain is safe and clean. This will be preceded by two auxiliary results clarifying the relationship between the marking of the refined net $\Omega(\boldsymbol{\Sigma})$ and the markings of the refining nets $\boldsymbol{\Sigma}$. The results stem from an observation that although in the general formulation of net refinement a token in a newly created place can be contributed by any of the places which were used to construct that place, for the kind of refinements considered in this chapter, the restrictions imposed on an SOS-operator box and on the tuples of boxes in its domain mean that at most one such box can actually contribute token(s).

Let Ω be an SOS-operator box and $\boldsymbol{\Sigma} \in \mathsf{dom}_\Omega$ be an Ω-tuple of boxes with the factorisation μ, fixed for the rest of this section. For every transition v in Ω, we define the set $\mathsf{mar}_\mu(v)$ of places of $\Omega(\boldsymbol{\Sigma})$ which are *markable* by v, in the following way:

$$\mathsf{mar}_\mu(v) = \begin{cases} \mathsf{SP}^{({}^\bullet v)}_{\mathsf{new}} & \text{if } v \in \mu_{\mathrm{e}}\\ \mathsf{SP}^{({}^\bullet v \cup v^\bullet)}_{\mathsf{new}} \cup \mathsf{ST}^v_{\mathsf{new}} & \text{if } v \in \mu_{\mathrm{d}}\\ \mathsf{SP}^{(v^\bullet)}_{\mathsf{new}} & \text{if } v \in \mu_{\mathrm{x}}\\ \emptyset & \text{if } v \in \mu_{\mathrm{s}}\,. \end{cases} \tag{7.2}$$

That is, $\mathsf{mar}_\mu(v)$ comprises all places into which the box Σ_v refining v can possibly insert tokens[4] and so, in particular, $\mathsf{mar}_\mu(v)$ is empty for a static

[4] In other words, Σ_v cannot insert a token into places outside $\mathsf{mar}_\mu(v)$.

Σ_v. From Prop. 7.1.1(4) it follows that a given place is markable by at most one transition, i.e., for all transitions v and w in Ω,

$$v \neq w \implies \mathsf{mar}_\mu(v) \cap \mathsf{mar}_\mu(w) = \emptyset . \tag{7.3}$$

Notice also that if $p \in \mathsf{mar}_\mu(v)$, then $\mathsf{trees}^v(p)$ is always nonempty. For our running example in Fig. 7.4, we will also assume that $\mu = (\{v_1\}, \{v_2\}, \emptyset, \{v_3\})$, the factorisation of the tuple $\boldsymbol{\Sigma}$ of Fig. 7.4, is the running factorisation. Then, we have $\mathsf{mar}_\mu(v_1) = \{p_1, p_2\}$, $\mathsf{mar}_\mu(v_2) = \{p_5, p_6, p_7\}$, and $\mathsf{mar}_\mu(v_3) = \emptyset$. Using the notion of markable places, we can relate the marking of $\Omega(\boldsymbol{\Sigma})$ to those of the nets in $\boldsymbol{\Sigma}$.

Lemma 7.1.1. Let p be a place in $\Omega(\boldsymbol{\Sigma})$. Then

$$M_{\Omega(\boldsymbol{\Sigma})}(p) = \begin{cases} \mathit{Tot}(M_{\Sigma_v}, \mathsf{trees}^v(p)) & \text{if } p \in \mathsf{mar}_\mu(v) \text{ for some } v \\ 0 & \text{otherwise} , \end{cases} \tag{7.4}$$

where $\mathit{Tot}(M, R) = \sum_{s \in R} M(s)$ is the *total number of tokens* on a finite set of places R.

Proof. First note that (7.4) is well formed due to (7.3). If $p \in \mathsf{ST}^v_{\mathsf{new}}$, for some v, then (7.4) follows directly from the definition of net refinement and (7.2). Suppose that $p \in \mathsf{SP}^s_{\mathsf{new}}$. If there is v such that $p \in \mathsf{mar}_\mu(v)$, then by Prop. 7.1.1(4) and (7.2),

$$w \in \Big(({}^\bullet s \cup s^\bullet) \cap (\mu_{\mathsf{e}} \cup \mu_{\mathsf{d}} \cup \mu_{\mathsf{x}}) \Big) \setminus \{v\} \implies s \in {}^\bullet w \setminus w^\bullet \ \wedge \ w \in \mu_{\mathsf{x}}$$

which in turn implies that $\mathit{Tot}(M_{\Sigma_w}, \mathsf{trees}^w(p)) = 0$, for all $w \in ({}^\bullet s \cup s^\bullet) \setminus \{v\}$. This and $M_\Omega(s) = 0$ yields $M_{\Omega(\boldsymbol{\Sigma})}(p) = \mathit{Tot}(M_{\Sigma_v}, \mathsf{trees}^v(p))$. If there is no v such that $p \in \mathsf{mar}_\mu(v)$, then $\mathit{Tot}(M_{\Sigma_w}, \mathsf{trees}^w(p)) = 0$, for all $w \in {}^\bullet s \cup s^\bullet$. This and $M_\Omega(s) = 0$ means that p is unmarked. □

Applying the above lemma to the places p_3 and p_6 in Fig. 7.4, we obtain $M_{\Omega_0(\boldsymbol{\Sigma})}(p_3) = 0$, since $p_3 \notin \mathsf{mar}_\mu(v_1) \cup \mathsf{mar}_\mu(v_2) \cup \mathsf{mar}_\mu(v_3)$, and

$$M_{\Omega_0(\boldsymbol{\Sigma})}(p_6) = \mathit{Tot}(\Sigma_{v_2}, \mathsf{trees}^{v_2}(p_6)) = M_{\Sigma_{v_2}}(q_{22}) = 1 .$$

Lemma 7.1.2. Let s be a place in Ω. Then all the places of $\mathsf{SP}^s_{\mathsf{new}}$ are marked in $\Omega(\boldsymbol{\Sigma})$ if and only if $s \in {}^\bullet\mu_{\mathsf{e}} \cup \mu_{\mathsf{x}}^\bullet$.

Proof. ($\Longrightarrow$) By Lemma 7.1.1 and $\mathsf{SP}^s_{\mathsf{new}} \neq \emptyset$ (see (4.10)), there is v such that $\mathsf{mar}_\mu(v) \cap \mathsf{SP}^s_{\mathsf{new}} \neq \emptyset$. If $v \in {}^\bullet s \setminus s^\bullet$ and $\Sigma_v \notin \mathsf{Box}^{\mathsf{x}}$, then there is $r \in \Sigma_v^\circ$ such that $M_{\Sigma_v}(r) = 0$. By Prop. 7.1.2(3), there is a place p in $\mathsf{SP}^s_{\mathsf{new}}$ such that $\mathsf{trees}^v(p) = \{r\}$. Hence, by Lemma 7.1.1, p is unmarked at $M_{\Omega(\boldsymbol{\Sigma})}$, a contradiction. Thus $v \in \mu_{\mathsf{x}}$. For $v \in s^\bullet \setminus {}^\bullet s$ the argument is symmetric, yielding $v \in \mu_{\mathsf{e}}$.

Suppose now that $v \in {}^\bullet s \cap s^\bullet$. Then, by the ex-exclusiveness of Σ_v (by (Dom1)) and without loss of generality, we may assume that ${}^\circ\Sigma_v \cap M_{\Sigma_v} = \emptyset$. If $\Sigma_v \notin \mathsf{Box}^{\mathsf{x}}$, then there is $r \in \Sigma_v^\circ$ such that $M_{\Sigma_v}(r) = 0$. By Prop. 7.1.2(4), there is a place p in $\mathsf{SP}^{s}_{\mathsf{new}}$ such that $\mathsf{trees}^{v}(p) = \{q, r\}$, where q is a place in ${}^{\circ}\Sigma_v$. Hence, by Lemma 7.1.1, p is unmarked at $M_{\Omega(\boldsymbol{\Sigma})}$, a contradiction. Thus $v \in \mu_{\mathsf{x}}$.

($\Longleftarrow$) If $s \in {}^{\bullet}\mu_{\mathsf{e}} \cup \mu_{\mathsf{x}}^{\bullet}$, then all the places in $\mathsf{SP}^{s}_{\mathsf{new}}$ are marked, which follows from (7.2), Lemma 7.1.1, and the definition of net refinement. □

Theorem 7.1.1. The marking of $\Omega(\boldsymbol{\Sigma})$ is safe and clean.

Proof. The marking is safe by Lemma 7.1.1, the safeness of the boxes in $\boldsymbol{\Sigma}$, and the ex-exclusiveness of the boxes Σ_v such that ${}^{\bullet}v \cap v^{\bullet} \neq \emptyset$ (by (Dom1)). To show that it is also clean, suppose that ${}^{\circ}\Omega(\boldsymbol{\Sigma}) \subseteq M_{\Omega(\boldsymbol{\Sigma})}$. From Lemma 7.1.2 it follows that ${}^{\circ}\Omega \subseteq {}^{\bullet}\mu_{\mathsf{e}} \cup \mu_{\mathsf{x}}^{\bullet}$. Hence, by Prop. 7.1.1(5), $\mu_{\mathsf{d}} = \emptyset$ and ${}^{\circ}\Omega = {}^{\bullet}\mu_{\mathsf{e}} \cup \mu_{\mathsf{x}}^{\bullet}$. It then follows from the safeness of $M_{\Omega(\boldsymbol{\Sigma})}$, (7.2), and Lemmata 7.1.1 and 7.1.2, that ${}^{\circ}\Omega(\boldsymbol{\Sigma}) = M_{\Omega(\boldsymbol{\Sigma})}$. The case $\Omega(\boldsymbol{\Sigma})^{\circ} \subseteq M_{\Omega(\boldsymbol{\Sigma})}$ is symmetric. □

A useful corollary of the results proved in this section is the following closed formula for the safe and clean marking of $\Omega(\boldsymbol{\Sigma})$:

$$M_{\Omega(\boldsymbol{\Sigma})} = \mathsf{SP}^{({}^{\bullet}\mu_{\mathsf{e}} \cup \mu_{\mathsf{x}}^{\bullet})}_{\mathsf{new}} \cup \bigcup_{v \in \mu_{\mathsf{d}}} \left\{ p \in \left(\mathsf{SP}^{({}^{\bullet}v \cup v^{\bullet})}_{\mathsf{new}} \cup \mathsf{ST}^{v}_{\mathsf{new}} \right) \,\middle|\, \mathsf{trees}^{v}(p) \cap M_{\Sigma_v} \neq \emptyset \right\}. \tag{7.5}$$

Applying it to the refinement in Fig. 7.4 yields the marking of $\Omega_0(\boldsymbol{\Sigma})$ (recall that the factorisation of $\boldsymbol{\Sigma}$ is $(\{v_1\}, \{v_2\}, \emptyset, \{v_3\})$):

$$\begin{aligned} M_{\Omega_0(\boldsymbol{\Sigma})} &= \mathsf{SP}^{({}^{\bullet}v_1)}_{\mathsf{new}} \cup \left\{ p \in \left(\mathsf{SP}^{({}^{\bullet}v_2 \cup v_2^{\bullet})}_{\mathsf{new}} \cup \mathsf{ST}^{v_2}_{\mathsf{new}} \right) \,\middle|\, \mathsf{trees}^{v_2}(p) \cap M_{\Sigma_{v_2}} \neq \emptyset \right\} \\ &= \mathsf{SP}^{s_1}_{\mathsf{new}} \cup \left\{ p \in \left(\mathsf{SP}^{\{s_2, s_4\}}_{\mathsf{new}} \cup \mathsf{ST}^{v_2}_{\mathsf{new}} \right) \,\middle|\, \mathsf{trees}^{v_2}(p) \cap \{q_{22}\} \neq \emptyset \right\} \\ &= \{p_1, p_2\} \cup \left\{ p \in \{p_5, p_7, p_6\} \,\middle|\, \mathsf{trees}^{v_2}(p) \cap \{q_{22}\} \neq \emptyset \right\} \\ &= \{p_1, p_2, p_6\}\,. \end{aligned}$$

Exercise 7.1.1. Show that, for every $v \in \mu_{\mathsf{e}}$, all places in $\mathsf{SP}^{(v^{\bullet} \setminus {}^{\bullet}v)}_{\mathsf{new}}$ are unmarked in $\Omega(\boldsymbol{\Sigma})$.

Exercise 7.1.2. Show that it is not always the case that all the places in $\mathsf{SP}^{({}^{\bullet}\mu_{\mathsf{x}})}_{\mathsf{new}}$ are unmarked in $\Omega(\boldsymbol{\Sigma})$ (or see the appendix for a counterexample).

7.2 Structured Operational Semantics of Composite Boxes

The generic rules of the operational semantics of a process algebra, (3.5) and (3.6), specify how the behaviour of a compound process term is related to the behaviour of its subterms. We will now take this idea and apply it to a completely different domain of compositional nets defined using SOS-operator boxes.

Let Ω be an SOS-operator box and $\boldsymbol{\Sigma}$ be an Ω-tuple in its domain, dom_Ω. When dealing with the operational semantics of the compositional box $\Omega(\boldsymbol{\Sigma})$, we shall use the notation

$$\Big(\Omega : \boldsymbol{\Sigma}\Big) \xrightarrow{U} \Big(\Omega : \boldsymbol{\Theta}\Big) \tag{7.6}$$

to mean that the boxes $\boldsymbol{\Sigma}$ can individually make moves which, when combined, yield step U and lead to new boxes $\boldsymbol{\Theta}$. More precisely, by definition, this will be the case whenever U is a finite set of transitions of $\Omega(\boldsymbol{\Sigma})$ and, for every transition v in Ω, $U \cap \mathsf{T}^v_{\mathsf{new}} = \{u_1, \ldots, u_k\}$ is a finite set of transitions such that

$$\Sigma_v \Big[\mathsf{trees}(u_1) + \cdots + \mathsf{trees}(u_k) \Big\rangle\, \Theta_v . \tag{7.7}$$

Consider, for example, the boxes of the running example depicted in Figs. 7.2 and 7.4. Then

$$\Big(\Omega_0 : \boldsymbol{\Sigma}\Big) \xrightarrow{\{w_1,w_3\}} \Big(\Omega_0 : \boldsymbol{\Theta}\Big)$$

where $\boldsymbol{\Theta}$ is the tuple of boxes shown in Fig. 7.5. Indeed, with $U = \{w_1, w_3\}$ we obtain

$$U \cap \mathsf{T}^{v_1}_{\mathsf{new}} = \{w_1\} \quad , \quad U \cap \mathsf{T}^{v_2}_{\mathsf{new}} = \{w_3\} \quad , \quad \text{and} \quad U \cap \mathsf{T}^{v_3}_{\mathsf{new}} = \emptyset .$$

Moreover, $\mathsf{trees}(w_1) = \{t_{11}, t_{12}\}$ and $\mathsf{trees}(w_3) = \{t_{22}\}$. And, finally,

$$\Sigma_{v_1} \Big[\{t_{11}, t_{12}\}\Big\rangle\, \Theta_{v_1} \quad , \quad \Sigma_{v_2} \Big[\{t_{22}\}\Big\rangle\, \Theta_{v_2} \quad , \quad \text{and} \quad \Sigma_{v_3} \Big[\emptyset\Big\rangle\, \Theta_{v_3} .$$

Notice that $\{w_1, w_3\}$ is also a valid step for the box $\Omega_0(\boldsymbol{\Sigma})$ and that we have $\Omega_0(\boldsymbol{\Sigma})\ [\{w_1, w_3\}\rangle\ \Omega_0(\boldsymbol{\Theta})$. We shall soon see that this is not accidental.

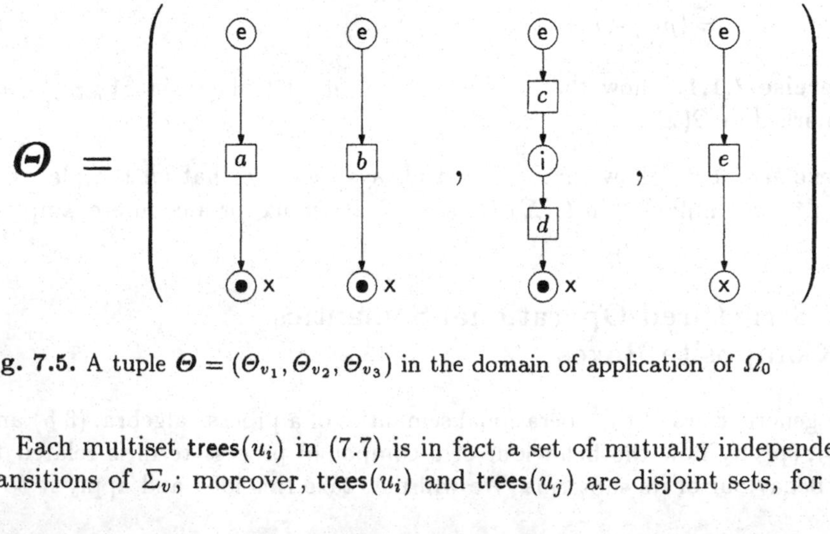

Fig. 7.5. A tuple $\boldsymbol{\Theta} = (\Theta_{v_1}, \Theta_{v_2}, \Theta_{v_3})$ in the domain of application of Ω_0

Each multiset $\mathsf{trees}(u_i)$ in (7.7) is in fact a set of mutually independent transitions of Σ_v; moreover, $\mathsf{trees}(u_i)$ and $\mathsf{trees}(u_j)$ are disjoint sets, for all distinct i and j. Both properties follow from the safeness of the boxes in $\boldsymbol{\Sigma}$ and Prop. 4.2.1(2). We can therefore denote the multiset sum in (7.7) as the disjoint union of k sets, $\mathsf{trees}(u_1) \uplus \cdots \uplus \mathsf{trees}(u_k)$. Notice also that $U \cap \mathsf{T}^v_{\mathsf{new}} \neq \emptyset$ (and so $\Sigma_v \neq \Theta_v$) only for finitely many v, since U is finite. Assuming that U is a multiset rather than a set would not add any new moves (7.6) since $\mathsf{T}^v_{\mathsf{new}} \cap \mathsf{T}^w_{\mathsf{new}} = \emptyset$ for $v \neq w$, and all boxes in $\boldsymbol{\Sigma}$ are safe. The above definition of operational semantics does not involve the *redo/skip* transitions, which are not refined but added only afterwards to obtain the transition systems of boxes.

Exercise 7.2.1. Show formally that the multiset sum in (7.7) is the disjoint union of k sets of mutually independent transitions, $\mathsf{trees}(u_1), \ldots, \mathsf{trees}(u_k)$.

As in Chap. 4, instead of expressing behaviours in terms of transitions, it is also possible to express them using actions, through the labelling function

$$\lambda_{\Omega(\boldsymbol{\Sigma})}(U) = \sum_{u \in U} \{\lambda_{\Omega(\boldsymbol{\Sigma})}(u)\} . \tag{7.8}$$

This returns multisets rather than sets since different transitions may have the same label. For the example above, this would yield $\lambda_{\Omega_0}(\{w_1, w_3\}) = \{f, d\}$, which is also here, by chance, a set. Finally, we need to observe that although each derivation (7.6) is underpinned by the derivations (7.7) for the boxes in $\boldsymbol{\Sigma}$, not all possible moves of the Σ_v's correspond to derivations captured by (7.6). For example, if we again take the running example, then

$$\Sigma_{v_1} \left[\{t_{11}\}\right\rangle \Sigma'_{v_1} \ , \quad \Sigma_{v_2} \left[\emptyset\right\rangle \Sigma'_{v_2} = \Sigma_{v_2} \ , \quad \text{and} \quad \Sigma_{v_3} \left[\emptyset\right\rangle \Sigma'_{v_3} = \Sigma_{v_3} \ .$$

Yet there is no move $(\Omega_0 \ : \ \boldsymbol{\Sigma}) \xrightarrow{U'} (\Omega_0 \ : \ \boldsymbol{\Sigma}')$ corresponding to it because there is no transition w in $\mathsf{T}^{v_1}_{\mathsf{new}}$ such that $\mathsf{trees}(w) = \{t_{11}\}$. This is due to the fact that the ϱ of Fig. 7.2 needs a synchronisation of t_{11} and t_{12}.

7.2.1 Soundness

What now follows is the (easier) half of the SOS-rule for boxes.

Theorem 7.2.1. Let $\boldsymbol{\Sigma}$ be a tuple in the domain of an SOS-operator box Ω and

$$\left(\Omega \ : \ \boldsymbol{\Sigma}\right) \xrightarrow{U} \left(\Omega \ : \ \boldsymbol{\Theta}\right) .$$

Then $\boldsymbol{\Theta}$ is also a tuple in the domain of Ω and $\Omega(\boldsymbol{\Sigma}) \; [U\rangle \; \Omega(\boldsymbol{\Theta})$.

Proof. Let μ be the factorisation of $\boldsymbol{\Sigma}$ and $V^{\delta\delta'} = \{v \in \mu_\delta \mid \Theta_v \in \mathsf{Box}^{\delta'}\}$, for $\delta, \delta' \in \{\mathrm{e}, \mathrm{d}, \mathrm{x}\}$. Then

$$\mu' = \begin{pmatrix} \mu_{\mathsf{e}}\backslash(V^{\mathrm{ed}} \cup V^{\mathrm{ex}}) \cup (V^{\mathrm{de}} \cup V^{\mathrm{xe}}) \\ \mu_{\mathsf{d}}\backslash(V^{\mathrm{de}} \cup V^{\mathrm{dx}}) \cup (V^{\mathrm{ed}} \cup V^{\mathrm{xd}}) \\ \mu_{\mathsf{x}}\backslash(V^{\mathrm{xe}} \cup V^{\mathrm{xd}}) \cup (V^{\mathrm{ex}} \cup V^{\mathrm{dx}}) \\ \mu_{\mathsf{s}} \end{pmatrix}$$

is the factorisation of Θ. Moreover, by multiple applications of all parts of Prop. 7.1.1(3) we obtain that $\mu' \in \mathsf{fact}_\Omega$ (note that $V^{\mathrm{ed}} \cup V^{\mathrm{ex}} \cup V^{\mathrm{de}} \cup V^{\mathrm{dx}} \cup V^{\mathrm{xe}} \cup V^{\mathrm{xd}}$ is a finite set since U is finite, and if $v \in V^{\mathrm{xe}} \cup V^{\mathrm{xd}}$, then Σ_v is non-x-directed and therefore, by (Dom2), v is reversible). We also observe that (Dom1) and (Dom2) are satisfied by $\boldsymbol{\Theta}$ since they were satisfied by $\boldsymbol{\Sigma}$. Hence $\boldsymbol{\Theta} \in \mathsf{dom}_\Omega$.

Let $U_v = U \cap \mathsf{T}^v_{\mathsf{new}}$, for every $v \in T_\Omega$ (recall that U_v and each $\mathsf{trees}(u)$, for $u \in U_v$, is a set). It follows from the definition of $\xrightarrow{U}$ and the enabling and execution rules in Sect. 4.2.3, that there is an Ω-tuple of boxes $\boldsymbol{\Delta}$ such that $\lfloor \boldsymbol{\Delta} \rfloor = \lfloor \boldsymbol{\Sigma} \rfloor$ and, for every $v \in T_\Omega$ and every $q \in S_{\Sigma_v}$,

$$M_{\Sigma_v}(q) = M_{\Delta_v}(q) + \sum_{u \in U_v} \sum_{x \in \mathsf{trees}(u)} W_{\Sigma_v}(q, x) \tag{7.9}$$

$$M_{\Theta_v}(q) = M_{\Delta_v}(q) + \sum_{u \in U_v} \sum_{x \in \mathsf{trees}(u)} W_{\Sigma_v}(x, q)\ . \tag{7.10}$$

(To see this, simply subtract one equation from the other.) We then observe that, since $M_{\Delta_v} \subseteq M_{\Sigma_v}$ for every $v \in T_\Omega$, and $\Omega(\boldsymbol{\Sigma})$ is defined, $\Omega(\boldsymbol{\Delta})$ is a well-defined application of net refinement (although it is not necessarily the case that $\boldsymbol{\Delta} \in \mathsf{dom}_\Omega$, since it could happen, for instance, that $M_{\Delta_v} = \emptyset$ while $v \in \mu_{\mathsf{d}}$). Moreover, it turns out that, for every $p \in S_{\Omega(\boldsymbol{\Sigma})}$,

$$M_{\Omega(\boldsymbol{\Sigma})}(p) = M_{\Omega(\boldsymbol{\Delta})}(p) + \sum_{u \in U} W_{\Omega(\boldsymbol{\Sigma})}(p, u) \tag{7.11}$$

$$M_{\Omega(\boldsymbol{\Theta})}(p) = M_{\Omega(\boldsymbol{\Delta})}(p) + \sum_{u \in U} W_{\Omega(\boldsymbol{\Sigma})}(u, p)\ . \tag{7.12}$$

To show (7.11), we consider two cases.

Case 1: $p = v \triangleleft q \in \mathsf{ST}^v_{\mathsf{new}}$. Then, by $\mathsf{trees}^v(p) = \{q\}$, (4.5), (4.7), (7.9), and $\mathsf{trees}(u)$ being sets, for all $u \in U_v$:

$$\begin{aligned} M_{\Omega(\boldsymbol{\Sigma})}(p) &= M_{\Sigma_v}(q) \\ &= M_{\Delta_v}(q) + \sum_{u \in U_v} \sum_{x \in \mathsf{trees}(u)} W_{\Sigma_v}(q, x) \\ &= M_{\Omega(\boldsymbol{\Delta})}(p) + \sum_{u \in U_v} W_{\Omega(\boldsymbol{\Sigma})}(p, u)\ . \end{aligned}$$

We then observe that from $\mathsf{trees}^z(p) = \emptyset$ (for $z \neq v$) and (4.7) it follows that $W_{\Omega(\boldsymbol{\Sigma})}(p, u) = 0$, for every $u \in U \backslash U_v$, so (7.11) holds.

Case 2: $p = s \lhd (\{v \lhd x_v\}_{v\in {}^\bullet s} + \{w \lhd e_w\}_{w\in s^\bullet}) \in \mathsf{SP}^s_{\mathsf{new}}$. Then we proceed as follows:

$$\begin{aligned}
M_{\Omega(\boldsymbol{\Sigma})}(p) &= \sum_{v\in {}^\bullet s} M_{\Sigma_v}(x_v) + \sum_{w\in s^\bullet} M_{\Sigma_w}(e_w) \\
&= \sum_{v\in {}^\bullet s} M_{\Delta_v}(x_v) + \sum_{v\in {}^\bullet s} \sum_{u\in U_v} \sum_{x\in \mathsf{trees}(u)} W_{\Sigma_v}(x_v, x) + \\
&\quad \sum_{w\in s^\bullet} M_{\Delta_w}(e_w) + \sum_{w\in s^\bullet} \sum_{u\in U_w} \sum_{x\in \mathsf{trees}(u)} W_{\Sigma_w}(e_w, x) \\
&= M_{\Omega(\boldsymbol{\Delta})}(p) + \sum_{h\in {}^\bullet s\cup s^\bullet} \sum_{u\in U_h} \sum_{z\in \mathsf{trees}^h(p)} \sum_{x\in \mathsf{trees}(u)} W_{\Sigma_h}(z, x) \\
&= M_{\Omega(\boldsymbol{\Delta})}(p) + \sum_{h\in {}^\bullet s\cup s^\bullet} \sum_{u\in U_h} W_{\Omega(\boldsymbol{\Sigma})}(p, u) ,
\end{aligned}$$

where the first equality follows from (4.5) and $M_\Omega(s) = 0$; the second from (7.9); the third from (4.5) and $M_\Omega(s) = 0$ and $x_v \neq e_w$ (for $v = w \in {}^\bullet s \cup s^\bullet$); and the last from (4.7) and $\mathsf{trees}(u)$ being sets. We then observe that from $\mathsf{trees}^z(p) = \emptyset$ (for $z \notin {}^\bullet s \cup s^\bullet$) and (4.7) it follows that $W_{\Omega(\boldsymbol{\Sigma})}(p, u) = 0$, for every $u \in U \backslash \bigcup_{h\in {}^\bullet s\cup s^\bullet} U_h$, so (7.11) holds.

We have shown that (7.11) holds, and the proof for (7.12) is similar.

Since U is finite, by (7.11), it is enabled at $M_{\Omega(\boldsymbol{\Sigma})}$. Let Δ be the box such that $\Omega(\boldsymbol{\Sigma})\ [U\rangle\ \Delta$. Then, by (7.11), (7.12), and U being a set, for every $p \in S_\Delta = S_{\Omega(\boldsymbol{\Sigma})} = S_{\Omega(\boldsymbol{\Theta})}$,

$$M_\Delta(p) = M_{\Omega(\boldsymbol{\Sigma})}(p) - \sum_{u\in U} W_{\Omega(\boldsymbol{\Sigma})}(p, u) + \sum_{u\in U} W_{\Omega(\boldsymbol{\Sigma})}(u, p) = M_{\Omega(\boldsymbol{\Theta})}(p).$$

This and $\lfloor \Omega(\boldsymbol{\Theta}) \rfloor = \lfloor \Delta \rfloor$ means that $\Omega(\boldsymbol{\Theta}) = \Delta$, which completes the proof.

□

7.2.2 Similarity Relation on Tuples of Boxes

A direct converse of Thm. 7.2.1 does not in general hold true. For consider the tuple of boxes $\boldsymbol{\Theta}$ in Fig. 7.5 and the box $\Omega_0(\boldsymbol{\Theta})$, shown in Fig. 7.6. We have $\Omega_0(\boldsymbol{\Theta})\ [\{w_4\}\rangle$, yet no nonempty move at all of the form

$$\left(\Omega_0 : \boldsymbol{\Theta}\right) \xrightarrow{U} \left(\Omega_0 : \boldsymbol{\Theta}'\right)$$

is possible. This is so because, when composing the nets, the tokens contributed by Θ_1 and Θ_2 are inserted into the composed net in such a way that they could all have been contributed by Θ_3 as well. More precisely, we have $\Omega_0(\boldsymbol{\Theta}) = \Omega_0(\boldsymbol{\Psi})$, where $\boldsymbol{\Psi}$ is the tuple of boxes depicted in Fig. 7.7. And the situation is now different since a move

$$\left(\Omega_0 : \boldsymbol{\Psi}\right) \xrightarrow{\{w_4\}} \left(\Omega_0 : \boldsymbol{\Psi}'\right)$$

$$\Omega_0(\boldsymbol{\Theta}) = \Omega_0(\boldsymbol{\Psi})$$

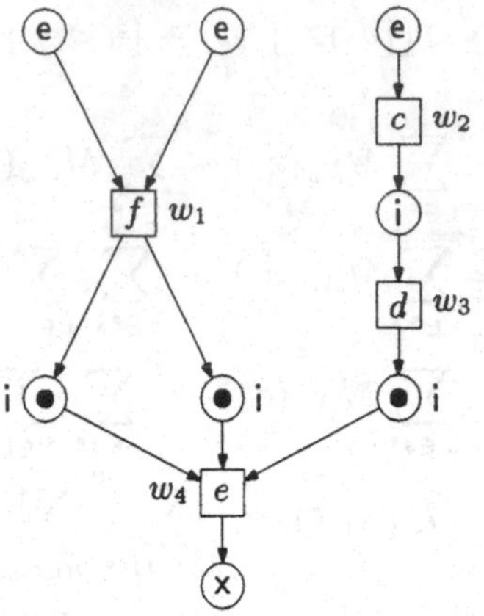

Fig. 7.6. Another application of Ω_0

is possible with $\Psi_1' = \Psi_1$, $\Psi_2' = \Psi_2$, and $\Psi_3' = \Psi_3$. A conclusion to be drawn from this example is that the markings in a tuple of boxes $\boldsymbol{\Sigma}$ may need to be rearranged before attempting to derive a move which is admitted by their composition $\Omega(\boldsymbol{\Sigma})$. Such a rearrangement is formalised by a similarity relation $\equiv_\Omega$ on Ω-tuples of boxes, defined next, which mirrors the redistribution of over- and underbars in the inaction rules of the operational semantics of PBC.

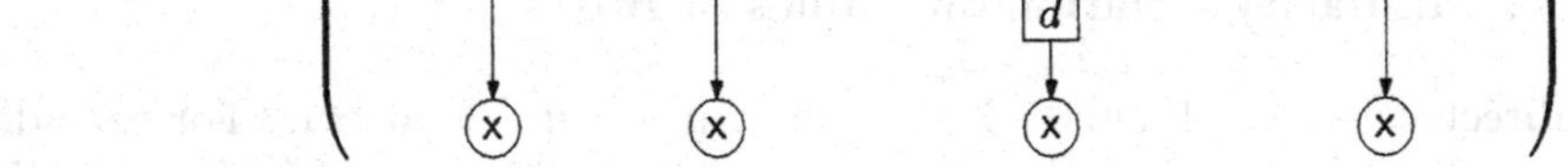

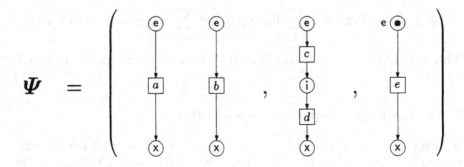

Fig. 7.7. A tuple $\boldsymbol{\Psi} = (\Psi_{v_1}, \Psi_{v_2}, \Psi_{v_3})$ in the domain of application of Ω_0

Let Ω be an SOS-operator box and $\boldsymbol{\Sigma}$ and $\boldsymbol{\Theta}$ be Ω-tuples of static and dynamic boxes whose factorisations are respectively μ and κ. Then $\boldsymbol{\Sigma} \equiv_\Omega \boldsymbol{\Theta}$ if μ and κ are factorisations of the same complex marking of Ω and $\lfloor \boldsymbol{\Sigma} \rfloor = \lfloor \boldsymbol{\Theta} \rfloor$ and $\Sigma_v = \Theta_v$, for every $v \in \mu_{\mathrm{d}} = \kappa_{\mathrm{d}}$. The last requirement is added because, as we will see, the rearrangement of tokens concerns only the entire entry and exit interfaces of the refining nets. It is clear that $\equiv_\Omega$ is an equivalence relation. We can further strengthen this by showing that it is closed in the domain of application of Ω, and that it relates tuples which yield the same boxes through refinement.

Proposition 7.2.1. If $\boldsymbol{\Sigma}$ is a tuple in the domain of an SOS-operator box Ω and $\boldsymbol{\Sigma} \equiv_\Omega \boldsymbol{\Theta}$, then $\boldsymbol{\Theta}$ is also a tuple in the domain of Ω and $\Omega(\boldsymbol{\Sigma}) = \Omega(\boldsymbol{\Theta})$. Moreover, if $\boldsymbol{\Sigma}$ and $\boldsymbol{\Theta}$ are tuples in the domain of Ω such that $\lfloor \boldsymbol{\Sigma} \rfloor = \lfloor \boldsymbol{\Theta} \rfloor$ and $\Omega(\boldsymbol{\Sigma}) = \Omega(\boldsymbol{\Theta})$, then $\boldsymbol{\Sigma} \equiv_\Omega \boldsymbol{\Theta}$.

Proof. Let μ and κ be respectively the factorisations of $\boldsymbol{\Sigma}$ and $\boldsymbol{\Theta}$.

By the definition of $\equiv_\Omega$, $\lfloor \boldsymbol{\Sigma} \rfloor = \lfloor \boldsymbol{\Theta} \rfloor$, which implies that $\boldsymbol{\Theta}$ satisfies (Dom1) and (Dom2) since $\boldsymbol{\Sigma}$ satisfies both conditions and $\Sigma_v = \Theta_v$, for every $v \in \mu_{\mathrm{d}} = \kappa_{\mathrm{d}}$. Hence $\boldsymbol{\Theta} \in \mathsf{dom}_\Omega$. The claim that $\Omega(\boldsymbol{\Sigma}) = \Omega(\boldsymbol{\Theta})$ follows from $\lfloor \Omega(\boldsymbol{\Sigma}) \rfloor = \Omega(\lfloor \boldsymbol{\Sigma} \rfloor) = \Omega(\lfloor \boldsymbol{\Theta} \rfloor) = \lfloor \Omega(\boldsymbol{\Theta}) \rfloor$, the fact that μ and κ are factorisations of the same marking of Ω and (7.5).

The second part can be shown in the following way. Let us denote

$$R_v^{\mathrm{e}} = {}^\circ\Sigma_v = {}^\circ\Theta_v \quad \text{and} \quad R_v^{\mathrm{x}} = \Sigma_v^\circ = \Theta_v^\circ \quad \text{and} \quad R_v = \ddot{\Sigma}_v = \ddot{\Theta}_v \ ,$$

for every $v \in T_\Omega$. We first observe that, for every $v \in T_\Omega$,

$$M_{\Sigma_v} \cap R_v \ = \ M_{\Theta_v} \cap R_v \ . \tag{7.13}$$

Indeed, let $q \in R_v$. By Prop. 7.1.2(1), there is $p \in S_{\Omega(\boldsymbol{\Sigma})} = S_{\Omega(\boldsymbol{\Theta})}$ such that $\mathsf{trees}^v(p) = \{q\}$ and so, by (7.5) and q being an internal place, $M_{\Sigma_v}(q) = M_{\Omega(\boldsymbol{\Sigma})}(p) = M_{\Omega(\boldsymbol{\Theta})}(p) = M_{\Theta_v}(q)$. We next show that for all $v \in T_\Omega$ and $\delta \in \{\mathrm{e}, \mathrm{x}\}$,

$$(M_{\Sigma_v} \backslash M_{\Theta_v}) \cap R_v^\delta \neq \emptyset \implies v \in \mu_\delta \ . \tag{7.14}$$

Suppose that $\delta = \mathrm{e}$ (the proof is symmetric for $\delta = \mathrm{x}$). Assume that $q \in (M_{\Sigma_v} \backslash M_{\Theta_v}) \cap R_v^{\mathrm{e}}$ (which means that $v \in \mu_{\mathrm{e}} \cup \mu_{\mathrm{d}}$) and $v \notin \mu_{\mathrm{e}}$. The latter means (since Σ_v is clean) that there is $q' \in {}^\circ\Sigma_v$ such that $M_{\Sigma_v}(q') = 0$. Take any $s \in {}^\bullet v$ (notice that $\mathsf{SP}^s_{\mathsf{new}} \subseteq \mathsf{mar}_\mu(v)$). If $s \notin v^\bullet$, then define $Z = \{q\}$ and $Z' = \{q'\}$. If $s \in v^\bullet$, then define $Z = \{q, r\}$ and $Z' = \{q', r\}$ where r is any place in Σ_v° (note that since Σ_v is ex-exclusive in such a case, we have $M_{\Sigma_v}(r) = 0$).

By Prop. 7.1.2(2,4), there is $p \in \mathsf{SP}^s_{\mathsf{new}}$ such that $\mathsf{trees}^v(p) = Z$. Moreover, by (4.5), $p \in M_{\Omega(\boldsymbol{\Sigma})}$. Thus, by $\Omega(\boldsymbol{\Sigma}) = \Omega(\boldsymbol{\Theta})$, (7.2), and Lemma 7.1.1, there must be a transition $w \in {}^\bullet s \cup s^\bullet$ such that $p \in \mathsf{SP}^s_{\mathsf{new}} \subseteq \mathsf{mar}_\kappa(w)$. We now consider two cases:

Case 1: $w = v$. Then there is $p' \in \mathsf{SP}^s_{\mathsf{new}}$ such that $\mathsf{trees}^v(p') = Z'$ and $\mathsf{trees}^z(p) = \mathsf{trees}^z(p')$, for $z \neq v$. We then observe that $M_{\Omega(\boldsymbol{\Sigma})}(p') = 0$. All these leads, by Lemma 7.1.1, to

$$\begin{aligned} Tot(M_{\Sigma_v}, Z) \ &= \ Tot(M_{\Theta_v}, Z) \ = 1 \\ Tot(M_{\Sigma_v}, Z') &= \ Tot(M_{\Theta_v}, Z') = 0 \end{aligned}$$

which is in contradiction with $M_{\Sigma_v}(q) = 1$ and $M_{\Sigma_v}(q') = M_{\Theta_v}(q) = 0$, as one can easily verify.

Case 2: $w \neq v$. Then there is $p' \in \mathsf{SP}^s_{\mathsf{new}}$ such that $\mathsf{trees}^v(p') = Z'$ and $\mathsf{trees}^w(p') = \mathsf{trees}^w(p)$. By Lemma 7.1.1, $M_{\Omega(\boldsymbol{\Theta})}(p') = M_{\Omega(\boldsymbol{\Theta})}(p)$. Yet $M_{\Omega(\boldsymbol{\Sigma})}(p) = 1 \neq 0 = M_{\Omega(\boldsymbol{\Sigma})}(p')$, a contradiction with $\Omega(\boldsymbol{\Sigma}) = \Omega(\boldsymbol{\Theta})$.

We have shown that (7.14) holds. Suppose now that $v \in \mu_{\mathrm{d}}$ and $M_{\Sigma_v} \neq M_{\Theta_v}$. We have, by (7.13), $M_{\Sigma_v} \cap R_v = M_{\Theta_v} \cap R_v$. Moreover, if

$$(M_{\Sigma_v} \backslash M_{\Theta_v}) \cap (R^{\mathrm{e}}_v \cup R^{\mathrm{x}}_v) \neq \emptyset ,$$

then by (7.14), $v \in \mu_{\mathrm{e}} \cup \mu_{\mathrm{x}}$, contradicting $v \in \mu_{\mathrm{d}}$. We therefore have

$$M_{\Sigma_v} \subseteq M_{\Theta_v} \quad \text{and} \quad M_{\Sigma_v} \cap R_v = M_{\Theta_v} \cap R_v \quad \text{and} \quad M_{\Sigma_v} \neq M_{\Theta_v} ,$$

so it must be the case that, for some $\delta \in \{\mathrm{e}, \mathrm{x}\}$,

$$(M_{\Theta_v} \backslash M_{\Sigma_v}) \cap R^{\delta}_v \neq \emptyset .$$

Hence, by the symmetric version of (7.14) (which clearly holds), we obtain $v \in \kappa_\delta$. If $\delta = \mathrm{e}$, then by (7.5), $\mathsf{SP}^{({}^\bullet v)}_{\mathsf{new}} \subseteq M_{\Omega(\boldsymbol{\Theta})} = M_{\Omega(\boldsymbol{\Sigma})}$. This and Lemma 7.1.2 means that ${}^\bullet v \subseteq {}^\bullet\mu_{\mathrm{e}} \cup \mu^\bullet_{\mathrm{x}}$, contradicting $v \in \mu_{\mathrm{d}}$ and Prop. 7.1.1(4). If $\delta = \mathrm{x}$, then by (7.5), $\mathsf{SP}^{(v^\bullet)}_{\mathsf{new}} \subseteq M_{\Omega(\boldsymbol{\Theta})} = M_{\Omega(\boldsymbol{\Sigma})}$ which, again by Lemma 7.1.2, means that $v^\bullet \subseteq {}^\bullet\mu_{\mathrm{e}} \cup \mu^\bullet_{\mathrm{x}}$, contradicting $v \in \mu_{\mathrm{d}}$ and Prop. 7.1.1(4). Hence $\Sigma_v = \Theta_v$ and therefore also $v \in \kappa_{\mathrm{d}}$. In a similar way, we can show that $v \in \kappa_{\mathrm{d}}$ implies $v \in \mu_{\mathrm{d}}$ and $\Sigma_v = \Theta_v$. Hence $\mu_{\mathrm{d}} = \kappa_{\mathrm{d}}$ and $\Sigma_v = \Theta_v$, for every $v \in \mu_{\mathrm{d}} = \kappa_{\mathrm{d}}$.

To complete the proof of $\boldsymbol{\Sigma} \equiv_\Omega \boldsymbol{\Theta}$, we still need to show that ${}^\bullet\mu_{\mathrm{e}} \cup \mu^\bullet_{\mathrm{x}} = {}^\bullet\kappa_{\mathrm{e}} \cup \kappa^\bullet_{\mathrm{x}}$. This, however, follows directly from Lemma 7.1.2 and $\Omega(\boldsymbol{\Sigma}) = \Omega(\boldsymbol{\Theta})$. □

The condition $\lfloor \boldsymbol{\Sigma} \rfloor = \lfloor \boldsymbol{\Theta} \rfloor$ in the second part of the last result cannot be omitted. For example, with our running example in Fig. 7.4, if we take Σ' to be Σ_{v_1} with the label of t_{11} changed to b, and Σ'' to be Σ_{v_1} with the label of t_{12} changed to a, then $\Omega_0(\Sigma', \Sigma_{v_2}, \Sigma_{v_3}) = \Omega_0(\Sigma'', \Sigma_{v_2}, \Sigma_{v_3})$, yet $(\Sigma', \Sigma_{v_2}, \Sigma_{v_3}) \equiv_\Omega (\Sigma'', \Sigma_{v_2}, \Sigma_{v_3})$ clearly does not hold.

7.2.3 Completeness

We are now ready to prove the second half of the SOS-rule for boxes. Together with Thm. 7.2.1, this will mean that for the class of SOS-operator boxes, the standard step sequence semantics of compositionally defined nets obeys a variant of the SOS-rule introduced originally for process algebras. The result is preceded by an auxiliary lemma which provides a characterisation of the enabling relation for the transitions in a refined net. Below, $\boldsymbol{\Sigma}$ is a tuple in the domain of an SOS-operator box Ω, and μ is the factorisation of $\boldsymbol{\Sigma}$.

Lemma 7.2.1. Let v be a transition in Ω and $U = \{u_1, \ldots, u_k\} \subseteq \mathsf{T}^v_{\mathsf{new}}$ be a nonempty set of transitions which is enabled in $\Omega(\boldsymbol{\Sigma})$. Moreover, let $\mathsf{trees}(u_l) = \{x_{l1}, \ldots, x_{lm_l}\}$, for every $l \leq k$, and

$$X = \{x_{11}, \ldots, x_{1m_1}, \ldots, x_{k1}, \ldots, x_{km_k}\} .$$

(1) For all $x \in X$ and $q \in {}^\bullet x$, $W_{\Sigma_v}(q,x) = 1$.
(2) For all $x_{li}, x_{hj} \in X$, if $(l,i) \neq (h,j)$, then ${}^\bullet x_{li} \cap {}^\bullet x_{hj} = \emptyset$ and $x_{li} \neq x_{hj}$.
(3) If $v \in \mu_{\mathsf{d}}$, then ${}^\bullet X \subseteq M_{\Sigma_v}$.
(4) If $v \in \mu_{\mathsf{e}} \cup \mu_{\mathsf{s}} \cup \mu_{\mathsf{x}}$, then

$$\left({}^\bullet X \subseteq {}^\circ\Sigma_v \wedge \mathsf{SP}^{({}^\bullet v)}_{\mathsf{new}} \subseteq M_{\Omega(\boldsymbol{\Sigma})}\right) \vee \left({}^\bullet X \subseteq \Sigma_v^\circ \wedge \mathsf{SP}^{(v^\bullet)}_{\mathsf{new}} \subseteq M_{\Omega(\boldsymbol{\Sigma})}\right).$$

Proof. (1),(2) We observe that if (1) does not hold for $x = x_{li}$, or if ${}^\bullet x_{li} \cap {}^\bullet x_{hj} = \emptyset$ (in (2)) does not hold for $l = h$, then, by the definition of net refinement, there is a place p in $\Omega(\boldsymbol{\Sigma})$ such that $W_{\Omega(\boldsymbol{\Sigma})}(p,u_l) \geq 2$, contradicting the safeness of $M_{\Omega(\boldsymbol{\Sigma})}$ (see Thm. 7.1.1) and $\{u_l\}$ being enabled in $\Omega(\boldsymbol{\Sigma})$. For $l \neq h$ in (2), we also have ${}^\bullet x_{li} \cap {}^\bullet x_{hj} = \emptyset$ since $M_{\Omega(\boldsymbol{\Sigma})}$ is safe, $\{u_l, u_h\} \subseteq U$ is enabled in $\Omega(\boldsymbol{\Sigma})$, and (the first part of) Prop. 7.1.5 holds. Moreover, in (2), $x_{li} \neq x_{hj}$ follows from ${}^\bullet x_{li} \cap {}^\bullet x_{hj} = \emptyset$ (which we have already shown) and the T-restrictedness of Σ_v.

Before proving the rest of the proposition, we observe that

$$q \in \ddot{\Sigma}_v \cap {}^\bullet X \implies M_{\Sigma_v}(q) > 0 \tag{7.15}$$

which follows from the fact that for such a q, by Prop. 7.1.4 and Lemma 7.1.1, we have $p = v \triangleleft q \in {}^\bullet U$ and $M_{\Omega(\boldsymbol{\Sigma})}(p) = M_{\Sigma_v}(q)$.

(3) By (7.15), we only need to show that if $q \in {}^\bullet X \cap ({}^\circ\Sigma_v \cup \Sigma_v^\circ)$, then $M_{\Sigma_v}(q) > 0$. To the contrary, assume that $M_{\Sigma_v}(q) = 0$ and, without loss of generality, that $q \in {}^\circ\Sigma_v$. Then, since $v \in \mu_{\mathsf{d}}$ and Σ_v is clean, there must be $r \in \Sigma_v^\circ$ such that $M_{\Sigma_v}(r) = 0$. It now follows from Prop. 7.1.2(2,4) that there is $p \in \mathsf{mar}_\mu(v)$ such that $\mathsf{trees}^v(p) = \{q, r\}$ or $\mathsf{trees}^v(p) = \{q\}$. We then notice that, by (7.2) and Lemma 7.1.1, $M_{\Omega(\boldsymbol{\Sigma})}(p) = 0$. Moreover, by Prop. 7.1.4, $p \in {}^\bullet U$, producing a contradiction with U being enabled in $\Omega(\boldsymbol{\Sigma})$.

(4) By (7.15), we have ${}^\bullet X \subseteq {}^\circ\Sigma_v \cup \Sigma_v^\circ$. We will first show that ${}^\bullet X \cap {}^\circ\Sigma_v \neq \emptyset$ implies $\mathsf{SP}^{({}^\bullet v)}_{\mathsf{new}} \subseteq M_{\Omega(\boldsymbol{\Sigma})}$. By Lemma 7.1.2, the latter always holds if $v \in \mu_{\mathsf{e}}$, so we may assume that $v \in \mu_{\mathsf{s}} \cup \mu_{\mathsf{x}}$ and $q \in {}^\bullet X \cap {}^\circ\Sigma_v$. To the contrary, suppose that $M_{\Omega(\boldsymbol{\Sigma})}(p) = 0$, for some $s \in {}^\bullet v$ and $p \in \mathsf{SP}^{s}_{\mathsf{new}}$. Let q' be such that $\{q'\} = {}^\circ\Sigma_v \cap \mathsf{trees}^v(p)$ (note that it may happen that $q = q'$, but it is always the case that $0 = M_{\Sigma_v}(q') \leq M_{\Sigma_v}(q)$). Then, by the definition of net refinement, there is $p' \in \mathsf{SP}^{s}_{\mathsf{new}}$ such that $\mathsf{trees}^v(p') = \mathsf{trees}^v(p) \setminus \{q'\} \cup \{q\}$ and $\mathsf{trees}^w(p') = \mathsf{trees}^w(p)$, for all $w \neq v$. Thus, by (4.5), $M_{\Omega(\boldsymbol{\Sigma})}(p') \leq M_{\Omega(\boldsymbol{\Sigma})}(p) = 0$. This, however, contradicts U being enabled in $\Omega(\boldsymbol{\Sigma})$ and $p' \in {}^\bullet U$ (by (4.7)).

In a similar way, we may show that if ${}^\bullet X \cap \Sigma_v^\circ \neq \emptyset$, then $\mathsf{SP}^{(v^\bullet)}_{\mathsf{new}} \subseteq M_{\Omega(\boldsymbol{\Sigma})}$.

Thus we only need to prove that it is impossible to have ${}^\bullet X \cap {}^\circ\Sigma_v \neq \emptyset \neq {}^\bullet X \cap \Sigma_v^\circ$. Indeed, if this was the case, then by what we have already shown, $\mathsf{SP}^{({}^\bullet v \cup v^\bullet)}_{\mathsf{new}} \subseteq M_{\Omega(\boldsymbol{\Sigma})}$ which, by Lemma 7.1.2, would mean that ${}^\bullet v \cup v^\bullet \subseteq {}^\bullet\mu_{\mathsf{e}} \cup \mu_{\mathsf{x}}^\bullet$. Moreover, ${}^\bullet v \cap v^\bullet = \emptyset$ since otherwise, by Prop. 7.1.2(4), (4.7), and (4.8), there would be $p \in \mathsf{SP}^{({}^\bullet v \cap v^\bullet)}_{\mathsf{new}}$ such that

$$\sum_{u \in U} W_{\Omega(\boldsymbol{\Sigma})}(p,u) \geq 2\,,$$

contradicting U being enabled at the safe marking $M_{\Omega(\boldsymbol{\Sigma})}$. Thus, since Ω is a box (and so is safe) we obtain that $v^\bullet = \emptyset$, a contradiction with Ω being T-restricted. □

The above lemma characterised two cases where a step U derived from the net refining a transition v can be executed in $\Omega(\boldsymbol{\Sigma})$. The first one is that the step is made possible by the marking of Σ_v and thus can be deduced using the rule (7.6); this is captured by part (3), and by part (4) with

$$({}^\bullet X \subseteq {}^\circ\Sigma_v \wedge v \in \mu_{\mathsf{e}}) \quad \text{or} \quad ({}^\bullet X \subseteq \Sigma_v^\circ \wedge v \in \mu_{\mathsf{x}}) .$$

Another case characterises those situations when (7.6) cannot be used; this is captured by part (4) with $v \in \mu_{\mathsf{s}}$ or

$$({}^\bullet X \subseteq {}^\circ\Sigma_v \wedge v \in \mu_{\mathsf{x}}) \quad \text{or} \quad ({}^\bullet X \subseteq \Sigma_v^\circ \wedge v \in \mu_{\mathsf{e}}) .$$

Notice that it applies to the tuple $\boldsymbol{\Theta}$ in Fig. 7.5 and $U = \{w_4\} \subseteq \mathsf{T}^{v_3}_{\mathsf{new}}$. Indeed, U is a step enabled in $\Omega_0(\boldsymbol{\Theta})$ (see Fig. 7.6), $\Theta_{v_3} \in \mathsf{Box}^{\mathsf{s}}$, ${}^\bullet\mathsf{trees}(w_4) = \{q_{31}\} \subseteq {}^\circ\Theta_{v_3}$, and

$$\mathsf{SP}^{({}^\bullet v_3)}_{\mathsf{new}} = \mathsf{SP}^{s_3}_{\mathsf{new}} \cup \mathsf{SP}^{s_4}_{\mathsf{new}} = \{p_3, p_4, p_7\} = M_{\Omega_0(\boldsymbol{\Theta})} .$$

Theorem 7.2.2. Let $\Omega(\boldsymbol{\Sigma})\ [U\rangle\ \Sigma'$. Then there are $\boldsymbol{\Theta}$ and $\boldsymbol{\Psi}$ in dom_Ω such that $\boldsymbol{\Sigma} \equiv_\Omega \boldsymbol{\Theta}$, $\Omega(\boldsymbol{\Psi}) = \Sigma'$ and

$$\Big(\Omega : \boldsymbol{\Theta}\Big) \xrightarrow{U} \Big(\Omega : \boldsymbol{\Psi}\Big) .$$

Proof. For every $v \in T_\Omega$, let $U_v = U \cap \mathsf{T}^v_{\mathsf{new}}$ and

$$X_v = \bigcup_{u \in U_v} \mathsf{trees}(u) .$$

Moreover, let $V = \{v \mid U_v \neq \emptyset\}$ and (M, Q) be the marking of Ω whose factorisation is the factorisation μ of $\boldsymbol{\Sigma}$. Finally, for $\delta \in \{\mathsf{e}, \mathsf{x}, \mathsf{s}\}$, let

$$Y_\delta = \{v \in V \cap \mu_\delta \mid {}^\bullet X_v \subseteq {}^\circ\Sigma_v\}$$
$$Z_\delta = \{v \in V \cap \mu_\delta \mid {}^\bullet X_v \subseteq \Sigma_v{}^\circ\} .$$

We now observe that

$$R = \left(\begin{array}{ccccc} \biguplus_{v \in Y_{\mathsf{e}}} {}^\bullet v & \uplus & \biguplus_{v \in Y_{\mathsf{x}}} {}^\bullet v & \uplus & \biguplus_{v \in Y_{\mathsf{s}}} {}^\bullet v \ \uplus \\ \biguplus_{v \in Z_{\mathsf{e}}} v^\bullet & \uplus & \biguplus_{v \in Z_{\mathsf{x}}} v^\bullet & \uplus & \biguplus_{v \in Z_{\mathsf{s}}} v^\bullet \end{array} \right) \subseteq M . \tag{7.16}$$

Indeed, the inclusion follows from Lemmata 7.1.2 and 7.2.1(4). The disjointness of the pre- and post-sets follows from the following argument: Take, for example, two different $v, w \in Y_{\mathsf{e}} \cup Y_{\mathsf{x}} \cup Y_{\mathsf{s}}$. Let $t \in U_v$ and $u \in U_w$ (clearly, $t \neq u$). Since t and u are simultaneously enabled in a safe marking $M_{\Omega(\boldsymbol{\Sigma})}$, ${}^\bullet t \cap {}^\bullet u = \emptyset$. Now, referring to the notation used in Prop. 7.1.6, it is the case that the two predicates In^{e}_t and In^{e}_u hold. Thus, by the first part of Prop. 7.1.6, ${}^\bullet v \cap {}^\bullet w = \emptyset$. Thus (7.16) is satisfied.

Hence, by Prop. 7.1.1(2) and (Dom2) (where the latter implies that each transition $v \in Z_e \cup Z_x \cup Z_s$ is reversible which guarantees that $(M \setminus v^\bullet, Q \cup \{v\})$ is a marking reachable from ${}^\circ\Omega$ or Ω°) we have that

$$(M', Q') = \Big(M \setminus R\,,\ Q \cup Y_e \cup Y_x \cup Y_s \cup Z_e \cup Z_x \cup Z_s\Big)$$

is a marking reachable from ${}^\circ\Omega$ or Ω°. Thus there is a factorisation $\xi \in \mathsf{fact}_\Omega$ of (M', Q'). It is then easy to see that

$$\kappa = \Big(\xi_e \cup Y_e \cup Y_x \cup Y_s\,,\ \mu_d\,,\ \xi_x \cup Z_e \cup Z_x \cup Z_s\,,\ \xi_s\Big)$$

is a valid factorisation of (M, Q). Let $\boldsymbol{\Theta}$ be the Ω-tuple of boxes such that $\lfloor \boldsymbol{\Theta} \rfloor = \lfloor \boldsymbol{\Sigma} \rfloor$, κ is the factorisation of $\boldsymbol{\Theta}$ and, for every $v \in \kappa_d = \mu_d$, $\Theta_v = \Sigma_v$, so that $\boldsymbol{\Theta} \equiv_\Omega \boldsymbol{\Sigma}$. Then, by Lemma 7.2.1(1,2,3), we have that

$$\Big(\Omega : \boldsymbol{\Theta}\Big) \xrightarrow{U} \Big(\Omega : \boldsymbol{\Psi}\Big)$$

for some $\boldsymbol{\Psi}$. Hence, by Thm. 7.2.1, we obtain $\Omega(\boldsymbol{\Theta})\ [U\rangle\ \Omega(\boldsymbol{\Psi})$ and, by Prop. 7.2.1, $\Omega(\boldsymbol{\Theta}) = \Omega(\boldsymbol{\Sigma})$. As a result, $\Sigma' = \Omega(\boldsymbol{\Psi})$. □

Various important consequences may be derived from Thms. 7.2.1 and 7.2.2. In particular, they imply that Requirement 1 is satisfied.

Theorem 7.2.3. $\Omega(\boldsymbol{\Sigma})$ is a static or dynamic box.

Proof. By Thm. 7.1.1, to show that all the markings reachable from $M_{\Omega(\boldsymbol{\Sigma})}$ are safe and clean, it suffices to show that if U is a step such that $\Omega(\boldsymbol{\Sigma})\ [U\rangle\ \Sigma'$, then there is a tuple $\boldsymbol{\Psi}$ in the domain of Ω such that $\Sigma' = \Omega(\boldsymbol{\Psi})$, which holds by Thm. 7.2.2 and Prop. 7.2.1. To deal with the markings reachable from ${}^\circ\Omega(\boldsymbol{\Sigma})$ or $\Omega(\boldsymbol{\Sigma})^\circ$, it suffices to observe that

$$\exists \boldsymbol{\Theta}, \boldsymbol{\Theta}' \in \mathsf{dom}_\Omega\ :\ \Omega(\boldsymbol{\Theta}) = \overline{\Omega(\boldsymbol{\Sigma})}\ \wedge\ \Omega(\boldsymbol{\Theta}') = \underline{\Omega(\boldsymbol{\Sigma})}\,. \tag{7.17}$$

To show that such a $\boldsymbol{\Theta}$ does indeed exist, we first observe that, since Ω is an SOS-operator box and Prop. 7.1.1(5) holds, there is a factorisation μ of the entry marking such that ${}^\bullet\mu_e \cup \mu_x^\bullet = {}^\circ\Omega$ and $\mu_d = \emptyset$. Define $\boldsymbol{\Theta}$ as the Ω-tuple of boxes such that $\lfloor \boldsymbol{\Theta} \rfloor = \lfloor \boldsymbol{\Sigma} \rfloor$ and μ is the factorisation of $\boldsymbol{\Theta}$. It follows directly from Lemma 7.1.2 that $\Omega(\boldsymbol{\Theta}) = \overline{\Omega(\boldsymbol{\Sigma})}$. The existence of $\boldsymbol{\Theta}'$ can be shown similarly. □

And, by inspecting the last proof, we obtain what may be considered as a result capturing Requirement 2.

Corollary 7.2.1. Let Σ' be a net derivable from $\overline{\Omega(\boldsymbol{\Sigma})}$ or $\underline{\Omega(\boldsymbol{\Sigma})}$. Then there is a tuple $\boldsymbol{\Theta}$ in the domain of Ω such that $\Sigma' = \Omega(\boldsymbol{\Theta})$. □

As a summary, SOS-compositionality in the domain of nets can be rendered by a rule similar to that commonly formulated for process algebras:

$$\frac{\left(\Omega : \Theta\right) \xrightarrow{U} \left(\Omega : \Theta'\right)}{\Omega(\Sigma) \left[U\right\rangle \Omega(\Sigma')} \quad \Sigma \equiv_\Omega \Theta \ , \ \Sigma' \equiv_\Omega \Theta' \tag{7.18}$$

Theorem 7.2.1 together with Prop. 7.2.1 verify its soundness, and Thm. 7.2.2 verifies its completeness.

The various net operations used by PBC can be accommodated within the meta-rule (7.18). For instance, in the $(OP_;)$ rules in Sect. 3.2.6, SL and SR are instances of (7.18) with $\Sigma = \Theta$ and $\Sigma' = \Theta'$, while IS2 is an instance of (7.18) with $U = \emptyset$. This follows from the fact that the standard PBC operator boxes are all SOS-operator boxes. Another useful corollary is a sufficient condition for a refinement to be ex-exclusive.

Theorem 7.2.4. Let Ω be an ex-exclusive SOS-operator box and $\Sigma \in \mathsf{dom}_\Omega$ be a tuple of boxes such that, for every $v \in T_\Omega$ satisfying

$$^\bullet v \cap {}^\circ\Omega \neq \emptyset \neq v^\bullet \cap \Omega^\circ \quad \text{or} \quad v^\bullet \cap {}^\circ\Omega \neq \emptyset \neq {}^\bullet v \cap \Omega^\circ ,$$

it is the case that Σ_v is ex-exclusive. Then $\Omega(\Sigma)$ is an ex-exclusive box.

Proof. Let (M, Q) be the marking of Ω with the same factorisation μ as Σ. From (7.17), Thm. 7.2.2 and Prop. 7.2.1, it follows that it suffices to show that ${}^\circ\Omega(\Sigma) \cap M_{\Omega(\Sigma)} = \emptyset$ or $\Omega(\Sigma)^\circ \cap M_{\Omega(\Sigma)} = \emptyset$. To the contrary, suppose that there are $p \in {}^\circ\Omega(\Sigma)$ and $p' \in \Omega(\Sigma)^\circ$ such that $p, p' \in M_{\Omega(\Sigma)}$. Then, by the definition of net refinement, there are $s \in {}^\circ\Omega$ and $s' \in \Omega^\circ$ such that $p \in \mathsf{SP}^s_{\mathsf{new}}$ and $p' \in \mathsf{SP}^{s'}_{\mathsf{new}}$. Moreover, by Lemma 7.1.1, there are v and v' such that $p \in \mathsf{mar}_\mu(v)$ and $p' \in \mathsf{mar}_\mu(v')$. We now consider two cases.

Case 1: $v = v'$. If $v \in \mu_{\mathsf{e}} \cup \mu_{\mathsf{x}}$, then $s, s' \in {}^\bullet\mu_{\mathsf{e}} \cup \mu_{\mathsf{x}}^\bullet$, contradicting Ω being ex-exclusive. Hence we may assume that $v \in \mu_{\mathsf{d}}$. If $s, s' \in {}^\bullet v \vee s, s' \in v^\bullet$, then by Prop. 7.1.1(3), we again obtain a contradiction with Ω being ex-exclusive. If neither $s, s' \in {}^\bullet v$ nor $s, s' \in v^\bullet$, we may assume without loss of generality, that $s \in {}^\bullet v$ and $s' \in v^\bullet$. Let $q \in {}^\circ\Sigma_v$ and $q' \in \Sigma_v^\circ$ be such that $q \in \mathsf{trees}^v(p)$ and $q' \in \mathsf{trees}^v(p')$. Then, since neither $s, s' \in {}^\bullet v$ nor $s, s' \in v^\bullet$, $\{q\} = \mathsf{trees}^v(p)$ and $\{q'\} = \mathsf{trees}^v(p')$. Hence $q, q' \in M_{\Sigma_v}$, contradicting Σ_v being ex-exclusive.

Case 2: $v \neq v'$. We first observe that the following hold:

- If $v \in \mu_{\mathsf{e}}$, then $p \in \mathsf{SP}^{({}^\bullet v)}_{\mathsf{new}}$ and $s \in {}^\bullet v \subseteq M$.
- If $v \in \mu_{\mathsf{x}}$, then $p \in \mathsf{SP}^{(v^\bullet)}_{\mathsf{new}}$ and $s \in v^\bullet \subseteq M$.
- If $v \in \mu_{\mathsf{d}}$, then $p \in \mathsf{SP}^{({}^\bullet v \cup v^\bullet)}_{\mathsf{new}}$ and $s \in {}^\bullet v \cup v^\bullet$; as a result $s \in {}^\bullet v$ and $(M \cup {}^\bullet v, \mu_{\mathsf{d}} \setminus \{v\})$ is reachable from ${}^\circ\Omega$ or Ω°, or $s \in v^\bullet$, and $(M \cup v^\bullet, \mu_{\mathsf{d}} \setminus \{v\})$ is reachable from ${}^\circ\Omega$ or Ω°.

A similar argument can be made for v' and s'. Hence, by $v \neq v'$ and $s \neq s'$, there is a marking (M', Q') reachable from ${}^\circ\Omega$ or Ω° such that $s, s' \in M'$, contradicting Ω being ex-exclusive. □

An important consequence of the last result is that it will allow us to identify a class of expressions whose corresponding boxes are guaranteed to be ex-exclusive. This result is similar, for a behavioural property, to Prop. 4.3.2, which plays the same role with respect to the structural properties of e- and x-directedness. Notice that, in the case of the standard PBC, $\Omega_{\|}$ is the only non-ex-exclusive operator box.

7.2.4 Solutions of Recursive Systems

The discussion in Chap. 6 identified several cases when certain restrictions, carefully placed on the kind of operator and plain boxes used in recursive net equations, guaranteed that their solutions would possess desirable behavioural properties. These results were all obtained by applying structural techniques based on S-invariants. In this chapter, we have developed an entirely different tool for achieving essentially the same effect; namely, Thm. 7.2.3 which provides us with a simple behavioural criterion for establishing the property of being a compound static or dynamic box. Although this result concerns net refinement, in view of our discussion in Chap. 5, one should expect that it can be lifted to the level of recursive net equations (recall that any solution of a recursive system can be approximated by a sequence of boxes generated by successive applications of net refinement starting from ex-boxes, and that many relevant, structural as well as behavioural, net properties are preserved through such approximations). In the next definition, as well as in the rest of this chapter, we will use $\mathsf{Box}^{\mathsf{xcl}}$, $\mathsf{Box}^{\mathsf{edir}}$, and $\mathsf{Box}^{\mathsf{xdir}}$ to denote, respectively, the sets of ex-exclusive, e-directed, and x-directed plain boxes.

Consider again the system of recursive equations (5.2) treated in Chap. 5. Let $\mathcal{X}^{\mathrm{wf}}$, $\mathcal{X}^{\mathrm{xcl}}$, $\mathcal{X}^{\mathrm{edir}}$, and $\mathcal{X}^{\mathrm{xdir}}$ be the largest subsets of $\mathcal{X}$ — called respectively the *well-formed*, *ex-exclusive*, *e-directed*, and *x-directed* variables[5] — such that the following hold, where we assume that a variable $X \in \mathcal{X}$ has the defining equation $X \stackrel{\mathrm{df}}{=} \Omega(\boldsymbol{\Delta})$:

(Var1) All ex-exclusive, e-directed, and x-directed variables are well-formed.

(Var2) If $X \in \mathcal{X}^{\mathrm{wf}}$, then Ω is an SOS-operator box and, for every $v \in T_\Omega$:
(i) $\Delta_v \in \mathsf{Box}^{\mathsf{s}} \cup \mathcal{X}^{\mathrm{wf}}$;
(ii) if ${}^\bullet v \cap v^\bullet \neq \emptyset$, then $\Delta_v \in \mathsf{Box}^{\mathsf{xcl}} \cup \mathcal{X}^{\mathrm{xcl}}$;

[5] In this and other syntactic definitions in Chaps. 7 and 8, the superscripts indicate the relevant properties of the corresponding boxes; more precisely, 'wf' indicates well-formedness, 'xcl' indicates ex-exclusiveness, 'e' indicates e-directedness, and 'x' indicates x-directedness.

(iii) if v is not reversible, then $\Delta_v \in \mathsf{Box}^{\mathrm{xdir}} \cup \mathcal{X}^{\mathrm{xdir}}$.

(Var3) If $X \in \mathcal{X}^{\mathrm{xcl}}$, then Ω is ex-exclusive and, for every $v \in T_\Omega$ satisfying

$$^\bullet v \cap {}^\circ\Omega \neq \emptyset \neq v^\bullet \cap \Omega^\circ \quad \text{or} \quad v^\bullet \cap {}^\circ\Omega \neq \emptyset \neq {}^\bullet v \cap \Omega^\circ ,$$

it is the case that $\Delta_v \in \mathsf{Box}^{\mathrm{xcl}} \cup \mathcal{X}^{\mathrm{xcl}}$.

(Var4) If $X \in \mathcal{X}^{\mathrm{edir}}$, then Ω is e-directed and, for every $v \in ({}^\circ\Omega)^\bullet$,
$\Delta_v \in \mathsf{Box}^{\mathrm{edir}} \cup \mathcal{X}^{\mathrm{edir}}$.

(Var5) If $X \in \mathcal{X}^{\mathrm{xdir}}$, then Ω is x-directed and, for every $v \in {}^\bullet(\Omega^\circ)$,

$$\Delta_v \in \mathsf{Box}^{\mathrm{xdir}} \cup \mathcal{X}^{\mathrm{xdir}}.$$

In the above, (Var2)(ii) and (Var2)(iii) encode (Dom1) and (Dom2), respectively; (Var3) encodes Thm. 7.2.4; and (Var4) together with (Var5) encode Prop. 4.3.2. Note that it is possible to formulate the above definition in terms of the 'largest subsets' since the constraints (Var1)—(Var5) are preserved by taking the union of sets of variables satisfying them.

It is the well-formed variables which are of primary interest to us as, thanks to the next result, any box solving them will always be static. In addition to the well-formed variables, we singled out three more classes: the ex-exclusive and x-directed variables (with corresponding ex-exclusive and x-directed solutions) are needed to reflect the conditions (Dom1) and (Dom2), whereas the e-directed variables are, strictly speaking, not necessary and have been added here for symmetry (the reader may find e-directed variables useful when attempting Exc. 7.2.6).

Theorem 7.2.5. The system of recursive equations (5.2) has at least one solution in the domain of plain boxes with empty markings. Moreover, the following hold for its every solution.

(1) If X is a well-formed variable, then the box corresponding to X is static.
(2) For $\delta \in \{\mathrm{xcl}, \mathrm{edir}, \mathrm{xdir}\}$, if $X \in \mathcal{X}^\delta$, then the box corresponding to X belongs to Box^δ.

Proof. Follows from the results obtained in Sect. 5.5, after taking into account: Props. 5.1.1(1,3,4,5,7,9) and 4.3.2; Thms. 5.2.6, 7.2.3, and 7.2.4; an observation that all ex-boxes are static boxes which are both ex-exclusive and ex-directed; and (Var1)—(Var5). □

Exercise 7.2.2. Show that being a well-formed, ex-exclusive, e-directed, or x-directed variable is not necessary for the corresponding box in a solution of (5.2) to be, respectively, static, ex-exclusive, e-directed, or x-directed.

7.2.5 Behavioural Restrictions

The conditions imposed on the domain of an SOS-operator box are sufficient for the main compositionality results contained in this chapter. However, they can still be seen as perhaps too weak due to not excluding some counterintuitive *behaviours* of composite boxes (this issue has already been discussed in Sect. 4.4.13 for the choice and iteration composition). We therefore formulate five additional conditions on an Ω-tuple $\boldsymbol{\Sigma}$ in the domain of application of an SOS-operator box Ω.

(Beh1) If $v \neq w$ and ${}^\bullet v \cap {}^\bullet w \neq \emptyset$, then $\Sigma_v, \Sigma_w \in \mathsf{Box}^{\mathsf{edir}}$.
(Beh2) If $v \neq w$ and $v^\bullet \cap w^\bullet \neq \emptyset$, then $\Sigma_v, \Sigma_w \in \mathsf{Box}^{\mathsf{xdir}}$.
(Beh3) If $v^\bullet \cap {}^\bullet w \neq \emptyset$, then $\Sigma_v \in \mathsf{Box}^{\mathsf{xdir}}$ or $\Sigma_w \in \mathsf{Box}^{\mathsf{edir}}$.
(Beh4) If ${}^\bullet v \cap \Omega^\circ \neq \emptyset$, then $\Sigma_v \in \mathsf{Box}^{\mathsf{edir}}$.
(Beh5) If $v^\bullet \cap {}^\circ\Omega \neq \emptyset$, then $\Sigma_v \in \mathsf{Box}^{\mathsf{xdir}}$.

With the above five conditions, we obtain the following domain restrictions for the PBC operators:

- $\Sigma \,\square\, \Theta$ Σ is ex-directed and Θ is ex-directed.
- $\langle \Sigma * \Theta]$ Σ is ex-exclusive and ex-directed, and Θ is e-directed.
- $[\Theta * \Sigma\rangle$ Σ is ex-exclusive and ex-directed, and Θ is x-directed.
- $\Sigma;\Theta$ Σ is x-directed, or Θ is e-directed.

Then one can see, for example, that a terminated behaviour of $\overline{\Sigma \,\square\, \Theta}$ is a terminated behaviour of $\overline{\Sigma}$ or $\overline{\Theta}$, whereas a terminated behaviour of $\overline{\langle \Sigma * \Theta]}$ is concatenated from zero or more terminated behaviours of $\overline{\Sigma}$, followed by a terminated behaviour of $\overline{\Theta}$.

Exercise 7.2.3. Show that if both Σ and Θ are ex-directed static boxes and $\overline{\Sigma \,\square\, \Theta}\,[U_1 \ldots U_k\rangle\, \Psi$, then one of (i) and (ii) in the following holds.

(i) There are steps $W_1, \ldots, W_k$ and and a box Σ' such that $\overline{\Sigma}\,[W_1 \ldots W_k\rangle\, \Sigma'$ and $\Psi = \Sigma' \,\square\, \Theta$ and, for every $i \leq k$,
$$U_i = \bigcup_{w \in W_i} \{(v^1_\square, \lambda_\Sigma(w)) \lhd w\} \,. \tag{7.19}$$
Moreover, $\Psi \in \mathsf{Box}^{\mathsf{x}}$ if and only if $\Sigma' \in \mathsf{Box}^{\mathsf{x}}$.

(ii) There are steps $W_1, \ldots, W_k$ and and a box Θ' such that $\overline{\Theta}\,[W_1 \ldots W_k\rangle\, \Theta'$ and $\Psi = \Sigma \,\square\, \Theta'$ and, for every $i \leq k$,
$$U_i = \bigcup_{w \in W_i} \{(v^2_\square, \lambda_\Theta(w)) \lhd w\} \,. \tag{7.20}$$
Moreover, $\Psi \in \mathsf{Box}^{\mathsf{x}}$ if and only if $\Theta' \in \mathsf{Box}^{\mathsf{x}}$.

Exercise 7.2.4. Show that the converse of the properties formulated in the previous exercise holds as well. First show that if $\overline{\Sigma}\ [W_1 \ldots W_k\rangle\ \Sigma'$, then $\overline{\Sigma \,\square\, \Theta}\,[U_1 \ldots U_k\rangle\ \Sigma' \,\square\, \Theta$ where, for every $i \leq k$, (7.19) holds. Then show that if $\overline{\Theta}\ [W_1 \ldots W_k\rangle\ \Theta'$, then $\overline{\Sigma \,\square\, \Theta}\ [U_1 \ldots U_k\rangle\ \Sigma \,\square\, \Theta'$ where, for every $i \leq k$, (7.20) is satisfied.

In the rest of this section, we will argue that (Beh1)—(Beh5) guarantee that the behaviours of a compound net are composed from behaviours of its components. It is limited, for the sake of simplicity of the presentation, to the case of operator boxes with one single token in each reachable real marking, of which all PBC control flow operators except parallel composition are special instances.

The appropriateness of the conditions (Beh1)—(Beh5). Let us consider an operator box Ω such that each real marking reachable from ${}^{\circ}\Omega$ or Ω° has exactly one token (in other words, we may assume that Ω is a *state machine* which means that $|{}^{\circ}\Omega| = |\Omega^{\circ}| = 1$ as well as $|{}^{\bullet}v| = |v^{\bullet}| = 1$, for every $v \in T_{\Omega}$) and that each transition of Ω is labelled by the identity relabelling, ϱ_{id}. Furthermore, let $\Omega(\boldsymbol{\Sigma})$ be a valid application of Ω to an Ω-tuple of static boxes $\boldsymbol{\Sigma}$ satisfying (Beh1)—(Beh5), and $U_1 \dots U_k$ be a nonempty step sequence composed of nonempty sets of transitions such that

$$\overline{\Omega(\boldsymbol{\Sigma})}\,[U_1 \dots U_k\rangle\, \Psi\,. \tag{7.21}$$

We aim at showing that the execution (7.21) can be seen as a combination of valid executions of Ω and the Σ_v's. To this end, we first observe that, by Thm. 7.2.2 and (7.17), there are Ω-tuples $\boldsymbol{\Delta}^1, \dots, \boldsymbol{\Delta}^k$ and $\boldsymbol{\Phi}^1, \dots, \boldsymbol{\Phi}^k$ in dom_{Ω} such that $\lfloor \boldsymbol{\Delta}^i \rfloor = \lfloor \boldsymbol{\Phi}^i \rfloor = \boldsymbol{\Sigma}$, for every $1 \le i \le k$, and

$$\forall 1 \le i \le k \;:\; \left(\Omega \,:\, \boldsymbol{\Delta}^i\right) \xrightarrow{U_i} \left(\Omega \,:\, \boldsymbol{\Phi}^i\right) \tag{7.22}$$

$$\forall 1 < i \le k \;:\; \Omega(\boldsymbol{\Delta}^i) \;=\; \Omega(\boldsymbol{\Phi}^{i-1}) \tag{7.23}$$

$$\Omega(\boldsymbol{\Delta}^1) \;=\; \overline{\Omega(\boldsymbol{\Sigma})} \quad \text{and} \quad \Omega(\boldsymbol{\Phi}^k) \;=\; \Psi\,. \tag{7.24}$$

Below, we will denote by μ^i and κ^i the factorisations of respectively $\boldsymbol{\Delta}^i$ and $\boldsymbol{\Phi}^i$. Notice that, by (7.23) and Prop. 7.2.1, for every $1 < i \le k$, the μ^i and κ^{i-1} are factorisations of the same complex marking of Ω, $\mathcal{M}_i = (M_i, Q_i)$. Moreover, we will denote by $\mathcal{M}_1 = (M_1, Q_1)$ and $\mathcal{M}_{k+1} = (M_{k+1}, Q_{k+1})$, the complex markings of Ω whose factorisations are respectively μ^1 and κ^k.

Our next observation is that since Ω is a state machine, for every $1 \le i \le k$, there is a *unique* transition $v_i \in T_{\Omega}$ which belongs to $\mu^i_{\mathsf{e}} \cup \mu^i_{\mathsf{d}} \cup \mu^i_{\mathsf{x}}$. Hence, each marking $\mathcal{M}_i$, for $1 \le i \le k+1$, is such that

$$|M_i| + |Q_i| = 1\,. \tag{7.25}$$

Moreover, for every $1 \le i \le k$, there is a nonempty set of transitions $V_i \subseteq T_{\Sigma_{v_i}}$ such that

$$U_i \;=\; \left\{ \left(v_i, \lambda_{\Sigma_{v_i}}(t)\right) \lhd t \;\middle|\; t \in V_i \right\}.$$

The nonemptiness of V_i follows from $U_i \ne \emptyset$. We finally observe that, by $V_i \ne \emptyset$ and (7.22) and the T-restrictedness of boxes, for every $1 \le i \le k$,

$$\left(v_i \in \mu^i_{\mathsf{x}} \Rightarrow \Sigma_{v_i} \notin \mathsf{Box}^{\mathsf{xdir}}\right) \wedge \left(v_i \in \kappa^i_{\mathsf{e}} \Rightarrow \Sigma_{v_i} \notin \mathsf{Box}^{\mathsf{edir}}\right). \tag{7.26}$$

Lemma 7.2.2. With the above notations, the following hold.

(1) ${}^\bullet v_1 = {}^\circ\Omega$ and $\Delta^1_{v_1} = \overline{\Sigma_{v_1}}$.
(2) For every $1 \le i < k$, we have either
  (a) $v_i = v_{i+1}$ and $\Phi^i_{v_i} = \Delta^{i+1}_{v_{i+1}}$, or
  (b) $v_i^\bullet = {}^\bullet v_{i+1}$ and $\Phi^i_{v_i} = \underline{\Sigma_{v_i}}$ and $\Delta^{i+1}_{v_{i+1}} = \overline{\Sigma_{v_{i+1}}}$
  (we do not exclude the possibility that $v_i = v_{i+1}$).
(3) $\Psi = \underline{\Omega(\Sigma)}$ if and only if $\Phi^k_{v_k} = \underline{\Sigma_{v_k}}$ and $v_k^\bullet = \Omega^\circ$.

Proof. (1) We have $M_{\Omega(\Delta^1)} = \mathsf{SP}^s_{\mathsf{new}}$ where $\{s\} = {}^\circ\Omega$. Hence, by Lemma 7.1.2 and (7.25), $\mathcal{M}_1 = (\{s\}, \emptyset)$ and $v_1 \in \mu^1_{\mathrm{e}} \cup \mu^1_{\mathrm{x}}$. If $v_1 \in \mu^1_{\mathrm{x}}$, then by (7.26), $\Sigma_{v_1} \notin \mathsf{Box}^{\mathrm{xdir}}$ which, together with $v_1^\bullet = \{s\}$, yields a contradiction with (Beh5). Hence $\mu^1_{\mathrm{e}} = \{v_1\}$ and ${}^\bullet v_1 = \{s\} = {}^\circ\Omega$.

Note: (Beh5) excludes the situation whereby, from the entry marking of the composed net, we can execute a behaviour originating from $\Sigma^\circ_{v_1}$ which bears no relationship to any possible behaviour of Ω (this intuitively corresponds to the backward reachability in Ω, which is not allowed).

(2) We consider two cases.

Case 1: $v_i \neq v_{i+1}$. Then μ^{i+1} and κ^i are two different factorisations of $\mathcal{M}_{i+1}$. By (7.25), there is $s \in S_\Omega$ such that $\mathcal{M}_{i+1} = (\{s\}, \emptyset)$, which in turn means that $v_i \in \kappa^i_{\mathrm{e}} \cup \kappa^i_{\mathrm{x}}$ and $v_{i+1} \in \mu^{i+1}_{\mathrm{e}} \cup \mu^{i+1}_{\mathrm{x}}$. We consider four subcases. Case 1(i): $v_i \in \kappa^i_{\mathrm{e}}$ and $v_{i+1} \in \mu^{i+1}_{\mathrm{e}}$. Then, by (7.26), $\Sigma_{v_i} \notin \mathsf{Box}^{\mathrm{edir}}$. On the other hand, ${}^\bullet v_i = \{s\} = {}^\bullet v_{i+1}$, producing a contradiction with (Beh1).

Note: With (Beh1), it is not possible to start the execution of Σ_{v_i} and later, without finishing it, to enter $\Sigma_{v_{i+1}}$ for a conflicting v_{i+1} because the initial state of Σ_{v_i} was reached.

Case 1(ii): $v_i \in \kappa^i_{\mathrm{x}}$ and $v_{i+1} \in \mu^{i+1}_{\mathrm{x}}$. Then, by (7.26), $\Sigma_{v_{i+1}} \notin \mathsf{Box}^{\mathrm{xdir}}$. On the other hand, $v_i^\bullet = \{s\} = v_{i+1}^\bullet$, producing a contradiction with (Beh2).

Note: (Beh2) prohibits to finish the execution of Σ_{v_i} and afterwards to enter $\Sigma_{v_{i+1}}$ from the rear.

Case 1(iii): $v_i \in \kappa^i_{\mathrm{e}}$ and $v_{i+1} \in \mu^{i+1}_{\mathrm{x}}$. Then, by (7.26), $\Sigma_{v_i} \notin \mathsf{Box}^{\mathrm{edir}}$ and $\Sigma_{v_{i+1}} \notin \mathsf{Box}^{\mathrm{xdir}}$. On the other hand, ${}^\bullet v_i = \{s\} = v_{i+1}^\bullet$, producing a contradiction with (Beh3).

Note: (Beh3) excludes a situation, combining the previous two problems, where we start the execution of Σ_{v_i} and later, without finishing it, enter $\Sigma_{v_{i+1}}$ from the rear.

Case 1(iv): $v_i \in \kappa^i_{\mathrm{x}}$ and $v_{i+1} \in \mu^{i+1}_{\mathrm{e}}$. Then $v_i^\bullet = \{s\} = {}^\bullet v_{i+1}$, and so (b) holds.

Case 2: $v_i = v_{i+1} (= v)$. If $\Phi^i_v = \Delta^{i+1}_v$, then (a) holds, so we assume that $\Phi^i_v \neq \Delta^{i+1}_v$. Then μ^{i+1} and κ^i are two different factorisations of $\mathcal{M}_{i+1}$. Hence,

by (7.25), there is $s \in S_\Omega$ such that $\mathcal{M}_{i+1} = (\{s\}, \emptyset)$, which in turn means that $v \in (\kappa_e^i \cup \kappa_x^i) \cap (\mu_e^{i+1} \cup \mu_x^{i+1})$ and ${}^\bullet v = \{s\} = v^\bullet$. Since $\Phi_v^i \neq \Delta_v^{i+1}$ we only need to consider two subcases.
Case 2(i): $v \in \kappa_e^i \cap \mu_x^{i+1}$. Then, by (7.26), $\Sigma_v \notin \mathsf{Box}^{\mathrm{edir}} \cup \mathsf{Box}^{\mathrm{xdir}}$. On the other hand, ${}^\bullet v = v^\bullet$, producing a contradiction with (Beh3).
Case 2(ii): $v \in \kappa_x^i \cap \mu_e^{i+1}$. This and ${}^\bullet v = v^\bullet$ means that (b) holds.

(3) If $\Psi = \underline{\Omega(\boldsymbol{\Sigma})}$, then $M_{\Omega(\boldsymbol{\Phi}^k)} = \mathsf{SP}^s_{\mathsf{new}}$ where $\{s\} = \Omega^\circ$. Hence, by Lemma 7.1.2 and (7.25), $\mathcal{M}_{k+1} = (\{s\}, \emptyset)$ and $v_k \in \kappa_e^k \cup \kappa_x^k$. If $v_k \in \kappa_e^k$, then by (7.26), $\Sigma_{v_k} \notin \mathsf{Box}^{\mathrm{edir}}$ which, together with ${}^\bullet v_k = \{s\}$, yields a contradiction with (Beh4). Hence $v_k \in \kappa_x^k$ and $v_k^\bullet = \{s\} = \Omega^\circ$.

Note: (Beh4) excludes the situation where we reach the exit marking, start the execution of Σ_{v_k} and return to the exit marking without finishing Σ_{v_k}, so that a completed history is not composed out of completed histories of the components corresponding to a completed history of Ω.

The reverse implication follows directly from (7.5). □

It is always possible to partition (in a unique way) the sequence of steps $V_1 \ldots V_k$ into the longest nonempty subsequences $\sigma_1, \ldots, \sigma_l$ so that $V_1 \ldots V_k = \sigma_1 \ldots \sigma_l$, and, for every $i \leq l$, if

$$\sigma_i = V_p V_{p+1} \ldots V_q ,$$

then it is the case that $v_p = v_{p+1} = \cdots = v_q (= w_i)$ and $\Phi_{w_i}^j = \Delta_{w_i}^{j+1}$, for every $p \leq j < q$. Finally,

Theorem 7.2.6. With the above notations, the following statements hold.

(1) $\overline{\Sigma_{w_i}} \, [\sigma_i\rangle \, \underline{\Sigma_{w_i}}$, for every $i < l$, and $\overline{\Sigma_{w_l}} \, [\sigma_l\rangle \, \Phi_{v_k}^k$.

(2) If $\Phi_{v_k}^k = \underline{\Sigma_{v_k}}$, then

$$\mathcal{M}_1 \left[\{w_1\} \ldots \{w_{l-1}\}\{w_l\} \right\rangle \mathcal{M}_{k+1} ;$$

otherwise

$$\mathcal{M}_1 \left[\{w_1\} \ldots \{w_{l-1}\}\{w_l\}^+ \right\rangle \mathcal{M}_{k+1} .$$

(3) $\Phi = \underline{\Omega(\boldsymbol{\Sigma})}$ if and only if $\mathcal{M}_{k+1} = (\Omega^\circ, \emptyset)$.

Proof. Follows with Lemma 7.2.2. □

This result shows that, as it was claimed, the original sequence $U_1 \ldots U_k$ can be decomposed into a behaviour of Ω (i.e., the sequence$\{w_1\}\{w_2\} \ldots$ in part (2) of the theorem), such that each w_i — except possibly the last one — corresponds to a full behaviour, σ_i, of Σ_{w_i} (cf. part (1) of the theorem). Part (3) of the theorem relates the full behaviours of $\Omega(\boldsymbol{\Sigma})$ and the full behaviours of Ω and the Σ_v's.

Exercise 7.2.5. Extend Thm. 7.2.6 to the remaining operator boxes.

The treatment of the recursive case can also be extended to accommodate the behavioural conditions (Beh1)—(Beh5).

Exercise 7.2.6. Formulate and prove a result similar to Thm. 7.2.5 in such a way that each solution of the system of recursive equations would also satisfy (Beh1)—(Beh5).

7.3 A Process Algebra and its Semantics

We now introduce an algebra of process expressions, called the *box algebra*, which is based on the class of operator boxes defined in the previous section. The box algebra is in fact a *meta-model* parameterised by two nonempty, possibly infinite, sets of Petri nets: a set ConstBox of static and dynamic plain boxes providing a denotational semantics of simple process expressions, and a disjoint set OpBox of SOS-operator boxes providing an interpretation for the connectives.

The only assumption made about the SOS-operator boxes in OpBox and the static boxes in ConstBox is that they have disjoint sets of simple root-only trees as their place and transition names, i.e., for all distinct static and/or operator boxes Σ and Θ in $\mathsf{OpBox} \cup \mathsf{ConstBox}$,

$$\begin{aligned} &S_\Sigma \cup T_\Sigma \subseteq \mathsf{P}_{\mathsf{root}} \cup \mathsf{T}_{\mathsf{root}} \\ &S_\Sigma \cap S_\Theta = T_\Sigma \cap T_\Theta \ = \ \emptyset \,. \end{aligned} \tag{7.27}$$

The first line of (7.27) will allow us to apply the results on solving recursive systems of equations on boxes developed in Chap. 5. It also implies that $\mathsf{P}_{\mathsf{root}}$ and $\mathsf{T}_{\mathsf{root}}$ must be sufficiently large. The second line of (7.27) will prove to be useful in defining a global independence relation on transitions in nets associated with process expressions. We do not require that the boxes in OpBox and ConstBox be finite.

Signature. We consider an algebra of process expressions over the signature

$$\mathsf{Const} \,\cup\, \left\{\overline{(.)}, \underline{(.)}\right\} \,\cup\, \left\{\mathsf{op}_\Omega \;\middle|\; \Omega \in \mathsf{OpBox}\right\}, \tag{7.28}$$

where Const is a fixed nonempty set of *constants* which will be modelled through the boxes in ConstBox, $\overline{(.)}$ and $\underline{(.)}$ are two unary operators, and each op_Ω is a connective of the algebra indexed by an SOS-operator box taken from the set OpBox. (The number of transitions in a finite operator box will be called the *arity* of the operator it defines.) Although we use the symbols $\overline{(.)}$ and $\underline{(.)}$ to denote both operations on boxes and process algebra connectives, it will always be clear from the context what the intended interpretation is (anyway, we shall see that these notations are strongly related).

The set of constants is partitioned into the *static* constants, $\mathsf{Const}^{\mathrm{s}}$, and *dynamic* constants, $\mathsf{Const}^{\mathrm{d}}$; moreover, there are two distinct disjoint subsets of $\mathsf{Const}^{\mathrm{d}}$, denoted by $\mathsf{Const}^{\mathrm{e}}$ and $\mathsf{Const}^{\mathrm{x}}$, and respectively called the *entry* and *exit* constants. We also identify three further, not necessarily disjoint subsets of Const, namely, $\mathsf{Const}^{\mathrm{xcl}}$, $\mathsf{Const}^{\mathrm{edir}}$, and $\mathsf{Const}^{\mathrm{xdir}}$, called the *ex-exclusive*, *e-directed*, and *x-directed* constants, respectively.

As in Chap. 3, we will also use a fixed set $\mathcal{X}$ of *process variables*.

Syntax. There are four classes of process expressions corresponding to previously introduced classes of plain boxes: the *entry*, *dynamic*, *exit*, and *static* expressions, denoted respectively by $\mathsf{Expr}^{\mathrm{e}}$, $\mathsf{Expr}^{\mathrm{d}}$, $\mathsf{Expr}^{\mathrm{x}}$, and $\mathsf{Expr}^{\mathrm{s}}$. Collectively, we will refer to them as the *box expressions*, $\mathsf{Expr}^{\mathrm{box}}$. We will also use a counterpart of the notion of the factorisation of a tuple of boxes. For an SOS-operator box Ω and an Ω-tuple of box expressions $\boldsymbol{D}$, we define the *factorisation* of $\boldsymbol{D}$ to be the quadruple $\mu = (\mu_{\mathrm{e}}, \mu_{\mathrm{d}}, \mu_{\mathrm{x}}, \mu_{\mathrm{s}})$ such that

$$\mu_\delta = \{v \in T_\Omega \mid D_v \in \mathsf{Expr}^\delta\}, \text{ for } \delta \in \{\mathrm{e},\mathrm{x},\mathrm{s}\}, \text{ and}$$
$$\mu_{\mathrm{d}} = \{v \in T_\Omega \mid D_v \in \mathsf{Expr}^{\mathrm{d}} \backslash (\mathsf{Expr}^{\mathrm{e}} \cup \mathsf{Expr}^{\mathrm{x}})\}\ .$$

The syntax for the box expressions $\mathsf{Expr}^{\mathrm{box}}$ is given as follows.

$$\begin{array}{llcl|c|c|c}
\mathsf{Expr}^{\mathrm{s}} & E & ::= & c^{\mathrm{s}} & X & \mathsf{op}_\Omega(\boldsymbol{E}) & \\
\mathsf{Expr}^{\mathrm{e}} & F & ::= & c^{\mathrm{e}} & \overline{E} & \mathsf{op}_\Omega(\boldsymbol{F}) & \\
\mathsf{Expr}^{\mathrm{x}} & G & ::= & c^{\mathrm{x}} & \underline{E} & \mathsf{op}_\Omega(\boldsymbol{G}) & \\
\mathsf{Expr}^{\mathrm{d}} & H & ::= & c^{\mathrm{d}} & F & G & \mathsf{op}_\Omega(\boldsymbol{H})\ ,
\end{array} \tag{7.29}$$

where $c^\delta \in \mathsf{Const}^\delta$, for $\delta \in \{\mathrm{e},\mathrm{x},\mathrm{s}\}$, and $c^{\mathrm{d}} \in \mathsf{Const}^{\mathrm{d}} \backslash (\mathsf{Const}^{\mathrm{e}} \cup \mathsf{Const}^{\mathrm{x}})$ are constants; $X \in \mathcal{X}$ is a process variable; $\Omega \in \mathsf{OpBox}$ is an SOS-operator box; and $\boldsymbol{E}$, $\boldsymbol{F}$, $\boldsymbol{G}$, and $\boldsymbol{H}$ are Ω-tuples[6] of box expressions. These tuples have to satisfy some conditions determined by the domain of application of the net operator induced by Ω, and so the factorisations of $\boldsymbol{E}$, $\boldsymbol{F}$, and $\boldsymbol{G}$ are respectively factorisations of the complex empty, entry, and exit marking of Ω, and the factorisation of $\boldsymbol{H}$ is a factorisation of a complex marking reachable from the entry or exit marking of Ω different from $({}^\circ\Omega, \emptyset)$ and $(\Omega^\circ, \emptyset)$.

The above syntax only reflects the first part of the definition of the domain of an operator box, dom_Ω, which stipulates that an Ω-tuple of boxes should have a factorisation belonging to fact_Ω. The remaining two conditions, (Dom1) and (Dom2), are not captured and their treatment will be given separately below. There are two reasons why we decided to proceed in this way.

---

[6] We allow operator boxes with infinitely many transitions, which may lead to a notational problem to specify such Ω-tuples explicitly when expressions are viewed as strings. This problem may be solved by viewing expressions as (syntax) trees rather than strings, as explained in Sect. 7.3.2.

The first is that in many cases (Dom1) and (Dom2) are already satisfied because of the specific properties of the operator boxes which parameterise the box algebra (more precisely, if all the operators are pure and have all transitions reversible); in particular, this is true of the extended PBC syntax defined in Sect. 3.4. And, in all such cases, (7.29) will be exactly what is needed. The second reason is that introducing (Dom1) and (Dom2) through a BNF-like notation would involve a significant number of syntactic classes corresponding to ex-exclusive boxes, x-directed boxes, etc., in combination with those already defined (cf. Exc. 7.3.1 and Sect. 8.1.3, where an example of syntactic definition will be given). We will instead introduce them through explicit conditions imposed on expressions satisfying the syntax (7.29).

Recursion. For every process variable $X \in \mathcal{X}$, there is a unique defining equation

$$X \stackrel{\mathrm{df}}{=} \mathsf{op}_\Omega(\boldsymbol{L}) \tag{7.30}$$

where $\Omega \in \mathsf{OpBox}$ and $\boldsymbol{L}$ is an Ω-tuple of process variables and static constants.

Syntactic Restrictions Resulting from (Dom1) and (Dom2). In the semantics of variables, we intend to use Thm. 7.2.5, and therefore need to define syntactic counterparts of well-formed, ex-exclusive, e-directed, and x-directed variables. To this end, we will now also have to consider (Dom1) and (Dom2).

Let $\mathsf{Expr}^{\mathrm{wf}}$, $\mathsf{Expr}^{\mathrm{xcl}}$, $\mathsf{Expr}^{\mathrm{edir}}$, and $\mathsf{Expr}^{\mathrm{xdir}}$ be the largest sets of expressions in $\mathsf{Expr}^{\mathrm{box}}$ — called respectively the *well-formed*, *ex-exclusive*, *e-directed*, and *x-directed* box expressions — such that the following hold:

(Expr1) All ex-exclusive, e-directed, and x-directed expressions are well formed.

(Expr2) If $\mathsf{op}_\Omega(\boldsymbol{D}) \in \mathsf{Expr}^{\mathrm{wf}}$, then for every $v \in T_\Omega$:
- (i) $D_v \in \mathsf{Expr}^{\mathrm{wf}}$;
- (ii) if ${}^\bullet v \cap v^\bullet \neq \emptyset$, then $D_v \in \mathsf{Expr}^{\mathrm{xcl}}$;
- (iii) if v is not reversible, then $D_v \in \mathsf{Expr}^{\mathrm{xdir}}$.

(Expr3) If $\mathsf{op}_\Omega(\boldsymbol{D}) \in \mathsf{Expr}^{\mathrm{xcl}}$, then Ω is ex-exclusive and, for every $v \in T_\Omega$ satisfying

$${}^\bullet v \cap {}^\circ\Omega \neq \emptyset \neq v^\bullet \cap \Omega^\circ \quad \text{or} \quad v^\bullet \cap {}^\circ\Omega \neq \emptyset \neq {}^\bullet v \cap \Omega^\circ ,$$

it is the case that $D_v \in \mathsf{Expr}^{\mathrm{xcl}}$.

(Expr4) If $\mathsf{op}_\Omega(\boldsymbol{D}) \in \mathsf{Expr}^{\mathrm{edir}}$, then Ω is e-directed and, for every $v \in ({}^\circ\Omega)^\bullet$, $D_v \in \mathsf{Expr}^{\mathrm{edir}}$.

(Expr5) If $\mathsf{op}_\Omega(\boldsymbol{D}) \in \mathsf{Expr}^{\mathrm{xdir}}$, then Ω is x-directed and, for every $v \in {}^\bullet(\Omega^\circ)$, $D_v \in \mathsf{Expr}^{\mathrm{xdir}}$.

(Expr6) If $\overline{E} \in \mathsf{Expr}^\delta$, or $\underline{E} \in \mathsf{Expr}^\delta$, or $X \in \mathsf{Expr}^\delta$ and $X \stackrel{\mathrm{df}}{=} E$, then $E \in \mathsf{Expr}^\delta$, for $\delta \in \{\mathrm{wf}, \mathrm{xcl}, \mathrm{edir}, \mathrm{xdir}\}$.

(Expr7) $\mathsf{Const} \subseteq \mathsf{Expr}^{\mathrm{wf}}$ and $\mathsf{Const}^\delta = \mathsf{Const} \cap \mathsf{Expr}^\delta$, for $\delta \in \{\mathrm{xcl}, \mathrm{edir}, \mathrm{xdir}\}$.

In the above, (Expr1)—(Expr5) correspond to (Var1)—(Var5) and, in particular: (Expr2)(ii) and (Expr2)(iii) encode (Dom1) and (Dom2), respectively; (Expr3) encodes Thm. 7.2.4; and (Expr4) together with (Expr5) encode Prop. 4.3.2. Moreover, (Expr6) and (Expr7) serve as a means for syntactic induction.

If all the operator boxes in OpBox are pure and have only reversible transitions, then (Expr2)(ii) and (Expr2)(iii) are vacuously satisfied and $\mathsf{Expr}^{\mathsf{wf}}$ is the entire set of box expressions defined by the syntax (7.29), i.e., $\mathsf{Expr}^{\mathsf{wf}} = \mathsf{Expr}^{\mathsf{box}}$. It is interesting to observe that this remark applies to the extended PBC syntax defined by (3.14) and (3.15) as well as to the DIY algebra introduced in the next section.

Exercise 7.3.1. The definition of $\mathsf{Expr}^{\mathsf{wf}}$, $\mathsf{Expr}^{\mathsf{xcl}}$, $\mathsf{Expr}^{\mathsf{edir}}$, and $\mathsf{Expr}^{\mathsf{xdir}}$ is not syntactic. Modify the syntax (7.29) in such a way that all these classes of expressions are given syntactically.

Exercise 7.3.2. Modify the definitions so that the conditions (Beh1)—(Beh5) are incorporated into the syntax of well-formed expressions.
Note: Chap. 8 will give a concrete syntax where this has been done.

Notice that the assumption that Expr^{δ}, for $\delta \in \{\mathrm{wf}, \mathrm{xcl}, \mathrm{edir}, \mathrm{xdir}\}$, are the largest sets satisfying (Expr1)—(Expr7) means that, for example, if E is a well-formed static expression, then $\overline{E}$ and $\underline{E}$ are well-formed entry and exit expressions, respectively.

7.3.1 A Running Example: the DIY Algebra

We will continue to use the running example based on the boxes depicted in Figs. 7.2 and 7.4, in order to construct a simple algebra of process expressions. More precisely, we will consider the *Do It Yourself (DIY)* algebra based on the following two sets of boxes:

$$\mathsf{OpBox} = \{\Omega_0\}$$

$$\mathsf{ConstBox} = \{\Phi_1, \Phi_{11}, \Phi_{12}\} \cup \{\Phi_2, \Phi_{21}, \Phi_{22}, \Phi_{23}\} \cup \{\Phi_3\}$$

where $\lfloor \Phi_{ij} \rfloor = \Phi_i = \lfloor \Sigma_{v_i} \rfloor$, for all i and j, and the following hold:

$$M_{\Phi_{11}} = \{q_{11}, q_{14}\}$$

$$M_{\Phi_{12}} = \{q_{13}, q_{12}\}$$

$$M_{\Phi_{2k}} = \{q_{2k}\} \qquad \text{(for } k = 1, 2, 3) .$$

The constants of the DIY algebra correspond to the boxes in ConstBox:

$$\mathsf{Const}^{\mathsf{e}} = \{c_{21}\} \qquad \mathsf{Const}^{\mathsf{s}} = \{c_1, c_2, c_3\}$$

$$\mathsf{Const}^{\mathsf{x}} = \{c_{23}\} \qquad \mathsf{Const}^{\mathsf{d}} = \{c_{11}, c_{12}, c_{21}, c_{22}, c_{23}\} .$$

Moreover, $\mathsf{Const} = \mathsf{Const}^{\mathsf{xcl}} = \mathsf{Const}^{\mathsf{edir}} = \mathsf{Const}^{\mathsf{xdir}}$.

The syntax of the DIY algebra is obtained by instantiating (7.29) with the concrete constants and operator introduced above. After taking into account the factorisations in fact_{Ω_0} (see (7.1)), we obtain

$$
\begin{array}{lrcl}
\mathsf{Expr}^{\mathsf{s}} & E & ::= & \mathsf{op}_{\Omega_0}(E,E,E) \mid c_1 \mid c_2 \mid c_3 \mid X \\
\mathsf{Expr}^{\mathsf{e}} & F & ::= & \mathsf{op}_{\Omega_0}(F,F,E) \mid c_{21} \mid \overline{E} \\
\mathsf{Expr}^{\mathsf{x}} & G & ::= & \mathsf{op}_{\Omega_0}(E,E,G) \mid c_{23} \mid \underline{E} \\
\mathsf{Expr}^{\mathsf{d}} & H & ::= & \mathsf{op}_{\Omega_0}(F,\widetilde{H},E) \mid \mathsf{op}_{\Omega_0}(\widetilde{H},F,E) \mid c_{11} \mid \\
 & & & \mathsf{op}_{\Omega_0}(F,G,E) \mid \mathsf{op}_{\Omega_0}(G,F,E) \mid c_{12} \mid \\
 & & & \mathsf{op}_{\Omega_0}(G,\widetilde{H},E) \mid \mathsf{op}_{\Omega_0}(\widetilde{H},G,E) \mid c_{22} \mid \\
 & & & \mathsf{op}_{\Omega_0}(\widetilde{H},\widetilde{H},E) \mid \mathsf{op}_{\Omega_0}(G,G,E) \mid F \mid \\
 & & & \mathsf{op}_{\Omega_0}(E,E,F) \mid \mathsf{op}_{\Omega_0}(E,E,\widetilde{H}) \mid G
\end{array}
$$

where $\widetilde{H}$ stands for an arbitrary expression in $\mathsf{Expr}^{\mathsf{d}} \setminus (\mathsf{Expr}^{\mathsf{e}} \cup \mathsf{Expr}^{\mathsf{x}})$. For instance, the first syntactic clause in the line defining (recursively) F comes about because $(\{v_1,v_2\},\emptyset,\emptyset,\{v_3\})$ factorises the entry marking $(\{s_1,s_2\},\emptyset)$ of Ω_0. Using the syntax we can see, for example, that $\mathsf{op}_{\Omega_0}(\overline{c_1},c_{22},c_3)$ is a valid dynamic expression of the DIY algebra. As we shall see, it corresponds to the net refinement $\Omega_0(\Sigma)$ of Fig. 7.4 discussed extensively earlier in this chapter.

The syntax for $\mathsf{Expr}^{\mathsf{d}}$ can be simplified. For example, we can replace $\mathsf{op}_{\Omega_0}(E,E,F)$ and $\mathsf{op}_{\Omega_0}(E,E,\widetilde{H})$ by $\mathsf{op}_{\Omega_0}(E,E,H)$ since $\mathsf{op}_{\Omega_0}(E,E,G)$ is part of the syntax for G, and G itself occurs in the syntax for $\mathsf{Expr}^{\mathsf{d}}$. This and other similar observations lead to

$$
\begin{array}{lrcl}
\mathsf{Expr}^{\mathsf{d}} & H & ::= & c_{11} \mid c_{12} \mid c_{22} \mid F \mid G \mid \\
 & & & \mathsf{op}_{\Omega_0}(H,H,E) \mid \mathsf{op}_{\Omega_0}(E,E,H)\,.
\end{array}
$$

This is not a mere chance since the syntax (7.29) can always be presented so that $\mathsf{Expr}^{\mathsf{d}}$ does not refer to $\widetilde{H}$. The interest in the form which employs $\widetilde{H}$ is that all the defining clauses are distinct, while this is not the case for the alternative form where $\mathsf{op}_{\Omega_0}(H,H,E)$ is also part of the definition of F.

Exercise 7.3.3. Prove that the syntax (7.29) can always be presented so that $\mathsf{Expr}^{\mathsf{d}}$ does not refer to $\widetilde{H}$.
Hint: If μ is a factorisation in fact_{Ω}, then so are the tuples $(\mu_{\mathsf{e}} \cup \mu_{\mathsf{d}},\emptyset,\mu_{\mathsf{x}},\mu_{\mathsf{s}})$ and $(\mu_{\mathsf{e}},\emptyset,\mu_{\mathsf{x}} \cup \mu_{\mathsf{d}},\mu_{\mathsf{s}})$ (cf. Prop. 7.1.1(3)).

7.3.2 Infinite Operators

In the case that each operator box Ω in OpBox has finitely many transitions, the meaning of the syntax (7.29) is clear; it simply defines four sets of finite strings, or terms. In the general case, however, we need to take into account Ω's with infinite transition sets (possibly uncountable, as allowed by the generalised parallel and choice composition), and one should ask what is meant by the syntax (7.29) and the expressions it generates. A possible answer is that expressions can, in general, be seen as trees and the syntax definition above as a definition of four sets of trees.

We will define such trees not just for the syntax (7.29), but even for a more general set of process expressions over the signature (7.28), denoted by Expr and referred to simply as *expressions*, which are defined by

$$C ::= c \;\Big|\; X \;\Big|\; \overline{C} \;\Big|\; \underline{C} \;\Big|\; \mathsf{op}_\Omega(\boldsymbol{C}) \tag{7.31}$$

where $c \in \mathsf{Const}$ is a constant, $X \in \mathcal{X}$ is a variable, and $\boldsymbol{C}$ is an Ω-tuple of expressions, for $\Omega \in \mathsf{OpBox}$. For instance, this allows one to write expressions such as $\overline{\overline{\overline{X}}}$, $\underline{\underline{\alpha}}$, $\overline{(\underline{\beta})}$, and $\underline{\alpha};\overline{\beta}$. Clearly, $\mathsf{Expr}^{\mathsf{box}}$ is a subset of Expr. We shall need a larger set of expressions Expr later on, in the definition of a similarity relation on process expressions and in their operational semantics.

To give a meaning to expressions defined by this syntax, we consider the set $\mathsf{ExprTrees}$ of all finite and infinite labelled trees τ, satisfying the following.

- Each node of τ is labelled by an element of the signature (7.28) or a variable in $\mathcal{X}$; moreover, each $x \in \mathsf{Const} \cup \mathcal{X}$ is identified with a single node tree in $\mathsf{ExprTrees}$ whose only node is labelled by x.
- If a node ξ is labelled by a constant or a variable, then ξ is a leaf.
- If a node ξ is labelled by $\overline{(.)}$ or $\underline{(.)}$, then ξ has exactly one child node to which it is connected by an unlabelled arc.
- If a node ξ is labelled by op_Ω, for some $\Omega \in \mathsf{OpBox}$, then its child nodes are $\{\xi_v \mid v \in T_\Omega\}$ and each ξ_v is connected to ξ by an arc labelled by v.

Then $\overline{(.)}$, $\underline{(.)}$, and $\mathsf{op}_\Omega(.)$ can be seen as mappings yielding trees in $\mathsf{ExprTrees}$. For example, $\overline{(.)} : \mathsf{ExprTrees} \to \mathsf{ExprTrees}$ is given by $\overline{C} = \overline{(.)} \lhd C$, with a slight abuse of notation since $\lhd$ was originally defined for name trees only.

The domain $(2^{\mathsf{ExprTrees}}, \subseteq)$ forms a complete lattice, with the set intersection as the 'meet' operation, and the set union as the 'join' operation. Consider now the syntax (7.31). It can be seen as defining a mapping

$$expr : 2^{\mathsf{ExprTrees}} \to 2^{\mathsf{ExprTrees}}$$

such that, for every $\mathcal{T} \subseteq \mathsf{ExprTrees}$, $expr(\mathcal{T})$ is equal to

$$\mathsf{Const} \cup \mathcal{X} \cup \mathcal{T} \cup \overline{\mathcal{T}} \cup \underline{\mathcal{T}} \cup$$

$$\bigcup_{\Omega \in \mathsf{OpBox}} \Big\{ \mathsf{op}_\Omega(\boldsymbol{C}) \;\Big|\; \boldsymbol{C} \text{ is an } \Omega\text{-tuple with values in } \mathsf{Const} \cup \mathcal{X} \cup \mathcal{T} \Big\} .$$

Clearly, *expr* is monotonic, hence it has a unique least fixpoint, denoted by Expr. Moreover, by an argument similar to that used in Sect. 5.2.6, there is a (possibly transfinite) ordinal α such that $\mathsf{Expr} = expr^{\alpha}(\emptyset)$, where for an ordinal β and $\mathcal{T} \subseteq \mathsf{ExprTrees}$,

$$expr^{\beta}(\mathcal{T}) = \begin{cases} \mathcal{T} & \text{if } \beta = 0 \\ expr(\bigcup_{\gamma<\beta} expr^{\gamma}(\mathcal{T})) & \text{if } \beta > 0 \,. \end{cases}$$

The above implies, in particular, that no tree in Expr has an infinite path from the root. It is then possible to associate with each expression C in Expr its *rank*, denoted by $\mathsf{rank}(C)$, which is the smallest ordinal β such that $C \in expr^{\beta}(\emptyset)$. It follows that the rank of all constants and variables is 1, $\mathsf{rank}(\overline{C}) = \mathsf{rank}(\underline{C}) = 1{+}\mathsf{rank}(C)$, and $\mathsf{rank}(\mathsf{op}_{\Omega}(\boldsymbol{C}))$ is the least ordinal β such that $\mathsf{rank}(C_v) < \beta$, for every $v \in T_{\Omega}$. The principle of transfinite induction means that we may use induction based on the rank of expressions. Such an inductive argument reduces to the standard induction on the depth of an expression if all the operator boxes in $\mathbf{OpBox}$ have finitely many transitions since then the rank of every expression is a natural number (a finite ordinal).

The meaning of the syntax (7.29) can now be explained. Consider its first line, i.e., the syntactic definition of a static expression. What it defines is the set $\mathsf{Expr}^{\mathsf{s}}$ of all expressions $E \in \mathsf{Expr}$ such that E is either a static constant, or a variable, or $E = \mathsf{op}_{\Omega}(\boldsymbol{E})$ where $\boldsymbol{E} \subseteq \mathsf{Expr}^{\mathsf{s}}$. By the transfinite induction argument, such a set exists and is unique. The definitions of Expr^{δ}, for $\delta \in \{\mathrm{e, x, d, wf, xcl, edir, xdir}\}$, all follow the same pattern.

Exercise 7.3.4. Show by induction on the rank of expressions that each subexpression (where, in general, a subexpression is a subtree of a tree in $\mathsf{ExprTrees}$) of an expression in Expr^{δ}, for $\delta \in \{\mathrm{box, s, wf}\}$, is also an expression in Expr^{δ}.

Properties. Several properties of box expressions can be proved by induction on the rank of expressions. In particular, it follows from the syntax (7.29) and

$$\mathsf{Const}^{\mathsf{s}} \cap \mathsf{Const}^{\mathsf{d}} = \mathsf{Const}^{\mathsf{e}} \cap \mathsf{Const}^{\mathsf{x}} = \emptyset$$

$$\mathsf{Const}^{\mathsf{e}} \cup \mathsf{Const}^{\mathsf{x}} \subseteq \mathsf{Const}^{\mathsf{d}}$$

that $\mathsf{Expr}^{\mathsf{e}}$ and $\mathsf{Expr}^{\mathsf{x}}$ are disjoint subsets of $\mathsf{Expr}^{\mathsf{d}}$; and that the static and dynamic expressions are disjoint sets, i.e.,

$$\begin{aligned} &\mathsf{Expr}^{\mathsf{s}} \cap \mathsf{Expr}^{\mathsf{d}} = \mathsf{Expr}^{\mathsf{e}} \cap \mathsf{Expr}^{\mathsf{x}} = \emptyset \\ &\mathsf{Expr}^{\mathsf{e}} \cup \mathsf{Expr}^{\mathsf{x}} \subseteq \mathsf{Expr}^{\mathsf{d}} \,. \end{aligned} \tag{7.32}$$

These can be compared to similar relationships in the domain of boxes:

$$\begin{aligned} &\mathsf{Box}^{\mathsf{s}} \cap \mathsf{Box}^{\mathsf{d}} = \mathsf{Box}^{\mathsf{e}} \cap \mathsf{Box}^{\mathsf{x}} = \emptyset \\ &\mathsf{Box}^{\mathsf{e}} \cup \mathsf{Box}^{\mathsf{x}} \subseteq \mathsf{Box}^{\mathsf{d}} \,. \end{aligned} \tag{7.33}$$

Notice that (7.32) implies that the factorisation of an Ω-tuple of box expressions is always a partition of the set T_Ω of transitions of Ω.

As in the domain of boxes and PBC expressions, it is convenient to have a notation for turning an expression in Expr into a corresponding static one. We again use $\lfloor . \rfloor$ to denote such an operation. To achieve the desired effect, we assume that (i) for each dynamic constant c there is a unique static constant $\lfloor c \rfloor$ satisfying

$$c \in \mathsf{Const}^\delta \iff \lfloor c \rfloor \in \mathsf{Const}^\delta ,$$

for $\delta \in \{\mathrm{xcl}, \mathrm{edir}, \mathrm{xdir}\}$; and (ii) if C is an expression, then $\lfloor C \rfloor$ is the static expression obtained by removing all occurrences of $\overline{(.)}$ and $\underline{(.)}$, and replacing every occurrence of each dynamic constant c by the corresponding static constant $\lfloor c \rfloor$. Formally,

$$\lfloor . \rfloor : \mathsf{Expr} \to \mathsf{Expr}$$

is a mapping defined by induction on the rank of expressions. If $\mathsf{rank}(C) = 1$, then $\lfloor C \rfloor = \lfloor c \rfloor$ for $C = c \in \mathsf{Const}^{\mathrm{d}}$, and $\lfloor C \rfloor = C$ otherwise. Moreover, for expressions with ranks higher than 1,

$$\lfloor \overline{C} \rfloor = \lfloor \underline{C} \rfloor = \lfloor C \rfloor \quad \text{and} \quad \lfloor \mathsf{op}_\Omega(\mathbf{C}) \rfloor = \mathsf{op}_\Omega(\lfloor \mathbf{C} \rfloor) . \tag{7.34}$$

Exercise 7.3.5. Show that $\lfloor C \rfloor$ is a static expression, for every expression C. Then show that, for all box expressions C and $\delta \in \{\mathrm{wf}, \mathrm{xcl}, \mathrm{edir}, \mathrm{xdir}\}$,

$$C \in \mathsf{Expr}^\delta \iff \lfloor C \rfloor \in \mathsf{Expr}^\delta .$$

The operators $\overline{(.)}$, $\underline{(.)}$ and $\lfloor . \rfloor$ can be applied in the usual way (i.e., elementwise) to sets as well as tuples of expressions. The same will be true of the semantical mapping box defined in the next section, and then also of the structural similarity relation $\equiv$. In what follows, the box expressions in $\mathsf{Const} \cup \overline{\mathsf{Const}^{\mathrm{s}}} \cup \underline{\mathsf{Const}^{\mathrm{s}}}$, i.e., those not involving any connective op_Ω nor a process variable (which hides an application of some op_Ω through its definition), will be called *flat*.

Exercise 7.3.6. Show that all flat expressions are well formed.

Running Example. In the DIY algebra, $\mathsf{op}_{\Omega_0}(c_1, c_2, \mathsf{op}_{\Omega_0}(c_1, c_2, \underline{c_3}))$ is a well-formed exit expression with rank 3. Moreover, we set $\lfloor c_{ij} \rfloor = c_i$, for every dynamic constant c_{ij}. Thus, for example,

$$\lfloor \mathsf{op}_{\Omega_0}(\overline{c_1}, c_{22}, c_3) \rfloor = \mathsf{op}_{\Omega_0}(c_1, c_2, c_3) .$$

7.3.3 Denotational Semantics

When dealing with the denotational semantics of the box algebra, we will first define the semantics of its constants, then of process variables, and finally, of compound box expressions. The semantics is given in the form of a mapping from well-formed box expressions to boxes, $\mathsf{box} : \mathsf{Expr}^{\mathsf{wf}} \to \mathsf{Box}$, defined by induction on the rank of box expressions.

Constants. Constant expressions are mapped onto constant boxes of corresponding types, i.e., for every constant c and $\delta \in \{\mathrm{e}, \mathrm{d}, \mathrm{x}, \mathrm{s}, \mathrm{xcl}, \mathrm{edir}, \mathrm{xdir}\}$,

$$c \in \mathsf{Const}^{\delta} \iff \mathsf{box}(c) \in \mathsf{Box}^{\delta} \cap \mathsf{ConstBox}\,. \tag{7.35}$$

We also make some additional assumptions which are not an inherent part of the definition of the denotational semantics. Their role is to ensure a consistency between the box semantics of dynamic constants and the corresponding static constants, as well as to guarantee that there are enough constants to model internal states of evolving static or dynamic constants.

Formally, it is assumed that, for every dynamic constant, the underlying box is the same as that for the corresponding static constant, and that it is reachable from the entry or exit marking of the latter; i.e., for every c in $\mathsf{Const}^{\mathsf{d}}$,

$$\big\lfloor \mathsf{box}(c) \big\rfloor = \mathsf{box}\big(\lfloor c \rfloor\big) \quad \text{and} \quad \mathsf{box}(c) \in \big[\overline{\mathsf{box}(\lfloor c \rfloor)}\big\rangle \cup \big[\underline{\mathsf{box}(\lfloor c \rfloor)}\big\rangle\,. \tag{7.36}$$

Moreover, for every non-entry and non-exit dynamic box reachable from an initially or terminally marked constant box, there is a corresponding dynamic constant, i.e., for every c in $\mathsf{Const}^{\mathsf{s}}$,

$$\big[\overline{\mathsf{box}(c)}\big\rangle \cup \big[\underline{\mathsf{box}(c)}\big\rangle \subseteq \mathsf{box}\big(\mathsf{Const}^{\mathsf{d}}\big) \cup \big\{\overline{\mathsf{box}(c)}, \underline{\mathsf{box}(c)}\big\}\,. \tag{7.37}$$

Assumption (7.37) is important for those constants of the algebra that are not simple multiactions, such as the data constants of Sect. 3.3.2.

Exercise 7.3.7. Show that for every entry or exit constant c, it is the case that $\mathsf{box}(c) = \overline{\lfloor \mathsf{box}(c) \rfloor}$ or $\mathsf{box}(c) = \underline{\lfloor \mathsf{box}(c) \rfloor}$, respectively.

Variables. With each defining equation $X \stackrel{\mathrm{df}}{=} \mathsf{op}_{\Omega}(\boldsymbol{L})$, we associate an equation on boxes:

$$X \stackrel{\mathrm{df}}{=} \Omega(\boldsymbol{\Psi})\,,$$

where $\Psi_v = L_v$ if L_v is a process variable (treated here as a box variable), and $\Psi_v = \mathsf{box}(L_v)$ if L_v is a static constant. This creates a system of net equations of the type considered in Chap. 5. Now, (Expr1)—(Expr7) and (7.35) have been set up in such a way that this system can be treated by Thm. 7.2.5 and, in particular, that we have $\mathcal{X}^{\delta} = \mathcal{X} \cap \mathsf{Expr}^{\delta}$, for $\delta \in \{\mathrm{wf}, \mathrm{xcl}, \mathrm{edir}, \mathrm{xdir}\}$, and that the conditions (Var1)—(Var5) are all satisfied. Thus, by Thm. 7.2.5, the system of equations we have just constructed has at least one solution in the domain of plain boxes with empty markings. We then fix *any* such solution sol (for instance, the maximal or a minimal one, following observations made in Chaps. 5 and 6) and define, for every well-formed process variable X, $\mathsf{box}(X) = \mathsf{sol}(X)$. As a result, for every well-formed variable X defined by the equation $X \stackrel{\mathrm{df}}{=} \mathsf{op}_\Omega(\boldsymbol{L})$,

$$\mathsf{box}(X) = \Omega(\mathsf{box}(\boldsymbol{L})) . \tag{7.38}$$

And, crucially, it follows from Thm. 7.2.5 that

$$\begin{aligned} &\mathsf{box}(\mathcal{X}^{\mathrm{wf}}) \subseteq \mathsf{Box}^{\mathrm{s}} \\ &\mathsf{box}(\mathcal{X}^{\delta}) \subseteq \mathsf{Box}^{\delta} \quad \text{for } \delta \in \{\mathrm{xcl}, \mathrm{edir}, \mathrm{xdir}\} . \end{aligned} \tag{7.39}$$

Compound Expressions. The definition of the semantical mapping box is completed by considering all the remaining static and dynamic well-formed expressions, following the syntax (7.29). The box mapping is a homomorphism; hence, for every static or dynamic expression $\mathsf{op}_\Omega(\boldsymbol{D})$ and every static expression E,

$$\begin{aligned} &\mathsf{box}(\mathsf{op}_\Omega(\boldsymbol{D})) = \Omega(\mathsf{box}(\boldsymbol{D})) \\ &\mathsf{box}(\overline{E}) \qquad\;\; = \overline{\mathsf{box}(E)} \\ &\mathsf{box}(\underline{E}) \qquad\;\; = \underline{\mathsf{box}(E)} . \end{aligned} \tag{7.40}$$

The above formulae are well formed since the ranks of the expressions to which box is applied appearing on the right-hand sides of the equality sign have strictly smaller ranks than the corresponding expressions on the left-hand sides.

Running Example. In the case of the DIY algebra, we define the box mapping by setting, for every static constant c_i, $\mathsf{box}(c_i) = \Phi_i$, and for every dynamic constant c_{ij}, $\mathsf{box}(c_{ij}) = \Phi_{ij}$. Other than that, we follow the general definitions. Thus, for example, the box in Fig. 7.4 can be derived as follows.

$$\begin{aligned} &\mathsf{box}\Big(\mathsf{op}_{\Omega_0}(\overline{c_1}, c_{22}, c_3)\Big) \\ &= \Omega_0\Big(\mathsf{box}(\overline{c_1}), \mathsf{box}(c_{22}), \mathsf{box}(c_3)\Big) && \text{(by line 1 of (7.40))} \\ &= \Omega_0\Big(\overline{\mathsf{box}(c_1)}, \Phi_{22}, \Phi_3\Big) && \text{(by line 2 of (7.40) and def. of constants)} \\ &= \Omega_0\Big(\overline{\Phi_1}, \Phi_{22}, \Phi_3\Big) && \text{(by def. of constants)} \\ &= \Omega_0\Big(\Sigma_{v_1}, \Sigma_{v_2}, \Sigma_{v_3}\Big) && \text{(by def. of the } \Sigma_{v_i}\text{'s)} \\ &= \Omega_0\Big(\boldsymbol{\Sigma}\Big) . && \text{(by def. of } \boldsymbol{\Sigma}\text{, cf. Fig. 7.4)} \end{aligned}$$

Properties. The semantical mapping always returns a box. Moreover, the assumed type consistency between constants and their denotations, (7.35), carries over to the remaining box expressions. Essentially, this means we have been able to capture syntactically the property of being a static, dynamic, entry, or exit box.

Theorem 7.3.1. Let D be a well-formed box expression.

(1) $\mathsf{box}(D)$ is a static or dynamic box.
(2) For every $\delta \in \{\mathrm{e},\mathrm{d},\mathrm{x},\mathrm{s}\}$, $D \in \mathsf{Expr}^{\delta}$ if and only if $\mathsf{box}(D) \in \mathsf{Box}^{\delta}$.
(3) For every $\delta \in \{\mathrm{xcl},\mathrm{edir},\mathrm{xdir}\}$, if $D \in \mathsf{Expr}^{\delta}$, then $\mathsf{box}(D) \in \mathsf{Box}^{\delta}$.

Proof. The proof will proceed by induction on the rank of D. Using (Expr2) and (Expr6), it is not hard to see (and the formal proof of this is left as an exercise to the reader — see Exc. 7.3.4) that $D \in \mathsf{Expr}^{\mathrm{wf}}$ implies that all subexpressions of D are well formed.

The base step corresponds to D being a constant or a well-formed variable. In the former case, (1,2,3) follow directly from (7.32), (7.33), (7.35), and (Expr7); and in the latter case from (7.39).

In the inductive step, we consider three cases.

Case 1: $D = \overline{E}$, for some static expression E. Then, by (7.29) and (7.32), $D \in \mathsf{Expr}^{\mathrm{e}} \subseteq \mathsf{Expr}^{\mathrm{d}}$ and $D \notin \mathsf{Expr}^{\mathrm{x}} \cup \mathsf{Expr}^{\mathrm{s}}$. Moreover, by the induction hypothesis and (7.40),

$$\mathsf{box}(D) = \overline{\mathsf{box}(E)} \in \overline{\mathsf{Box}^{\mathrm{s}}} = \mathsf{Box}^{\mathrm{e}} \subseteq \mathsf{Box}^{\mathrm{d}}$$

and, by (7.33), $\mathsf{box}(D) \notin \mathsf{Box}^{\mathrm{x}} \cup \mathsf{Box}^{\mathrm{s}}$. Thus (1) and (2) are satisfied. Moreover, (3) is also satisfied, by $\mathsf{box}(E) = \lfloor \mathsf{box}(D) \rfloor$, (Expr6) (which implies that $E \in \mathsf{Expr}^{\delta}$), and the induction hypothesis. Hence the result holds in this case.

Case 2: $D = \underline{E}$, for some static expression E. Then we proceed similarly as in Case 1.

Case 3: $D = \mathsf{op}_{\Omega}(\boldsymbol{D})$. Let μ be the factorisation of $\boldsymbol{D}$. By the induction hypothesis, for every $v \in T_{\Omega}$, $\mathsf{box}(D_v)$ is a static or dynamic box; moreover, for all $\delta \in \{\mathrm{e},\mathrm{d},\mathrm{x},\mathrm{s}\}$ and $v \in \mu_{\delta}$,

$$\mathsf{box}(D_v) \subseteq \mathsf{Box}^{\delta} \tag{7.41}$$

which means that μ is also the factorisation of $\mathsf{box}(\boldsymbol{D})$. Furthermore, by (Expr2) and the induction hypothesis, $\mathsf{box}(\boldsymbol{D})$ is an Ω-tuple satisfying (Dom1) and (Dom2). This and $\mu \in \mathsf{fact}_{\Omega}$ means that $\mathsf{box}(\boldsymbol{D}) \in \mathsf{dom}_{\Omega}$ and so, by (7.40), well-formedness and Thm. 7.2.3, we obtain that $\mathsf{box}(D)$ is a static or dynamic box. We then consider subcases depending on the type of marking of Ω whose factorisation is μ.

If μ is the factorisation of the empty marking, then $\mu_{\mathrm{e}} = \mu_{\mathrm{d}} = \mu_{\mathrm{x}} = \emptyset$. Thus $D \in \mathsf{Expr}^{\mathrm{s}}$ and, by (7.5), (7.40), and (7.41), $\mathsf{box}(D) \in \mathsf{Box}^{\mathrm{s}}$. Moreover, by (7.32) and (7.33), $D \notin \mathsf{Expr}^{\delta}$ and $\mathsf{box}(D) \notin \mathsf{Box}^{\delta}$, for $\delta \in \{\mathrm{e},\mathrm{d},\mathrm{x}\}$. Hence (1) and (2) are satisfied.

If μ is a factorisation of ${}^{\circ}\Omega$, then by (7.29) and (7.32), $D \in \mathsf{Expr}^{\mathrm{e}} \subseteq \mathsf{Expr}^{\mathrm{d}}$ and $D \notin \mathsf{Expr}^{\mathrm{x}} \cup \mathsf{Expr}^{\mathrm{s}}$. Moreover, by Prop. 7.1.1(4), ${}^{\bullet}\mu_{\mathrm{e}} \cup \mu_{\mathrm{x}}^{\bullet} = {}^{\circ}\Omega$ and $\mu_{\mathrm{d}} = \emptyset$. Hence, by (7.5), (7.40), and (7.41), $\mathsf{box}(D) \in \mathsf{Box}^{\mathrm{e}} \subseteq \mathsf{Box}^{\mathrm{d}}$. The latter and (7.33) further implies $\mathsf{box}(D) \notin \mathsf{Box}^{\mathrm{x}} \cup \mathsf{Box}^{\mathrm{s}}$. Hence (1) and (2) are satisfied. If μ is a factorisation of Ω°, we proceed in a similar way.

If μ is a factorisation of a complex marking reachable from ${}^\circ\Omega$ or Ω°, but different from ${}^\circ\Omega$ and Ω°, then, by (7.29) and (7.32),

$$D \in \mathsf{Expr}^{\mathsf{d}} \backslash (\mathsf{Expr}^{\mathsf{e}} \cup \mathsf{Expr}^{\mathsf{x}}) \quad \text{and} \quad D \notin \mathsf{Expr}^{\mathsf{s}} .$$

Clearly, by (7.40) and (7.41), $\mathsf{box}(D) \notin \mathsf{Box}^{\mathsf{s}}$. Suppose $\mathsf{box}(D) \in \mathsf{Box}^{\mathsf{e}} \cup \mathsf{Box}^{\mathsf{x}}$. Then, by (7.40), (7.41), and Lemma 7.1.2, ${}^\circ\Omega \subseteq {}^\bullet\mu_{\mathsf{e}} \cup \mu_{\mathsf{x}}^\bullet$ or $\Omega^\circ \subseteq {}^\bullet\mu_{\mathsf{e}} \cup \mu_{\mathsf{x}}^\bullet$. Hence, by Prop. 7.1.1(4), μ is a factorisation of ${}^\circ\Omega$ or Ω°, a contradiction. Thus $\mathsf{box}(D) \in \mathsf{Box}^{\mathsf{d}} \backslash (\mathsf{Box}^{\mathsf{e}} \cup \mathsf{Box}^{\mathsf{x}})$, and so (1) and (2) are satisfied.

We have shown that (1) and (2) are satisfied. We then observe that (3) follows from (Expr3)—(Expr5), the induction hypothesis, $\mathsf{box}(D) = \Omega(\mathsf{box}(\boldsymbol{D}))$, Thm. 7.2.4 and Prop. 4.3.2. □

It may be observed that the families of boxes involved in Thm. 7.3.1(3) are defined using transitions, hence they are sensitive to the relabellings annotating transitions in operator boxes. On the other hand, Thm. 7.3.1(2) involves families of boxes which are defined purely through place markings, allowing a stronger formulation of the result. Therefore, unlike Thm. 7.3.1(2), Thm. 7.3.1(3) states implication rather than equivalence. For example, a non-e-directed operator box yields an e-directed box if it is refined with plain boxes which contain no transitions at all.

The semantic translation commutes with removing the over- and underbars and replacing dynamic constants by static ones. Moreover, in a box generated from an expression, either all transitions are single-node trees, or none is.

Proposition 7.3.1. Let D be a dynamic and E a static well-formed box expression.

(1) $\lfloor \mathsf{box}(D) \rfloor = \mathsf{box}(\lfloor D \rfloor)$.
(2) Exactly one of the following holds:
  (a) $T_{\mathsf{box}(E)} \subseteq \mathsf{T}_{\mathsf{root}}$ and $E \in \mathsf{Const}^{\mathsf{s}}$.
  (b) $T_{\mathsf{box}(E)} \cap \mathsf{T}_{\mathsf{root}} = \emptyset$ and $\mathsf{box}(E) = \mathsf{box}(\mathsf{op}_\Omega(\boldsymbol{E}))$, for some well-formed static expression $\mathsf{op}_\Omega(\boldsymbol{E})$.

Proof. The first part follows from Prop. 4.3.3, (7.36) and by a straightforward induction on the rank of D. The second part follows from (7.27), (7.35), and the following observation: if $E = X$ and the defining equation for X is $X \stackrel{\mathrm{df}}{=} \mathsf{op}_\Omega(\boldsymbol{L})$, then by (7.38) and (7.40), $\mathsf{box}(E) = \mathsf{box}(X) = \Omega(\mathsf{box}(\boldsymbol{L})) = \mathsf{box}(\mathsf{op}_\Omega(\boldsymbol{L}))$. □

A result similar to Prop. 7.3.1(2) does not hold for the place set $S_{\mathsf{box}(E)}$, due to the possible presence of isolated places in an operator box.

7.3.4 Structural Similarity Relation on Expressions

We now define a structural similarity relation on box expressions, $\equiv$. It provides a partial identification of box expressions with the same denotational semantics, equating those expressions which can be shown denotationally equal by purely syntactic means. The way in which it will be defined resembles to some extent the definition of the similarity relation for (tuples of) boxes in Sect. 7.2.2. This should not be too surprising since the denotational semantics of the box algebra is based on the algebra of boxes presented earlier in this chapter.

The reason why we introduced, in (7.31), a larger set of expressions than was necessary to define the domain of box expressions is that the rules of the structural equivalence (and, later, operational semantics) should act as term rewriting rules. That is, whenever a (well-formed) box expression can match one side of such a rule, then it should be guaranteed that the other side is a (well-formed) box expression too. To be able to express and prove such a property, we need to allow for more general expressions such as

$$\overline{\overline{X}} \;,\quad \underline{\underline{\alpha}} \;,\quad \overline{(\underline{\beta})} \;,\quad \text{and} \quad \underline{\alpha};\overline{\beta}\,.$$

In the definition of the structural similarity on box expressions, we need to take into account the way in which boxes were associated with process variables and possible factorisations of the operator boxes (since the relation we are going to define can be seen as a counterpart of the similarity relations $\equiv_\Omega$ on Ω-tuples of boxes). Formally, we define $\equiv$ to be the least binary relation on expressions in Expr such that (7.42)—(7.50) hold.

Reflexivity, Symmetry, and Transitivity. For all expressions D, H and J,

$$D \equiv D \qquad \frac{D \equiv H}{H \equiv D} \qquad \frac{D \equiv J\;,\; J \equiv H}{D \equiv H} \tag{7.42}$$

Flat Expressions. For all flat expressions D and H satisfying $\mathsf{box}(D) = \mathsf{box}(H)$,

$$D \equiv H \tag{7.43}$$

Process Variables. For every defining equation $X \stackrel{\mathrm{df}}{=} \mathsf{op}_\Omega(\boldsymbol{L})$,

$$X \equiv \mathsf{op}_\Omega(\boldsymbol{L}) \tag{7.44}$$

Entry Expressions. For every SOS-operator box Ω in OpBox and every factorisation μ of ${}^\circ\Omega$,

$$\mathsf{op}_\Omega(\boldsymbol{D}) \equiv \overline{\mathsf{op}_\Omega(\boldsymbol{H})} \tag{7.45}$$

where $\boldsymbol{D}$ and $\boldsymbol{H}$ are Ω-tuples of expressions[7] such that, for every $v \in T_\Omega$,

[7] Notice that here and later it is not required that $\boldsymbol{D}$ and $\boldsymbol{H}$ be Ω-tuples of *box* expressions.

$$D_v = \begin{cases} \overline{H_v} \text{ if } & v \in \mu_e \\ \underline{H_v} \text{ if } & v \in \mu_x \\ H_v \text{ otherwise} . \end{cases}$$

Exit Expressions. For every SOS-operator box Ω in OpBox and every factorisation μ of Ω°,

$$\boxed{\mathsf{op}_\Omega(\boldsymbol{D}) \equiv \underline{\mathsf{op}_\Omega(\boldsymbol{H})}} \tag{7.46}$$

where $\boldsymbol{D}$ and $\boldsymbol{H}$ are as in (7.45).

Equivalent Factorisations. For every SOS-operator box Ω in OpBox, for every complex marking reachable from the entry or exit marking of Ω different from ${}^\circ\Omega$ and Ω°, and for every pair of different factorisations μ and κ of that marking,

$$\boxed{\mathsf{op}_\Omega(\boldsymbol{D}) \equiv \mathsf{op}_\Omega(\boldsymbol{H})} \tag{7.47}$$

where $\boldsymbol{D}$ and $\boldsymbol{H}$ are Ω-tuples of expressions for which there is an Ω-tuple of expressions $\boldsymbol{C}$ such that, for every $v \in T_\Omega$,

$$D_v = \begin{cases} \overline{C_v} \text{ if } & v \in \mu_e \\ \underline{C_v} \text{ if } & v \in \mu_x \\ C_v \text{ otherwise} \end{cases} \quad \text{and} \quad H_v = \begin{cases} \overline{C_v} \text{ if } & v \in \kappa_e \\ \underline{C_v} \text{ if } & v \in \kappa_x \\ C_v \text{ otherwise} . \end{cases} \tag{7.48}$$

Substitutivity. For all expressions D and H,

$$\boxed{\frac{D \equiv H}{\overline{D} \equiv \overline{H}} \qquad \frac{D \equiv H}{\underline{D} \equiv \underline{H}}} \tag{7.49}$$

and, for every SOS-operator box Ω in OpBox,

$$\boxed{\frac{\boldsymbol{D} \equiv \boldsymbol{H}}{\mathsf{op}_\Omega(\boldsymbol{D}) \equiv \mathsf{op}_\Omega(\boldsymbol{H})}} \tag{7.50}$$

where $\boldsymbol{D}$ and $\boldsymbol{H}$ are Ω-tuples of expressions.

The tuple $\boldsymbol{C}$ used in the formulation of (7.47) intuitively corresponds to the common (and thus unchanging) part of the tuples $\boldsymbol{D}$ and $\boldsymbol{H}$. For example, in the DIY algebra, since $(\{v_3\}, \emptyset, \emptyset, \{v_1, v_2\})$ and $(\emptyset, \emptyset, \{v_1, v_2\}, \{v_3\})$ are factorisations of the same marking reachable from ${}^\circ\Omega_0$, we shall have

$$\mathsf{op}_{\Omega_0}(c_1, c_2, \overline{c_3}) \equiv \mathsf{op}_{\Omega_0}(\underline{c_1}, \underline{c_2}, c_3)$$
$$\mathsf{op}_{\Omega_0}(\overline{c_1}, \overline{c_2}, \overline{c_3}) \equiv \mathsf{op}_{\Omega_0}\left(\underline{(\overline{c_1})}, \underline{(\overline{c_2})}, c_3\right) .$$

Notice that for infinite operator boxes, the context rule IREC defined in Sect. 3.2.15 is too weak to effect 'infinitely broad' substitutions, which are possible under the general substitutivity rules, since it only allows us to replace instances of one recursive variable at a time.

Infinite Operators. As it was the case for the syntax of the box algebra, the meaning of the similarity relation $\equiv$ is clear if we only consider operator boxes OpBox with finite transition sets. However, since we also allow infinite operator boxes, some explanations are needed to clarify what is meant by the 'the least binary relation on expressions such that (7.42)—(7.50) hold'. We start by observing that the domain $(2^{\mathsf{Expr}\times\mathsf{Expr}},\subseteq)$ forms a complete lattice, with component-wise set intersection as the 'meet' operation, and component-wise set union as the 'join' operation, and that the rules (7.42)—(7.50) define a monotonic mapping

$$sim : 2^{\mathsf{Expr}\times\mathsf{Expr}} \to 2^{\mathsf{Expr}\times\mathsf{Expr}},$$

where $sim(\mathcal{E})$ is $\mathcal{E}$ together with all pairs in Expr × Expr which can be derived from the pairs in $\mathcal{E}$ using a single application of any of the rules (7.42)—(7.50).

By following the same line of reasoning as before, we conclude that sim has a unique least fixpoint, denoted by $\equiv$, and equal to $sim^{\alpha'}(\emptyset)$, for some ordinal α' (such that $sim^{\alpha'}(\emptyset) = sim^{\beta}(\emptyset)$, for any $\beta \geq \alpha'$). And, as before, we can define the *rank* of a pair (D, H) belonging to $\equiv$, denoted by $\mathsf{rank}(D, H)$, as the least ordinal β such that $(D, H) \in sim^{\beta}(\emptyset)$. It follows from the principle of transfinite induction that we can apply induction on the rank of elements of $\equiv$, which reduces to the standard induction if all operator boxes in OpBox have finite transition sets.

Notice that $\mathsf{rank}(D, H)$ is not related in any obvious way to $\mathsf{rank}(D)$ and $\mathsf{rank}(H)$. For example, whatever the rank of D, $\mathsf{rank}(D, D) = 1$. Conversely, in the DIY algebra, if $X_0 \stackrel{\text{df}}{=} \Omega_0(c_1, c_1, c_1)$, $Y_0 \stackrel{\text{df}}{=} \Omega_0(c_1, c_1, c_1)$ and, for every $i \geq 1$,

$$X_i \stackrel{\text{df}}{=} \Omega_0(X_{i-1}, X_{i-1}, X_{i-1}) \quad \text{and} \quad Y_i \stackrel{\text{df}}{=} \Omega_0(Y_{i-1}, Y_{i-1}, Y_{i-1}),$$

then $\mathsf{rank}(X_n, Y_n) = 3 \cdot n + 3$, for $n \geq 0$, while $\mathsf{rank}(X_n) = \mathsf{rank}(Y_n) = 1$.

The next two exercises can be solved by a straightforward application of (transfinite) induction on the rank of pairs belonging to $\equiv$.

Exercise 7.3.8. Show that $\equiv$ is an equivalence relation.

Exercise 7.3.9. Show that if E and F are structurally equivalent static expressions, then no derivation for $E \equiv F$ can use the rules (7.45), (7.46), (7.47), and (7.49).

Properties. The structural similarity relation is closed in the domain of box expressions and preserves their types.

Theorem 7.3.2. Let D and H be expressions such that $D \equiv H$. Then $D \in \mathsf{Expr}^{\delta}$ if and only if $H \in \mathsf{Expr}^{\delta}$, for every $\delta \in \{\mathrm{e, d, x, s, wf, xcl, edir, xdir, box}\}$.

Proof. The proof proceeds by induction on $\mathsf{rank}(D, H)$. In the base step, $D \equiv H$ is directly derived from the first part of (7.42) or one of (7.43)—(7.47). If $D = H$, then there is nothing to prove; for (7.43) the result follows from

(7.35) and (Expr7); and for (7.44) from $D, H \in \mathsf{Expr}^{\mathsf{s}}$, (7.32) and (Expr6). We will now show that the result holds if $D \equiv H$ has been derived directly from (7.45) or (7.47) (the proof for (7.46) is similar to that for (7.45)).

Suppose that $D = \mathsf{op}_\Omega(\boldsymbol{D})$ and $H = \mathsf{op}_\Omega(\boldsymbol{H})$ are as in (7.45). If D is a box expression, then by the syntax (7.29), $\boldsymbol{D}$ is an Ω-tuple of box expressions such that the factorisation κ of $\boldsymbol{D}$ belongs to fact_Ω. Moreover, we have $\mu_{\mathrm{e}} \subseteq \kappa_{\mathrm{e}}$ and $\mu_{\mathrm{x}} \subseteq \kappa_{\mathrm{x}}$, and so ${}^\circ\Omega = {}^\bullet\mu_{\mathrm{e}} \cup \mu_{\mathrm{x}}^\bullet \subseteq {}^\bullet\kappa_{\mathrm{e}} \cup \kappa_{\mathrm{x}}^\bullet$. Hence, by Prop. 7.1.1(5), we have $\mu = \kappa$, $\mu_{\mathrm{d}} = \emptyset$, $\boldsymbol{H}$ is a tuple of static expressions, and D is an entry expression. Thus, by $\boldsymbol{H} \subseteq \mathsf{Expr}^{\mathsf{s}}$, H is also an entry expression. Conversely, if H is a box expression, then H must be an entry expression and, as before, $\boldsymbol{H}$ must be a tuple of static expressions. Thus D is also an entry expression. The part of the result for $\delta \in \{\mathrm{wf}, \mathrm{xcl}, \mathrm{edir}, \mathrm{xdir}\}$ follows from (Expr2)—(Expr6) and $D_v \in \{H_v, \overline{H_v}, \underline{H_v}\}$, for every $v \in T_\Omega$, and the fact that the Expr^δ's (for $\delta \in \{\mathrm{wf}, \mathrm{xcl}, \mathrm{edir}, \mathrm{xdir}\}$) are the largest sets satisfying (Expr1)—(Expr7).

Suppose that $D = \mathsf{op}_\Omega(\boldsymbol{D})$ and $H = \mathsf{op}_\Omega(\boldsymbol{H})$ are as in (7.47). Assume that D is a box expression (if H is a box expression, the argument is symmetric). Let ϕ be the factorisation of $\boldsymbol{D}$. By (7.48), since D is a box expression and $\boldsymbol{D}_{\mu_{\mathrm{e}}} = \overline{\boldsymbol{C}_{\mu_{\mathrm{e}}}}$ and $\boldsymbol{D}_{\mu_{\mathrm{x}}} = \underline{\boldsymbol{C}_{\mu_{\mathrm{x}}}}$, we have that $\boldsymbol{C}$ is an Ω-tuple of box expressions, $\mu_{\mathrm{e}} \subseteq \phi_{\mathrm{e}}$, $\mu_{\mathrm{x}} \subseteq \phi_{\mathrm{x}}$,

$$\boldsymbol{C}_{\mu_{\mathrm{e}}} \cup \boldsymbol{C}_{\mu_{\mathrm{x}}} \cup \boldsymbol{C}_{\mu_{\mathrm{s}}} \subseteq \mathsf{Expr}^{\mathsf{s}} \quad , \quad \boldsymbol{D}_{\mu_{\mathrm{e}}} \subseteq \overline{\mathsf{Expr}^{\mathsf{s}}} \quad , \quad \text{and} \quad \boldsymbol{D}_{\mu_{\mathrm{x}}} \subseteq \underline{\mathsf{Expr}^{\mathsf{s}}} \; .$$

(Note that, for an Ω-tuple of expressions $\boldsymbol{Z}$ and $V \subseteq T_\Omega$, we denote by $\boldsymbol{Z}_V$ the standard *restriction* of $\boldsymbol{Z}$ to V.)
We now observe that $\boldsymbol{H}$ is a tuple of box expressions. Indeed, suppose that H_v is not a box expression. Then, by (7.48) and $\boldsymbol{C}$ being a tuple of box expressions, $v \in \kappa_{\mathrm{e}} \cup \kappa_{\mathrm{x}}$. Thus, by $\boldsymbol{C}_{\mu_{\mathrm{e}}} \cup \boldsymbol{C}_{\mu_{\mathrm{x}}} \cup \boldsymbol{C}_{\mu_{\mathrm{s}}} \subseteq \mathsf{Expr}^{\mathsf{s}}$, it must be the case that $v \in \mu_{\mathrm{d}}$. On the other hand, since μ and κ are factorisations of the same marking, we have $({}^\bullet\mu_{\mathrm{e}} \cup \mu_{\mathrm{x}}^\bullet) \cap ({}^\bullet v \cup v^\bullet) \neq \emptyset$. But this contradicts 7.1.1(4). Thus, $\boldsymbol{H}$ is a tuple of box expressions whose factorisation, given by

$$\psi = \Big((\phi_{\mathrm{e}} \backslash \mu_{\mathrm{e}}) \cup \kappa_{\mathrm{e}} \, , \, \phi_{\mathrm{d}} \, , \, (\phi_{\mathrm{x}} \backslash \mu_{\mathrm{x}}) \cup \kappa_{\mathrm{x}} \, , \, \phi_{\mathrm{s}} \cup (\mu_{\mathrm{e}} \backslash \kappa_{\mathrm{e}}) \cup (\mu_{\mathrm{x}} \backslash \kappa_{\mathrm{x}}) \Big) \, ,$$

is a factorisation of the same marking as ϕ. Hence $H = \mathsf{op}_\Omega(\boldsymbol{H})$ is a box expression and $D \in \mathsf{Expr}^\delta$ if and only if $H \in \mathsf{Expr}^\delta$, for every $\delta \in \{\mathrm{e}, \mathrm{d}, \mathrm{x}, \mathrm{s}\}$. The part of the result for $\delta \in \{\mathrm{wf}, \mathrm{xcl}, \mathrm{edir}, \mathrm{xdir}\}$ follows from (Expr2)—(Expr6) and $D_v, H_v \in \{C_v, \overline{C_v}, \underline{C_v}\}$, for every $v \in T_\Omega$, and the fact that the Expr^δ's (for $\delta \in \{\mathrm{wf}, \mathrm{xcl}, \mathrm{edir}, \mathrm{xdir}\}$) are the largest sets satisfying (Expr1)—(Expr7).

In the inductive step, we consider four cases.

Case 1: $D \equiv H$ and $\mathsf{rank}(H, D) < \mathsf{rank}(D, H)$. Then, by the induction hypothesis, the result holds for $H \equiv D$. Clearly, it then also holds for $D \equiv H$.

Case 2: There is J such that $D \equiv J$, $J \equiv H$, and

$$\max\{\mathsf{rank}(D, J), \mathsf{rank}(J, H)\} < \mathsf{rank}(D, H) \, .$$

Then, by the induction hypothesis, the result holds for both $D \equiv J$ and $J \equiv H$. Clearly, it then also holds for $D \equiv H$.

Case 3: $D = \overline{D_0}$, $H = \overline{H_0}$, $D_0 \equiv H_0$, and $\mathsf{rank}(D_0, H_0) < \mathsf{rank}(D, H)$. Suppose D is a box expression (if H is a box expression, the argument is symmetric). Then D_0 is a static expression. Hence, by the induction hypothesis, H_0 is also a static expression. Thus $D, H \in \mathsf{Expr}^{\mathrm{e}}$.
If $D = \underline{D_0}$ and $H = \underline{H_0}$ the proof is similar.

Case 4: $D = \mathsf{op}_\Omega(\boldsymbol{D})$, $H = \mathsf{op}_\Omega(\boldsymbol{H})$, $\boldsymbol{D} \equiv \boldsymbol{H}$ and, for every $v \in T$, $\mathsf{rank}(D_v, H_v) < \mathsf{rank}(D, H)$. Suppose D is a box expression (if H is a box expression, the argument is symmetric). Then $\boldsymbol{D}$ is an Ω-tuple of box expressions and, by the induction hypothesis, $\boldsymbol{H}$ is an Ω-tuple of box expressions with the same factorisation as $\boldsymbol{D}$. Hence H is a box expression and $D \in \mathsf{Expr}^\delta$ if and only if $H \in \mathsf{Expr}^\delta$, for every $\delta \in \{\mathrm{e}, \mathrm{d}, \mathrm{x}, \mathrm{s}\}$.

Finally, we observe for the Cases 1–4, that the part of the result for $\delta \in \{\mathrm{wf}, \mathrm{xcl}, \mathrm{edir}, \mathrm{xdir}\}$ follows from (Expr2)—(Expr5), the induction hypothesis and the fact that the Expr^δ's (for $\delta \in \{\mathrm{wf}, \mathrm{xcl}, \mathrm{edir}, \mathrm{xdir}\}$) are the largest sets satisfying (Expr1)—(Expr7). □

Exercise 7.3.10. Show that for all static expressions, D and H, $D \equiv H$ if and only if either $D = H \in \mathsf{Const}^{\mathrm{s}}$, or there is $\Omega \in \mathsf{OpBox}$ and two Ω-tuples of static expressions, $\boldsymbol{D} \equiv \boldsymbol{H}$, such that $D = \mathsf{op}_\Omega(\boldsymbol{D})$ or $D \stackrel{\mathrm{df}}{=} \mathsf{op}_\Omega(\boldsymbol{D})$, and $H = \mathsf{op}_\Omega(\boldsymbol{H})$ or $H \stackrel{\mathrm{df}}{=} \mathsf{op}_\Omega(\boldsymbol{H})$.

Exercise 7.3.11. Show that if D and H are structurally equivalent expressions in Expr^δ, where $\delta \in \{\mathrm{box}, \mathrm{s}, \mathrm{wf}\}$, then every derivation for $D \equiv H$ can only use expressions in Expr^δ.

The structural equivalence on expressions and the structural similarity on tuples of boxes are related in the following way.

Corollary 7.3.1. If $\mathsf{op}_\Omega(\boldsymbol{D})$ and $\mathsf{op}_\Omega(\boldsymbol{H})$ are well-formed box expressions satisfying the conditions in the rule for equivalent factorisations, (7.47), extended to factorisations of ${}^\circ\Omega$ and Ω° (i.e., we allow μ and κ to be factorisations of the entry or exit marking of Ω), then $\mathsf{box}(\boldsymbol{D}) \equiv_\Omega \mathsf{box}(\boldsymbol{H})$.

Proof. An inspection of the proof of Thm. 7.3.2 reveals that the factorisations of $\boldsymbol{D}$ and $\boldsymbol{H}$ are factorisations of the same marking of Ω. Thus, by Thm. 7.3.1, the factorisations of $\mathsf{box}(\boldsymbol{D})$ and $\mathsf{box}(\boldsymbol{H})$ are factorisations of the same marking of Ω. Moreover, by Prop. 7.3.1(1) and $\lfloor \boldsymbol{D} \rfloor = \lfloor \boldsymbol{C} \rfloor = \lfloor \boldsymbol{H} \rfloor$, we have $\lfloor \mathsf{box}(\boldsymbol{D}) \rfloor = \lfloor \mathsf{box}(\boldsymbol{H}) \rfloor$. And, if $v \in \mu_{\mathrm{d}} = \kappa_{\mathrm{d}}$, then $D_v = C_v = H_v$. Hence $\mathsf{box}(\boldsymbol{D}) \equiv_\Omega \mathsf{box}(\boldsymbol{H})$. □

A central property of the structural similarity relation is preceded by an auxiliary lemma.

Lemma 7.3.1. If D is an entry or exit expression, then $D \equiv \overline{\lfloor D \rfloor}$ or $D \equiv \underline{\lfloor D \rfloor}$, respectively.

Proof. We proceed by induction on $\mathsf{rank}(D)$. In the base step, D is a dynamic constant, and $D \equiv \overline{\lfloor D \rfloor}$ or $D \equiv \underline{\lfloor D \rfloor}$ follows from (7.35), (7.36), and (7.43). In the inductive step we consider four cases.

Case 1: $D = \overline{E}$ and E is a static expression. Then $E = \lfloor D \rfloor$ and $D = \overline{E} = \overline{\lfloor D \rfloor} \equiv \overline{\lfloor D \rfloor}$ since $\equiv$ is reflexive.

Case 2: $D = \underline{E}$ and E is a static expression. We then proceed similarly as in Case 1.

Case 3: $D = \mathsf{op}_\Omega(\boldsymbol{D})$ and the factorisation μ of the Ω-tuple of box expressions $\boldsymbol{D}$ is a factorisation of the entry marking of Ω and for each $v \in T_\Omega$, $\mathsf{rank}(D_v) < \mathsf{rank}(D)$. By the induction hypothesis and (7.50), we have $\mathsf{op}_\Omega(\boldsymbol{D}) \equiv \mathsf{op}_\Omega(\boldsymbol{H})$ where, for every $v \in T_\Omega$,

$$H_v = \begin{cases} \overline{\lfloor D_v \rfloor} & \text{if } \ v \in \mu_{\mathsf{e}} \\ \underline{\lfloor D_v \rfloor} & \text{if } \ v \in \mu_{\mathsf{x}} \\ \lfloor D_v \rfloor & \text{otherwise}\ . \end{cases}$$

Notice that $\lfloor D_v \rfloor = D_v$ for $v \notin \mu_{\mathsf{e}} \cup \mu_{\mathsf{x}}$ since $\mu_{\mathsf{d}} = \emptyset$ as μ is a factorisation of ${}^\circ\Omega$. We may now apply (7.45) to conclude that

$$\mathsf{op}_\Omega(\boldsymbol{H}) \equiv \overline{\mathsf{op}_\Omega(\lfloor \boldsymbol{D} \rfloor)}$$

which, together with (7.34) and (7.42), yields $D \equiv \overline{\lfloor D \rfloor}$.

Case 4: $D = \mathsf{op}_\Omega(\boldsymbol{D})$ and the factorisation of the Ω-tuple of box expressions $\boldsymbol{D}$ is a factorisation of the exit marking of Ω. We then proceed similarly as in Case 3. □

Theorem 7.3.3. Let D and H be well-formed box expressions. Then $D \equiv H$ if and only if $\lfloor D \rfloor \equiv \lfloor H \rfloor$ and $\mathsf{box}(D) = \mathsf{box}(H)$.

Proof. ($\Longrightarrow$) The proof proceeds by induction on $\mathsf{rank}(D, H)$. In the base step we consider six cases.

Case 1: $D = H$. Then $\mathsf{box}(D) = \mathsf{box}(H)$ and $\lfloor D \rfloor \equiv \lfloor H \rfloor$ follows from the reflexivity of $\equiv$.

Case 2: D and H are flat expressions and $\mathsf{box}(D) = \mathsf{box}(H)$. Then $\lfloor D \rfloor$ and $\lfloor H \rfloor$ are static constants and, by Prop. 7.3.1(1), $\mathsf{box}(\lfloor D \rfloor) = \mathsf{box}(\lfloor H \rfloor)$. Hence, by (7.36), $\lfloor D \rfloor \equiv \lfloor H \rfloor$.

Case 3: $D = X$, $H = \mathsf{op}_\Omega(\boldsymbol{L})$, and $X \stackrel{\mathrm{df}}{=} \mathsf{op}_\Omega(\boldsymbol{L})$. Then $D \equiv H$, $\lfloor D \rfloor = D$, and $\lfloor H \rfloor = H$, and so $\lfloor D \rfloor \equiv \lfloor H \rfloor$. Moreover, $\mathsf{box}(D) = \mathsf{box}(H)$ follows from (7.38) and (7.40).

Case 4: $D = \mathsf{op}_\Omega(\boldsymbol{D})$ and $H = \overline{\mathsf{op}_\Omega(\boldsymbol{H})}$ where $\mathsf{op}_\Omega(\boldsymbol{D})$ and $\mathsf{op}_\Omega(\boldsymbol{H})$ are as in (7.45). Then $\lfloor D \rfloor = \mathsf{op}_\Omega(\boldsymbol{H}) = \lfloor H \rfloor$, so $\lfloor D \rfloor \equiv \lfloor H \rfloor$ follows from the reflexivity of $\equiv$. Also, H is an entry expression, so by Thm. 7.3.2, D is also an entry expression. Moreover, by Prop. 7.3.1(1) and $\lfloor D \rfloor = \lfloor H \rfloor$,

$$\lfloor \mathsf{box}(D) \rfloor = \mathsf{box}(\lfloor D \rfloor) = \mathsf{box}(\lfloor H \rfloor) = \lfloor \mathsf{box}(H) \rfloor\ .$$

Hence $\mathsf{box}(D) = \mathsf{box}(H) = \overline{\mathsf{box}(\mathsf{op}_\Omega(\boldsymbol{H}))}$.

Case 5: $D = \mathsf{op}_\Omega(\boldsymbol{D})$ and $H = \underline{\mathsf{op}_\Omega(\boldsymbol{H})}$ where $\mathsf{op}_\Omega(\boldsymbol{D})$ and $\mathsf{op}_\Omega(\boldsymbol{H})$ are as in (7.46). Then we proceed similarly as in Case 4.

Case 6: $D = \mathsf{op}_\Omega(\boldsymbol{D})$ and $H = \mathsf{op}_\Omega(\boldsymbol{H})$ where $\mathsf{op}_\Omega(\boldsymbol{D})$ and $\mathsf{op}_\Omega(\boldsymbol{H})$ are as in (7.47). Then $\lfloor D_v \rfloor = \lfloor H_v \rfloor$ for each v, so $\lfloor D \rfloor \equiv \lfloor H \rfloor$ follows from the reflexivity of $\equiv$. Moreover, by Cor. 7.3.1, $\mathsf{box}(\boldsymbol{D}) \equiv_\Omega \mathsf{box}(\boldsymbol{H})$ which, by (7.40) and Prop. 7.2.1, means that $\mathsf{box}(D) = \mathsf{box}(H)$.

In the inductive step, we consider four cases.

Case 1: $H \equiv D$ and $\mathsf{rank}(H, D) < \mathsf{rank}(D, H)$. Then, by the induction hypothesis, $\lfloor H \rfloor \equiv \lfloor D \rfloor$ and $\mathsf{box}(H) = \mathsf{box}(D)$. Hence $\lfloor D \rfloor \equiv \lfloor H \rfloor$, by the symmetry of $\equiv$.

Case 2: There is J such that $D \equiv J$, $J \equiv H$, and

$$\max\{\mathsf{rank}(D, J), \mathsf{rank}(J, H)\} < \mathsf{rank}(D, H) .$$

Then, by the induction hypothesis, $\lfloor D \rfloor \equiv \lfloor J \rfloor$ and $\lfloor J \rfloor \equiv \lfloor H \rfloor$ and $\mathsf{box}(D) = \mathsf{box}(J) = \mathsf{box}(H)$. Hence $\lfloor D \rfloor \equiv \lfloor H \rfloor$, by the transitivity of $\equiv$.

Case 3: $D = \overline{D_0}$, $H = \overline{H_0}$, $D_0 \equiv H_0$, and $\mathsf{rank}(D_0, H_0) < \mathsf{rank}(D, H)$. By the induction hypothesis, $\lfloor D_0 \rfloor \equiv \lfloor H_0 \rfloor$ and $\mathsf{box}(D_0) = \mathsf{box}(H_0)$. Hence $\lfloor D \rfloor \equiv \lfloor H \rfloor$ since $\lfloor D \rfloor = \lfloor D_0 \rfloor$ and $\lfloor H \rfloor = \lfloor H_0 \rfloor$. Moreover,

$$\mathsf{box}(D) = \overline{\mathsf{box}(D_0)} = \overline{\mathsf{box}(H_0)} = \mathsf{box}(H) .$$

If $D = \underline{D_0}$ and $H = \underline{H_0}$, then the proof is similar.

Case 4: $D = \mathsf{op}_\Omega(\boldsymbol{D})$, $H = \mathsf{op}_\Omega(\boldsymbol{H})$, $\boldsymbol{D} \equiv \boldsymbol{H}$, and, for every $v \in T$, $\mathsf{rank}(D_v, H_v) < \mathsf{rank}(D, H)$. Then, by the induction hypothesis, $\lfloor \boldsymbol{D} \rfloor \equiv \lfloor \boldsymbol{H} \rfloor$ and $\mathsf{box}(\boldsymbol{D}) = \mathsf{box}(\boldsymbol{H})$. Hence, by (7.34) and (7.50),

$$\lfloor D \rfloor = \mathsf{op}_\Omega(\lfloor \boldsymbol{D} \rfloor) \equiv \mathsf{op}_\Omega(\lfloor \boldsymbol{H} \rfloor) = \lfloor H \rfloor .$$

Finally, we obtain

$$\mathsf{box}(D) = \mathsf{box}(\mathsf{op}_\Omega(\boldsymbol{D})) = \mathsf{box}(\mathsf{op}_\Omega(\boldsymbol{H})) = \mathsf{box}(H)$$

which follows from $\mathsf{box}(\boldsymbol{H}) = \mathsf{box}(\boldsymbol{D})$ and (7.40).

($\Longleftarrow$) By Thm. 7.3.1 and $\mathsf{box}(D) = \mathsf{box}(H)$, it suffices to consider the following four cases.

Case 1: $D, H \in \mathsf{Expr}^{\mathsf{s}}$. Then $D = \lfloor D \rfloor \equiv \lfloor H \rfloor = H$.

Case 2: $D, H \in \mathsf{Expr}^{\mathsf{e}}$. Then, by Lemma 7.3.1, $D \equiv \overline{\lfloor D \rfloor}$ and $H \equiv \overline{\lfloor H \rfloor}$. Hence, by (7.49) and the symmetry of $\equiv$, we have

$$D \equiv \overline{\lfloor D \rfloor} \equiv \overline{\lfloor H \rfloor} \equiv H .$$

Thus, by transitivity of $\equiv$, $D \equiv H$.

Case 3: $D, H \in \mathsf{Expr}^{\mathsf{x}}$. Then we proceed similarly as in Case 2.

Case 4: $D, H \in \mathsf{Expr}^{\mathsf{d}} \backslash (\mathsf{Expr}^{\mathsf{e}} \cup \mathsf{Expr}^{\mathsf{x}})$. Then we proceed by (transfinite) induction on $\max\{\mathsf{rank}(D), \mathsf{rank}(H)\}$. In the base step, D and H are dynamic constants, and $D \equiv H$ follows from (7.43). In the inductive step, we have $D = \mathsf{op}_\Omega(\boldsymbol{D})$ and $H = \mathsf{op}_\Omega(\boldsymbol{H})$ where $\lfloor \boldsymbol{D} \rfloor \equiv \lfloor \boldsymbol{H} \rfloor$ and

$$\max\{\mathsf{rank}(D_v), \mathsf{rank}(H_v)\} < \max\{\mathsf{rank}(D), \mathsf{rank}(H)\} ,$$

for every $v \in T_\Omega$. Since $\mathsf{box}(\lfloor \boldsymbol{D} \rfloor) = \mathsf{box}(\lfloor \boldsymbol{H} \rfloor)$ and $\mathsf{box}(D) = \mathsf{box}(H)$, by Prop. 7.2.1, $\mathsf{box}(\boldsymbol{D}) \equiv_\Omega \mathsf{box}(\boldsymbol{H})$. Let μ and κ be factorisations of respectively $\mathsf{box}(\boldsymbol{D})$ and $\mathsf{box}(\boldsymbol{H})$, and so $\mu_\mathrm{d} = \kappa_\mathrm{d}$. (Then, by Thm. 7.3.1(2), μ and κ are factorisations of, respectively, $\boldsymbol{D}$ and $\boldsymbol{H}$.) We have $\mathsf{box}(\boldsymbol{D}_{\mu_\mathrm{d}}) = \mathsf{box}(\boldsymbol{H}_{\mu_\mathrm{d}})$; hence, by $\lfloor \boldsymbol{D}_{\mu_\mathrm{d}} \rfloor = \lfloor \boldsymbol{H}_{\mu_\mathrm{d}} \rfloor$ and the induction hypothesis, $\boldsymbol{D}_{\mu_\mathrm{d}} \equiv \boldsymbol{H}_{\mu_\mathrm{d}}$. Moreover, by Lemma 7.3.1,

$$\boldsymbol{D}_{\mu_\mathrm{e}} \equiv \overline{\lfloor \boldsymbol{D}_{\mu_\mathrm{e}} \rfloor}\,,\quad \boldsymbol{D}_{\mu_\mathrm{x}} \equiv \underline{\lfloor \boldsymbol{D}_{\mu_\mathrm{x}} \rfloor}\,,\quad \boldsymbol{H}_{\kappa_\mathrm{e}} \equiv \overline{\lfloor \boldsymbol{H}_{\kappa_\mathrm{e}} \rfloor}\,,\quad \text{and } \boldsymbol{H}_{\kappa_\mathrm{x}} \equiv \underline{\lfloor \boldsymbol{H}_{\kappa_\mathrm{x}} \rfloor}.$$

Then one can show $D \equiv H$ by applying (7.47), (7.49), and (7.50). □

We have seen that $\equiv$ is a sound equivalence notion from the point of view of the denotational semantics. It is also complete in the sense that $\mathsf{box}(D) = \mathsf{box}(H)$ implies $D \equiv H$ provided that $\lfloor D \rfloor \equiv \lfloor H \rfloor$. One might ask whether the latter condition could be left out. The answer is no and in the DIY algebra a counterexample is provided by two expressions, $D = \overline{X}$ and $H = \overline{Y}$, where X and Y are variables defined by $X \stackrel{\mathrm{df}}{=} \Omega_0(c_1, c_1, X)$ and $Y \stackrel{\mathrm{df}}{=} \Omega_0(c_1, c_1, Y)$.

Thus the structural equivalence on expressions is complete for all those expressions whose underlying static expressions are structurally equivalent. And, as we shall see later, this will be enough for our purposes since all expressions derivable from a box expression through the rules of operational semantics will satisfy this property.

Running Example. The DIY algebra gives rise to five specific rules of the structural equivalence relation (we omit here their symmetric, hence redundant, counterparts). The first two are derived from (7.43):

$$\boxed{\overline{c_2} \equiv c_{21} \qquad \underline{c_2} \equiv c_{23}}$$

The next two are derived from respectively (7.45) and (7.46):

$$\boxed{\mathsf{op}_{\Omega_0}(\overline{E}, \overline{F}, G) \equiv \overline{\mathsf{op}_{\Omega_0}(E, F, G)} \qquad \mathsf{op}_{\Omega_0}(E, F, \underline{G}) \equiv \underline{\mathsf{op}_{\Omega_0}(E, F, G)}}$$

Finally, there is a single instance of (7.47):

$$\boxed{\mathsf{op}_{\Omega_0}(\underline{E}, \underline{F}, G) \equiv \mathsf{op}_{\Omega_0}(E, F, \overline{G})}$$

Using these rules, one can derive $\mathsf{op}_{\Omega_0}(\underline{c_1}, c_{23}, c_3) \equiv \mathsf{op}_{\Omega_0}(c_1, c_2, \overline{c_3})$, as shown below:

$$\mathsf{op}_{\Omega_0}(\underline{c_1}, c_{23}, c_3) \equiv \mathsf{op}_{\Omega_0}(c_1, c_2, \overline{c_3})$$

$$\overbrace{\mathsf{op}_{\Omega_0}(\underline{c_1}, c_{23}, c_3) \equiv \mathsf{op}_{\Omega_0}(\underline{c_1}, \underline{c_2}, c_3) \qquad \mathsf{op}_{\Omega_0}(\underline{c_1}, \underline{c_2}, c_3) \equiv \mathsf{op}_{\Omega_0}(c_1, c_2, \overline{c_3})}$$

$$\overbrace{\underline{c_1} \equiv \underline{c_1} \qquad c_{23} \equiv \underline{c_2} \qquad c_3 \equiv c_3}$$

7.3.5 Transition-based Operational Semantics

In developing the operational semantics of the box algebra, we will follow the following steps: first (in this section), we introduce operational rules based on transitions of nets which provide the denotational semantics of box expressions. Based on these, we will formulate our key consistency result (in Sect. 7.3.6). And, finally, we will formulate the label-based rules (Sect. 7.3.7) and partial order semantics (Sect. 7.3.8), together with derived consistency results. Note that this deviates slightly from the path taken in the treatment of PBC, where the label-based rules were given and the other two were omitted. The transition-based rules for PBC will be given in Chap. 8.

Transition Trees. Consider the set of all transition trees in the boxes derived through the box mapping:

$$\mathsf{T}^{\mathsf{box}}_{\mathsf{tree}} = \bigcup_{D \in \mathsf{Expr}^{\mathsf{wf}}} \mathcal{T}_{\mathsf{box}(D)} =_{\mathsf{prop.7.3.1(1)}} \bigcup_{E \in \mathsf{Expr}^{\mathsf{s}} \cap \mathsf{Expr}^{\mathsf{wf}}} \mathcal{T}_{\mathsf{box}(E)} \, .$$

Every $t \in \mathsf{T}^{\mathsf{box}}_{\mathsf{tree}}$ has a unique label, $\mathsf{lab}(t)$, in all boxes associated with box expressions in which it occurs. More precisely, if $t \in \mathsf{T}_{\mathsf{root}}$, then this follows from (7.27) and (7.36), and if t has the form $t = (v, \alpha) \lhd R$, then by the definition of net refinement which underpins the semantical mapping box, $\mathsf{lab}(t) = \alpha$. To see that the same transition may belong to different boxes in $\mathsf{box}(\mathsf{Expr}^{\mathsf{s}})$, consider $E = \mathsf{op}_{\Omega_0}(c_1, c_2, c_2)$ and $F = \mathsf{op}_{\Omega_0}(c_1, c_2, c_3)$ in DIY. Then the transition $t = (v_1, f) \lhd \{t_{11}, t_{12}\}$ belongs to both $\mathsf{box}(E)$ and $\mathsf{box}(F)$, yet $\mathsf{box}(E) \neq \mathsf{box}(F)$.

Exercise 7.3.12. Show that in the DIY algebra, each transition tree in $\mathsf{T}^{\mathsf{box}}_{\mathsf{tree}}$ has one of the following four forms:

$$\begin{array}{c}
(v_{n_1}, a_{ij}) \lhd \cdots \lhd (v_{n_k}, a_{ij}) \lhd t_{ij} \\
(v_{m_1}, a) \lhd \cdots \lhd (v_{m_{k_1}}, a) \lhd t_{11} \\
(v_{l_1}, b) \lhd \cdots \lhd (v_{l_{k_2}}, b) \lhd t_{12} \\
(v_{n_1}, f) \lhd \cdots \lhd (v_{n_k}, f) \lhd (v_1, f) \lhd \left\{ \begin{array}{c} (v_{m_1}, a) \lhd \cdots \lhd (v_{m_{k_1}}, a) \lhd t_{11} \\ (v_{l_1}, b) \lhd \cdots \lhd (v_{l_{k_2}}, b) \lhd t_{12} \end{array} \right\}
\end{array}$$

where $k, k_1, k_2 \geq 0$, $t_{ij} \in \{t_{21}, t_{22}, t_{31}\}$, $v_{n_i} \in \{v_1, v_2, v_3\}$, $v_{m_i}, v_{l_i} \in \{v_2, v_3\}$, $a_{21} = c$, $a_{22} = d$, and $a_{31} = e$.

SOS Semantics Rules. The operational semantics we will now define has moves of the form $D \xrightarrow{U} H$ such that D and H are box expressions and $U \in \mathsf{tlab}_{\mathsf{sr}}$, where $\mathsf{tlab}_{\mathsf{sr}}$ is the set of all finite subsets of $\mathsf{T}^{\mathsf{box}}_{\mathsf{tree}} \cup \{\mathsf{skip}, \mathsf{redo}\}$. The idea here is that U is a valid step for the boxes associated with D and H, after augmenting them with the skip and redo transitions. We will denote, for every such set,

$$\mathsf{lab}(U) = \sum_{t \in U} \left\{ \mathsf{lab}(t) \right\} ,$$

assuming that $\mathsf{lab}(\mathsf{skip}) = \mathsf{skip}$ and $\mathsf{lab}(\mathsf{redo}) = \mathsf{redo}$; as before, $\mathsf{lab}(U)$ is a finite multiset of labels. The derivation system is defined in four stages. First, we define the rules for skip and redo, next for the inaction rules, then we treat the flat expressions and, finally, introduce the derivation rule for each connective op_Ω. Formally, we define a ternary relation $\longrightarrow$ which is the least relation comprising all $(D, U, H) \in \mathsf{Expr} \times \mathsf{tlab_{sr}} \times \mathsf{Expr}$ such that (7.51)—(7.55) below hold. Notice that we use $D \xrightarrow{U} H$ to denote $(D, U, H) \in\longrightarrow$.

skip and redo. There are two basic rules governing the applications of the moves involving skip and redo. For every static expression E,

$$\overline{E} \xrightarrow{\{\mathsf{skip}\}} \underline{E} \qquad \underline{E} \xrightarrow{\{\mathsf{redo}\}} \overline{E} \tag{7.51}$$

Inaction Rules. There is a single basic inaction rule,

$$\frac{D \equiv H}{D \xrightarrow{\emptyset} H} \tag{7.52}$$

where D and H are box expressions. Moreover, an empty move can be absorbed by the move which precedes or follows it:

$$\frac{D \xrightarrow{\emptyset} J \xrightarrow{U} H}{D \xrightarrow{U} H} \qquad \frac{D \xrightarrow{U} J \xrightarrow{\emptyset} H}{D \xrightarrow{U} H} \tag{7.53}$$

where D, J, and H are expressions. The inaction rules render the view, adopted in Chaps. 3 and 4, that empty moves do not modify the state of a system, but rather change the view on it.

Axioms for Flat Expressions. For all flat expressions D and H and every $U \in \mathsf{tlab_{sr}}$ such that $\mathsf{box}(D)\ [U\rangle\ \mathsf{box}(H)$,

$$D \xrightarrow{U} H \tag{7.54}$$

Inference Rules for Operators. For every SOS-operator box Ω in OpBox, there is an inference rule:

$$\frac{\forall v \in T_\Omega : D_v \xrightarrow{U_v^1 \uplus \cdots \uplus U_v^{k_v}} H_v}{\mathsf{op}_\Omega(\boldsymbol{D}) \xrightarrow{U} \mathsf{op}_\Omega(\boldsymbol{H})}\ \mathit{cond} \tag{7.55}$$

where $\boldsymbol{D}$ and $\boldsymbol{H}$ are Ω-tuples of expressions,

$$U = \bigcup_{v \in T_\Omega} \{(v, \alpha_v^1) \lhd U_v^1, \ldots, (v, \alpha_v^{k_v}) \lhd U_v^{k_v}\} ,$$

and *cond* is

$$\left(\sum_{v\in T_\Omega} k_v < \infty\right) \wedge \left(\forall v \in T_\Omega\ \forall i \le k_v : \left(\mathsf{lab}(U_v^i), \alpha_v^i\right) \in \lambda_\Omega(v)\right) .$$

The finiteness of U means that finitely many k_v's are nonzero. Notice that for the other ones we do not require that $D_v = H_v$ but only that

$$D_v \xrightarrow{\emptyset} H_v .$$

In such a case, it will follow from Cor. 7.3.2 that $D_v \equiv H_v$.

Since **skip** and **redo** never occur in any $\lambda_\Omega(v)$, the last rule will never be applied if any of the sets U_v^i contains **skip** or **redo**. Similarly, it will never generate any of these two special labels.

Exercise 7.3.13. Show that if $D \xrightarrow{U} H$ and $\mathsf{skip} \in U$, then $U = \{\mathsf{skip}\}$, and if $D \xrightarrow{U} H$ and $\mathsf{redo} \in U$, then $U = \{\mathsf{redo}\}$.

Infinite Operators. The meaning of the operational semantics rules is again clear if all boxes in OpBox have finite transition sets; in general, however, we need to treat infinite operator boxes as well. Following what is now an established practice, we consider the domain

$$(2^{\mathsf{ExprTrees}\times\mathsf{tlab}_{sr}\times\mathsf{ExprTrees}}, \subseteq)$$

which forms a complete lattice with the $\cap$ and $\cup$ operations. And the inference rules (7.51)—(7.55) define a monotonic mapping

$$opsem : 2^{\mathsf{ExprTrees}\times\mathsf{tlab}_{sr}\times\mathsf{ExprTrees}} \to 2^{\mathsf{ExprTrees}\times\mathsf{tlab}_{sr}\times\mathsf{ExprTrees}}$$

whose least fixed point, $\longrightarrow$, is equal to $opsem^{\alpha''}(\emptyset)$, for some ordinal α''. We also define, for every $(D, U, H) \in \longrightarrow$, its *rank*, denoted by $\mathsf{rank}(D, U, H)$, defined as the smallest ordinal β such that $(D, U, H) \in opsem^{\beta}(\emptyset)$. Thus we are in a position to apply (transfinite) induction on the rank of the triples in $\longrightarrow$. If all operator boxes in OpBox have finite transition sets, the standard induction can be used.

Corollary 7.3.2. If D and H are expressions such that $D \xrightarrow{\emptyset} H$, then $D \equiv H$.

Proof. Follows by a straightforward induction on $\mathsf{rank}(D, \emptyset, H)$. □

Factorisability Revisited. We now briefly return to the relevance of the factorisability property. Consider the nonfactorisable operator box N shown in Fig. 7.1. To see that this operator does not have a complete operational semantics in the above sense, try defining a quaternary operator $\mathsf{op}_N(E_1, E_2, E_3, E_4)$ from it, with four arguments (where $E_1 - E_4$ are supposed to describe the same behaviour as transitions $v_1 - v_4$ of the operator box) and the following set of equations, obtained from the generic rules in Sect. 7.3.4:

$$\overline{\mathsf{op}_N(E_1,E_2,E_3,E_4)} \equiv \mathsf{op}_N(\overline{E_1},\overline{E_2},E_3,E_4)$$
$$\mathsf{op}_N(\underline{E_1},\overline{E_2},E_3,E_4) \equiv \mathsf{op}_N(E_1,\overline{E_2},\overline{E_3},E_4)$$
$$\mathsf{op}_N(\underline{E_1},\underline{E_2},E_3,E_4) \equiv \mathsf{op}_N(E_1,E_2,\overline{E_3},\overline{E_4})$$
$$\mathsf{op}_N(E_1,\underline{E_2},\underline{E_3},E_4) \equiv \mathsf{op}_N(E_1,E_2,\underline{E_3},\overline{E_4})$$
$$\mathsf{op}_N(E_1,E_2,\underline{E_3},\underline{E_4}) \equiv \underline{\mathsf{op}_N(E_1,E_2,E_3,E_4)}\ .$$

At first sight, the equations seem to describe the control flow within the operator box. However, if we consider

$$E = \mathsf{op}_N\Big(a\ ,\ (b;b')\ ,\ (c;c')\ ,\ d\Big)\ ,$$

then the box corresponding to the expression $\overline{E}$ can execute the sequence of labels, $abcb'd$, but this cannot be derived from the generic evolution rules, and the associated labelled evolutions, using the above set of equations. Essentially, the third equation can never be applied in this case.

Properties. A move of the operational semantics transforms a box expression into another box expression with structurally equivalent underlying static expression, and the move generated is a valid step for the corresponding boxes. We interpret this result as one establishing the soundness of the operational semantics of box expressions.

Theorem 7.3.4. Let D be a well-formed box expression and $D \xrightarrow{U} H$. Then H is a well-formed box expression such that $\mathsf{box}(D)_{\mathsf{sr}}\ [U\rangle\ \mathsf{box}(H)_{\mathsf{sr}}$ and $\lfloor D \rfloor \equiv \lfloor H \rfloor$.

Proof. The proof proceeds by induction on $\mathsf{rank}(D,U,H)$. In the base step, we consider three cases.

Case 1: $D = \overline{E} \xrightarrow{\{\mathsf{skip}\}} \underline{E} = H$ or $D = \underline{E} \xrightarrow{\{\mathsf{redo}\}} \overline{E} = H$ where E is a well-formed static expression. Then both D and H are well-formed box expressions, $\lfloor D \rfloor = E = \lfloor H \rfloor$ and

$$\mathsf{box}(\overline{E})_{\mathsf{sr}}\ [\{\mathsf{skip}\}\rangle\ \mathsf{box}(\underline{E})_{\mathsf{sr}} \quad \text{or} \quad \mathsf{box}(\underline{E})_{\mathsf{sr}}\ [\{\mathsf{redo}\}\rangle\ \mathsf{box}(\overline{E})_{\mathsf{sr}}$$

follows from $\mathsf{box}(\overline{E})_{\mathsf{sr}} = \overline{\mathsf{box}(E)_{\mathsf{sr}}}$ and $\mathsf{box}(\underline{E})_{\mathsf{sr}} = \underline{\mathsf{box}(E)_{\mathsf{sr}}}$ and the definition of $(.)_{\mathsf{sr}}$.

Case 2: $D \equiv H$ and $U = \emptyset$. Then, by Thms. 7.3.2 and 7.3.3, H is a well-formed box expression such that $\lfloor D \rfloor \equiv \lfloor H \rfloor$ and $\mathsf{box}(D) = \mathsf{box}(H)$. Hence $\mathsf{box}(D)_{\mathsf{sr}}\ [\emptyset\rangle\ \mathsf{box}(H)_{\mathsf{sr}}$.

Case 3: D and H are flat expressions and $\mathsf{box}(D)\ [U\rangle\ \mathsf{box}(H)$. Then, by (7.36), $\mathsf{box}(\lfloor D \rfloor) = \mathsf{box}(\lfloor H \rfloor)$. Hence, by (7.43), $\lfloor D \rfloor \equiv \lfloor H \rfloor$. Clearly, $\mathsf{box}(D)_{\mathsf{sr}}\ [U\rangle\ \mathsf{box}(H)_{\mathsf{sr}}$ also holds.

In the inductive step we consider two cases.

Case 1: There is J such that $D \xrightarrow{\emptyset} J \xrightarrow{U} H$ and

$$\max\{\mathsf{rank}(D,\emptyset,J), \mathsf{rank}(J,U,H)\} < \mathsf{rank}(D,U,H)\ .$$

(If $D \xrightarrow{U} J \xrightarrow{\emptyset} H$, then we proceed similarly.) By the induction hypothesis, both J and H are well-formed box expressions such that $\lfloor D \rfloor \equiv \lfloor J \rfloor \equiv \lfloor H \rfloor$ and $\mathsf{box}(D)_{\mathsf{sr}}\ [\emptyset\rangle\ \mathsf{box}(J)_{\mathsf{sr}}\ [U\rangle\ \mathsf{box}(H)_{\mathsf{sr}}$. Hence, by the transitivity of $\equiv$ and $\mathsf{box}(D) = \mathsf{box}(J)$, $\lfloor D \rfloor \equiv \lfloor H \rfloor$, and $\mathsf{box}(D)_{\mathsf{sr}}\ [U\rangle\ \mathsf{box}(H)_{\mathsf{sr}}$.

Case 2: $D = \mathsf{op}_\Omega(\boldsymbol{D})$ and $H = \mathsf{op}_\Omega(\boldsymbol{H})$ where $\mathsf{op}_\Omega(\boldsymbol{D})$, $\mathsf{op}_\Omega(\boldsymbol{H})$, and U are as in (7.55); moreover, for every $v \in T_\Omega$, $\mathsf{rank}(D_v, U_v, H_v) < \mathsf{rank}(D, U, H)$. From D being a well-formed box expression (which implies that $\boldsymbol{D} \subseteq \mathsf{Expr}^{\mathsf{wf}}$), the induction hypothesis (which implies that, for each v,

$$\lfloor D_v \rfloor \equiv \lfloor H_v \rfloor \quad \text{and} \quad \mathsf{box}(D_v)\ [U_v^1 \uplus \ldots \uplus U_v^{k_v}\rangle\ \mathsf{box}(H_v)$$

since no U_v^i can contain $\mathsf{skip}, \mathsf{redo}$), and Thm. 7.3.1, it follows that $\boldsymbol{D}$ and $\boldsymbol{H}$ are Ω-tuples of well-formed box expressions, $\mathsf{box}(\boldsymbol{D}) \in \mathsf{dom}_\Omega$, and

$$\Big(\Omega : \mathsf{box}(\boldsymbol{D})\Big) \xrightarrow{U} \Big(\Omega : \mathsf{box}(\boldsymbol{H})\Big) .$$

Hence, by Thm. 7.2.1, $\mathsf{box}(\boldsymbol{H}) \in \mathsf{dom}_\Omega$ and $\Omega(\mathsf{box}(\boldsymbol{D}))\ [U\rangle\ \Omega(\mathsf{box}(\boldsymbol{H}))$; notice that the finiteness of the set U is needed to reach this conclusion. Moreover, by $\mathsf{box}(\boldsymbol{H}) \in \mathsf{dom}_\Omega$ and Thm. 7.3.1, H is a box expression which is well formed since D is well formed and $\lfloor \boldsymbol{D} \rfloor \equiv \lfloor \boldsymbol{H} \rfloor$ and Thm. 7.3.2 holds. Hence

$$\mathsf{box}(D) = \mathsf{box}(\mathsf{op}_\Omega(\boldsymbol{D}))\ [U\rangle\ \mathsf{box}(\mathsf{op}_\Omega(\boldsymbol{H})) = \mathsf{box}(H) .$$

Finally, we obtain

$$\lfloor D \rfloor = \mathsf{op}_\Omega(\lfloor \boldsymbol{D} \rfloor) \equiv \mathsf{op}_\Omega(\lfloor \boldsymbol{H} \rfloor) = \lfloor H \rfloor$$

by $\lfloor \boldsymbol{D} \rfloor \equiv \lfloor \boldsymbol{H} \rfloor$ and (7.50). □

We have seen that if $D \xrightarrow{U} H$, then $D \in \mathsf{Expr}^{\mathsf{wf}}$ implies $H \in \mathsf{Expr}^{\mathsf{wf}}$. It is, however, not always the case that $H \in \mathsf{Expr}^{\mathsf{wf}}$ implies $D \in \mathsf{Expr}^{\mathsf{wf}}$.

Exercise 7.3.14. Show this.
Hint: SOS-operator boxes are safe under the forward execution, but not necessarily under the backward execution, and not all transitions are reversible; see also the appendix.

The next result states completeness of the operational semantics of box expressions.

Theorem 7.3.5. Let D be a well-formed box expression and Σ be a net such that $\mathsf{box}(D)_{\mathsf{sr}}\ [U\rangle\ \Sigma_{\mathsf{sr}}$. Then there is a well-formed box expression H such that $\mathsf{box}(H) = \Sigma$ and $D \xrightarrow{U} H$.

Proof. If $U = \emptyset$, then by (7.52) and $\Sigma = \mathsf{box}(D)$, the result holds for $H = D$. If $U \neq \emptyset$, then we first observe that either $U = \{\mathsf{skip}\}$, or $U = \{\mathsf{redo}\}$, or $U \subseteq \mathsf{T}^{\mathsf{box}}_{\mathsf{tree}}$ (see Excs. 4.2.2 and 4.2.5, and Thm. 7.3.1(1)). Suppose $U = \{\mathsf{skip}\}$ (the case $U = \{\mathsf{redo}\}$ is similar). Then $\mathsf{box}(D) \in \mathsf{Box}^{\mathsf{e}}$ and $\Sigma = \underline{\mathsf{box}(\lfloor D \rfloor)}$. By

Thm. 7.3.1, $D \in \mathsf{Expr}^{\mathsf{e}}$. Hence, by Lemma 7.3.1, $D \equiv \overline{\lfloor D \rfloor}$. Thus, by (7.51) and (7.52),

$$D \xrightarrow{\emptyset} \overline{\lfloor D \rfloor} \xrightarrow{\{\mathsf{skip}\}} \underline{\lfloor D \rfloor} .$$

Hence, by (7.53), $D \xrightarrow{\{\mathsf{skip}\}} \underline{\lfloor D \rfloor}$ and we can take $H = \underline{\lfloor D \rfloor}$ since

$$\mathsf{box}(H) = \mathsf{box}(\underline{\lfloor D \rfloor}) = \underline{\mathsf{box}(\lfloor D \rfloor)} = \Sigma .$$

If $U \subseteq \mathsf{T}^{\mathsf{box}}_{\mathsf{tree}}$, then $\mathsf{box}(D)\ [U\rangle\ \Sigma$ and we proceed by induction on the maximal depth h of the transition trees in U (such an inductive argument is valid since U is a finite set of finite trees).

In the base step ($h = 0$), by Prop. 7.3.1(2), D is a flat expression. Hence, from (7.36), (7.37), and (7.54) it follows that there is a flat expression H such that

$$D \xrightarrow{U} H$$

and $\mathsf{box}(D)\ [U\rangle\ \mathsf{box}(H)$. The latter further implies that $\mathsf{box}(H) = \Sigma$.

In the inductive step ($h > 0$), we may assume that $D = \mathsf{op}_\Omega(\boldsymbol{D})$, for some Ω and $\boldsymbol{D}$.
Indeed, since D is not a flat expression (which follows from Prop. 7.3.1(2)), we can use (7.44), (7.45), (7.46), and (7.49) to find D' such that $D \equiv D' = \mathsf{op}_\Omega(\boldsymbol{D})$. And, after proving the result for D', extend it to D using the inaction rules, (7.52) and (7.53).
Hence, by Thm. 7.3.1, we have $\mathsf{box}(\boldsymbol{D}) \in \mathsf{dom}_\Omega$ and $\Omega(\mathsf{box}(\boldsymbol{D}))\ [U\rangle\ \Sigma$. From Thm. 7.2.2 it follows that there are Ω-tuples in the domain of Ω, $\boldsymbol{\Theta}$, and $\boldsymbol{\Psi}$, such that $\mathsf{box}(\boldsymbol{D}) \equiv_\Omega \boldsymbol{\Theta}$ and $\Sigma = \Omega(\boldsymbol{\Psi})$ and

$$\left(\Omega : \boldsymbol{\Theta}\right) \xrightarrow{U} \left(\Omega : \boldsymbol{\Psi}\right) . \tag{7.56}$$

We next observe that there is a well-formed box expression $\mathsf{op}_\Omega(\boldsymbol{F})$ such that $\mathsf{op}_\Omega(\boldsymbol{D}) \equiv \mathsf{op}_\Omega(\boldsymbol{F})$ and $\mathsf{box}(\boldsymbol{F}) = \boldsymbol{\Theta}$. That $\mathsf{op}_\Omega(\boldsymbol{F})$ exists can be shown in the following way. Let μ and κ be respectively the factorisations of $\mathsf{box}(\boldsymbol{D})$ and $\boldsymbol{\Theta}$ which are factorisations of the same complex marking of Ω (notice that, by Thm. 7.3.1, μ is also the factorisation of $\boldsymbol{D}$). By Lemma 7.3.1 and (7.50), we may assume $\boldsymbol{D}_{\mu_{\mathsf{e}}} \subseteq \overline{\mathsf{Expr}^{\mathsf{s}}}$ and $\boldsymbol{D}_{\mu_{\mathsf{x}}} \subseteq \underline{\mathsf{Expr}^{\mathsf{s}}}$. Let $\boldsymbol{C}$ be the Ω-tuple of well-formed expressions such that, for every $t \in T_\Omega$,

$$D_v = \begin{cases} \overline{C_v} & \text{if} \quad v \in \mu_{\mathsf{e}} \\ \underline{C_v} & \text{if} \quad v \in \mu_{\mathsf{x}} \\ C_v & \text{otherwise} . \end{cases}$$

Then, by (7.47), $\mathsf{op}_\Omega(\boldsymbol{D}) \equiv \mathsf{op}_\Omega(\boldsymbol{F})$ where, for every $v \in T_\Omega$,

$$F_v = \begin{cases} \overline{C_v} & \text{if} \quad v \in \kappa_{\mathsf{e}} \\ \underline{C_v} & \text{if} \quad v \in \kappa_{\mathsf{x}} \\ C_v & \text{otherwise} . \end{cases}$$

Hence, by Thm. 7.3.2, $\mathsf{op}_\Omega(\boldsymbol{F})$ is a well-formed box expression which satisfies $\mathsf{box}(\boldsymbol{F}) = \boldsymbol{\Theta}$. Indeed, $\lfloor \mathsf{box}(\boldsymbol{F}) \rfloor = \lfloor \mathsf{box}(\boldsymbol{D}) \rfloor = \lfloor \boldsymbol{\Theta} \rfloor$, the factorisation of $\boldsymbol{F}$ is κ, and, by $\mu_\mathsf{d} = \kappa_\mathsf{d}$, $\mathsf{box}(\boldsymbol{F}_{\kappa_\mathsf{d}}) = \boldsymbol{\Theta}_{\kappa_\mathsf{d}}$.

Define, for every $v \in T_\Omega$, $U_v = U \cap \mathsf{T}^v_{\mathsf{new}} = \{u^1_v, \ldots, u^{k_v}_v\}$. By (7.56), we have, for every $v \in T$,

$$\Theta_v \left[\mathsf{trees}(u^1_v) \uplus \cdots \uplus \mathsf{trees}(u^{k_v}_v)\right\rangle \Psi_v$$

and only finitely many k_v are greater than zero. Thus, by the induction hypothesis and the $U = \emptyset$ case, for every $v \in T$, there is a well-formed box expression H_v such that $\mathsf{box}(H_v) = \Psi_v$ and

$$F_v \xrightarrow{\mathsf{trees}(u^1_v) \uplus \cdots \uplus \mathsf{trees}(u^{k_v}_v)} H_v \ .$$

Hence, by (7.55), $\mathsf{op}_\Omega(\boldsymbol{F}) \xrightarrow{U} \mathsf{op}_\Omega(\boldsymbol{H})$. And, by Thm. 7.3.4, we obtain that $H = \mathsf{op}_\Omega(\boldsymbol{H})$ is a well-formed box expression such that

$$\mathsf{box}(\mathsf{op}_\Omega(\boldsymbol{F}))\ [U\rangle\ \mathsf{box}(H) \ .$$

Thus, by Thm. 7.3.3 and $D \equiv \mathsf{op}_\Omega(\boldsymbol{F})$, we have $\mathsf{box}(D)\ [U\rangle\ \mathsf{box}(H)$, and so $\mathsf{box}(H) = \Sigma$. □

Running Example. The operational semantics of flat expressions is given in DIY by

$$\begin{array}{cccc}
\overline{c_1} \xrightarrow{\{t_{11}\}} c_{12} & \overline{c_2} \xrightarrow{\{t_{21}\}} c_{22} & \overline{c_1} \xrightarrow{\{t_{12}\}} c_{11} & c_{21} \xrightarrow{\{t_{21}\}} c_{22} \\
\overline{c_1} \xrightarrow{\{t_{11},t_{12}\}} \underline{c_1} & c_{22} \xrightarrow{\{t_{22}\}} \underline{c_2} & c_{12} \xrightarrow{\{t_{12}\}} \underline{c_1} & c_{22} \xrightarrow{\{t_{22}\}} c_{23} \\
c_{11} \xrightarrow{\{t_{11}\}} \underline{c_1} & \overline{c_3} \xrightarrow{\{t_{31}\}} \underline{c_3} & &
\end{array}$$

and the inference rule for the only operator box, Ω_0, can be formulated in the following way:

$$\frac{D \xrightarrow{\{t_1,u_1,\ldots,t_k,u_k,x_1,\ldots,x_l\}} D' \ , \ H \xrightarrow{\{y_1,\ldots,y_m\}} H' \ , \ G \xrightarrow{\{z_1,\ldots,z_n\}} G'}{\mathsf{op}_{\Omega_0}(D,G,H) \xrightarrow{U} \mathsf{op}_{\Omega_0}(D',G',H')}$$

where $k,l,m,n \geq 0$; $\{\mathsf{lab}(t_i), \mathsf{lab}(u_i)\} = \{a,b\}$, for every $i \leq k$; $\mathsf{lab}(x_i) \notin \{a,b\}$, for every $i \leq l$; and U is given by

$$U = \left\{ \begin{array}{ccc}
(v_1,f) \lhd \{t_1,u_1\} & ,\ \ldots\ , & (v_1,f) \lhd \{t_k,u_k\} \\
(v_1,\mathsf{lab}(x_1)) \lhd x_1 & ,\ \ldots\ , & (v_1,\mathsf{lab}(x_l)) \lhd x_l \\
(v_2,\mathsf{lab}(y_1)) \lhd y_1 & ,\ \ldots\ , & (v_2,\mathsf{lab}(y_m)) \lhd y_m \\
(v_3,\mathsf{lab}(z_1)) \lhd z_1 & ,\ \ldots\ , & (v_3,\mathsf{lab}(z_n)) \lhd z_n
\end{array} \right\} .$$

Exercise 7.3.15. Show that if the above rule is applied to a well-formed expression, then $n = 0$ or $k = l = m = 0$.

For example, the following is a valid sequence of three moves in the DIY algebra:

$$\begin{aligned}\overline{\mathsf{op}_{\Omega_0}(c_1, c_2, c_3)} &\xrightarrow{\{w_1,w_2\}} \mathsf{op}_{\Omega_0}(\underline{c_1}, c_{22}, c_3) \\ &\xrightarrow{\{w_3\}} \mathsf{op}_{\Omega_0}(c_1, c_2, \overline{c_3}) \\ &\xrightarrow{\{w_4\}} \underline{\mathsf{op}_{\Omega_0}(c_1, c_2, c_3)} \,.\end{aligned}$$

A derivation for the first move is shown below.

$$\overbrace{\overline{\mathsf{op}_{\Omega_0}(c_1, c_2, c_3)} \xrightarrow{\emptyset} \mathsf{op}_{\Omega_0}(\overline{c_1}, \overline{c_2}, c_3) \qquad \overbrace{\overline{c_1} \xrightarrow{\{t_{11},t_{12}\}} \underline{c_1} \qquad \overline{c_2} \xrightarrow{\{t_{21}\}} c_{22} \qquad c_3 \xrightarrow{\emptyset} c_3}^{\mathsf{op}_{\Omega_0}(\overline{c_1}, \overline{c_2}, c_3) \xrightarrow{\{\,(v_1,f)\triangleleft\{t_{11},t_{12}\}\,,\,(v_2,c)\triangleleft t_{21}\,\}} \mathsf{op}_{\Omega_0}(\underline{c_1}, c_{22}, c_3)}}^{\overline{\mathsf{op}_{\Omega_0}(c_1, c_2, c_3)} \xrightarrow{\{w_1,w_2\}} \mathsf{op}_{\Omega_0}(\underline{c_1}, c_{22}, c_3)}$$

7.3.6 Consistency of the Two Semantics

The consistency between the denotational and operational semantics of box expressions will be formulated in terms of the transition systems they generate (see also Sects. 3.2.2 and 4.2.4). This will be possible since, thanks to Thms. 7.3.4 and 7.3.5, we are now in a position to compare transition systems generated by a well-formed box expression and the corresponding net.

Let G be a well-formed dynamic expression. We will use $[G\rangle_{\mathsf{sr}}$ to denote the least set of expressions containing G such that if $H \in [G\rangle_{\mathsf{sr}}$ and

$$H \xrightarrow{U} C\,,$$

for some $U \in \mathsf{tlab}_{\mathsf{sr}}$, then $C \in [G\rangle_{\mathsf{sr}}$. The *full transition system* of G is defined as $\mathsf{fts}_G = (V, L_{\mathsf{sr}}, A, [G]_{\equiv})$ where $V = \{[H]_{\equiv} \mid H \in [G\rangle_{\mathsf{sr}}\}$ is the set of states, $L = \mathsf{tlab}_{\mathsf{sr}}$ is the set of arc labels, and

$$A = \{([H]_{\equiv}, U, [C]_{\equiv}) \in V \times \mathsf{tlab}_{\mathsf{sr}} \times V \mid H \xrightarrow{U} C\}$$

is the set of arcs. For a well-formed static expression E, $\mathsf{fts}_E = \mathsf{fts}_{\overline{E}}$.

Let Σ be a box generated by a well-formed dynamic expression. The *full transition system* of Σ is defined as $\mathsf{fts}_\Sigma = (V, L, A, v_0)$ where

$$V = \{\Theta \mid \mathsf{skip}, \mathsf{redo} \notin T_\Theta \ \wedge\ \Theta_{\mathsf{sr}} \in [\Sigma_{\mathsf{sr}}\rangle\,\}$$

is the set of states, $v_0 = \Sigma$ is the initial state, $L = \mathsf{tlab}_{\mathsf{sr}}$ is the set of arc labels, and

$$A = \{(\Theta, U, \Psi) \in V \times \mathsf{tlab_{sr}} \times V \mid \Theta_{\mathsf{sr}}\ [U\rangle\ \Psi_{\mathsf{sr}}\}$$

is the set of arcs. In other words, fts_Σ is the reachability graph of Σ_{sr} with all references to skip and redo in the nodes of the graph erased. For a box Σ generated by a well-formed static expression, $\mathsf{fts}_\Sigma = \mathsf{fts}_{\overline{\Sigma}}$.

We now state a fundamental result which demonstrates that the operational and denotational semantics of a well-formed box expression capture the same behaviour, in arguably the strongest sense.

Theorem 7.3.6. For every well-formed box expression D, the relation

$$isom \ = \ \left\{ \left([H]_\equiv, \mathsf{box}(H) \right) \middle| \, [H]_\equiv \text{ is a node of } \mathsf{fts}_D \right\}$$

is an isomorphism between fts_D and $\mathsf{fts}_{\mathsf{box}(D)}$, and the relation

$$isom' \ = \ \left\{ \left([H]_\equiv, \mathsf{box}(H) \right) \middle| \, [H]_\equiv \text{ is a node of } \mathsf{fts}_{\lfloor D \rfloor} \right\}$$

is an isomorphism between $\mathsf{fts}_{\lfloor D \rfloor}$ and $\mathsf{fts}_{\mathsf{box}(\lfloor D \rfloor)}$.

Proof. By Thms. 7.3.3 and 7.3.4, *isom* is a well-defined injective mapping. The rest of the result follows from Prop. 7.2.1 and Thms. 7.3.4, 7.3.5 and 7.3.1. □

Running Example. In the DIY algebra, Thm. 7.3.6 can be illustrated by taking $D = \mathsf{op}_{\Omega_0}(\overline{c_1}, c_{22}, c_3)$ and the corresponding box shown in Fig. 7.4. Figure 7.8 depicts fts_D and $\mathsf{fts}_{\mathsf{box}(D)}$.

7.3.7 Label-based Operational Semantics

The operational semantics based on transition names and captured by full transition systems is very expressive; in particular, we will see in Sect. 7.3.8 that it contains enough information to retrieve partial order semantics of nets corresponding to box expressions. However, it may often be sufficient to record only the labels of executed transitions, in the usual style of process algebras and also that used in the operational semantics of PBC in Chap. 3. Such a treatment can be accommodated within the scheme developed so far. First, we retain the structural similarity relation $\equiv$ on box expressions without any change. Next, we define moves of the form

$$D \xrightarrow{\Gamma} H$$

where D and H are expressions as before, and Γ is a finite multiset in

$$\mathsf{mlab_{sr}} = \mathsf{mult}(\mathsf{Lab} \cup \{\mathsf{skip}, \mathsf{redo}\}) \ .$$

We keep the rules (7.51)—(7.53) unchanged (with $\emptyset$ now denoting the empty multiset of labels, {skip} and {redo} denoting singleton multisets, and U being changed to Γ), but modify the axioms for flat expressions and inference rules for operator boxes, in the following way.

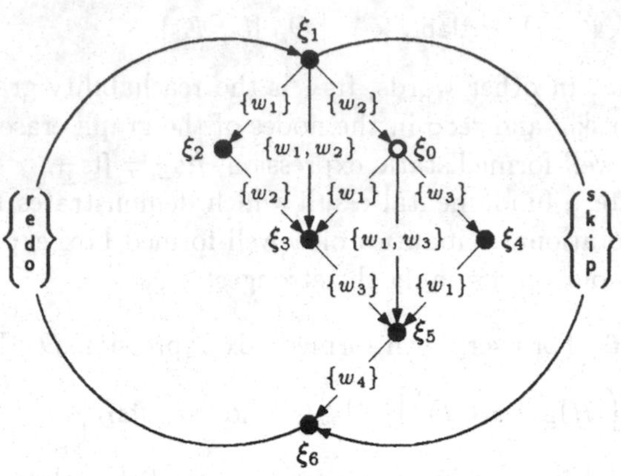

fts_D

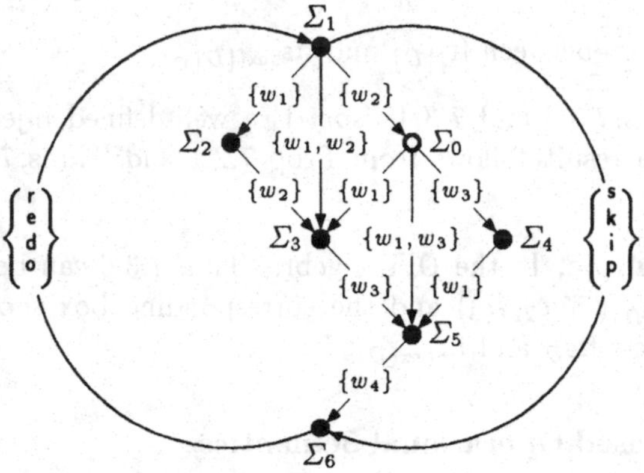

$\mathsf{fts}_{\mathsf{box}(D)}$

$$\xi_0 = \left\{ \mathsf{op}_{\Omega_0}(\overline{c_1}, c_{22}, c_3) \right\} \qquad M_{\Sigma_0} = \{p_1, p_2, p_6\}$$

$$\xi_1 = \left\{ \overline{\mathsf{op}_{\Omega_0}(c_1, c_2, c_3)}, \mathsf{op}_{\Omega_0}(\overline{c_1}, \overline{c_2}, c_3), \mathsf{op}_{\Omega_0}(\overline{c_1}, c_{21}, c_3) \right\} \qquad M_{\Sigma_1} = \{p_1, p_2, p_5\}$$

$$\xi_2 = \left\{ \mathsf{op}_{\Omega_0}(\underline{c_1}, \overline{c_2}, c_3), \mathsf{op}_{\Omega_0}(\underline{c_1}, c_{21}, c_3) \right\} \qquad M_{\Sigma_2} = \{p_3, p_4, p_5\}$$

$$\xi_3 = \left\{ \mathsf{op}_{\Omega_0}(\underline{c_1}, c_{22}, c_3) \right\} \qquad M_{\Sigma_3} = \{p_3, p_4, p_6\}$$

$$\xi_4 = \left\{ \mathsf{op}_{\Omega_0}(\overline{c_1}, c_{23}, c_3), \mathsf{op}_{\Omega_0}(\overline{c_1}, \underline{c_2}, c_3) \right\} \qquad M_{\Sigma_4} = \{p_1, p_2, p_7\}$$

$$\xi_5 = \left\{ \mathsf{op}_{\Omega_0}(\underline{c_1}, \underline{c_2}, c_3), \mathsf{op}_{\Omega_0}(\underline{c_1}, c_{23}, c_3), \mathsf{op}_{\Omega_0}(c_1, c_2, \overline{c_3}) \right\} \qquad M_{\Sigma_5} = \{p_3, p_4, p_7\}$$

$$\xi_6 = \left\{ \underline{\mathsf{op}_{\Omega_0}(c_1, c_2, c_3)}, \mathsf{op}_{\Omega_0}(c_1, c_2, \underline{c_3}) \right\} \qquad M_{\Sigma_6} = \{p_8\}$$

Fig. 7.8. Isomorphic full transition systems

For all flat expressions D and H and every nonempty $\Gamma \in \mathsf{mlab}_{\mathsf{sr}}$ such that $\mathsf{box}(D)\ [\Gamma\rangle_{\mathsf{lab}}\ \mathsf{box}(H)$,

$$\boxed{D \xrightarrow{\Gamma} H} \tag{7.57}$$

and for every SOS-operator box Ω:

$$\boxed{\frac{\forall v \in T_\Omega\ :\ D_v \xrightarrow{\Gamma_v^1+\cdots+\Gamma_v^{k_v}} H_v}{\mathsf{op}_\Omega(\boldsymbol{D}) \xrightarrow{\sum_{v\in T_\Omega}\{\alpha_v^1\}+\cdots+\{\alpha_v^{k_v}\}} \mathsf{op}_\Omega(\boldsymbol{H})}\ \mathit{cond}} \tag{7.58}$$

where $\boldsymbol{D}$ and $\boldsymbol{H}$ are Ω-tuples of expressions and *cond* is

$$\left(\sum_{v\in T_\Omega} k_v < \infty\right) \wedge \left(\forall v \in T_\Omega\ \forall i \le k_v\ :\ \left(\Gamma_v^i, \alpha_v^i\right) \in \lambda_\Omega(v)\right)\ .$$

Properties. The two types of operational semantics are strongly related; in essence, each label-based move is a transition-based move with only transitions labels being recorded.

Proposition 7.3.2. Let D be a well-formed box expression and $\Gamma \in \mathsf{mlab}_{\mathsf{sr}}$. Then $D \xrightarrow{\Gamma} H$ if and only if there is $U \in \mathsf{tlab}_{\mathsf{sr}}$ such that $D \xrightarrow{U} H$ and $\mathsf{lab}(U) = \Gamma$.

Proof. Both implications can be proved by a straightforward induction on the rank of $D \xrightarrow{\Gamma} H$ and $D \xrightarrow{U} H$, respectively. □

The results concerning the transition-based operational semantics directly extend to the label-based one. Let D be a well-formed box expression. In view of Prop. 7.3.2, the label-based operational semantics of D is faithfully captured by the *transition system* of D, denoted by ts_D and defined as fts_D with each arc label U changed to $\mathsf{lab}(U)$ (which is consistent with the definition in Sect. 3.2). Using it, the consistency result for the label-based operational semantics can be formulated in the following way.

Theorem 7.3.7. For every well-formed box expression D,

$$\mathit{isom}\ =\ \left\{\left([H]_\equiv, \mathsf{box}(H)\right)\,\middle|\, [H]_\equiv \text{ is a node of } \mathsf{ts}_D\right\}$$

is an isomorphism between ts_D and $\mathsf{ts}_{\mathsf{box}(D)}$.

Proof. Follows from Thm. 7.3.6, Prop. 7.3.2, and the fact that the structural similarity relation is the same for the transition- and label-based operational semantics. □

Running Example. For the DIY algebra, the label-based operational semantics of flat expressions is given by

$$
\boxed{
\begin{array}{llll}
\overline{c_1} \xrightarrow{\{a\}} c_{12} & \overline{c_2} \xrightarrow{\{c\}} c_{22} & \overline{c_1} \xrightarrow{\{b\}} c_{11} & c_{21} \xrightarrow{\{c\}} c_{22} \\
\overline{c_1} \xrightarrow{\{a,b\}} \underline{c_1} & c_{22} \xrightarrow{\{d\}} \underline{c_2} & c_{12} \xrightarrow{\{b\}} \underline{c_1} & c_{22} \xrightarrow{\{d\}} c_{23} \\
c_{11} \xrightarrow{\{a\}} \underline{c_1} & \overline{c_3} \xrightarrow{\{e\}} \underline{c_3} & &
\end{array}
}
$$

and the inference rule for the only operator box can be formulated in the following way:

$$
\boxed{
\frac{D_1 \xrightarrow{k\cdot\{a,b\}+\Gamma_1} H_1 \ , \ D_2 \xrightarrow{\Gamma_2} H_2 \ , \ D_3 \xrightarrow{\Gamma_3} H_3}{\mathsf{op}_{\Omega_0}(D_1, D_2, D_3) \xrightarrow{k\cdot\{f\}+\Gamma_1+\Gamma_2+\Gamma_3} \mathsf{op}_{\Omega_0}(H_1, H_2, H_3)} \quad a, b \notin \Gamma_1
}
$$

An application of these rules is illustrated on the move

$$\overline{\mathsf{op}_{\Omega_0}(c_1, c_2, c_3)} \xrightarrow{\{c,f\}} \mathsf{op}_{\Omega_0}(\underline{c_1}, c_{22}, c_3)$$

for which a derivation looks as follows:

$$\overline{\mathsf{op}_{\Omega_0}(c_1, c_2, c_3)} \xrightarrow{\{c,f\}} \mathsf{op}_{\Omega_0}(\underline{c_1}, c_{22}, c_3)$$

$$\overbrace{\overline{\mathsf{op}_{\Omega_0}(c_1, c_2, c_3)} \xrightarrow{\emptyset} \mathsf{op}_{\Omega_0}(\overline{c_1}, \overline{c_2}, c_3) \qquad \mathsf{op}_{\Omega_0}(\overline{c_1}, \overline{c_2}, c_3) \xrightarrow{\{c,f\}} \mathsf{op}_{\Omega_0}(\underline{c_1}, c_{22}, c_3)}$$

$$\overbrace{\overline{c_1} \xrightarrow{\{a,b\}} \underline{c_1} \qquad \overline{c_2} \xrightarrow{\{c\}} c_{22} \qquad c_3 \xrightarrow{\emptyset} c_3}$$

7.3.8 Partial Order Semantics of Box Expressions

In Sect. 6.4.1, we considered a possible way of giving partial order semantics to nets of the PBC model using occurrence nets and processes. Here, we aim at achieving the same for box expressions. But occurrence nets are not well suited for dealing with process expressions for the simple reason that the latter do not support an explicit notion of place or local state. We will therefore use Mazurkiewicz traces — a model of partial order behaviour based solely on transitions, which is equivalent to the occurrence net approach for safe nets (see [5]), which is our concern here.

Mazurkiewicz Traces. Let A be a set and $\mathsf{ind} \subseteq A \times A$ be an irreflexive symmetric relation on A. Intuitively, A is meant to represent the set of possible events in a concurrent system, and ind an *independence* relation which identifies events that can be executed concurrently. With every sequence $\sigma = A_1 \ldots A_k$, where each A_i is a finite subset of A such that $(a, b) \in \mathsf{ind}$ for all distinct $a, b \in A_i$, one can associate a partial order, denoted by $\mathsf{poset}_{\mathsf{ind}}(\sigma)$, in the following way.

The set of *event occurrences* of σ, occ_σ, comprises all pairs $(a, l) \in A \times \mathbf{N}$ such that $a \in A_1 \cup \ldots \cup A_k$ and l ranges between 1 and the number of times a occurs within σ, $1 \leq l \leq |\{i \mid a \in A_i\}|$. Moreover, we denote by $\mathsf{idx}_{(a,l)}$ the index m such that A_m contains the l-th occurrence of a, i.e., $a \in A_m$ and $l = |\{i \mid a \in A_i \wedge i \leq m\}|$. We then define a *precedence* relation on occ_σ, $\prec_\sigma$, by stipulating that $(a, l) \prec_\sigma (b, n)$ if $(a, b) \notin \mathsf{ind}$ and $\mathsf{idx}_{(a,l)} < \mathsf{idx}_{(b,n)}$. Then

$$\mathsf{poset}_{\mathsf{ind}}(\sigma) = (\mathsf{occ}_\sigma, \prec_\sigma^*)$$

where $\prec_\sigma^*$ is the reflexive transitive closure of $\prec_\sigma$.

Exercise 7.3.16. Show that $\mathsf{poset}_{\mathsf{ind}}(\sigma)$ is a partial order.

Consider now a safe labelled net Σ, and take A to be its transition set, $A = T_\Sigma$, and ind to be its independence relation, $\mathsf{ind} = \mathsf{ind}_\Sigma$ (see Sect. 4.2.3). Then, with every finite step sequence σ of Σ, $\Sigma\ [\sigma\rangle\ \Theta$, we can associate a partial order, $\mathsf{poset}_{\mathsf{ind}_\Sigma}(\sigma)$, in the way described above. One of the crucial properties of the partial order semantics is that every sequence of sets of transitions consistent with $\mathsf{poset}_{\mathsf{ind}_\Sigma}(\sigma)$ is also a valid step sequence. More precisely, if $\omega = U_1 \ldots U_k$ is a sequence of finite sets of transitions of Σ such that $(t, u) \in \mathsf{ind}_\Sigma$, for every U_i and all distinct $t, u \in U_i$, then $\mathsf{poset}_{\mathsf{ind}_\Sigma}(\omega) = \mathsf{poset}_{\mathsf{ind}_\Sigma}(\sigma)$ implies that ω is a valid step sequence for Σ leading to the same marked net as σ, i.e., $\Sigma\ [\omega\rangle\ \Theta$. For example, the step sequence $\sigma = \{t_0, t_2\}\emptyset\{t_1, t_2\}\{t_2\}\emptyset$ of the labelled net Σ_0 in Fig. 4.3 generates a partial order $\mathsf{poset}_{\mathsf{ind}_{\Sigma_0}}(\sigma)$ whose Hasse diagram, giving the precedence relation, is shown on the left-hand side of Fig. 7.9. This partial order is normally represented as a *labelled* partial order with nodes being labelled just by transitions, as shown on the right-hand side of Fig. 7.9.

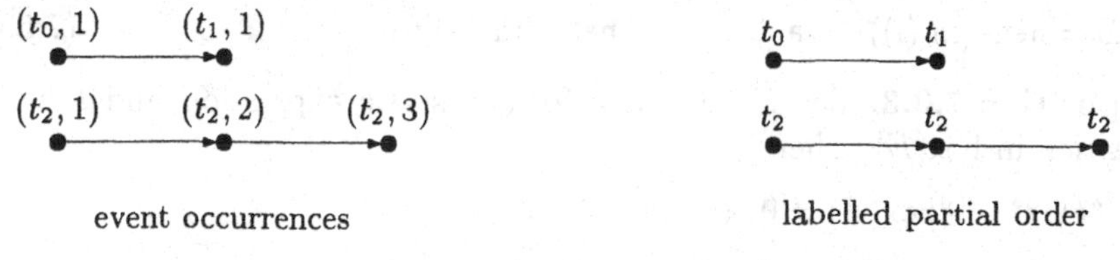

Fig. 7.9. Hasse diagrams of partial orders

The presence of a path between two nodes in such a graph is interpreted as *causality*, and the lack of ordering as *concurrency*. Thus, in the example above, the occurrences of t_0 and t_1 are causally related, and both are concurrent with the three occurrences of t_2. Whenever $\Sigma\ [\sigma\rangle\ \Theta$, we will write $\Sigma\ [\mathsf{poset}_{\mathsf{ind}_\Sigma}(\sigma)\rangle_{\mathsf{po}}\ \Theta$ to indicate that Θ can be derived from Σ by executing the poset within the brackets $[\ldots\rangle_{\mathsf{po}}$. There is a very close relationship between such partial orders and the partial orders obtained from processes in the sense of Sect. 6.4.1, when only their events are considered (see [5]).

Connectedness Properties of Transition Trees. We will now investigate connectedness between transitions in nets generated by box expressions. It will turn out that the name trees of these transitions contain information as to when two transitions are adjacent to a common place, and the relevant information is independent of the specific net to which the two transitions happen to belong.

We can now again take advantage of the assumption made in (7.27) that the operator and static constant boxes have disjoint transition sets. This allows one to identify, for every transition t in the boxes $\mathsf{OpBox} \cup (\mathsf{ConstBox} \cap \mathsf{Box}^{\mathsf{s}})$, a unique box in $\mathsf{OpBox} \cup (\mathsf{ConstBox} \cap \mathsf{Box}^{\mathsf{s}})$ to which t belongs; we will denote it by net_t. For example, in the DIY algebra, $\mathsf{net}_{v_1} = \Omega_0$ and $\mathsf{net}_{t_{22}} = \Phi_2$.

To start with, we can capture the property of 'being connected to an entry or exit place'. By definition, a transition tree $t \in \mathsf{T}^{\mathsf{box}}_{\mathsf{tree}}$ belongs to the set Con^{e} or Con^{x}, if (i) or (ii) below holds.

(i) There is a plain box $\Sigma \in \mathsf{ConstBox} \cap \mathsf{Box}^{\mathsf{s}}$ such that $t \in T_\Sigma$ and we have, respectively:

$$({}^\bullet t \cup t^\bullet) \cap {}^\circ\Sigma \neq \emptyset \quad \text{or} \quad ({}^\bullet t \cup t^\bullet) \cap \Sigma^\circ \neq \emptyset\,.$$

(ii) There is an operator box $\Omega \in \mathsf{OpBox}$ such that $t = (v,\alpha) \lhd Q$ and $v \in T_\Omega$, and we have, respectively:

$$\left({}^\bullet v \cap {}^\circ\Omega \neq \emptyset \neq Q \cap Con^{\mathrm{e}} \;\vee\; v^\bullet \cap {}^\circ\Omega \neq \emptyset \neq Q \cap Con^{\mathrm{x}}\right) \quad \text{or}$$
$$\left({}^\bullet v \cap \Omega^\circ \neq \emptyset \neq Q \cap Con^{\mathrm{e}} \;\vee\; v^\bullet \cap \Omega^\circ \neq \emptyset \neq Q \cap Con^{\mathrm{x}}\right).$$

The sets Con^{e} and Con^{x} are well defined since transition trees are always finite. Notice also that, by (7.27), we have

$$\Sigma = \mathsf{net}_t \text{ (in (i))} \qquad \text{and} \qquad \Omega = \mathsf{net}_v \text{ (in (ii))}\,. \tag{7.59}$$

Proposition 7.3.3. Let E be a well-formed static expression and t be a transition in $\mathsf{box}(E)$. Then

$$({}^\bullet t \cup t^\bullet) \cap {}^\circ\mathsf{box}(E) \neq \emptyset \iff t \in Con^{\mathrm{e}}$$
$$({}^\bullet t \cup t^\bullet) \cap \mathsf{box}(E)^\circ \neq \emptyset \iff t \in Con^{\mathrm{x}}\,.$$

Proof. We only prove the first equivalence, the proofs of the other one being symmetric. We proceed by induction on the depth h of t. In the base step ($h = 0$), by Prop. 7.3.1(2) and (7.27), $\mathsf{box}(E) = \mathsf{net}_t \in \mathsf{ConstBox} \cap \mathsf{Box}^{\mathsf{s}}$. Hence the equivalence holds by the definition of Con^{e} and (7.59).

In the inductive step ($h > 0$), we may assume that $E = \mathsf{op}_\Omega(\boldsymbol{E})$, by Prop. 7.3.1(2), which means that $t = (v,\alpha) \lhd Q$; and so, by (7.27), $\Omega = \mathsf{net}_v$. By Prop. 7.1.7, $({}^\bullet t \cup t^\bullet) \cap {}^\circ\mathsf{box}(E) \neq \emptyset$ if and only if

$$^\bullet v \cap {}^\circ\Omega \neq \emptyset \neq ({}^\bullet Q \cup Q^\bullet) \cap {}^\circ\mathsf{box}(E_v) \quad \text{or}$$

$$v^\bullet \cap {}^\circ\Omega \neq \emptyset \neq ({}^\bullet Q \cup Q^\bullet) \cap \mathsf{box}(E_v)^\circ .$$

Hence, the equivalence holds by $\Omega = \mathsf{net}_v$, (7.59), and the induction hypothesis. □

Global Independence Relation. By definition, a pair of transition trees $(t,u) \in \mathsf{T}^{\mathsf{box}}_{\mathsf{tree}} \times \mathsf{T}^{\mathsf{box}}_{\mathsf{tree}}$ belongs to the set $\mathsf{ind}^{\mathsf{box}}$ if (i) or (ii) below holds.

(i) There is a plain box $\Sigma \in \mathsf{ConstBox} \cap \mathsf{Box}^{\mathsf{s}}$ such that $(t,u) \in \mathsf{ind}_\Sigma$.
(ii) There is an operator box $\Omega \in \mathsf{OpBox}$ such that $t = (v,\alpha) \lhd Q$, $u = (w,\beta) \lhd R$, and $v, w \in T_\Omega$, and we have

$$\begin{aligned} ^\bullet v \cap w^\bullet \neq \emptyset &\implies Q \cap Con^{\mathsf{e}} = \emptyset \vee R \cap Con^{\mathsf{x}} = \emptyset \\ v^\bullet \cap {}^\bullet w \neq \emptyset &\implies Q \cap Con^{\mathsf{x}} = \emptyset \vee R \cap Con^{\mathsf{e}} = \emptyset \\ v \neq w \wedge {}^\bullet v \cap {}^\bullet w \neq \emptyset &\implies Q \cap Con^{\mathsf{e}} = \emptyset \vee R \cap Con^{\mathsf{e}} = \emptyset \\ v \neq w \wedge v^\bullet \cap w^\bullet \neq \emptyset &\implies Q \cap Con^{\mathsf{x}} = \emptyset \vee R \cap Con^{\mathsf{x}} = \emptyset . \end{aligned}$$

Moreover, if $v = w$, then $Q \times R \subseteq \mathsf{ind}^{\mathsf{box}}$.

Notice also that similarly as before, by (7.27), we have

$$\Sigma = \mathsf{net}_t \text{ (in (i))} \quad \text{and} \quad \Omega = \mathsf{net}_v = \mathsf{net}_w \text{ (in (ii))} . \tag{7.60}$$

The relation we just defined can be thought of as a global independence relation for the boxes generated by the denotational semantics.

Theorem 7.3.8. For every well-formed static expression E,

$$\mathsf{ind}_{\mathsf{box}(E)} = (T_{\mathsf{box}(E)} \times T_{\mathsf{box}(E)}) \cap \mathsf{ind}^{\mathsf{box}} .$$

Proof. Let t and u be transition trees in $\mathsf{box}(E)$. We proceed by induction on $h = \max\{\mathsf{depth}(t), \mathsf{depth}(u)\}$.

In the base step ($h = 0$), by Prop. 7.3.1(2) and (7.27),

$$\mathsf{box}(E) = \mathsf{net}_t = \mathsf{net}_u \in \mathsf{ConstBox} \cap \mathsf{Box}^{\mathsf{s}} .$$

Hence, by the definition of $\mathsf{ind}^{\mathsf{box}}$ and (7.60), $(t,u) \in \mathsf{ind}_{\mathsf{box}(E)}$ if and only if $(t,u) \in \mathsf{ind}^{\mathsf{box}}$.

In the inductive step ($h > 0$), we may assume that $E = \mathsf{op}_\Omega(\boldsymbol{E})$, by Prop. 7.3.1(2). Thus $t = (v,\alpha) \lhd Q$ and $u = (w,\beta) \lhd R$, and so, by (7.27), $\Omega = \mathsf{net}_v = \mathsf{net}_w$. We now consider two cases.

Case 1: $v = w$. We have $Q \cup R \subseteq \mathsf{box}(E_v)$. By Prop. 7.1.5, $(t,u) \in \mathsf{ind}_{\mathsf{box}(E)}$ if and only if $Q \times R \subseteq \mathsf{ind}_{\mathsf{box}(E_v)}$ and, moreover, if $^\bullet v \cap v^\bullet \neq \emptyset$, then (also by Prop. 7.3.3)

$$\left(Q \cap Con^{\mathsf{e}} = \emptyset \vee R \cap Con^{\mathsf{x}} = \emptyset\right) \wedge \left(Q \cap Con^{\mathsf{x}} = \emptyset \vee R \cap Con^{\mathsf{e}} = \emptyset\right) .$$

Thus, by $\Omega = \mathsf{net}_v = \mathsf{net}_w$, (7.60), and the induction hypothesis, $(t,u) \in \mathsf{ind}_{\mathsf{box}(E)}$ if and only if $(t,u) \in \mathsf{ind}^{\mathsf{box}}$.

Case 2: $v \neq w$. We proceed similarly as in Case 1, using Prop. 7.1.6 in place of Prop. 7.1.5. □

Consistency Result. With the global independence relation $\mathsf{ind}^{\mathsf{box}}$, we can introduce partial order behaviours directly into the box algebra. All we need to do is say what a partial-order execution associated with a step sequence generated by a box expression is. Let D and H be two well-formed box expressions, and $U_1, \ldots, U_k$ $(k \geq 0)$ be subsets of $\mathsf{T}^{\mathsf{box}}_{\mathsf{tree}}$ such that there are well-formed box expressions $D_0, D_1, \ldots, D_k$ satisfying $D = D_0$, $H = D_k$ and, for every $i < k$,

$$D_i \xrightarrow{U_{i+1}} D_{i+1} \, .$$

We will call $\pi = \mathsf{poset}_{\mathsf{ind}^{\mathsf{box}}}(U_1 \ldots U_k)$ a *partial order execution* from D to H and denote $D \xrightarrow{\pi} H$. The consistency result obtained for transition-based operational semantics can then be lifted to the level of partial order executions.

Theorem 7.3.9. Let D and H be well-formed box expressions, and π be a partial order.

(1) If $D \xrightarrow{\pi} H$, then $\mathsf{box}(D)\ [\pi\rangle_{\mathsf{po}}\ \mathsf{box}(H)$.
(2) If $\mathsf{box}(D)\ [\pi\rangle_{\mathsf{po}}\ \Sigma$, then there is a well-formed box expression J such that $\mathsf{box}(J) = \Sigma$ and $D \xrightarrow{\pi} J$.

Proof. Follows from Thms. 7.3.4, 7.3.5, and 7.3.8, after making an easy observation that for any partial order π, there is at least one step sequence consistent with it (one can simply perform a topological sort and consider singleton steps). □

The last result demonstrated that partial-order executions associated with a given well-formed box expression are exactly the same as those generated by the corresponding box. Indirectly, this indicates that the device of name trees together with the global independence relation faithfully capture the causality and independence in a system's behaviour.

7.4 Literature and Background

The results presented in this chapter extend those contained in [68] by allowing non-pure and non-ex-directed operator boxes with possibly infinitely many transitions. The consistency results have also been strengthened since they are now formulated in terms of transition system isomorphism rather than bisimulation equivalence.

8. PBC and Other Process Algebras

In this chapter, we apply the theory developed in Chap. 7 to various concrete algebras, viz. PBC, CCS, TCSP, and COSY. We demonstrate that the theory we have developed lends itself to these — at first sight — very different setups. In particular, we argue that it is sufficiently flexible and extensible to describe the key semantic properties of the above algebras.

In Sect. 8.1, we take a fresh look at PBC and argue that it can be seen as an instance of the box algebra, and so it inherits all coherence and completeness properties obtained for the general framework in Chap. 7. This applies, in particular, to the generalised PBC defined in Chap. 3. We also present a PBC syntax extended by two non-ex-directed operator boxes for iteration and show that the resulting model is also an instance of the general scheme. In Sect. 8.2, we explain why CCS, TCSP, and COSY may all be regarded as instances of the box algebra as well.

8.1 (Generalised) PBC is a Box Algebra

We will now see how various parts of the syntax and semantics of the general box algebra are realised by the standard PBC. In Sect. 8.1.1, we will exclude the ternary iteration construct since it does not directly fit the general framework, and can be treated in two different ways, corresponding to the two operator boxes ($\Omega_{[**]}$ and $\Omega_{[*]}$) which can be used to model the net semantics of iteration. We will show how the theory can be extended to yield a safe translation of the ternary iteration operator in Sect. 8.1.2. In Sect. 8.1.3, we will introduce more general iteration constructs whose arguments must satisfy certain constraints to yield safe translations, and we will then also describe a second way of treating the ternary iteration.

8.1.1 PBC Without Loops

Syntax and Denotational Semantics. The syntax of PBC we are interested in was given in Sect. 3.4. It is not difficult to see that it conforms to that of the general box algebra in (7.29), after making some straightforward simplifications and adjustments, such as changing the mode of application of the operators (e.g., using infix notation instead of functional one). The operator boxes used to define the denotational semantics of PBC expressions are all ex-directed pure SOS-operator boxes with reversible transitions (so the standard PBC syntax contains only well-formed expressions), and the set of (ex-directed and ex-exclusive) constants is given by $\mathsf{Const}^{\mathsf{s}} = \mathsf{Lab}_{\mathsf{PBC}} \cup \{[x_{()}] \mid x \text{ is a data variable}\}$, $\mathsf{Const}^{\mathsf{e}} = \mathsf{Const}^{\mathsf{x}} = \emptyset$, and $\mathsf{Const}^{\mathsf{d}} = \{[\overline{x_{(u)}}] \mid x \text{ is a data variable and } u \in \mathit{Type}(x)\}$. The semantical mapping from the PBC expressions to boxes, $\mathsf{box}_{\mathsf{PBC}}$, was defined in the same way as the box mapping in Chap. 7.

Structural Equivalence Relation. The equation system used in Chap. 3 to define the structural equivalence on PBC expressions is an instance of that defined for the general box algebra, with some notational modifications and simplifications. In particular, when formulating the PBC rules, we have consciously used the concrete syntax; thus, for example, we have $\underline{E}; F \equiv E; \overline{F}$ for static expressions E and F, rather than the (generic) $\underline{D}; H \equiv D; \overline{H}$ for any D and H. Below we indicate the relationship between the general rules and those specific to PBC:

Rule	Instance(s) in PBC
(7.42)	IN1, IN2, ILN, IRN
(7.45)	IPAR1, IC1L, IC1R, IS1, ISY1, IR1, IRS1, ISC1, IGSY1, IGRS1, IGSC1
(7.46)	IPAR2, IC2L, IC2R, IS3, ISY2, IR2, IRS2, ISC2, IGSY2, IGRS2, IGSC2
(7.47)	IS2

Moreover, for the dynamic PBC expressions, the context rule IREC is equivalent to a combination of (7.44) and the two general substitutivity rules, (7.49) and (7.50), because all PBC operator boxes are finite. And, since the transition system semantics of static expressions is that of the entry (dynamic) expressions, the structural equivalence of PBC expressions is in effect the same as that of the generic algebra (see also Exc. 3.2.51).

Operational Semantics. We have seen that with nonessential modifications the structural similarity relation of PBC (without iteration) conforms to that of the general box algebra. The transition-based operational semantics rules are defined now, by applying the general box algebra scheme.

Axiomatising flat PBC expressions is straightforward. The labelled version of (7.54) leads to the rules AR, DAT1, DAT2, and DAT3. In the transition-based framework,

RAc	$\overline{\alpha} \xrightarrow{\{v_\alpha\}} \underline{\alpha}$

and, for every constant representing a data variable x and $u, v \in \mathit{Type}(x)$:

RDa1	$\underline{[\overline{x_{()}}]} \xrightarrow{\{v_x^{\bullet u}\}} [\overline{x_{(u)}}]$
RDa2	$[\overline{x_{(u)}}] \xrightarrow{\{v_x^{uv}\}} [\overline{x_{(v)}}]$
RDa3	$[\overline{x_{(u)}}] \xrightarrow{\{v_x^{u\bullet}\}} \underline{[x_{()}]}$

The rules SKP and RDO are those given by (7.51). Figure 8.1 specifies the inference rules for the transition-based operational semantics of the various PBC operators (other than iteration), deduced directly from (7.55). In the figure, D, H, D', and H' are dynamic or static PBC expressions, and U, U', and U_i are finite sets of transition trees. The notation $(v, f) \oplus U$ where v is a transition name, f a relabelling defined for action particles (lifted to PBC multiactions in the usual way, see Sect. 3.2.11), and $U = \{t_1, \ldots, t_k\}$ a finite set of transition trees, is a set of transition trees defined as

$$(v, f) \oplus U = \Big\{ \big(v, f(\mathsf{lab}(t_1))\big) \lhd t_1 \,, \ldots , \big(v, f(\mathsf{lab}(t_k))\big) \lhd t_k \Big\} \,.$$

Moreover, $v \oplus U = (v, id) \oplus U$ where id is the identity relabelling. Note that in Fig. 8.1 we could have used the other two synchronisation schemes discussed in Chaps. 3 and 4, by suitably replacing the relabellings $\varrho_{\mathsf{sy}\, a}$ and $\varrho_{\mathsf{sy}\, A}$.

The label-based derivation rules have all been given in Chap. 3. Let us verify, for instance, that the labelled version of RSq is equivalent to the two rules, SR and SL, given in Chap. 3. First, we need to look closer at the syntax of legal PBC expressions. The SOS-operator box $\Omega_;$ has the following six factorisations of nonempty markings reachable from its entry or exit marking:

$$(\{v_;^1\}, \emptyset, \emptyset, \emptyset) \quad (\emptyset, \{v_;^1\}, \emptyset, \emptyset) \quad (\emptyset, \emptyset, \{v_;^1\}, \emptyset)$$
$$(\{v_;^2\}, \emptyset, \emptyset, \emptyset) \quad (\emptyset, \{v_;^2\}, \emptyset, \emptyset) \quad (\emptyset, \emptyset, \{v_;^2\}, \emptyset) \,.$$

Hence RSq can be seen as giving rise to the following two different labelled rules, where this time D, D' denote dynamic PBC expressions and E, E' denote static ones (note that by Thms. 7.3.1 and 7.3.4 and the T-restrictedness of boxes, static expressions can only execute empty steps).

$$\mathrm{RS}_1 \quad \frac{D \xrightarrow{\Gamma} D' \,,\; E \xrightarrow{\emptyset} E'}{D;E \xrightarrow{\Gamma} D';E'} \qquad\qquad \mathrm{RS}_2 \quad \frac{E \xrightarrow{\emptyset} E' \,,\; D \xrightarrow{\Gamma'} D'}{E;D \xrightarrow{\Gamma} E';D'}$$

Thus, if we take $E' = E$ (which is always possible since $E \xrightarrow{\emptyset} E$ for every expression E), we obtain the rules SR and SL. And, conversely, SR and SL together with ILN, (3.8), and IREC, yield RS_1 and RS_2.

Exercise 8.1.1. Discuss in a similar way other operational rules in Fig. 8.1.

RSq

$$\frac{D \xrightarrow{U} D' \;,\; H \xrightarrow{U'} H'}{D;H \xrightarrow{v_;^1 \oplus U \,\cup\, v_;^2 \oplus U'} D';H'}$$

RC

$$\frac{D \xrightarrow{U} H \;,\; H \xrightarrow{U'} H'}{D \,\square\, H \xrightarrow{v_\square^1 \oplus U \,\cup\, v_\square^2 \oplus U'} D' \,\square\, H'}$$

RCC

$$\frac{D \xrightarrow{U} H \;,\; D' \xrightarrow{U'} H'}{D \| H \xrightarrow{v_\|^1 \oplus U \,\cup\, v_\|^2 \oplus U'} D' \| H'}$$

RRe

$$\frac{D \xrightarrow{U} H}{D[f] \xrightarrow{(v_{[f]}, f) \oplus U} H[f]}$$

RRs

$$\frac{D \xrightarrow{U} H \quad \text{where} \quad \forall u \in U : a, \widehat{a} \notin \mathsf{lab}(u)}{D \;\mathsf{rs}\; a \xrightarrow{v_{\mathsf{rs}\,a} \oplus U} H \;\mathsf{rs}\; a}$$

RSy

$$\frac{D \xrightarrow{U_1 \uplus \ldots \uplus U_k} H \quad \text{where} \quad \forall i : (\mathsf{lab}(U_i), \alpha_i) \in \varrho_{\mathsf{sy}\,a}}{D \;\mathsf{sy}\; a \xrightarrow{\{(v_{\mathsf{sy}\,a}, \alpha_1) \lhd U_1, \ldots, (v_{\mathsf{sy}\,a}, \alpha_k) \lhd U_k\}} H \;\mathsf{sy}\; a}$$

RSc

$$\frac{D \xrightarrow{U_1 \uplus \ldots \uplus U_k} H \quad \text{where} \quad \forall i : (\mathsf{lab}(U_i), \alpha_i) \in \varrho_{[a:]}}{[a : D] \xrightarrow{\{(v_{[a:]}, \alpha_1) \lhd U_1, \ldots, (v_{[a:]}, \alpha_k) \lhd U_k\}} [a : H]}$$

GRRs

$$\frac{D \xrightarrow{U} H \quad \text{where} \quad \forall u \in U : (A \cup \widehat{A}) \cap \mathsf{lab}(u) = \emptyset}{D \;\mathsf{rs}\; A \xrightarrow{v_{\mathsf{rs}\,A} \oplus U} H \;\mathsf{rs}\; A}$$

GRSy

$$\frac{D \xrightarrow{U_1 \uplus \ldots \uplus U_k} H \quad \text{where} \quad \forall i : (\mathsf{lab}(U_i), \alpha_i) \in \varrho_{\mathsf{sy}\,A}}{D \;\mathsf{sy}\; A \xrightarrow{\{(v_{\mathsf{sy}\,A}, \alpha_1) \lhd U_1, \ldots, (v_{\mathsf{sy}\,A}, \alpha_k) \lhd U_k\}} H \;\mathsf{sy}\; A}$$

GRSc

$$\frac{D \xrightarrow{U_1 \uplus \ldots \uplus U_k} H \quad \text{where} \quad \forall i : (\mathsf{lab}(U_i), \alpha_i) \in \varrho_{[A:]}}{[A : D] \xrightarrow{\{(v_{[A:]}, \alpha_1) \lhd U_1, \ldots, (v_{[A:]}, \alpha_k) \lhd U_k\}} [A : H]}$$

Fig. 8.1. Inference rules for the transition-based semantics of PBC (without iteration)

For syntactically legal PBC expressions it would, of course, be possible to use only one format of operational semantic rules (as well as structural equivalence rules) instead of two equivalent but different ones. However, we have found the generic rules used in Chap. 7 better for carrying out formal manipulations, whereas those used in Chap. 3 are more intuitive and better tailored for the concrete syntax of PBC.

8.1.2 Safe Translation of the Ternary PBC Iteration

We deferred the discussion of the iteration construct to a separate section as it requires more careful handling than other PBC operators. The general scheme has to be extended since $[E * F * E']$ is a ternary operator, unlike the operator box $\Omega_{[**]}$ used to give the (1-safe) denotational net semantics of iteration. However, this will not prevent us from applying the theory developed in Chap. 7. Our plan is first to add to the PBC algebra without iteration the 6-ary operator box $\Omega_{[**]}$ and the corresponding 6-ary operator $\mathsf{op}_{\Omega_{[**]}}$, claim the consistency results proved for the general box algebra, and only after that treat the ternary iteration operation. In doing so, we shall be able to transfer the results obtained in Chap. 7 into the PBC algebra.

We start by adding $\Omega_{[**]}$ to the set of PBC operators considered so far in this chapter, which is something we can clearly do as $\Omega_{[**]}$ is an SOS-operator box (which is pure, ex-directed, and has all transitions reversible), assuming the following ordering of its transitions:

$$v^{10}_{[**]}\,,\ v^{20}_{[**]}\,,\ v^{30}_{[**]}\,,\ v^{11}_{[**]}\,,\ v^{21}_{[**]}\,,\ v^{31}_{[**]}\ .$$

We shall call the resulting algebra of expressions PBC0.

According to the general scheme, $\Omega_{[**]}$ contributes to the syntax of PBC0 in the following way (below we elided the syntactic rules for expressions not involving $\Omega_{[**]}$):

$$
\begin{array}{lll}
\mathsf{Expr}^{\mathsf{s}}\ E & ::= & \dots \\
 & & \mathsf{op}_{\Omega_{[**]}}(E,E,E,E,E,E) \\
\mathsf{Expr}^{\mathsf{e}}\ F & ::= & \dots \\
 & & \mathsf{op}_{\Omega_{[**]}}(F,E,E,E,E,E) \mid \mathsf{op}_{\Omega_{[**]}}(E,E,E,F,E,E) \\
\mathsf{Expr}^{\mathsf{x}}\ G & ::= & \dots \\
 & & \mathsf{op}_{\Omega_{[**]}}(E,E,G,E,E,E) \mid \mathsf{op}_{\Omega_{[**]}}(E,E,E,E,E,G) \\
\mathsf{Expr}^{\mathsf{d}}\ H & ::= & \dots \\
 & & \mathsf{op}_{\Omega_{[**]}}(H,E,E,E,E,E) \mid \mathsf{op}_{\Omega_{[**]}}(E,H,E,E,E,E) \mid \\
 & & \mathsf{op}_{\Omega_{[**]}}(E,E,H,E,E,E) \mid \mathsf{op}_{\Omega_{[**]}}(E,E,E,H,E,E) \mid \\
 & & \mathsf{op}_{\Omega_{[**]}}(E,E,E,E,H,E) \mid \mathsf{op}_{\Omega_{[**]}}(E,E,E,E,E,H)\ .
\end{array}
$$

The structural equivalence rules derived from $\Omega_{[**]}$, (7.45), and (7.46) are given below where, for an $\Omega_{[**]}$-tuple of static expressions $\boldsymbol{E}$ and a transition $v^{ij}_{[**]}$ of $\Omega_{[**]}$, we denote by $\boldsymbol{E}[ij \leftarrow D]$ the $\Omega_{[**]}$-tuple obtained from $\boldsymbol{E}$ by replacing the expression corresponding to $v^{ij}_{[**]}$ by a dynamic expression D.

$$\begin{aligned} &\mathsf{op}_{\Omega_{[\bullet\bullet]}}\left(\boldsymbol{E}\left[10 \leftarrow \overline{E_{10}}\right]\right) \equiv \mathsf{op}_{\Omega_{[\bullet\bullet]}}\left(\boldsymbol{E}\left[11 \leftarrow \overline{E_{11}}\right]\right) \equiv \overline{\mathsf{op}_{\Omega_{[\bullet\bullet]}}(\boldsymbol{E})} \\ &\mathsf{op}_{\Omega_{[\bullet\bullet]}}\left(\boldsymbol{E}\left[30 \leftarrow \underline{E_{30}}\right]\right) \equiv \mathsf{op}_{\Omega_{[\bullet\bullet]}}\left(\boldsymbol{E}\left[31 \leftarrow \underline{E_{31}}\right]\right) \equiv \underline{\mathsf{op}_{\Omega_{[\bullet\bullet]}}(\boldsymbol{E})} \end{aligned} \tag{8.1}$$

There are six rules derived from $\Omega_{[**]}$ and (7.47), after omitting those which can be derived using the transitivity and symmetry of $\equiv$:

$$\begin{aligned} &\mathsf{op}_{\Omega_{[\bullet\bullet]}}\left(\boldsymbol{E}\left[10 \leftarrow \underline{E_{10}}\right]\right) \equiv \mathsf{op}_{\Omega_{[\bullet\bullet]}}\left(\boldsymbol{E}\left[21 \leftarrow \underline{E_{21}}\right]\right) \equiv \\ &\mathsf{op}_{\Omega_{[\bullet\bullet]}}\left(\boldsymbol{E}\left[20 \leftarrow \overline{E_{20}}\right]\right) \equiv \mathsf{op}_{\Omega_{[\bullet\bullet]}}\left(\boldsymbol{E}\left[30 \leftarrow \overline{E_{30}}\right]\right) \\ &\mathsf{op}_{\Omega_{[\bullet\bullet]}}\left(\boldsymbol{E}\left[11 \leftarrow \underline{E_{11}}\right]\right) \equiv \mathsf{op}_{\Omega_{[\bullet\bullet]}}\left(\boldsymbol{E}\left[20 \leftarrow \underline{E_{20}}\right]\right) \equiv \\ &\mathsf{op}_{\Omega_{[\bullet\bullet]}}\left(\boldsymbol{E}\left[21 \leftarrow \overline{E_{21}}\right]\right) \equiv \mathsf{op}_{\Omega_{[\bullet\bullet]}}\left(\boldsymbol{E}\left[31 \leftarrow \overline{E_{31}}\right]\right) \end{aligned} \tag{8.2}$$

We have one instance of the rule (7.50):

$$\frac{\forall (i,j) \in \{1,2,3\} \times \{0,1\} \ : D_{ij} \equiv H_{ij}}{\mathsf{op}_{\Omega_{[\bullet\bullet]}}(\boldsymbol{D}) \equiv \mathsf{op}_{\Omega_{[\bullet\bullet]}}(\boldsymbol{H})} \tag{8.3}$$

and two operational semantics rules, for the transition-based and label-based operational semantics, derived from (7.55) and (7.58), respectively:

$$\frac{\forall (i,j) \in \{1,2,3\} \times \{0,1\} \ : D_{ij} \xrightarrow{U_{ij}} H_{ij}}{\mathsf{op}_{\Omega_{[\bullet\bullet]}}(\boldsymbol{D}) \xrightarrow{\bigcup_{(i,j)\in\{1,2,3\}\times\{0,1\}} v^{ij}_{[\bullet\bullet]} \oplus U_{ij}} \mathsf{op}_{\Omega_{[\bullet\bullet]}}(\boldsymbol{H})} \tag{8.4}$$

$$\frac{\forall (i,j) \in \{1,2,3\} \times \{0,1\} \ : D_{ij} \xrightarrow{\Gamma_{ij}} H_{ij}}{\mathsf{op}_{\Omega_{[\bullet\bullet]}}(\boldsymbol{D}) \xrightarrow{\sum_{(i,j)\in\{1,2,3\}\times\{0,1\}} \Gamma_{ij}} \mathsf{op}_{\Omega_{[\bullet\bullet]}}(\boldsymbol{H})} \tag{8.5}$$

All the results obtained in Chap. 7 hold in the PBC0 model. In particular, we obtain the following instance of the consistency results, Thms. 7.3.6 and 7.3.7, for the transition systems generated by process expressions and the corresponding static and dynamic boxes.

Theorem 8.1.1. For every box expression D in PBC0,

$$isom = \left\{ \left([H]_{\equiv}, \mathsf{box}(H) \right) \Big| \, [H]_{\equiv} \text{ is a node of } \mathsf{fts}_D \right\}$$

is an isomorphism between fts_D and $\mathsf{fts}_{\mathsf{box}(D)}$, as well as between ts_D and $\mathsf{ts}_{\mathsf{box}(D)}$. □

$\Omega_{[**]}$-symmetric PBC Expressions. In the next step we restrict PBC0 to a submodel of $\Omega_{[**]}$-*symmetric expressions*, PBC1. It comprises all PBC0 expressions D such that, if

$$D' = \mathsf{op}_{\Omega_{[**]}}(D_{10}, D_{20}, D_{30}, D_{11}, D_{21}, D_{31})$$

is a subexpression of D, then $\lfloor D_{i0} \rfloor = \lfloor D_{i1} \rfloor$, for $i = 1, 2, 3$. The same holds if D' appears in a defining equation for a process variable.

We restrict the rules (8.3), (8.4), and (8.5), by requiring that $\lfloor D_{i0} \rfloor = \lfloor D_{i1} \rfloor$ and $\lfloor H_{i0} \rfloor = \lfloor H_{i1} \rfloor$, for $i = 1, 2, 3$ (note that replacing recursion variables by their defining equations can break the symmetry). We will denote the resulting rules by (8.3'), (8.4'), and (8.5'), respectively. The rules (8.1) and (8.2) do not need similar restriction since when one side of any such rule is matched by an $\Omega_{[**]}$-symmetric expression, then the other side is an $\Omega_{[**]}$-symmetric expression as well. We will use $\equiv_1$ and $\longrightarrow_1$ to denote the resulting structural similarity and operational semantics, respectively. Both relations enjoy properties similar to those which are true of their counterparts in the box algebra.

Proposition 8.1.1. Let D be a box expression in PBC1, $U \in \mathsf{tlab}_{\mathsf{sr}}$, and $\Gamma \in \mathsf{mlab}_{\mathsf{sr}}$.

(1) $D \equiv_1 H$ if and only if $D \xrightarrow{\emptyset}_1 H$.
(2) If $D \equiv_1 H$ or $D \xrightarrow{U}_1 H$ or $D \xrightarrow{\Gamma}_1 H$, then H is a box expression in PBC1 and $\lfloor D \rfloor \equiv_1 \lfloor H \rfloor$.
(3) $D \xrightarrow{\Gamma}_1 H$ if and only if there is $U' \in \mathsf{tlab}_{\mathsf{sr}}$ such that $D \xrightarrow{U'}_1 H$ and $\mathsf{lab}(U') = \Gamma$.
(4) If H is a box expression in PBC1 such that both $\mathsf{box}(D) = \mathsf{box}(H)$ and $\lfloor D \rfloor \equiv_1 \lfloor H \rfloor$, then $D \equiv_1 H$.

Proof. Similar to the proofs of the corresponding results for $\equiv$ and $\longrightarrow$, i.e., Thms. 7.3.3 and 7.3.4, Prop. 7.3.2, and Cor. 7.3.2. □

The next group of properties relates the similarity and operational semantics relations in PBC0 and PBC1.

Proposition 8.1.2. Let D and H be box expressions in PBC1, $U \in \mathsf{tlab}_{\mathsf{sr}}$, and $\Gamma \in \mathsf{mlab}_{\mathsf{sr}}$.

(1) If $D \equiv_1 H$ or $D \xrightarrow{U}_1 H$ or $D \xrightarrow{\Gamma}_1 H$, then respectively,

$$D \equiv H \text{ or } D \xrightarrow{U} H \text{ or } D \xrightarrow{\Gamma} H .$$

(2) If $D \xrightarrow{U} J$, then there is a box expression C in PBC1 such that

$$D \xrightarrow{U}_1 C \equiv J .$$

(3) If $D \xrightarrow{\Gamma} J$, then there is a box expression C in PBC1 such that

$$D \xrightarrow{\Gamma}_1 C \equiv J\ .$$

Proof. (1) Follows from the fact that (8.3'), (8.4'), and (8.5') are restricted versions of the rules (8.3), (8.4), and (8.5), respectively.

(2) The proof proceeds by induction on $\mathsf{rank}(D, U, J)$. Its only nonstandard part is when in the inductive step we have

$$D = \mathsf{op}_{\Omega_{[\bullet\bullet]}}(D_{10}, D_{20}, D_{30}, D_{11}, D_{21}, D_{31}) \text{ where } \lfloor D_{i0} \rfloor = \lfloor D_{i1} \rfloor \text{ for all } i$$
$$J = \mathsf{op}_{\Omega_{[\bullet\bullet]}}(H_{10}, H_{20}, H_{30}, H_{11}, H_{21}, H_{31})$$
$$D_{ij} \xrightarrow{U_{ij}} H_{ij} \text{ and } \mathsf{rank}(D_{ij}, U_{ij}, H_{ij}) < \mathsf{rank}(D, U, J) \text{ for all } i \text{ and } j$$
$$U = \bigcup_{(i,j)\in\{1,2,3\}\times\{0,1\}} v^{ij}_{[**]} \oplus U_{ij}\ .$$

Since in every factorisation μ of a marking reachable from the entry or exit marking of $\Omega_{[**]}$, there is exactly one transition in $\mu_e \cup \mu_x \cup \mu_d$, we may assume without loss of generality that, for all $(i,j) \in (\{1,2,3\} \times \{1,0\}) \setminus \{(1,0)\}$, D_{ij} is a static expression. Then also H_{ij} is a static expression and $U_{ij} = \emptyset$ which follows from D_{ij} being static and Thms. 7.3.1(2) and 7.3.4. By the induction hypothesis, there is a box expression C' in PBC1 such that $D_{10} \xrightarrow{U_{10}}_1 C' \equiv H_{10}$. Define

$$C = \mathsf{op}_{\Omega_{[\bullet\bullet]}}(C', D_{20}, D_{30}, \lfloor C' \rfloor, D_{21}, D_{31})\ .$$

We now observe that $\lfloor C' \rfloor \equiv D_{11}$ which follows from $D_{11} = \lfloor D_{10} \rfloor$ and $\lfloor D_{10} \rfloor \equiv_1 \lfloor C' \rfloor$ (by Prop. 8.1.1(2)). Moreover, by $D_{ij} \xrightarrow{\emptyset} H_{ij}$ and Cor. 7.3.2, we have $H_{ij} \equiv D_{ij}$, for all $(i,j) \in (\{1,2,3\} \times \{0,1\}) \setminus \{(1,0)\}$. Hence, by (8.3), $C \equiv J$. Furthermore, $D \xrightarrow{U}_1 C$, by (8.4) and Prop. 8.1.1(1).

(3) Follows directly from (2) and Prop. 8.1.1(3). □

We now import into PBC1 the consistency results stated for PBC0. Below, for every box expression D in PBC1, we use fts^1_D and ts^1_D to denote the transition systems defined in exactly the same way as fts_D and fts_D, but using $\equiv_1$ and $\longrightarrow_1$ instead of $\equiv$ and $\longrightarrow$, respectively.

Theorem 8.1.2. For every box expression D in PBC1,

$$isom = \left\{ \left([H]_{\equiv_1}, \mathsf{box}(H) \right) \middle| [H]_{\equiv_1} \text{ is a node of } \mathsf{fts}^1_D \right\}$$

is an isomorphism between fts^1_D and $\mathsf{fts}_{\mathsf{box}(D)}$, as well as between ts^1_D and $\mathsf{ts}_{\mathsf{box}(D)}$.

Proof. Follows from Props. 8.1.1 and 8.1.2, and Thm. 8.1.1. □

Abridged PBC. We will now make a crucial step towards incorporating the ternary iteration. We introduce a syntax driven by a mapping η from $\Omega_{[**]}$-symmetric PBC expressions into *abridged* PBC expressions. The mapping is defined homomorphically for all operators other than $\mathsf{op}_{\Omega_{[**]}}$, and leaves constants and variables unchanged. For $\mathsf{op}_{\Omega_{[**]}}$ it is defined so that if $D = \mathsf{op}_{\Omega_{[**]}}(D_{10}, D_{20}, D_{30}, D_{11}, D_{21}, D_{31})$ is a box expression in PBC1, then

$$\eta(D) = \begin{cases} \Big[\eta(D_{10}) * \eta(D_{20}) * \eta(D_{30})\Big] & \text{if} \quad D \in \mathsf{Expr}^{\mathsf{s}} \\ \Big[\eta(D_{10}) * \eta(D_{20}) * \eta(D_{30})\Big]_0 & \text{if} \quad D_{i0} \in \mathsf{Expr}^{\mathsf{d}} \text{ for some } i \\ \Big[\eta(D_{11}) * \eta(D_{21}) * \eta(D_{31})\Big]_1 & \text{if} \quad D_{i1} \in \mathsf{Expr}^{\mathsf{d}} \text{ for some } i \,. \end{cases}$$

The mapping is well defined since, for the reason already stated in the proof of Prop. 8.1.2, at most one D_{ij} can be a non-static expression. The reverse translation η^{-1} is also well defined since η is injective (notice the index 0 or 1 attached to each dynamic ternary iteration construct). Thus η bijectively maps box expressions in PBC1 into another set of expressions, which we denote by PBC2. The syntax of PBC2 is that of PBC1 with one change, namely the part contributed by the iteration construct changes to the following (here and later $i \in \{0, 1\}$):

$\mathsf{Expr}^{\mathsf{s}}$	E ::= ...	$[E * E * E]$			
$\mathsf{Expr}^{\mathsf{e}}$	F ::= ...	$[F * E * E]_i$			
$\mathsf{Expr}^{\mathsf{x}}$	G ::= ...	$[E * E * G]_i$			
$\mathsf{Expr}^{\mathsf{d}}$	H ::= ...	$[H * E * E]_i$	$[E * H * E]_i$	$[E * E * H]_i$.	

The equations of the structural similarity relation (8.1) and (8.2) are also modified:

$$\begin{array}{|ll|} \hline [E * F * E'] & \equiv [\overline{E} * F * E']_i \\ [\underline{E} * F * E']_i & \equiv [E * \overline{F} * E']_i \\ [E * \overline{F} * E']_i & \equiv [E * \underline{F} * E']_{1-i} \\ [E * \underline{F} * E']_{1-i} & \equiv [E * F * \overline{E'}]_i \\ [E * F * \underline{E'}]_i & \equiv \underline{[E * F * E']} \\ \hline \end{array}$$

The rule (8.3) is split into two:

$$\begin{array}{|c|} \hline \dfrac{\forall j \in \{1,2,3\} \;:\; D_j \equiv H_j}{[D_1 * D_2 * D_3] \;\equiv\; [H_1 * H_2 * H_3]} \\ \dfrac{\forall j \in \{1,2,3\} \;:\; D_j \equiv H_j}{[D_1 * D_2 * D_3]_i \;\equiv\; [H_1 * H_2 * H_3]_i} \\ \hline \end{array} \qquad (8.6)$$

while the two derivation rules of the operational semantics, (8.4) and (8.5), are modified in the following way.

$$\frac{\forall j \in \{1,2,3\} \, : \, D_j \xrightarrow{U_j} H_j}{[D_1 * D_2 * D_3]_i \xrightarrow{\bigcup_{j\in\{1,2,3\}} v^{ji}_{[\bullet\bullet]} \oplus U_j} [H_1 * H_2 * H_3]_i}$$

$$\frac{\forall j \in \{1,2,3\} \, : \, D_j \xrightarrow{\Gamma_j} H_j}{[D_1 * D_2 * D_3]_i \xrightarrow{\Gamma_1+\Gamma_2+\Gamma_3} [H_1 * H_2 * H_3]_i}$$

We will use $\equiv_2$ and $\longrightarrow_2$ to denote the resulting structural similarity and operational semantics relations, respectively.

PBC2 inherits the consistency properties of PBC1 since the former is nothing but an η-based representation of the latter. Below, for an expression D in PBC2, we define $\mathsf{box}(D) = \mathsf{box}(\eta^{-1}(D))$, and use fts^2_D and ts^2_D to denote transition systems defined in exactly the same way as fts_D and ts_D using $\equiv_2$ and $\longrightarrow_2$ instead of $\equiv$ and $\longrightarrow$, respectively.

Theorem 8.1.3. For every box expression D in PBC2,

$$isom \;=\; \left\{ \left([H]_{\equiv_2}, \mathsf{box}(H)\right) \,\middle|\, [H]_{\equiv_2} \text{ is a node of } \mathsf{fts}^2_D \right\}$$

is an isomorphism between fts^2_D and $\mathsf{fts}_{\mathsf{box}(D)}$, as well as between ts^2_D and $\mathsf{ts}_{\mathsf{box}(D)}$.

Proof. Follows from Thm. 8.1.2 and the fact that η is a bijection between PBC1 and PBC2 which preserves the structural equivalence and operational semantics relations. □

There is nothing we would like to do further with the transition-based operational semantics of PBC2. Thus we take PBC2 to be the (extended) standard PBC algebra with the transition-based operational semantics. However, for the label-based operational semantics, one more simplification is possible leading straight to the original PBC model of Chap. 3.

The Original PBC. The idea behind the final transformation step stems from the observation that the index i in an expression $[D_1 * D_2 * D_3]_i$ is irrelevant a far as the label-based operational semantics is concerned, in the sense that the transition systems of $[D_1 * D_2 * D_3]_i$ and $[D_1 * D_2 * D_3]_{1-i}$ are isomorphic. Hence, each such index may be removed from the syntax for this type of semantics. This results in PBC3, the last model based on PBC that we consider, with the syntax given as in Sect. 3.4. The equations of the structural similarity relation are turned into the rules IIT1, IIT2a, IIT2b, IIT2c, and IIT3 of Chap. 3. The second rule in (8.6) is dropped as it reduces to the first one after removing the index i, and the label-based operational semantics rule becomes

$$\frac{\forall j \in \{1,2,3\} : D_j \xrightarrow{\Gamma_j} H_j}{[D_1 * D_2 * D_3] \xrightarrow{\Gamma_1+\Gamma_2+\Gamma_3} [H_1 * H_2 * H_3]}$$

Note that the above is essentially a condensed presentation of the rules IT1, IT2, and IT3 of Chap. 3.

For every expression D in PBC3, let ts^3_D be the transition systems defined in the same way as ts_D, using the similarity relation, $\equiv_3$, and operational semantics, $\longrightarrow_3$, of PBC3. Moreover, let $\mathsf{box}(D) = \mathsf{box}(H)$ for any H in PBC2 such that D is H after removing the index i from each instance of the iteration construct. It is easy to see that $\mathsf{box}(D)$ is unique up to net isomorphism.

As far as the consistency result is concerned, we cannot carry forward the result obtained for PBC2 for the simple reason that the transition systems which we would like to compare, ts^3_D and $\mathsf{ts}_{\mathsf{box}(D)}$, may have different numbers of nodes. For example, if $D = \overline{[a * \mathsf{stop} * \mathsf{stop}]}$, then ts^3_D has three nodes, whereas $\mathsf{ts}_{\mathsf{box}(D)}$ has four, as shown in Fig. 8.2. However, symmetries present in the transition systems of PBC3 expressions mean that ts^3_D and ts^2_D are strongly equivalent, and so are ts^3_D and $\mathsf{ts}_{\mathsf{box}(D)}$.

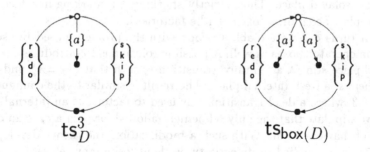

Fig. 8.2. Transition systems of $\overline{[a * \mathsf{stop} * \mathsf{stop}]}$ and of the corresponding box

Theorem 8.1.4. For every expression D in PBC3, $\mathsf{ts}^3_D \approx \mathsf{ts}_{\mathsf{box}(D)}$.

Proof. Let ξ be a mapping which takes a PBC2 expression and removes the index i from each instance of the iteration construct. Moreover, let H be the expression in PBC2 obtained from D by replacing each occurrence $[D_1 * D_2 * D_3]$ of the dynamic iteration construct by $[D_1 * D_2 * D_3]_0$.

Define $\mathcal{BS}$ as the relation comprising all pairs $([\xi(G)]_{\equiv_3}, [G]_{\equiv_2})$ such that $[G]_{\equiv_2}$ is a node in ts^2_H. One can see that $\mathcal{BS}$ is a strong bisimulation for ts^3_D and ts^2_H. Hence $\mathsf{ts}^3_D \approx \mathsf{ts}^2_H$ and, by Thm. 8.1.3, ts^3_D and $\mathsf{ts}_{\mathsf{box}(D)}$ are strongly equivalent transition systems, $\mathsf{ts}^3_D \approx \mathsf{ts}_{\mathsf{box}(D)}$. □

Thus PBC3, which is nothing but the original PBC of Chap. 3 with the label-based operational semantics, can indeed be treated as an instance of the general box algebra, and its denotational and operational semantics are equivalent in the sense of generating strongly equivalent label-based transition systems (Thm. 8.1.4). As already indicated, for the transition-based PBC we can adopt the PBC2 model; its denotational and operational semantics are equivalent in the sense of generating isomorphic transition-based and label-based transition systems (Thm. 8.1.3).

Generalised Control Flow Operators. We finally discuss the generalised control flow operators dealt with in Sects. 3.3.3 and 4.4.11, for whom the corresponding operator boxes are shown in Fig. 4.40. Referring to that figure, we observe that all operator boxes are ex-directed static boxes having all transitions reversible. Thus the only point which needs discussion is their factorisability. Clearly, $\Omega_{\square I}$ and $\Omega_{;J}$ and $\Omega_{\|J}$ are all factorisable. $\Omega_{\|I}$ is not factorisable, and there are at least two ways of dealing with this problem. One is to use its version with the two isolated places removed, which is factorisable and thus satisfies the definition of an SOS-operator box. Intuitively, this corresponds to taking the (unique) minimal solution of an unguarded recursive equation $X \stackrel{\mathrm{df}}{=} \Omega_{\|}(\mathsf{N}_\alpha, X)$ rather than the maximal one. Another way of dealing with the nonfactorisability of $\Omega_{\|I}$ relates to the observation that the problem stems from the fact that there are entry or exit places which are isolated. This is also true for the generalised sequential operator box $\Omega_{;I}$ with one isolated place. Thus, strictly speaking, no marking in which these isolated places contain a token can be factorised.

As it turns out we are able to cope with the problem posed by isolated places in operator boxes as well. A possible solution is to introduce, for every isolated place s in Ω, a 'dummy' transition v_s such that $\{s\} = v_s^\bullet$ and ${}^\bullet v_s = \{s'\}$ where s' a fresh internal place. The result is similar to the configuration in Fig. 7.3, where a dead transition was used to factor out an internal place. Then we stipulate that the only refinement allowed for such a v_s is an ex-box with no behaviour at all. With such a modification, the theory developed in Chap. 7 can be applied in its entirety, without losing any relevant property. To conclude, all generalised operator boxes in Fig. 4.40 can be accommodated within the generic framework developed in Chap. 7.

Exercise 8.1.2. Define transition-based derivation rules for the generalised control flow operators.

8.1.3 PBC with Generalised Loops

In the PBC adopted at the end of Chap. 3, all the operator boxes are pure and ex-directed and have only reversible transitions. It can also be checked that all refinements corresponding by the standard PBC syntax satisfy the behavioural properties (Beh1)—(Beh5); this in itself provided a strong motivation for considering such a syntax. However, allowing generalised loop constructs which are modelled by non-ex-directed and non-pure operator boxes can be highly advantageous; in particular, when modelling guarded while loops in the programming language discussed in the next chapter.

In this section, we will introduce a PBC with generalised loops (for static expressions only, leaving out the syntax of dynamic expressions which can be easily derived) containing two generalised iteration constructs, $\langle *]$ and $[* \rangle$. Since the relevant aspects of non-ex-exclusiveness and non-ex-directedness are associated with control flow rather than communication interface operators, for the purposes of the discussion in this section, we may treat the latter as equivalent. Hence we shall shorten the syntax by arbitrarily selecting only one communication interface operator, viz. basic relabelling. Similar comment applies to constant boxes; since all PBC constant boxes are both ex-exclusive and ex-directed, they are all equivalent as far as these two properties are concerned. We will therefore include just the basic action. In the next chapter, we will define new plain boxes modelling process constants, which will also be ex-directed and ex-exclusive; as a result, the discussion presented in this section will still be applicable.

We first give, in Fig. 8.3, a syntax incorporating the conditions (Dom1) and (Dom2); notice that the latter is trivially satisfied since all operator boxes we use have only reversible transitions. It is assumed that X^{wf} and X^{xcl} are recursion variables defined by $X^{\mathrm{wf}} \stackrel{\mathrm{df}}{=} E^{\mathrm{wf}}$ and $X^{\mathrm{xcl}} \stackrel{\mathrm{df}}{=} E^{\mathrm{xcl}}$, respectively.

$$
\begin{array}{llllll}
E^{\mathrm{wf}} ::= X^{\mathrm{wf}} & \mid E^{\mathrm{wf}}[f] & \mid E^{\mathrm{xcl}} & \mid E^{\mathrm{wf}} \,\square\, E^{\mathrm{wf}} & \mid E^{\mathrm{wf}} \| E^{\mathrm{wf}} & \mid \\
\qquad \langle E^{\mathrm{xcl}} * E^{\mathrm{wf}}] & \mid [E^{\mathrm{wf}} * E^{\mathrm{xcl}}\rangle & & & & \\
E^{\mathrm{xcl}} ::= X^{\mathrm{xcl}} & \mid E^{\mathrm{xcl}}[f] & \mid \alpha & \mid E^{\mathrm{xcl}} \,\square\, E^{\mathrm{xcl}} & \mid E^{\mathrm{wf}} ; E^{\mathrm{wf}} & \mid \\
\qquad \langle E^{\mathrm{xcl}} * E^{\mathrm{xcl}}] & \mid [E^{\mathrm{xcl}} * E^{\mathrm{xcl}}\rangle & \mid [E^{\mathrm{wf}} * E^{\mathrm{wf}} * E^{\mathrm{wf}}] & & &
\end{array}
$$

Fig. 8.3. PBC syntax with two generalised loops incorporating (Dom1)—(Dom2)

In the above, E^{wf} and E^{xcl} define the well-formed and ex-exclusive expressions, respectively. The syntax is an instance of the general box algebra, thus all results formulated in the previous chapter hold for it. Moreover, the syntax in Fig. 8.3 is sufficient to model the concurrent programming language discussed in the next chapter.

The syntax in Fig. 8.3 allows expressions for whom the corresponding boxes do not satisfy the behavioural conditions (Beh1)—(Beh5). We therefore introduce a final syntax, presented in Fig. 8.4, which identifies a subset of expressions generated by Fig. 8.3 where these conditions have been incorporated through (Expr1)—(Expr7), by means of Prop. 4.3.2 and Thm. 7.2.4, assuming the ternary iteration is modelled by $\Omega_{[**]}$. The syntax — more restrictive but still sufficient for the modelling of the programming language introduced in the next chapter — is based on eight syntactic classes E^β and each recursive variable X^β is defined by $X^\beta \stackrel{\mathrm{df}}{=} E^\beta$. It conforms to the general box algebra scheme and, moreover, all refinements it specifies satisfy the behavioural conditions, (Beh1)—(Beh5). In particular, with reference to the notation used in Sect. 7.3.3, it may be checked that all the expressions E^{wf} are well formed, and all the expressions E^{x} are x-directed. Similarly, for instance, $E^{\mathrm{e:x:xcl}}$ are expressions which are e-directed, x-directed, and ex-exclusive.

Exercise 8.1.3. Analyse the necessity of the constraints imposed on the first and third arguments of the ternary iteration operator in Fig. 8.4, and possibly add and justify extra clauses relaxing them.

Exercise 8.1.4. Verify that the boxes corresponding to the expressions E^{xcl}, E^{e}, and E^{x} are all ex-exclusive, e-directed, and x-directed, respectively. Moreover, show that all refinements corresponding to the expressions in E^{wf} satisfy (Beh1)—(Beh5).

We can therefore conclude that all consistency results formally established in Chap. 7 for the general box algebra hold also for the extended PBC given by the syntax in Fig. 8.4.

The discussion presented in this section could be repeated so as to yield — as mentioned earlier on — a second method of treating the ternary operator $[E*F*E']$ using $\Omega_{[*]}$ instead of $\Omega_{[**]}$. Note that although $\Omega_{[*]}$ is not pure, it is ex-directed. Hence to obtain a safe translation, it suffices to ensure that F be ex-exclusive. Figure 8.5 shows the resulting syntax for the static expressions. Notice that since all the expressions it generates are ex-directed, the syntax satisfies the behavioural conditions (Beh1)—(Beh5).

Exercise 8.1.5. Give the syntax for the two generalised loops together with the ternary iteration modelled by $\Omega_{[*]}$ (but see also the Appendix).

8.2 Other Process Algebras

In this section we outline how CCS, TCSP, and COSY could be treated in the general compositionality framework provided by the box algebra. In what follows, by a *simple operator box* we will mean a two-place one-transition operator box as shown in Sects. 4.4.5–4.4.8. We will denote such an operator box by $\Omega{:}\varrho$, where ϱ is the relabelling of its only transition; for instance, $\Omega_{\mathsf{rs}\,a}$ may also be denoted as $\Omega{:}\varrho_{\mathsf{rs}\,a}$.

$$
\begin{array}{lcl|l|l|l}
E^{\mathrm{wf}} & ::= & X^{\mathrm{wf}} & E^{\mathrm{wf}}[f] & E^{\mathrm{e}} & [E^{\mathrm{x}} * E^{\mathrm{e:x:xcl}}\rangle \\
 & & E^{\mathrm{xcl}} & E^{\mathrm{wf}} \| E^{\mathrm{wf}} & E^{\mathrm{x}} & \langle E^{\mathrm{e:x:xcl}} * E^{\mathrm{e}}] \\
E^{\mathrm{e}} & ::= & X^{\mathrm{e}} & E^{\mathrm{e}}[f] & E^{\mathrm{e:xcl}} & E^{\mathrm{e:x}} \\
 & & E^{\mathrm{e}} \| E^{\mathrm{e}} & [E^{\mathrm{e:x}} * E^{\mathrm{e:x:xcl}}\rangle & & \\
E^{\mathrm{x}} & ::= & X^{\mathrm{x}} & E^{\mathrm{x}}[f] & E^{\mathrm{x:xcl}} & E^{\mathrm{e:x}} \\
 & & E^{\mathrm{x}} \| E^{\mathrm{x}} & \langle E^{\mathrm{e:x:xcl}} * E^{\mathrm{e:x}}] & & \\
E^{\mathrm{xcl}} & ::= & X^{\mathrm{xcl}} & E^{\mathrm{xcl}}[f] & E^{\mathrm{e:xcl}} & \langle E^{\mathrm{e:x:xcl}} * E^{\mathrm{e:xcl}}] \\
 & & E^{\mathrm{x}} ; E^{\mathrm{wf}} & E^{\mathrm{wf}} ; E^{\mathrm{e}} & E^{\mathrm{x:xcl}} & [E^{\mathrm{x:xcl}} * E^{\mathrm{e:x:xcl}}\rangle \\
E^{\mathrm{e:x}} & ::= & X^{\mathrm{e:x}} & E^{\mathrm{e:x}}[f] & E^{\mathrm{e:x:xcl}} & E^{\mathrm{e:x}} \| E^{\mathrm{e:x}} \\
 & & E^{\mathrm{e:x}} \Box E^{\mathrm{e:x}} & & & \\
E^{\mathrm{e:xcl}} & ::= & X^{\mathrm{e:xcl}} & E^{\mathrm{e:xcl}}[f] & E^{\mathrm{e:x:xcl}} & [E^{\mathrm{e:x:xcl}} * E^{\mathrm{e:x:xcl}}\rangle \\
 & & E^{\mathrm{e}} ; E^{\mathrm{e}} & E^{\mathrm{e:x}} ; E^{\mathrm{wf}} & & \\
E^{\mathrm{x:xcl}} & ::= & X^{\mathrm{x:xcl}} & E^{\mathrm{x:xcl}}[f] & E^{\mathrm{e:x:xcl}} & \langle E^{\mathrm{e:x:xcl}} * E^{\mathrm{e:x:xcl}}] \\
 & & E^{\mathrm{x}} ; E^{\mathrm{x}} & E^{\mathrm{wf}} ; E^{\mathrm{e:x}} & & \\
E^{\mathrm{e:x:xcl}} & ::= & X^{\mathrm{e:x:xcl}} & E^{\mathrm{e:x:xcl}}[f] & \alpha & E^{\mathrm{e:x:xcl}} \Box E^{\mathrm{e:x:xcl}} \\
 & & E^{\mathrm{e:x}} ; E^{\mathrm{x}} & E^{\mathrm{e}} ; E^{\mathrm{e:x}} & [E^{\mathrm{e:x}} * E^{\mathrm{e:x}} * E^{\mathrm{e:x}}] &
\end{array}
$$

Fig. 8.4. PBC syntax with two generalised loops incorporating (Dom1)—(Dom2) and (Beh1)—(Beh5)

$$
\begin{array}{lcl|l|l|l|l|l}
E^{\mathrm{wf}} & ::= & X^{\mathrm{wf}} & E^{\mathrm{wf}}[f] & E^{\mathrm{wf}} \Box E^{\mathrm{wf}} & E^{\mathrm{wf}} \| E^{\mathrm{wf}} & E^{\mathrm{xcl}} & \\
E^{\mathrm{xcl}} & ::= & X^{\mathrm{xcl}} & E^{\mathrm{xcl}}[f] & E^{\mathrm{xcl}} \Box E^{\mathrm{xcl}} & E^{\mathrm{wf}} ; E^{\mathrm{wf}} & \alpha & [E^{\mathrm{wf}} * E^{\mathrm{xcl}} * E^{\mathrm{wf}}]
\end{array}
$$

Fig. 8.5. PBC syntax with ternary generalised loop incorporating (Dom1)—(Dom2) and (Beh1)—(Beh5)

8.2.1 CCS

To model CCS processes we assume that $\mathsf{Lab} = Act \cup \{\tau\}$ is the set of CCS labels and $\widehat{\ }: Act \to Act$ is a bijection on Act satisfying $\widehat{\widehat{a}} = a$ and $\widehat{a} \neq a$, for all $a \in Act$. We then define five simple operator boxes, $\Omega{:}restr(b)$, $\Omega{:}relab(h)$, $\Omega{:}left$, $\Omega{:}right$, and $\Omega{:}syn(CCS)$, where $b \in Act$ is a label and $h : \mathsf{Lab} \to \mathsf{Lab}$ is a mapping on labels commuting with $\widehat{(.)}$ and preserving τ, defined by the following relabellings:

$$\begin{aligned}
restr(b) &= \Big\{ \big(\{a\}, a\big) \;\Big|\; a \in (\mathsf{Lab} \setminus \{b\}) \Big\} \\
relab(h) &= \Big\{ \big(\{a\}, h(a)\big) \;\Big|\; a \in \mathsf{Lab} \Big\} \\
left &= \Big\{ \big(\{a\}, a^L\big) \;\Big|\; a \in \mathsf{Lab} \Big\} \\
right &= \Big\{ \big(\{a\}, a^R\big) \;\Big|\; a \in \mathsf{Lab} \Big\} \\
syn(CCS) &= \Big\{ \big(\{a^L, \widehat{a}^R\}, \tau\big) \;\Big|\; a \in Act \Big\} \\
&\quad \cup \Big\{ \big(\{a^L\}, a\big) \;\Big|\; a \in \mathsf{Lab} \Big\} \cup \Big\{ \big(\{a^R\}, a\big) \;\Big|\; a \in \mathsf{Lab} \Big\} .
\end{aligned}$$

It is assumed that Lab is extended by the labels a^L and a^R, for $a \in \mathsf{Lab}$, but neither a^L nor a^R are allowed in the syntax of CCS expressions (hence also not in the domain or codomain of h); they are auxiliary symbols used to model correctly the semantics of CCS's parallel composition. Below, we give the Petri box denotational semantics of CCS, through the full translation ϕ_{CCS} for the CCS processes conforming to the syntax (2.1):

$$\begin{aligned}
\phi_{CCS}(\mathsf{nil}) &= \mathsf{stop} \\
\phi_{CCS}(a.E) &= \mathsf{op}_{\Omega_;}\big(\mathsf{N}_a, \phi_{CCS}(E)\big) \\
\phi_{CCS}(E{+}F) &= \mathsf{op}_{\Omega_\square}\big(\phi_{CCS}(E), \phi_{CCS}(F)\big) \\
\phi_{CCS}(E|F) &= \mathsf{op}_{\Omega:syn(CCS)}\left(\mathsf{op}_{\Omega_\|}\left(\begin{array}{c}\mathsf{op}_{\Omega:left}\big(\phi_{CCS}(E)\big) \\ \mathsf{op}_{\Omega:right}\big(\phi_{CCS}(F)\big)\end{array}\right)\right) \\
\phi_{CCS}(E[h]) &= \mathsf{op}_{\Omega:relab(h)}\big(\phi_{CCS}(E)\big) \\
\phi_{CCS}(E \setminus a) &= \mathsf{op}_{\Omega:restr(a)}\big(\phi_{CCS}(E)\big) .
\end{aligned}$$

Exercise 8.2.1. Show that the translation given by $\phi_{CCS}(E)$ preserves the interleaving semantics of CCS expressions given in [81]. More precisely, show that the transition system of E is strongly bisimilar with that of $\overline{\phi_{CCS}(E)}$ after (i) deleting all the arcs labelled by steps of cardinality greater than one; (ii) ignoring all the moves involving the redo/skip actions; and (iii) replacing each arc label $\{a\}$ by a.

8.2.2 TCSP

The other two models employ synchronisation mechanisms different from those used in CCS and PBC. We will not provide a full translation of TCSP and COSY into a box algebra, leaving this task as an exercise for the reader. Rather, we shall briefly explain how to handle parallel composition in these two models.

The parallel composition in TCSP is a binary operator, $E\|_A F$, where $A \subseteq \mathsf{Lab}$ is a set of actions on which $\|_A$ enforces synchronisation. The resulting process can execute an action $a \in A$ if it can be executed simultaneously by the two component processes. The actions outside A can be executed autonomously by the two component processes. Such a synchronisation discipline can be modelled similarly as in CCS. This time, however, we do not assume that the set of labels, Lab, has any special properties. The relevant fragment of the transformation ϕ_{TCSP} from TCSP expressions to box algebra expressions is modelled as follows:

$$\phi_{TCSP}(E\|_A F) = \mathsf{op}_{\Omega:TCSP(A)}\left(\mathsf{op}_{\Omega_{\|}}\begin{pmatrix}\mathsf{op}_{\Omega:left}\left(\phi_{TCSP}(E)\right)\\ \mathsf{op}_{\Omega:right}\left(\phi_{TCSP}(F)\right)\end{pmatrix}\right),$$

where $\Omega{:}TCSP(A)$ is a simple operator box with the relabelling

$$TCSP(A) = \left\{\left(\{a^L, a^R\}, a\right) \,\middle|\, a \in A\right\} \cup$$
$$\left\{\left(\{a^L\}, a\right) \,\middle|\, a \in (\mathsf{Lab}\backslash A)\right\} \cup \left\{\left(\{a^R\}, a\right) \,\middle|\, a \in (\mathsf{Lab}\backslash A)\right\}.$$

As before, we assume that the set of labels is temporarily extended by labels a^L and a^R, for $a \in A$, which are not allowed in the TCSP syntax.

Exercise 8.2.2. Complete the translation for TCSP processes.

8.2.3 COSY

The parallel composition in COSY is based on multiway synchronisation. Consider a path program

$$prog = \mathsf{program}\ \ path_1\ \ \ldots\ \ path_n\ \ \mathsf{endprogram}\ .$$

It can execute a if it is executed simultaneously in all the paths $path_i$ in which a occurs. To model such a synchronisation mechanism, we take the generalised parallel composition operator box $\Omega_{\|J}$, shown in Fig. 4.40. Let $A \subseteq \mathsf{Lab}$ be the set of labels occurring in *prog* and, for every $a \in A$, let $index(a)$ be the set of all indices i such that a occurs in $path_i$. We define a simple operator $\Omega{:}Index$ with the relabelling being given as follows:

$$Index = \bigcup_{a \in A}\left\{\left(\bigcup_{i \in index(a)} \{a^i\}\ ,\ a\right)\right\}$$

and n simple operators, $\Omega{:}index_i$, with the relabellings

$$index_i = \left\{ \left(\{a\}, a^i \right) \,\middle|\, a \in A \right\}, \text{ for } i = 1, \ldots, n\,.$$

Then the relevant fragment of the transformation ϕ_{COSY} from COSY programs to box algebra expressions is given by

$$\phi_{COSY}(prog) \;=\; \mathsf{op}_{\Omega:Index}\left(\mathsf{op}_{\Omega_{\|J}} \begin{pmatrix} \mathsf{op}_{\Omega:index_1}\left(\phi_{COSY}(path_1)\right) \\ \cdots \\ \mathsf{op}_{\Omega:index_n}\left(\phi_{COSY}(path_n)\right) \end{pmatrix} \right).$$

Exercise 8.2.3. Complete the translation for COSY path programs and provide a similar translation for ACP [2].

8.3 Literature and Background

Other Petri net denotational semantics of process algebras may be found, for example, in [44] (for CCS), in [84] (for TCSP) and in [43] (for ACP).

It is possible to consider equivalences in the domain of PBC expressions other than the structural equivalence. For example, two static expressions are *duplication equivalent* if the corresponding static boxes are also duplication equivalent. The resulting equivalence was studied in [55, 56] where its sound and complete axiomatisation for a variant of the PBC syntax was given.

9. A Concurrent Programming Language

In this chapter, we discuss the use of PBC in giving the semantics of a concurrent programming language. We define an example language, called *Razor*, and use PBC (compositionally) as a semantic domain for *Razor*. This also induces, by the Petri net semantics of PBC given previously, a consistent compositional net semantics of *Razor*.

The language *Razor* is a mini-Pascal-like [101] block-structured imperative language supporting both shared data parallelism, meaning that it is possible to express concurrent processes operating on a common set of variables, and buffered communication, meaning that processes can communicate via shared buffers (we shall allow buffers of arbitrary capacity). *Razor* is an extension of Dijkstra's guarded command language [34] as well as a partial extension of *occam* [76] (excluding *occam*'s various priority operators).

Loops and choices (selection statements) of *Razor* are such that an inner loop is always 'guarded' and may never occur right at the start of an enclosing choice or loop. This avoids undesirable semantic situations relating to generalised loops, and thus we may use generalised PBC expressions defined in Sect. 8.1.3 in order to give the semantics of *Razor*.

We start, in Sect. 9.1, by describing the syntax of *Razor*. In Sect. 9.2, we give its semantics in terms of PBC and thus, implicitly, in terms of Petri nets. More precisely, we define a mapping pbc_{Razor} which associates with every *Razor* program a PBC expression (and thus, through $\mathsf{box}_{\mathrm{PBC}}$, also a net). One of the characteristic properties of this mapping is that it is compositional, i.e., it enjoys the homomorphism property of mapping operators of the programming language to operations on PBC expressions (and thus to operator nets). In Sect. 9.3, we illustrate the translation on examples. In Sect. 9.4 we describe an extended *Razor* supporting procedures and, in Sect. 9.5, we apply the general theory developed in the previous chapters in order to infer some properties of the semantics of (extended) *Razor*. Finally, in Sect. 9.6, we discuss possible applications.

9.1 Syntax of *Razor*

The syntax of *Razor*, specified in Fig. 9.1, should be understood as being prototypical for a whole class of similar languages which may, for instance, differ in terms of the set of arithmetic or logical operators they use. In a concrete implemented language, additional control flow operators may be needed, or some of the operators given in Fig. 9.1 may not be needed at all. The theory developed so far is flexible and should be able to cope easily with a wide range of such variations.

In this section, we explain the syntactical clauses in the order of their appearance, adding also restrictions that are not captured by Fig. 9.1 (note that some standard items in the syntax, such as variable identifier *id* and channel identifier *chanid*, are not specified explicitly). It is unavoidable that in this explanation, the intended (intuitive) semantics of *Razor* is anticipated informally. The formal semantics is presented later, in Sect. 9.2.

$$
\begin{array}{lll}
1.\ \mathit{Program} & ::= & \mathit{Block} \\
2.\ \mathit{Block} & ::= & \mathbf{begin}\ \mathit{Body}\ \mathbf{end} \\
3.\ \mathit{Body} & ::= & \mathit{Decl};\ \mathit{Body} \mid \mathit{Com} \\
4.\ \mathit{Decl} & ::= & \mathbf{var}\ \mathit{id} : \mathit{Set} \mid \mathbf{var}\ \mathit{chanid} : \mathbf{chan}\ \mathit{Capacity}\ \mathbf{of}\ \mathit{Set} \\
5.\ \mathit{Com} & ::= & \mathit{Block} \mid \llbracket \mathit{Act} \rrbracket \mid \mathit{Com}_1; \mathit{Com}_2 \mid \mathit{Com}_1 \| \mathit{Com}_2 \mid \\
 & & \mathbf{if}\ \mathit{GC}_1 \ \square \ \ldots \ \square \ \mathit{GC}_m\ \mathbf{fi} \mid \mathbf{do}\ \mathit{GC}_1 \ \square \ \ldots \ \square \ \mathit{GC}_m\ \mathbf{od} \\
6.\ \mathit{Act} & ::= & \mathit{id}{:=}\mathit{Expr} \mid \mathit{chanid}?x \mid \mathit{chanid}!\mathit{Expr} \\
7.\ \mathit{GC} & ::= & \mathit{GC}; \mathit{Com} \mid \llbracket \mathit{Bexpr} \rrbracket \mid \llbracket \mathit{Bexpr}; \mathit{Act} \rrbracket \\
8.\ \mathit{Expr} & ::= & \mathit{id} \mid \mathit{Const} \mid \mathit{Expr}\ \mathit{Binop}\ \mathit{Expr} \mid \mathit{Unop}\ \mathit{Expr} \mid \mathit{Bexpr} \\
9.\ \mathit{Bexpr} & ::= & \mathit{boolid} \mid \mathbf{false} \mid \mathbf{true} \mid \\
 & & \mathit{Bexpr}\ \mathit{Boolop}\ \mathit{Bexpr} \mid \neg\ \mathit{Bexpr} \mid \mathit{Expr}\ \mathit{Relop}\ \mathit{Expr} \\
10.\ \mathit{Binop} & ::= & + \mid - \mid * \mid \mathbf{mod} \mid \mathbf{div} \\
11.\ \mathit{Unop} & ::= & + \mid - \\
12.\ \mathit{Boolop} & ::= & \wedge \mid \vee \mid \rightarrow \\
13.\ \mathit{Relop} & ::= & = \mid \neq \mid < \mid \leq
\end{array}
$$

Fig. 9.1. Syntax of *Razor*

9.1.1 Programs and Blocks

As in other block-structured imperative languages, a program is simply a block. We require that it be closed, in the sense defined below.

A block starts with a declaration part (which may be empty) and ends with a command part (which must be nonempty). We require *declaration unambiguity*, in the sense that all variables declared in the declaration part of a block must be different.

Blocks may be nested, since the syntax of *Com* allows it to be a block. A block *Bl* is *closed* if every program variable occurrence inside its *Com* part is declared in *Bl*, or in some block nested within *Bl* enclosing that occurrence. The usual *scope rule*, which we adopt, applies in cases in which two or more blocks contain declarations of the same variable. The rule says that the declaration which is significant for the occurrence of a variable is the one which appears in the innermost block enclosing that occurrence. By declaration unambiguity, such a declaration (if it exists) is unique, and if a block is closed, then every occurrence of a variable in its *Com* part is declared. In particular, in a program — since it is closed — every variable occurrence has exactly one significant declaration, namely the one of the innermost enclosing block that declares it.

9.1.2 Declarations

It is assumed that *id* is an identifier for a plain program variable and *chanid* is an identifier for a channel variable. Variable and channel identifiers obey the same syntax: they are strings of letters and digits, starting with a letter. This is a standard convention which is not represented explicitly in the syntax of Fig. 9.1. We may (and will) assume that the set of variable identifiers and the set of channel identifiers are disjoint.

The syntactic entity *Set* is called the *type* of a variable *id* or the *base type* of the channel variable *chanid*; it will be denoted by *Type*(*id*) and *Type*(*chanid*), respectively. *Set* is an arbitrary finite or denumerably infinite set, and may either be specified directly (e.g., enumerated as $\{0, 1\}$ or $\{\mathbf{false}, \mathbf{true}\}$; in the former case, the variable is called 'binary', and in the latter case, 'Boolean') or by a 'type identifier' (e.g., $\mathbf{N}$ may be viewed as a type identifier for the set $\{0, 1, 2, \ldots\}$). Concrete languages such as Pascal allow considerable syntactic flexibility in the way types may be specified, but for our purposes we do not need to elaborate on this.

The syntactic entity *Capacity* is a constant in the set $\mathbf{N} \cup \{\omega\}$, called the *capacity* of the channel. A channel is used for the communication between parallel processes, and *Capacity* specifies the maximal number of values it may store. If the capacity is 0, then such a channel may be used in order to realise a 'handshake' communication as in CSP [58] and *occam*. A nonzero integer capacity k defines a bounded FIFO (first-in-first-out, like a UNIX pipe) buffer which may hold up to k items. The infinite capacity ω defines an unbounded FIFO buffer which may hold arbitrarily many values.

9.1.3 Commands and Actions

The intuitive meaning of the brackets $[\![\ldots]\!]$ is that they enclose *atomic actions*. The meaning of $x{:=}Expr$ is standard; it denotes the assignment of the value given by *Expr* to variable x. Thus, for instance, the *semaphore* [35] action 'signal' $V(x)$, with a semaphore variable x, can be encoded as $[\![x{:=}x{+}1]\!]$ (or as $[\![x{:=}1]\!]$ if x is a binary semaphore). The other two actions, $c?x$ and $c!Expr$, denote respectively the removal of a value from the channel c followed atomically by the assignment of this value to the variable x; and the insertion of a value given by the expression *Expr* to the channel c.

The constructs $Com_1; Com_2$ and $Com_1 \| Com_2$ denote sequential composition and parallel composition, respectively. We will assume that the former binds more strongly than the latter. Thus, $Com_1; Com_2 \| Com_3$ is equivalent to $(Com_1; Com_2) \| Com_3$ rather than $Com_1; (Com_2 \| Com_3)$. As usual, bracketing can override this precedence rule, in order to make syntactic descriptions unambiguous or more comprehensible. Some brackets may be redundant by algebraic rules. Strictly speaking, for instance, $(Com_1; Com_2); Com_3$ differs from $Com_1; (Com_2; Com_3)$, but after giving the semantic definitions we may use the associativity of the sequential composition in order to omit the brackets altogether.

The **if** ... **fi** and **do** ... **od** clauses denote choice and loop (with an inner choice within the body), respectively, as in Dijkstra's guarded command language. Informally, **if** ... **fi** means: 'choose one of the enclosed guarded commands that can be executed', and **do** ... **od** means: 'repeatedly choose one of the enclosed guarded commands, as long as possible'. The index m in $GC_1 \,\Box\, \ldots \,\Box\, GC_m$ may be any natural number, including zero, in which case the two commands become **if fi** and **do od**; the semantics of the two latter constructs will be discussed in detail in Sect. 9.2.3.

9.1.4 Guarded Commands

The syntactic entity *GC* is called a *guarded command*, and the Boolean expression *Bexpr* in its first part is called a 'guard'. More formally, we define *guard*(*GC*) inductively, as follows:

$$\begin{aligned} guard(GC; Com) &= guard(GC) \\ guard([\![Bexpr]\!]) &= Bexpr \\ guard([\![Bexpr; Act]\!]) &= Bexpr \;. \end{aligned}$$

Thus, for example,

$$\begin{aligned} guard([\![(x{>}0) \,;\, (x{:=}x{-}1)]\!]) &= (x{>}0) \\ guard([\![(x{=}1) \,;\, (x{:=}0)]\!]) &= (x{=}1) \\ guard([\![x{=}0]\!]) &= (x{=}0) \;. \end{aligned}$$

The idea is that a guard can be 'chosen' or 'passed' — and thus, the atomic command it is guarding executed — only if it evaluates to **true**. For instance, the atomic semaphore action $P(x)$, which is defined informally as 'wait until x is greater than 0, then decrement x', can be encoded as **if** ⟦ $(x{>}0)$; $(x{:=}x{-}1)$ ⟧ **fi** (or as **if** ⟦ $(x{=}1)$; $(x{:=}0)$ ⟧ **fi**, if x is binary). This expresses that the positive test for $x > 0$ (respectively, $x = 1$) and the subsequent change of x have to be viewed and implemented as an indivisible action. As another example, a simple 'wait until $x = 0$' can be encoded in the language as **if** ⟦ $x{=}0$ ⟧ **fi**.

Note that by the syntax of *GC*, choice clauses **if** ... **fi** and loop clauses **do** ... **od** may not start directly with other loops, since an inner loop is always guarded by a nonlooping clause of the form ⟦ *Bexpr* ⟧ or ⟦ *Bexpr*; *Act* ⟧, which has to be executed atomically.

9.1.5 Expressions and Operators

Expr denotes an *expression* and *Bexpr* denotes a *Boolean expression.* For expressions, we allow several binary and unary *operators*, *Binop* and *Unop*, as well as *constants*, *Const*. A constant is any member of a (base type) set. For example, if both x and y are variables of type $\{0, 1\}$, we may consider the constant 1 and the binary operator '−'; then the assignment $x{:=}1{-}y$ is well defined. For any given value (0 or 1) of y, it yields a well-defined new value of x (1 or 0, respectively). As another example, **false** is a constant belonging to the Boolean type {**false**, **true**}.

Boolean expressions *Bexpr* are special. To serve as a Boolean expression by itself, a variable *boolid* (line 9 of the syntax) has to be of Boolean type. We allow the usual binary Boolean operators *Boolop*, the (only) usual unary Boolean operator (¬), and few selected relational operators *Relop*. For instance, $x{\neq}y$ is a Boolean expression, provided x and y are variables of the same type. The only reason for treating Boolean expressions separately is that in guarded commands (line 7) we require Boolean expressions, rather than just any expressions, to be present.

The 'expressions' and 'constants' as defined in this section pertain to *Razor* programs and must not be confused with the '(PBC or box algebra) expressions' and '(process) constants' used elsewhere in this book.

9.1.6 Syntactic Variations

As a 'syntactic sugar', we will allow explicit initialisations of variables, according to the scheme:

Decl ::= ... | **var** x : *Set* (**init** *Expr*) .

Such a declaration is translated into one without the **init** clause which is replaced by an additional assignment ⟦ $x{:=}$*Expr* ⟧ at the start of the *Com* part of the block. If there are more than one initialisation clauses, then the corresponding assignments are inserted in the same order as the declarations. Usually, the initialisation expressions *Expr* are simply constants.

Sometimes we will also use **var** $x, y : Set$ as an abbreviation of the declarations **var** $x : Set$; **var** $y : Set$.

Exercise 9.1.1. Carry out the expansion of the program

begin **var** $x : \{1,2\}$ (**init** 1);
  begin **var** $y : \{1,2\}$ (**init** x);
    var $x : \{1,2\}$ (**init** 2);
    ...
  end
end

9.2 Semantics of *Razor*

Let P be a *Razor* program fragment, which could be a program, block, body, declaration, etc., defined by lines 1–7 of Fig. 9.1. We will now define a mapping pbc_{Razor} associating with P a PBC expression, $\mathsf{pbc}_{Razor}(P)$, and thus also a box, $\mathsf{box}_{\mathrm{PBC}}(\mathsf{pbc}_{Razor}(P))$. In defining this mapping, we will resort to basic process expressions and operators defined previously in this book, except for the translation of channel declarations and modelling of atomic actions and guards, where we will provide new basic PBC processes (i.e., constants in the process algebra sense). Channel constants are analogous to the data constants already discussed in Sects. 2.8, 3.3.2, and 4.4.10. The semantics of atomic actions and guards will be expressed using simple (static) constant expressions, $\Box_{\mathcal{A}}$, where $\mathcal{A}$ is a set of PBC multiactions. The operational semantics of $\Box_{\mathcal{A}}$ is given by

$$(OP_{[\Box_{\mathcal{A}}]}) \quad \boxed{\overline{\Box_{\mathcal{A}}} \xrightarrow{\{\alpha\}} \underline{\Box_{\mathcal{A}}} \qquad \text{for } \alpha \in \mathcal{A}}$$

and the net associated with $\Box_{\mathcal{A}}$ is the plain static box $\mathsf{N}_{\Box_{\mathcal{A}}}$ shown in Fig. 9.2.

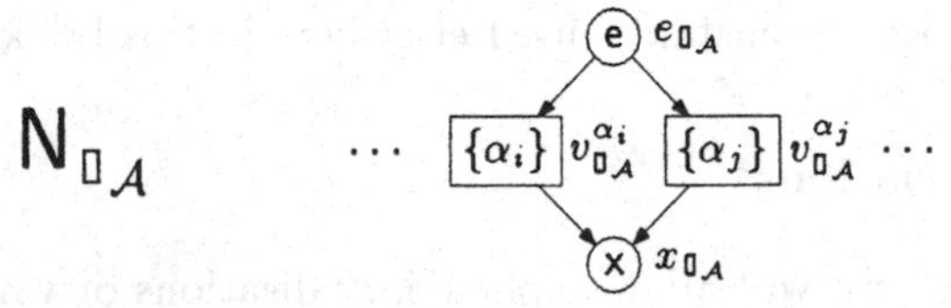

Fig. 9.2. Static box representing the constant $\Box_{\mathcal{A}}$ where $\mathcal{A} = \{\ldots, \alpha_i, \alpha_j, \ldots\}$

Both the operational and net semantics of $\Box_{\mathcal{A}}$ are equivalent to that of the generalised choice $\Box_{\alpha\in\mathcal{A}}\{\alpha\}$ and, moreover, $\Box_{\emptyset}$ is equivalent to stop, and $\Box_{\{\alpha\}}$ to α. When discussing examples, we will take advantage of this correspondence. We do not use the potentially infinitary generalised choice operator since we would then need to re-evaluate the theory to convince the reader that its actual usage in the modelling of *Razor* programs is unproblematic. By the above device, generalised choice is not necessary for giving the semantics of *Razor*.

In Sects. 9.2.1–9.2.4, we proceed top-down, by induction on the syntax of *Razor*. The whole translation has 16 parts, (1)—(16), of which we give explicitly the first 14, cf. formulae (1)—(14) below, and leave the remaining two as an exercise for the reader.

9.2.1 Programs, Blocks, and Declarations

Following an idea due to Milner [80, 81], the main device in describing a block is to juxtapose (in parallel) the nets for its declarations (the *Decl* part) and the net for the rest of its body (the *Com* part), and then to enclose both of them within scoping brackets, as in

$$[\,A\,:\,(\mathit{Decl}\parallel \mathit{Com})\,]\,,$$

where A is a suitable set of action particles. In a particular realisation of this idea, we will associate with every data variable x and channel variable c a set of action particles. Whenever the declaration part of a block is translated, we will use the 'unhatted' versions of these particles, while whenever one of the commands inside the block is translated, we will use their 'hatted' versions; thus, on the left-hand side of the parallel operator, only unhatted actions appear, while on the right-hand side, only hatted actions appear. After scoping, all action particles pertaining to variables or channels of the present block disappear, while those pertaining to enclosing blocks remain. In this way, local variables are made invisible outside the block effecting the application of the scope rule. We also need a termination mechanism, since the block may be nested inside a loop, and upon subsequent entries to the loop, variables are not supposed to retain any values from the previous iterations.

Let *Decl* be the declaration **var** x : *Set* of a data variable x. Then, by definition, x_{ab} and $\widehat{x_{ab}}$, for all $a,b \in \mathit{Type}(x) \cup \{\bullet\} = \mathit{Set} \cup \{\bullet\}$, are PBC action particles associated with it. The particle x_{ab} as well as its conjugate particle $\widehat{x_{ab}}$ represent the change of x's value from a to b, and $\bullet$ stands for an 'undefined value' (see Sects. 2.8 and 3.3.2). Moreover, we define

$$\delta(\mathit{Decl}) = \{\, x_{ab} \mid a,b \in \mathit{Type}(x) \cup \{\bullet\}\,\} \quad \text{and} \quad \tau(\mathit{Decl}) = \Box_{\mathcal{A}(\mathit{Decl})}\,,$$

where $\mathcal{A}(\mathit{Decl}) = \{\{\widehat{x_{a\bullet}}\} \mid a \in \mathit{Type}(x) \cup \{\bullet\}\}$. Thus $\delta(\mathit{Decl})$ is the set of all 'unhatted' action particles that pertain to the declaration of x. This will be the set over which scoping is carried out. The process constant $\tau(\mathit{Decl})$ specifies a choice over all 'hatted' actions whose particles can lead to the termination of the variable (resulting in an undefined value). Such a construct will be used for the termination of a block in which the declaration occurs.

Recall from Sects. 3.3.2 and 4.4.10 that the semantics of *Decl* will be described by an expression $[x_{()}]$, which uses the action particles just defined. Actually, $[x_{()}]$ depends on the set *Set* that occurs in *Decl*, and in order to be completely precise in circumstances when more than one such set is used, we will parameterise $[x_{()}]$ by this set. Hence we will use the notation $[x_{()}^{Set}]$, and if *Set* is clear from the context, we will use $[x_{()}]$ as an abbreviation.

Let *Decl* be the declaration **var** c : **chan** k **of** *Set* of a channel c with the basic type $\mathit{Type}(c) = \mathit{Set}$ and capacity $k \in \mathbf{N} \cup \{\omega\}$. The capacity can range from $k = 0$ (handshake synchronisation) to $k \in \mathbf{N}\backslash\{0\}$ (bounded FIFO buffer of capacity k) to $k = \omega$ (unbounded FIFO buffer). We define the *value domain* $\mathit{Type}^k(c)$ of c as the set of all *finite* sequences of elements of $\mathit{Type}(c)$, of length at most k; in particular,

$$\mathit{Type}^0(c) = \{\varepsilon\},\ \mathit{Type}^1(c) = \mathit{Type}(c) \cup \{\varepsilon\},\ \text{and}\ \mathit{Type}^\omega(c) = \mathit{Type}(c)^*\ .$$

Again, we define PBC action particles associated with channel variable c, namely $c_{!a}$ and $c_{?a}$, for $a \in \mathit{Type}(c)$, and $c_\bullet$, as well as their 'hatted' versions, $\widehat{c_{!a}}$ and $\widehat{c_{?a}}$, for $a \in \mathit{Type}(c)$, and $\widehat{c_\bullet}$. Moreover,

$$\delta(\mathit{Decl}) \ = \ \{\, c_{!a} \mid a \in \mathit{Type}(c)\,\} \ \cup \ \{\, c_{?a} \mid a \in \mathit{Type}(c)\,\} \ \cup \ \{c_\bullet\}$$

and $\tau(\mathit{Decl}) = \{\widehat{c_\bullet}\}$. The particles $c_{!a}$ and $\widehat{c_{!a}}$ represent the fact that value a is sent to (and then inserted, capacity permitting, at the rear of) channel c; $c_{?a}$ and $\widehat{c_{?a}}$ represent the fact that value a is read and removed (if present) from the front of channel c; and $c_\bullet$ together with $\widehat{c_\bullet}$ represent termination.

We then introduce a new family of basic processes, $[c_{()}]$, modelling channels c in the same style as plain variables are modelled by $[x_{()}]$. To be exact, $[c_{()}]$ depends also on k and *Set*, whence we will use the notation $[c_{()}^{k,Set}]$, unless k and *Set* are clear from the context. The semantics of the constant expression $[c_{()}]$ will be defined in Sect. 9.2.2.

We are now in a position to define the semantics of blocks (and hence *Razor* programs), as follows:

$$\begin{aligned}
&\mathsf{pbc}_{Razor}(\textbf{begin}\ \mathit{Body}\ \textbf{end}) \\
&= \mathsf{pbc}_{Razor}(\mathit{Body}) && (1) \\
&\mathsf{pbc}_{Razor}(\mathit{Decl};\mathit{Body}) \\
&= \Big[\ \delta(\mathit{Decl})\ :\ \Big(\ \mathsf{pbc}_{Razor}(\mathit{Decl}) \parallel (\,\mathsf{pbc}_{Razor}(\mathit{Body}); \tau(\mathit{Decl})\,)\ \Big)\Big] && (2) \\
&\mathsf{pbc}_{Razor}\Big(\ \textbf{var}\ x : \mathit{Set}\ \Big) \\
&= [x_{()}^{Set}]\ \square\ \{x_{\bullet\bullet}\} && (3) \\
&\mathsf{pbc}_{Razor}(\,\textbf{var}\ c : \textbf{chan}\ k\ \textbf{of}\ \mathit{Set}\,) \\
&= [c_{()}^{k,Set}]\ \square\ \{c_\bullet\}\ . && (4)
\end{aligned}$$

Note that in line (2), we used the generalised scoping operator

$$[\, \delta(Decl) \; : \; \ldots \,]$$

introduced in Sect. 3.3.3, whose well-definedness relies on the commutativity of scoping (which itself results from the fact that noninterfering synchronisation and restriction commute, see Exc. 3.2.31). The set $\delta(Decl)$ may even be infinite if the type of the variable in question is infinite too.

Line (1) is self-explanatory. Line (2) translates declarations sequentially in the order of their occurrence, in the sense that it creates a nested expression containing as its innermost part the *Com* part of the block *Body*, which is enclosed by pairs of parallel compositions of declarations (one by one, innermost being the last one) and scopings over the action particles of the declaration. The scopings synchronise the 'unhatted' particles on the left-hand side of the parallel operator with the 'hatted' particles on its right-hand side.

Lines (3) and (4) introduce the 'unhatted' particles of the declarations. The expression $[x_{()}^{Set}]$ in line (3) controls the possible changes of the value of variable x, from its initial undefined value. The expression $[c_{()}^{k,Set}]$ in line (4) plays a similar role for channels. The extra actions $\ldots \, \square \, \{x_{\bullet\bullet}\}$ (line (3)) and $\ldots \, \square \, \{c_{\bullet}\}$ (line (4)) cover the possibility that a variable or channel may be created at the block entry, and then destroyed at the block exit without being accessed by any executed command.

Exercise 9.2.1. Use Thm. 4.4.1 and the net semantics of channels described in Sect. 9.2.2 to demonstrate that the scopings (synchronisations) described above do not introduce infinitely many transitions in the net modelling a *Razor* program, unless there are already infinitely many transitions in the component nets (due to infinite type sets).

9.2.2 Basic Channel Processes

In this section we assume declarations **var** c : **chan** k **of** *Set* to be given, and augment PBC with a family of static constants $[c_{()}^{k,Set}]$ (which we abbreviate $[c_{()}]$ since k and *Set* will be clear from the context). These constants are used in the semantics of channel declarations, as described in the previous section, in line (4).

Asynchronous Channels. We first assume that $k \neq 0$. Then c may store up to k elements of $Type(c)$ as values (or arbitrarily many if $k = \omega$). The label-based operational semantics of the channel expression $[c_{()}]$ is given by the following rules, where $a \in Type(c)$ and $w, v \in Type^k(c)$, and $[\, \overline{c_{(w)}} \,]$ is a dynamic constant.

$$(OP_{[c_{()}]})\quad \begin{array}{|llcll|}\hline \text{ACHAN1} & \underline{[c_{()}]} & \xrightarrow{\{\{c!a\}\}} & [\,\overline{c_{(a)}}\,] & \\ \text{ACHAN2a} & [\,\overline{c_{(w)}}\,] & \xrightarrow{\{\{c!a\}\}} & [\,\overline{c_{(aw)}}\,] & \text{if } \;|w| < k \\ \text{ACHAN2b} & [\,\overline{c_{(w)}}\,] & \xrightarrow{\{\{c?a\}\}} & [\,\overline{c_{(v)}}\,] & \text{if } \; w = va \\ \text{ACHAN3} & [\,\overline{c_{(w)}}\,] & \xrightarrow{\{\{c_\bullet\}\}} & \underline{[c_{()}]} & \\ \hline \end{array}$$

That is, an empty asynchronous channel c can be initialised by being sent a value (ACHAN1), repeatedly filled or emptied provided the relevant conditions are met (ACHAN2a and ACHAN2b, respectively), and terminated (ACHAN3). Notice that for $k = \omega$ the condition of ACHAN2a is void. The following derivation illustrates the use of these rules for a channel c with the declaration **var** c : **chan** 2 **of** $\{0, 1\}$:

$$\begin{array}{lll} \overline{[c_{()}]} \xrightarrow{\{\{c!0\}\}} [\,\overline{c_{(0)}}\,] & \text{ACHAN1} \\ & (0 \text{ is sent to } c \text{ which now stores } 0) \\ \xrightarrow{\{\{c!1\}\}} [\,\overline{c_{(10)}}\,] & \text{ACHAN2a } (w = 0 \text{ and } |0| = 1 < 2) \\ & (1 \text{ is sent to } c \text{ which now stores } 10) \\ \xrightarrow{\{\{c?0\}\}} [\,\overline{c_{(1)}}\,] & \text{ACHAN2b } (w = 10 \text{ and } v = 1) \\ & (0 \text{ is removed from } c \text{ which now stores } 1) \\ \xrightarrow{\{\{c_\bullet\}\}} \underline{[c_{()}]} & \text{ACHAN3} \\ & (c \text{ terminates}). \end{array}$$

We still need to provide a Petri net translation of the static constant $[c_{()}]$. Consider, as a special case, the declaration **var** c : **chan** 1 **of** $\{0, 1\}$, of a binary channel of capacity 1. Figure 9.3 represents the box $\mathsf{N}_{[c_{()}]}$ associated with $[c_{()}]$. Notice that e_c is the 'undefined' state of the buffer while i_c^ε is the state in which it stores the empty sequence of values. Keeping in mind that according to line (4) of the semantics, $[c_{()}]$ is always used in choice with $\{c_\bullet\}$, e_c and i_c^ε could be identified if losing e-directedness would not be a problem. This identification would correspond to the fact that the empty sequence is a natural initial value for any channel. In Fig. 9.3, the middle row consists of exactly three places because the binary channel of capacity 1 can store exactly one out of the three sequences: 0 or ε or 1. The three dynamic plain boxes corresponding to the dynamic constants $[\,\overline{c_{(0)}}\,]$, $[\,\overline{c_{(\varepsilon)}}\,]$, and $[\,\overline{c_{(1)}}\,]$, are defined as $\mathsf{N}_{[c_{()}]}$ with a single token in the place i_c^0, i_c^ε, and i_c^1, respectively.

Exercise 9.2.2. Formulate a Petri net translation for an arbitrary $k > 0$ and arbitrary *Type*(c) ensuring behavioural conformity with the rules $(OP_{[c_{()}]})$. Also, formulate the transition-based operational semantics for the corresponding channel constants.

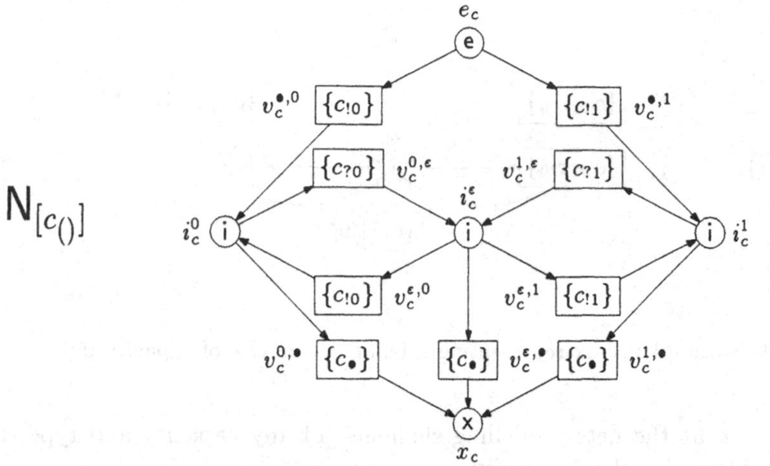

Fig. 9.3. Static plain box representing a binary channel c of capacity 1

Synchronous Channels. We next assume that $k = 0$. This means that c may not actually store any value, but rather effects a handshake synchronisation (instantaneous transmission of values) between a sender and a receiver. Then the operational semantics of the channel expression $[c_{()}]$ is given by the following rules, where $a \in \mathit{Type}(c)$ and $[\overline{c_{()}}]$ is a dynamic constant:

$$(OP'_{[c_{()}]}) \qquad \begin{array}{|lcl|} \text{SCHAN1} & \overline{[c_{()}]} \xrightarrow{\{\{c!_a, c?_a\}\}} [\,\overline{c_{()}}\,] \\ \text{SCHAN2} & [\,\overline{c_{()}}\,] \xrightarrow{\{\{c!_a, c?_a\}\}} [\,\overline{c_{()}}\,] \\ \text{SCHAN3} & [\,\overline{c_{()}}\,] \xrightarrow{\{\{c_\bullet\}\}} \underline{[c_{()}]} \end{array}$$

That is, a synchronous channel c can perform an initial handshake (SCHAN1), then repeatedly perform further handshakes (SCHAN2), and finally terminate (SCHAN3). Again, the possibility that the channel terminates without ever being used is modelled as part of the semantics of a channel declaration in line (4) of Sect. 9.2.1.

Figure 9.4 describes the box $\mathsf{N}_{[c_{()}]}$ representing $[c_{()}]$ for the channel declared by **var** c : **chan** 0 **of** $\{0,1\}$. This box can informally be viewed as obtained from the previous one (in Fig. 9.3) by folding the three places in the middle row into a single central place, which corresponds to the fact that previously $\mathit{Type}^1(c) = \{\varepsilon, 0, 1\}$ had three elements while now $\mathit{Type}^0(c) = \{\varepsilon\}$ has only one. The dynamic plain box corresponding to the dynamic constant $[\overline{c_{()}}]$ is defined as $\mathsf{N}_{[c_{()}]}$ with a single token in the place i_c.

Exercise 9.2.3. Formulate a Petri net translation for $k = 0$ and an arbitrary (countable) $\mathit{Type}(c)$ ensuring behavioural conformity with the rules $(OP'_{[c_{()}]})$. Also, formulate the transition-based operational semantics for the corresponding channel constants.

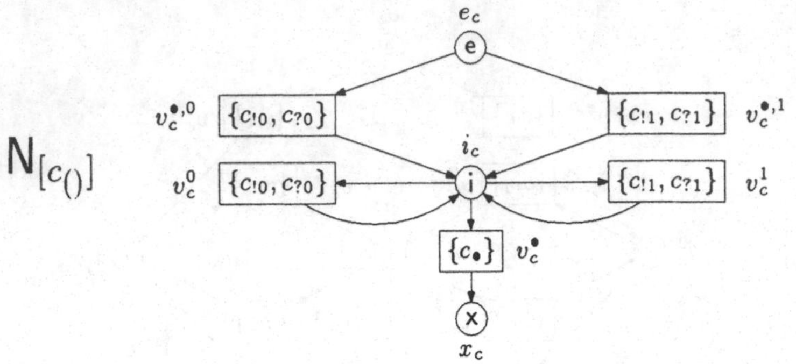

Fig. 9.4. Static plain box representing a binary channel c of capacity 0

Notice that the nets modelling channels (of any capacity and type) are both ex-directed and ex-exclusive.

9.2.3 Command Connectives

The semantics of the command connectives is defined in the following way:

$$\begin{aligned}&\mathsf{pbc}_{Razor}(Com_1; Com_2)\\ &\quad = \mathsf{pbc}_{Razor}(Com_1) ; \mathsf{pbc}_{Razor}(Com_2)\end{aligned} \tag{5}$$

$$\begin{aligned}&\mathsf{pbc}_{Razor}(Com_1 \| Com_2)\\ &\quad = \mathsf{pbc}_{Razor}(Com_1) \| \mathsf{pbc}_{Razor}(Com_2)\end{aligned} \tag{6}$$

$$\begin{aligned}&\mathsf{pbc}_{Razor}(\mathbf{if}\ GC_1 \,\square \ldots \square\, GC_m\ \mathbf{fi})\\ &\quad = \mathsf{pbc}_{Razor}(GC_1) \,\square\, \ldots \,\square\, \mathsf{pbc}_{Razor}(GC_m)\end{aligned} \tag{7}$$

$$\begin{aligned}&\mathsf{pbc}_{Razor}(\mathbf{do}\ GC_1 \,\square \ldots \square\, GC_m\ \mathbf{od})\\ &\quad = \left\langle \begin{array}{l}(\mathsf{pbc}_{Razor}(GC_1) \,\square\, \ldots \,\square\, \mathsf{pbc}_{Razor}(GC_m))\\ * \ \mathsf{pbc}_{Razor}([\![\neg guard(GC_1) \wedge \ldots \wedge \neg guard(GC_m)]\!])\end{array}\right]\end{aligned} \tag{8}$$

$$\begin{aligned}&\mathsf{pbc}_{Razor}(GC; Com)\\ &\quad = \mathsf{pbc}_{Razor}(GC) ; \mathsf{pbc}_{Razor}(Com) .\end{aligned} \tag{9}$$

Lines (5)—(7) and (9) translate sequence, parallel composition, and **if** ... **fi** commands into the corresponding PBC operators. Note that in line (7), as well as in line (8), we use the associativity of the choice operator to avoid nested parentheses. In line (8), the *Razor* loop is translated into the generalised loop construct $\langle F * E]$ where F corresponds to the body of the loop and E corresponds to its termination. F is a direct translation of the loop body, while E is the translation of the conjunction of all negated guards. Thus, termination can occur (only) if no guard is passable. Note that we assume here that the evaluation of all the guards is done atomically; in a nonatomic discipline — not considered here — already tested guards may change their truth values while others are being examined. The generalised iteration operator $\langle F * E]$ is described in Sect. 3.3.1 (see also Sect. 8.1.3), and its operator box $\Omega_{\langle *]}$ is shown in Fig. 4.44. Referring to the discussion and results in Sect. 8.1, we will show in Sect. 9.5 that $\langle F * E]$ can be used 'safely' in line (8). For $m = 1$, lines (7) and (8) reduce to the following:

$$\mathsf{pbc}_{Razor}(\textbf{if } GC_1 \textbf{ fi}) \quad = \mathsf{pbc}_{Razor}(GC_1)$$
$$\mathsf{pbc}_{Razor}(\textbf{do } GC_1 \textbf{ od}) = \langle\, \mathsf{pbc}_{Razor}(GC_1) * \mathsf{pbc}_{Razor}([\![\, \neg guard(GC_1)\,]\!])\,] \,.$$

The special case of $m = 0$ in lines (7) and (8) is treated as follows. On the right-hand side of line (7), we then get an empty choice. Since choice composition is commutative and associative (cf. Excs. 3.2.6 and 4.4.2), the translation should result in a unit of choice. Thus it will be taken to be the stop static constant (with no rules of the operational semantics) and, in terms of nets, the ex-box of Fig. 4.35 (cf. Exc. 4.4.21). Thus, colloquially, '**if fi** equals deadlock'. On the right-hand side of line (8), we then get stop as the loop body, and as the termination expression $\mathsf{pbc}_{Razor}([\![\, \textbf{true}\,]\!])$ (due to the empty conjunction). The latter, as we will see, corresponds to the 'silent' box expression $\emptyset$ and so the body of such a loop can never be executed, but its termination action can always be executed. Thus, colloquially, '**do od** equals **skip** — a terminating action with no external effect'.

9.2.4 Actions and Guarded Commands

In order to complete the definition of the mapping pbc_{Razor}, we still need to specify its effect for the following seven clauses:

(10)	$[\![\, Bexpr \,]\!]$	(Boolean expression)
(11)	$[\![\, x{:=}Expr \,]\!]$	(assignment)
(12)	$[\![\, Bexpr\, ;\, x{:=}Expr \,]\!]$	(guarded assignment)
(13)	$[\![\, c?x \,]\!]$	(receive action)
(14)	$[\![\, c!Expr \,]\!]$	(send action)
(15)	$[\![\, Bexpr\, ;\, c?x \,]\!]$	(guarded receive action)
(16)	$[\![\, Bexpr\, ;\, c!Expr \,]\!]$	(guarded send action) .

The definitions all follow the same pattern. Before giving them, we need to define a few auxiliary notions which are standard in the semantics of programming languages (see [3]).

For a given expression *Expr*, let *Var*(*Expr*) be the set of variables occurring in *Expr*. For instance, $Var(1{-}y)$ equals $\{y\}$ and $Var(x{\neq}y)$ equals $\{x, y\}$. Let $x^1, \ldots, x^m$ be typed variables, *Expr* be an expression with $Var(Expr) \subseteq \{x^1, \ldots, x^m\}$, and $a_1, \ldots, a_m$ be members of the types of, respectively, $x^1, \ldots, x^m$. Then the expression

$$Expr[x^1, \ldots, x^m \leftarrow a_1, \ldots, a_m]$$

is defined to be the same as *Expr*, except that every occurrence of each x^j is replaced by a_j. Note that the resulting expression is variable-free. For example, $(1-y)[y \leftarrow 0]$ equals $(1-0)$ and $(x \neq y)[x \leftarrow 0, y \leftarrow 0]$ equals $(0 \neq 0)$. Finally, let *Expr* be any variable-free expression, i.e., $Var(Expr) = \emptyset$. Then *Expr* can be 'evaluated', i.e., a partial function $eval(.)$ exists which may yield a value of *Expr*. For instance, $eval(1-0) = 1$, $eval(0 \neq 0) = \mathbf{false}$, and $eval(1\ \mathbf{div}\ 0)$ is undefined. In giving the semantics of *Razor*, we shall assume that a variable-free expression can always be evaluated, otherwise the *Razor* program is deemed semantically incorrect and has no true semantics. That is, we put aside the error handling aspects of real implementation, as these are orthogonal to the issues we are concerned with in this book.

Boolean Expression: $[\![$ *Bexpr* $]\!]$. Let $Var(Bexpr) = \{x^1, \ldots, x^m\}$ and let $\mathcal{A}(BExpr)$ be a set of multiactions defined as follows:

$$\left\{ \left\{ \widehat{x^1_{a_1 a_1}}, \ldots, \widehat{x^m_{a_m a_m}} \right\} \middle| \begin{array}{l} a_1 \in Type(x^1) \ \wedge \ \ldots \ \wedge \ a_m \in Type(x^m) \\ eval(Bexpr[x^1, \ldots, x^m \leftarrow a_1, \ldots, a_m]) = \mathbf{true} \end{array} \right\}.$$

Then the translation for $[\![$ *Bexpr* $]\!]$ is

$$\mathsf{pbc}_{Razor}([\![\, Bexpr \,]\!]) \ = \ \Box_{\mathcal{A}(Bexpr)} \,. \tag{10}$$

That is, $\mathsf{pbc}_{Razor}([\![\, Bexpr \,]\!])$ consists of a choice of multiactions (using the previously defined action particles pertaining to the free variables) describing possible values of the free variables, such that the expression evaluates to **true**. For instance, assuming that x and y are declared by **var** $x, y : \{0,1\}$, we obtain:

$$\mathsf{pbc}_{Razor}([\![\, y{=}0 \,]\!]) = (\,\{\widehat{y_{00}}\}\,)$$
$$\mathsf{pbc}_{Razor}([\![\, x{\neq}y \,]\!]) = (\,\{\widehat{x_{00}}, \widehat{y_{11}}\} \,\Box\, \{\widehat{x_{11}}, \widehat{y_{00}}\}\,)\,.$$

Also, consider $Bexpr = \mathbf{false}$. Then $m = 0$ because there are no free variables, and since $eval(\mathbf{false}) \neq \mathbf{true}$ and

$$\mathcal{A}(\mathbf{false}) = \{\{\,\} \mid eval(\mathbf{false}) = \mathbf{true}\} = \emptyset\,,$$

we obtain an empty choice, which is equal to the stop box expression. Consider, on the other hand, $Bexpr = \mathbf{true}$. Then, since $eval(\mathbf{true}) = \mathbf{true}$ and

$$\mathcal{A}(\mathbf{true}) = \{\{\,\} \mid eval(\mathbf{true}) = \mathbf{true}\} = \{\emptyset\}\,,$$

we obtain the 'silent' box expression $\emptyset$.

Assignment: $[\![$ *x:=Expr* $]\!]$. The translation is given by

$$\mathsf{pbc}_{Razor}([\![\, x{:=}Expr \,]\!]) \ = \ \Box_{\mathcal{A}(x:=Expr)}\,, \tag{11}$$

where the set of multiactions $\mathcal{A}(x{:=}Expr)$ is defined as follows. If $x \notin Var(Expr)$, then $\mathcal{A}(x{:=}Expr)$ is

$$\left\{ \left\{ \widehat{x_{ab}}, \widehat{x^1_{a_1 a_1}}, \ldots, \widehat{x^m_{a_m a_m}} \right\} \middle| \begin{array}{l} a \in \mathit{Type}(x) \cup \{\bullet\} \ \wedge \ b \in \mathit{Type}(x) \\ a_1 \in \mathit{Type}(x^1) \ \wedge \ \ldots \ \wedge \ a_m \in \mathit{Type}(x^m) \\ \mathit{eval}(\mathit{Expr}[x^1, \ldots, x^m \leftarrow a_1, \ldots, a_m]) = b \end{array} \right\}$$

provided that $\mathit{Var}(\mathit{Expr}) = \{x^1, \ldots, x^m\}$, and if $x \in \mathit{Var}(\mathit{Expr})$, then the set of multiactions $\mathcal{A}(x{:=}\mathit{Expr})$ is

$$\left\{ \left\{ \widehat{x_{ab}}, \widehat{x^1_{a_1 a_1}}, \ldots, \widehat{x^m_{a_m a_m}} \right\} \middle| \begin{array}{l} a \in \mathit{Type}(x) \ \wedge \ b \in \mathit{Type}(x) \\ a_1 \in \mathit{Type}(x^1) \ \wedge \ \ldots \ \wedge \ a_m \in \mathit{Type}(x^m) \\ \mathit{eval}\left(\mathit{Expr} \left[\begin{array}{c} x, x^1, \ldots, x^m \\ \leftarrow a, a_1, \ldots, a_m \end{array} \right] \right) = b \end{array} \right\}$$

provided that $\mathit{Var}(\mathit{Expr}) = \{x, x^1, \ldots, x^m\}$. For instance, with the same declaration of x and y as above, $\mathsf{pbc}_{Razor}([\![\, x{:=}1{-}y \,]\!])$ is

$$\{\widehat{x_{\bullet 1}}, \widehat{y_{00}}\} \,\square\, \{\widehat{x_{01}}, \widehat{y_{00}}\} \,\square\, \{\widehat{x_{11}}, \widehat{y_{00}}\} \,\square\, \{\widehat{x_{\bullet 0}}, \widehat{y_{11}}\} \,\square\, \{\widehat{x_{00}}, \widehat{y_{11}}\} \,\square\, \{\widehat{x_{10}}, \widehat{y_{11}}\} \,.$$

The case that $a = \bullet$ occurs if x_{ab} describes the very first access to x inside the block in which x is declared. It is assumed this never happens if x occurs in an expression on the right of :=.

Guarded Assignment: $[\![\, \mathit{Bexpr} \,;\, x{:=}\mathit{Expr} \,]\!]$. The translation is

$$\mathsf{pbc}_{Razor}([\![\, \mathit{Bexpr} \,;\, x{:=}\mathit{Expr} \,]\!]) \;=\; \square_{\mathcal{A}(\mathit{Bexpr}\,;\,x:=\mathit{Expr})} \,, \tag{12}$$

where $\mathcal{A}(\mathit{Bexpr}\,;\,x{:=}\mathit{Expr})$ is defined in the following way. If $x \notin \mathit{Var}(\mathit{Expr}) \cup \mathit{Var}(\mathit{Bexpr})$, then $\mathcal{A}(\mathit{Bexpr}\,;\,x{:=}\mathit{Expr})$ is

$$\left\{ \left\{ \widehat{x_{ab}}, \widehat{x^1_{a_1 a_1}}, \ldots, \widehat{x^m_{a_m a_m}} \right\} \middle| \begin{array}{l} a \in \mathit{Type}(x) \cup \{\bullet\} \ \wedge \ b \in \mathit{Type}(x) \\ a_1 \in \mathit{Type}(x^1) \ \wedge \ \ldots \ \wedge \ a_m \in \mathit{Type}(x^m) \\ \mathit{eval}\left(\mathit{Bexpr} \left[\begin{array}{c} x^1, \ldots, x^m \\ \leftarrow a_1, \ldots, a_m \end{array} \right] \right) = \mathbf{true} \\ \mathit{eval}\left(\mathit{Expr} \left[\begin{array}{c} x^1, \ldots, x^m \\ \leftarrow a_1, \ldots, a_m]) \end{array} \right] \right) = b \end{array} \right\}$$

provided that $\mathit{Var}(\mathit{Expr}) \cup \mathit{Var}(\mathit{Bexpr}) = \{x^1, \ldots, x^m\}$; moreover, if we have $x \in \mathit{Var}(\mathit{Expr}) \cup \mathit{Var}(\mathit{Bexpr})$, then $\mathcal{A}(\mathit{Bexpr}\,;\,x{:=}\mathit{Expr})$ is given by

$$\left\{ \left\{ \widehat{x_{ab}}, \widehat{x^1_{a_1 a_1}}, \ldots, \widehat{x^m_{a_m a_m}} \right\} \middle| \begin{array}{l} a \in \mathit{Type}(x) \ \wedge \ b \in \mathit{Type}(x) \\ a_1 \in \mathit{Type}(x^1) \ \wedge \ \ldots \ \wedge \ a_m \in \mathit{Type}(x^m) \\ \mathit{eval}\left(\mathit{Bexpr} \left[\begin{array}{c} x, x^1, \ldots, x^m \\ \leftarrow a, a_1, \ldots, a_m \end{array} \right] \right) = \mathbf{true} \\ \mathit{eval}\left(\mathit{Expr} \left[\begin{array}{c} x, x^1, \ldots, x^m \\ \leftarrow a, a_1, \ldots, a_m \end{array} \right] \right) = b \end{array} \right\}$$

provided that $Var(Expr) \cup Var(Bexpr) = \{x, x^1, \ldots, x^m\}$. For instance, still with the same declarations of x and y,

$$\mathsf{pbc}_{Razor}([\![\, y{=}0; x{:=}1{-}y \,]\!]) \;=\; (\{\widehat{x_{\bullet 1}}, \widehat{y_{00}}\} \,\Box\, \{\widehat{x_{01}}, \widehat{y_{00}}\} \,\Box\, \{\widehat{x_{11}}, \widehat{y_{00}}\}) \,.$$

In the special case that $Bexpr = \mathbf{true}$, line (12) yields

$$\mathsf{pbc}_{Razor}([\![\, \mathbf{true}; x{:=}Expr \,]\!]) \;=\; \mathsf{pbc}_{Razor}([\![\, x{:=}Expr \,]\!]) \,.$$

Hence we may use $[\![\, \mathbf{true}; x{:=}Expr \,]\!]$ and $[\![\, x{:=}Expr \,]\!]$ interchangeably, or consider the latter as an abbreviation of the former.

Receive Action: $[\![\, c?x \,]\!]$**.** The translation is simply

$$\mathsf{pbc}_{Razor}([\![\, c?x \,]\!]) \;=\; \Box_{\mathcal{A}(c?x)} \,, \tag{13}$$

where the set of multiactions $\mathcal{A}(c?x)$ is defined by

$$\left\{ \left\{\widehat{c?b}, \widehat{x_{ab}}\right\} \,\middle|\, \begin{array}{l} b \in Type(x) \cap Type(c) \\ a \in Type(x) \cup \{\bullet\} \end{array} \right\} .$$

For instance, with binary x, and with c declared as **var** c : **chan** k **of** $\{0, 1\}$, $\mathsf{pbc}_{Razor}([\![\, c?x \,]\!])$ is given by

$$\{\widehat{c?0}, \widehat{x_{\bullet 0}}\} \,\Box\, \{\widehat{c?0}, \widehat{x_{00}}\} \,\Box\, \{\widehat{c?0}, \widehat{x_{10}}\} \,\Box\, \{\widehat{c?1}, \widehat{x_{\bullet 1}}\} \,\Box\, \{\widehat{c?1}, \widehat{x_{01}}\} \,\Box\, \{\widehat{c?1}, \widehat{x_{11}}\} \,.$$

Send Action: $[\![\, c!Expr \,]\!]$**.** Let $Var(Expr) = \{x^1, \ldots, x^m\}$ and let the set of multiactions $\mathcal{A}(c!Expr)$ be given by

$$\left\{ \left\{\widehat{c!b}, \widehat{x^1_{a_1 a_1}}, \ldots, \widehat{x^m_{a_m a_m}}\right\} \,\middle|\, \begin{array}{l} b \in Type(x) \cap Type(c) \\ a_1 \in Type(x^1) \;\wedge\; \ldots \;\wedge\; a_m \in Type(x^m) \\ eval(Expr[x^1, \ldots, x^m \leftarrow a_1, \ldots, a_m]) = b \end{array} \right\} .$$

Then the translation for $[\![\, c!Expr \,]\!]$ is

$$\mathsf{pbc}_{Razor}([\![\, c!Expr \,]\!]) \;=\; \Box_{\mathcal{A}(c!Expr)} \,. \tag{14}$$

For instance, $\mathsf{pbc}_{Razor}([\![\, c!0 \,]\!]) = \{\widehat{c!0}\}$. Moreover, with binary x and y, and with a channel identifier c declared as **var** c : **chan** k **of** $\{\mathbf{true}, \mathbf{false}\}$, we have that $\mathsf{pbc}_{Razor}([\![\, c!(x{\neq}y) \,]\!])$ is

$$\{\widehat{c!\mathbf{false}}, \widehat{x_{00}}, \widehat{y_{00}}\} \,\Box\, \{\widehat{c!\mathbf{true}}, \widehat{x_{00}}, \widehat{y_{11}}\} \,\Box\, \{\widehat{c!\mathbf{true}}, \widehat{x_{11}}, \widehat{y_{00}}\} \,\Box\, \{\widehat{c!\mathbf{false}}, \widehat{x_{11}}, \widehat{y_{11}}\} .$$

Exercise 9.2.4. Specify the remaining parts, (15) and (16), of the semantics for the clauses $[\![\, Bexpr \,;\, c?x \,]\!]$ and $[\![\, Bexpr \,;\, c!Expr \,]\!]$, respectively.

9.3 Three *Razor* Programs

In this section, we give three examples in order to illustrate: the block structure (program P_1), shared variable parallelism (program P_2), and channel communication (program P_3).

The first program, P_1, exemplifies the impact of block structure and scoping rule:

$$\begin{array}{l} P_1 : \textbf{begin}\ \ \textbf{var}\ x : \{0\}\ (\textbf{init}\ 0);\ \textbf{var}\ y : \{0,1\}; \\ \qquad\qquad \textbf{begin}\ \ \textbf{var}\ x : \{1\}\ (\textbf{init}\ 1); \\ \qquad\qquad\qquad [\![\, y{:=}x \,]\!] \\ \qquad\qquad \textbf{end} \\ \qquad \textbf{end} \end{array}$$

First we expand P_1 in such a way that it corresponds strictly to the *Razor* syntax, calling it also P_1 because no confusion may arise:

$$\begin{array}{l} P_1 : \textbf{begin}\ \ \textbf{var}\ x : \{0\}; \\ \qquad\qquad \textbf{var}\ y : \{0,1\}; \\ \qquad\qquad [\![\, x{:=}0 \,]\!]; \\ \qquad\qquad \textbf{begin}\ \ \textbf{var}\ x : \{1\}; \\ \qquad\qquad\qquad [\![\, x{:=}1 \,]\!]; \\ \qquad\qquad\qquad [\![\, y{:=}x \,]\!] \qquad (*) \\ \qquad\qquad \textbf{end} \\ \qquad \textbf{end} \end{array}$$

In line $(*)$, we have an occurrence of x, but there are two declarations of x. The scope rule applies at this point, determining that the declaration which is significant for x at $(*)$ is the inner one, and thus $\mathit{Type}(x) = \{1\}$ at that point. The PBC expression $\mathsf{pbc}_{\mathit{Razor}}(P_1)$ is shown in Fig. 9.5. The text after each % is a comment which does not belong to the expression.

Exercise 9.3.1. Derive $\mathsf{box}_{\mathsf{PBC}}(\mathsf{pbc}_{\mathit{Razor}}(P_1))$.
Hint: It has 17 places (three for the outer x, four for y, three for the inner x, and seven for the flow of control) and 13 transitions (two for $x{:=}0$, two for $x{:=}1$, two for $y{:=}x$, two for the termination of the inner x, three for the termination of y, and two for the termination of the outer x).

The next example, P_2, shows how shared data are treated:

$$\begin{array}{l} P_2 : \textbf{begin}\ \ \textbf{var}\ x : \{0,1\}; \\ \qquad\qquad [\![\, x{:=}0 \,]\!] \parallel [\![\, x{:=}1 \,]\!] \\ \qquad \textbf{end} \end{array}$$

The PBC expression $\mathsf{pbc}_{\mathit{Razor}}(P_2)$ is shown in Fig. 9.6 whilst Fig. 9.7 shows the corresponding box. The central e-place and one of the x-places belong to the declaration of x. The e-place on the left-hand side and on the right-hand side belongs, respectively, to the atomic actions $[\![\, x{:=}0 \,]\!]$ and $[\![\, x{:=}1 \,]\!]$. The x-places of these two commands have changed into internal places because of

$[\,\{x_{\bullet\bullet}, x_{\bullet 0}, x_{0\bullet}, x_{00}\}:$	% declaration of outer x
$\quad ((\,[x_{()}^{\{0\}}] \Box \{x_{\bullet\bullet}\}\,) \parallel$	% data constant for outer x
	% with $\mathit{Type}(x) = \{0\}$
$\quad ($	
$\qquad [\,\{y_{\bullet\bullet}, y_{\bullet 0}, y_{\bullet 1}, y_{0\bullet}, y_{1\bullet}, y_{00}, y_{01}, y_{10}, y_{11}\}:$	% declaration of y
$\qquad ((\,[y_{()}^{\{0,1\}}] \Box \{y_{\bullet\bullet}\}\,) \parallel$	% data constant for y
	% with $\mathit{Type}(y) = \{0,1\}$
$\qquad ($	
$\qquad\quad ((\,\{\widehat{x_{\bullet 0}}\} \Box \{\widehat{x_{00}}\}\,);$	% $[\![x{:=}0]\!]$
$\qquad\quad [\,\{x_{\bullet\bullet}, x_{\bullet 1}, x_{1\bullet}, x_{11}\}:$	% declaration of inner x
$\qquad\qquad ((\,[x_{()}^{\{1\}}] \Box \{x_{\bullet\bullet}\}\,) \parallel$	% data constant for inner x
	% with $\mathit{Type}(x) = \{1\}$
$\qquad\qquad\quad ((\,\{\widehat{x_{\bullet 1}}\} \Box \{\widehat{x_{11}}\}\,);$	% $[\![x{:=}1]\!]$
$\qquad\qquad\quad (\,\{\widehat{y_{\bullet 1}}, \widehat{x_{11}}\} \Box \{\widehat{y_{01}}, \widehat{x_{11}}\} \Box \{\widehat{y_{11}}, \widehat{x_{11}}\}\,)$	% $[\![y{:=}x]\!]$; note: at this
	% point $\mathit{Type}(x) = \{1\}$
$\qquad\qquad\quad);\ (\,\{\widehat{x_{\bullet\bullet}}\} \Box \{\widehat{x_{1\bullet}}\}\,)$	% termination of inner x
$\qquad\qquad)$	
$\qquad\quad]$	
$\qquad\quad);\ (\,\{\widehat{y_{0\bullet}}\} \Box \{\widehat{y_{1\bullet}}\} \Box \{\widehat{y_{\bullet\bullet}}\}\,)$	% termination of y
$\qquad\quad)$	
$\qquad)$	
$\qquad];\ (\,\{\widehat{x_{0\bullet}}\} \Box \{\widehat{x_{\bullet\bullet}}\}\,)$	% termination of outer x
$\quad)$	
$\quad)$	
$]$	

Fig. 9.5. PBC expression $\mathsf{pbc}_{\mathit{Razor}}(P_1)$

the sequential composition with the termination clause. The other x-place at the bottom of Fig. 9.6 belongs to the latter.

The third program, P_3, exemplifies synchronous channel-based communication between parallel processes:

P_3 : **begin** **var** c : **chan** 0 **of** $\{0\}$;
  var x : $\{0\}$;
  $[\![c!0]\!] \parallel [\![c?x]\!]$
end

The PBC expression $\mathsf{pbc}_{\mathit{Razor}}(P_3)$ is shown in Fig. 9.8.

Exercise 9.3.2. Derive $\mathsf{box}_{\mathsf{PBC}}(\mathsf{pbc}_{\mathit{Razor}}(P_3))$.

$$
\begin{array}{ll}
[\,\{x_{\bullet\bullet}, x_{\bullet 0}, x_{\bullet 1}, x_{0\bullet}, x_{1\bullet}, x_{00}, x_{01}, x_{10}, x_{1\bullet}\} \; : & \%\ \text{declaration of } x \\
\quad (& \\
\quad\quad ([x_{()}^{\{0,1\}}] \,\Box\, \{x_{\bullet\bullet}\}) \parallel & \%\ \text{data constant for } x \\
 & \%\ \text{with } \mathit{Type}(x) = \{0,1\} \\
\quad\quad (& \\
\quad\quad (\{\widehat{x_{\bullet 0}}\} \,\Box\, \{\widehat{x_{00}}\} \,\Box\, \{\widehat{x_{10}}\}) \parallel & \%\ [\![x{:=}0]\!] \\
\quad\quad (\{\widehat{x_{\bullet 1}}\} \,\Box\, \{\widehat{x_{01}}\} \,\Box\, \{\widehat{x_{11}}\}) & \%\ [\![x{:=}1]\!] \\
\quad\quad); (\{\widehat{x_{0\bullet}}\} \,\Box\, \{\widehat{x_{1\bullet}}\} \,\Box\, \{\widehat{x_{\bullet\bullet}}\}) & \%\ \text{termination of } x \\
\quad) & \\
]
\end{array}
$$

Fig. 9.6. PBC expression $\mathsf{pbc}_{Razor}(P_2)$

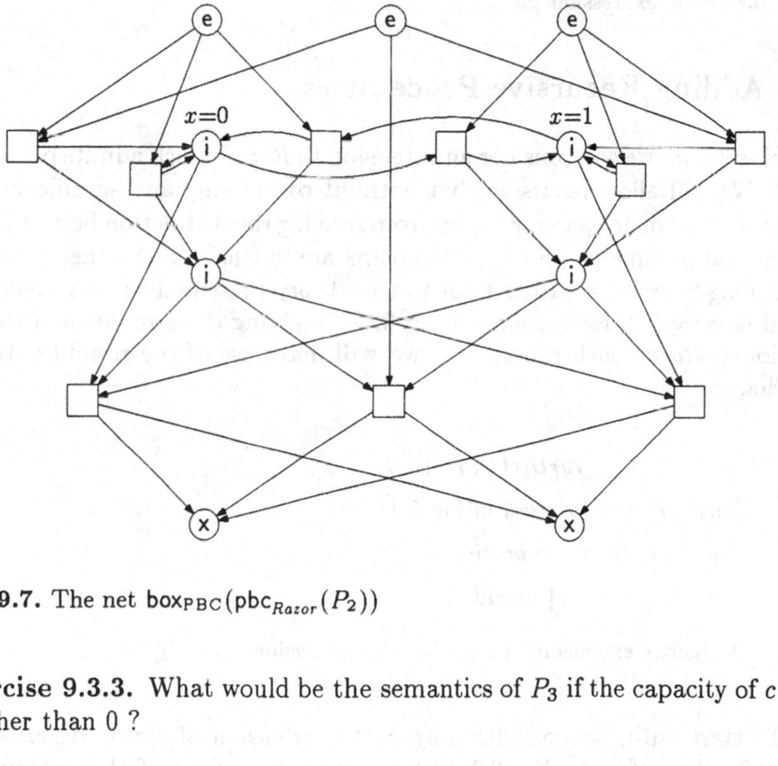

Fig. 9.7. The net $\mathsf{box}_{\mathrm{PBC}}(\mathsf{pbc}_{Razor}(P_2))$

Exercise 9.3.3. What would be the semantics of P_3 if the capacity of c was 1 rather than 0 ?

Exercise 9.3.4. Analyse a program with a guard of the form $[\![\, x{=}0; c?x \,]\!]$.

Exercise 9.3.5. The above examples do not explain why the termination part $\tau(Decl)$ in the semantics of a block is really necessary. Provide a formal motivation for the presence of this construct.
Hint: Consider a program in which a block is nested inside a loop that may be executed more than once.

$$
\begin{array}{ll}
[\,\{c_{!0}, c_{?0}, c_{\bullet}\} : & \%\ \text{declaration of } c \\
\quad (\,(\,[c^{0,\{0\}}_{()}] \,\square\, \{c_{\bullet}\}\,)\ \| & \%\ \text{data constant for } c \\
\quad (& \\
\qquad [\,\{x_{\bullet\bullet}, x_{\bullet 0}, x_{0\bullet}, x_{00}\} : & \%\ \text{declaration of } x \\
\qquad\quad (\,(\,[x^{\{0\}}_{()}] \,\square\, \{x_{\bullet\bullet}\}\,)\ \| & \%\ \text{data constant for } x \\
\qquad\qquad (\,\{\widehat{c_{!0}}\} \| (\{\widehat{c_{?0}}, \widehat{x_{\bullet 0}}\} \,\square\, \{\widehat{c_{?0}}, \widehat{x_{00}}\}) & \%\ [\![\, c!0 \,]\!] \| [\![\, c?x \,]\!] \\
\qquad\qquad);\ (\,\{\widehat{x_{0\bullet}}\} \,\square\, \{\widehat{x_{\bullet\bullet}}\}\,) & \%\ \text{termination of } x \\
\qquad\quad) & \\
\qquad];\ \{\widehat{c_{\bullet}}\} & \%\ \text{termination of } c \\
\quad) & \\
\ \ ) & \\
] &
\end{array}
$$

Fig. 9.8. PBC expression $\mathsf{pbc}_{\mathit{Razor}}(P_3)$

9.4 Adding Recursive Procedures

In this section, we will consider an extension to *Razor* which admits procedure calls. We will allow recursion, but without considering any specific mechanisms for parameter passing, apart from making the distinction between local and global parameters. Such mechanisms are intrinsic to the theory of programming languages, rather than to the theory presented in this book, and could be adapted, for instance, from [57]. In giving the semantics of the extension of *Razor* with procedures, we will make use of the results obtained in Chap. 5.

Decl	::=	*VarDecl* | *ProcDecl*
VarDecl	::=	(as *Decl* in Fig. 9.1)
ProcDecl	::=	**proc** *procid* : *Block*
Com	::=	... | *procid*

Fig. 9.9. Syntax extensions for *Razor* with procedures

To start with, we consider a syntactic extension of *Razor* described in Fig. 9.9. This refers to Fig. 9.1 and specifies those lines of the syntax that are added or modified for the present purposes. The extensions are twofold. First, in the declaration part, we may have the declaration of a procedure with name *procid* and procedure body *Block*. Second, in the command part, we may have the invocation of a procedure. We may (and will) assume that the set of procedure names is disjoint from the set of variable identifiers and the set of channel identifiers, as well as that no procedure name is defined more than once. The latter property can be ensured, if necessary, by systematic renaming; it will allow us to associate a (possibly recursive) variable definition to each procedure name. In order to distinguish between variable and channel declarations, and procedure declarations, we use lower case letters for the former and upper case letters for the latter. For instance, consider the following program:

$$
\begin{array}{ll}
P_4 : \textbf{begin} & \textbf{var}\ x : \{0\}; \\
 & \textbf{proc}\ X : \textbf{begin}\ [\![x{:=}0]\!]\ \textbf{end}; \\
 & X \\
\textbf{end} &
\end{array}
$$

The main program consists only of the invocation of procedure X in its command part, where X is declared as simply setting the value of x to 0. In our translation, we will aim at associating with P_4 a PBC expression, as follows:

$$
\begin{array}{rl}
\mathsf{pbc}_{Razor}(P_4) = & [\ \{x_{\bullet\bullet}, x_{\bullet 0}, x_{0\bullet}, x_{00}\}\ : \\
 & \quad ([x_{()}] \,\Box\, \{x_{\bullet\bullet}\}) \parallel (X; (\{\widehat{x_{0\bullet}}\} \,\Box\, \{\widehat{x_{\bullet\bullet}}\})) \\
 &] \\
X \stackrel{\mathrm{df}}{=} & \{\widehat{x_{00}}\} \,\Box\, \{\widehat{x_{\bullet 0}}\}\ .
\end{array}
\tag{9.1}
$$

Here, the expression given in the first three lines contains a recursion variable, X, whose defining equation is given in the fourth line. In this way, the declaration of procedure X in P_4 is separated from the main body of P_4.

Static vs. Dynamic Binding. Before we give the translation in general, we need to discuss the scope of identifiers. Indeed, even with the restrictions we imposed on the naming of variables, channels, and procedures, we still need to clarify the binding of variables, as exemplified in the following program:

$$
\begin{array}{lll}
P_5 : \textbf{begin} & \textbf{var}\ x : \{0\}; & \\
 & \textbf{proc}\ X : \textbf{begin}\ [\![x{:=}0]\!]\ \textbf{end}; & \\
 & \textbf{begin} & \textbf{var}\ x : \{0\}; \\
 & & X \\
 & \textbf{end} & \\
\textbf{end} & &
\end{array}
$$

The invocation of X in the fourth line yields an access to variable x, but which of the two x's is meant, that of line 1 or that of line 3? *Static binding* would declare the x of the first line to be the one accessed, while *dynamic binding* would identify the x of the third line as the accessed variable. Normally, in a block-structured language, static rather than dynamic binding is used (dynamic binding is more common in interpreted macroprocessing languages), and this is also what we shall adopt. This means that in order to find the variable x accessed by X, we would need to search through all declarations in blocks that enclose the procedure body of X (not its invocation), in order to find a suitable declaration of x; and if there was none, the program would not be closed.

But we always may, systematically and without loss of generality, remove multiple declarations of the same variable (even in different blocks) by appropriate name substitutions. In P_5, for instance, we may distinguish the two declarations of x by substituting a variable $x1$ consistently for the first and a variable $x2$ consistently for the second. The occurrence of x in the procedure body of X belongs to $x1$ by static binding, and is therefore substituted by $x1$, rather than $x2$:

P_5' : **begin** **var** $x1 : \{0\}$;
  proc X : **begin** $[\![x1{:=}0]\!]$ **end**;
  begin **var** $x2 : \{0\}$;
    X
  end
end

Note that such syntactic renaming of variables is only possible for the static binding discipline because in the case of dynamic binding, different invocations in different blocks may need to be renamed differently, but since there is always only one procedure body, this may be impossible.

Since we have committed ourselves to static binding, we may (and will) from now on assume that any variable has *at most* one declaration (and *exactly* one, if the program is closed). This excludes program P_5, but allows P_5', where the static binding of P_5 is expressed syntactically. This does not foreclose the simultaneous coexistence of various variables (or channels) with the same identifier, if they are declared in a truly recursive procedure, but they all will have the same type (and capacity).

Semantics. Let P be any program fragment, possibly with procedures, that conforms to the restrictions given above, i.e., every variable identifier is declared at most once. We will associate with P two objects:

$\mathsf{pbc}_{Razor}(P)$: the PBC expression of P

$\mathsf{env}_{Razor}(P)$: the PBC environment of P.

Here, $\mathsf{pbc}_{Razor}(P)$ is an expression as before, which may also contain recursion variables, while $\mathsf{env}_{Razor}(P)$ is a set of equations with definitions for such recursion variables. Consider P_4 above; then $\mathsf{pbc}_{Razor}(P_4)$ comprises the first three lines of (9.1), while $\mathsf{env}_{Razor}(P)$ comprises just one equation, $X \overset{\text{df}}{=} \{\widehat{x_{00}}\} \,\square\, \{\widehat{x_{\bullet 0}}\}$, i.e., the last line of (9.1).

The compositional definition of pbc_{Razor} and env_{Razor} given in Fig. 9.10 incorporates and expands each part, (1)—(16), of Sect. 9.2 into two or more parts, as necessary. Only the significant new parts are shown. We then apply this definition to program P_4. The expression $\mathsf{pbc}_{Razor}(P_4)$ is derived as before, and it contains X as the recursion variable (see the first three lines of (9.1)). The environment $\mathsf{env}_{Razor}(P_4)$ is derived as follows:

(1) from Sect. 9.2.1

$\mathsf{env}_{Razor}(\textbf{begin}\ Body\ \textbf{end}) \ = \ \mathsf{env}_{Razor}(Body)$ (*i*)

$\mathsf{pbc}_{Razor}(VarDecl;\ Body) \ = $ as in (2) with *Decl* changed to *VarDecl*

$\mathsf{env}_{Razor}(VarDecl;\ Body) \ = \ \mathsf{env}_{Razor}(Body)$ (*ii*)

$\mathsf{pbc}_{Razor}(ProcDecl;\ Body) \ = \ \mathsf{pbc}_{Razor}(Body)$

$\mathsf{env}_{Razor}(ProcDecl;\ Body) \ = \ \mathsf{env}_{Razor}(ProcDecl) \cup \mathsf{env}_{Razor}(Body)$ (*ii'*)

$\mathsf{pbc}_{Razor}(\textbf{proc}\ procid : Block) \ = $ not applicable (see the first line of (*ii'*))

$\mathsf{env}_{Razor}(\textbf{proc}\ procid : Block) \ =$

$$\{\, procid \overset{\mathrm{df}}{=} \mathsf{pbc}_{Razor}(Block)\,\} \cup \mathsf{env}_{Razor}(Block) \qquad (ii'')$$

(3) from Sect. 9.2.1

$\mathsf{env}_{Razor}(\textbf{var}\ x : Set) \ = \ \emptyset$ (*iii*)

(4) from Sect. 9.2.1

$\mathsf{env}_{Razor}(\textbf{var}\ c : \textbf{chan}\ k\ \textbf{of}\ Set) \ = \ \emptyset$ (*iv*)

(5) from Sect. 9.2.3

$\mathsf{env}_{Razor}(Com_1;\ Com_2) \ = \ \mathsf{env}_{Razor}(Com_1) \cup \mathsf{env}_{Razor}(Com_2)$ (*v*)

similar to (*v*) and based on (6)—(8) in Sect. 9.2.3 (*vi*)—(*viii*)

(9) from Sect. 9.2.3

$\mathsf{env}_{Razor}(GC;\ Com) \ = \ \mathsf{env}_{Razor}(GC) \cup \mathsf{env}_{Razor}(Com)$ (*ix*)

(10) from Sect. 9.2.4

$\mathsf{env}_{Razor}(\llbracket\, Bexpr\,\rrbracket) \ = \emptyset$ (*x*)

(11) from Sect. 9.2.4

$\mathsf{env}_{Razor}(\llbracket\, x{:=}Expr\,\rrbracket) \ = \ \emptyset$ (*xi*)

similar to (*x*)—(*xi*) and based on (12)—(16) in Sect. 9.2.4 (*xii*)—(*xvi*)

$\mathsf{pbc}_{Razor}(X) \ = \ X$

$\mathsf{env}_{Razor}(X) \ = \ \emptyset$ (*xvii*)

Fig. 9.10. Semantic definitions for recursive *Razor*

$$
\begin{aligned}
&\mathsf{env}_{Razor}(P_4)\\
&= (\text{by } (i))\\
&\quad \mathsf{env}_{Razor}(\textbf{var } x : \{0\};\ \textbf{proc } X : \textbf{begin} [\![x{:=}0]\!] \textbf{ end};\ X)\\
&= (\text{by } (ii))\\
&\quad \mathsf{env}_{Razor}(\textbf{proc } X : \textbf{begin} [\![x{:=}0]\!] \textbf{ end};\ X)\\
&= \text{by } (ii')\\
&\quad \mathsf{env}_{Razor}(\textbf{proc } X : \textbf{begin} [\![x{:=}0]\!] \textbf{ end}) \cup \mathsf{env}_{Razor}(X)\\
&= (\text{by } (ii'') \text{ and } (xvii))\\
&\quad \{ X \stackrel{\text{df}}{=} \mathsf{pbc}_{Razor}(\textbf{begin} [\![x{:=}0]\!] \textbf{ end}) \}\\
&\quad \cup\ \mathsf{env}_{Razor}(\textbf{begin} [\![x{:=}0]\!] \textbf{ end}) \cup \emptyset\\
&= (\text{by } (i))\\
&\quad \{ X \stackrel{\text{df}}{=} \mathsf{pbc}_{Razor}([\![x{:=}0]\!]) \} \cup \mathsf{env}_{Razor}([\![x{:=}0]\!])\\
&= (\text{by } (xi) \text{ and } \mathit{Type}(x) = \{0\})\\
&\quad \{ X \stackrel{\text{df}}{=} \{\widehat{x_{00}}\} \,\square\, \{\widehat{x_{\bullet 0}}\} \}\ .
\end{aligned}
$$

That is, we obtain the last line of (9.1).

In this section, we have described parameterless, possibly recursive, procedures. The mechanism we used is essentially macro substitution after resolving static binding. This mechanism would actually correspond to dynamic binding in programs such as P_5. However, since we require static binding to be resolved syntactically, the difference between the two forms of binding vanishes: given the translation of P_5 into P_5', dynamic binding in P_5' is the same as static binding in P_5.

9.5 Some Consequences of the Theory

In this section, we state some fundamental properties of the Petri net translation of *Razor* programs that are consequences of the general theory presented in the previous chapters of this book.

For the purpose of this section, let P be any (possibly recursive) *Razor* program fragment as in Sect. 9.2,

$$\mathcal{E}_P = \mathsf{pbc}_{Razor}(P)$$

its corresponding expression, and

$$\mathcal{B}_P = \mathsf{box}_{\mathrm{PBC}}(\mathcal{E}_P)$$

the corresponding box (Petri net). It is clear that $\mathcal{E}_P$ is a PBC expression, possibly employing generalised PBC operators, including the generalised iteration operator $\langle F * E]$, as well as the data and channel constants.

The basic result of this section, which is actually a straightforward consequence of the results of Chap. 8, states that $\mathcal{E}_P$ is a PBC expression of the (generalised) form E^{wf} defined in Sect. 8.1. This implies, in particular, that *Razor* can never give rise to expressions such as

$$\alpha \,\square\, \langle \beta * \gamma] \quad \text{or} \quad \langle\langle \alpha * \beta] * \gamma] \quad \text{or} \quad \langle(\alpha \| \beta) * \gamma]$$

in which loops are nested at the beginning of choices or other loops, or a loop body starts with a parallel composition. As we have already discussed in this book, allowing such expressions may have undesirable consequences in that, while Petri net and SOS semantics are still equivalent (in the case of the first two expressions), they may not exactly correspond to the intuition normally associated with, for instance, the choice operator, or that they may lead to non-safeness.

Theorem 9.5.1. $\mathcal{E}_P$ is of the form E^{wf} (see Fig. 8.4 in Sect. 8.1.3).

Proof. In this proof, we will use the notation $E \in E^{\delta}$ to denote that an expression E is of the form E^{δ} in Fig. 8.4. We will also prove a somewhat stronger result, namely, we will show that

- if P is an atomic action, guarded command, or declaration, then it is the case that $\mathcal{E}_P \in E^{\mathrm{e:x:xcl}}$,
- if P is a loop, selection, or sequential composition statement, then it is the case that $\mathcal{E}_P \in E^{\mathrm{x:xcl}}$,
- otherwise, we have $\mathcal{E}_P \in E^{\mathrm{x}}$ (this includes the case when $P = X$ is a recursion variable).

The proof proceeds inductively by inspection of the semantic definitions (i)—$(xvii)$ of Fig. 9.10 in the reverse order.

Rule $(xvii)$: $P = X$ and $\mathcal{E}_P = X$. Then, by the defining equation for X in (ii''), the induction hypothesis, Fig. 8.4 for E^{x}, and the fact that X^{δ} is the largest set of hierarchical variables such that

$$X^{\delta} \stackrel{\mathrm{df}}{=} E$$

with $E \in E^{\delta}$ for the various possible values of δ, we obtain $\mathcal{E}_P \in E^{\mathrm{x}}$.

Rules (xvi)—(x): This is the first base case of the inductive proof, in which P is of the form $[\![\ \ldots\ ]\!]$ and $\mathcal{E}_P$ is of the form $\square_{\mathcal{A}}$. Hence, by applying Fig. 8.4 for $E^{\mathrm{e:x:xcl}}$ and recalling that $\square_{\mathcal{A}}$ is a constant which is both ex-directed and ex-exclusive, we obtain $\mathcal{E}_P \in E^{\mathrm{e:x:xcl}}$.

Rule (ix): $P = GC; Com$ and $\mathcal{E}_P = \mathcal{E}_{GC}; \mathcal{E}_{Com}$. Then, by the induction hypothesis and Fig. 8.4 for E^{x} and $E^{\mathrm{x:xcl}}$, we have $\mathcal{E}_{GC} \in E^{\mathrm{e:x:xcl}}$ and $\mathcal{E}_{Com} \in E^{\mathrm{x}}$ (notice that Com may be a variable). Thus, by Fig. 8.4 for $E^{\mathrm{e:x}}$ and $E^{\mathrm{e:x:xcl}}$, we obtain $\mathcal{E}_P \in E^{\mathrm{e:x:xcl}}$.

Rule $(viii)$: $P =$ **do** $GC_1 \,\square\, \ldots \,\square\, GC_m$ **od** and $\mathcal{E}_P$ is of the form $\langle\, \square_{i=1}^{m} \mathcal{E}_j * \square_{\mathcal{A}}]$. Then, by the induction hypothesis, $\mathcal{E}_j \in E^{\mathrm{e:x:xcl}}$, for every i. Thus, after applying $m-1$ times Fig. 8.4 for $E^{\mathrm{e:x:xcl}}$, we have $\square_{i=1}^{m} \mathcal{E}_j \in E^{\mathrm{e:x:xcl}}$. Moreover, $\square_{\mathcal{A}} \in E^{\mathrm{e:x:xcl}}$, by Fig. 8.4 for $E^{\mathrm{e:x:xcl}}$. Hence $\mathcal{E}_P \in E^{\mathrm{x:xcl}}$, by Fig. 8.4 for $E^{\mathrm{x:xcl}}$.

Rule (vii): $P =$ **if** $GC_1 \,\square\, \ldots \,\square\, GC_m$ **fi** and $\mathcal{E}_P$ is of the form $\square_{i=1}^{m} \mathcal{E}_j$. Then, by the induction hypothesis, $\mathcal{E}_j \in E^{\mathrm{e:x:xcl}}$, for every i. Thus $\mathcal{E}_P \in E^{\mathrm{e:x:xcl}}$, after applying $m-1$ times Fig. 8.4 for $E^{\mathrm{e:x:xcl}}$.

Rule (vi): $P = P_1 \| P_2$ and $\mathcal{E}_P = \mathcal{E}_1 \| \mathcal{E}_2$. Then, by the induction hypothesis and Fig. 8.4 for E^{x} and $E^{\mathsf{x:xcl}}$, we have $\mathcal{E}_1, \mathcal{E}_2 \in E^{\mathsf{x}}$. Then, by Fig. 8.4 for E^{x}, we obtain $\mathcal{E}_P \in E^{\mathsf{x}}$.

Rule (v): $P = P_1; P_2$ and $\mathcal{E}_P = \mathcal{E}_1; \mathcal{E}_2$. Then, by the induction hypothesis and Fig. 8.4 for E^{x} and $E^{\mathsf{x:xcl}}$, we have $\mathcal{E}_1, \mathcal{E}_2 \in E^{\mathsf{x}}$. Then, by Fig. 8.4 for $E^{\mathsf{x:xcl}}$, we obtain $\mathcal{E}_P \in E^{\mathsf{x:xcl}}$.

Rule (iv): This is the second base case of the inductive proof, in which P is a channel declaration and $\mathcal{E}_P = \gamma \,\square\, (\, \square_{\mathcal{A}})$ where $\gamma \in E^{\mathsf{e:x:xcl}}$ and $\square_{\mathcal{A}} \in E^{\mathsf{e:x:xcl}}$. Then, by Fig. 8.4 for $E^{\mathsf{e:x:xcl}}$, we obtain $\mathcal{E}_P \in E^{\mathsf{e:x:xcl}}$.

Rule (iii): This is the third base case of the inductive proof, in which a variable declaration can be treated similarly as the channel declaration above.

Rule (ii): Then P is a block and $\mathcal{E}_P = [C : (F \| (E; \square_{\mathcal{A}}))]$. Then, by the induction hypothesis and Fig. 8.4 for E^{x} and $E^{\mathsf{x:xcl}}$, we have $F, E \in E^{\mathsf{x}}$, $\square_{\mathcal{A}} \in E^{\mathsf{e:x:xcl}}$, and both A and C may be infinite. Thus $(E; \square_{\mathcal{A}}) \in E^{\mathsf{x}}$. Moreover, by Fig. 8.4 for E^{x}, we obtain $(F \| (E; \square_{\mathcal{A}})) \in E^{\mathsf{x}}$. Finally, we obtain that $\mathcal{E}_P \in E^{\mathsf{x}}$, by Fig. 8.4 for E^{x} applied to scoping rather than relabelling.

Rule (ii'): Directly by definition and the induction hypothesis.

Rule (ii"): Not relevant.

Rule (i): Directly by definition and the induction hypothesis. □

Corollary 9.5.1. The labelled net $\mathcal{B}_P$ is T-restricted, ex-restricted, and x-directed; moreover, it is safe and clean under the initial marking. In particular, it is a plain static box.

Proof. Follows from Thm. 9.5.1 in combination with the results proved in the previous two chapters; in particular, Thm. 7.3.1. □

It is not difficult to see that $\mathcal{B}_P$ may not be e-directed (since a loop may occur initially). However, any violations of e-directedness notwithstanding, the operational and Petri net semantics of P are fully consistent with each other. What is more, choices and loops behave in the way they (intuitively) should, as implied by the behavioural conditions (Beh1)—(Beh5).

Corollary 9.5.2. The operational semantics associated with $\mathcal{E}_P$ and the Petri net semantics associated with $\mathcal{B}_P$ are fully consistent in the sense of Thms. 7.3.6, 7.3.7, and 7.3.9. Moreover, they are consistent with the intuition behind the composition operators expressed by the conditions (Beh1)—(Beh5).

Proof. Follows from Thm. 9.5.1 and Cor. 9.5.1, as well as the discussion in Chaps. 7 and 8. □

Summarising, the last two corollaries imply that, even though the generalised iteration operator has been used in the semantics of *Razor*, we still have the following desirable properties:

- The box associated with a *Razor* program is safe and clean.

- The operational and denotational semantics of a *Razor* program, as defined implicitly above, are strongly consistent as specified in Chap. 7, as well as consistent with the intuition behind the composition operators captured by the conditions (Beh1)—(Beh5).

Furthermore, we have the following additional properties (referring again to P, $\mathcal{E}_P$, and $\mathcal{B}_P$ as defined above).

Corollary 9.5.3. If P contains no infinite types and no recursion, then $\mathcal{B}_P$ is a finite box.

Proof. Follows using Exc. 9.2.1. □

Corollary 9.5.4. $\mathcal{B}_P$ is a simple net which means that the weight function always returns 0 or 1 (see Sect. 4.2.3).

Proof. All the direct translations (of declarations and atomic actions) lead to simple nets.

Non-simpleness could be introduced if there was a place in a net with two input transitions or two output transitions, one of which contained a 'hatted' action particle and the other a corresponding 'unhatted' action particle. However, the translation of *Razor* programs is such that matching hatted and unhatted action particles can only be found on different sides of a parallel composition operator, and are never connected by a common input or output place.

Non-simpleness could also be introduced if there was a skewed transition (for instance, a transition with two output places, one being an e-place and the other being an x-place) inside the body of a loop. Such transitions fail to be present, however, in the translations of declarations and atomic actions. Moreover, as a consequence of Exc. 4.3.2, they cannot be created by the semantic constructions.

From what we have already said and the fact that simpleness is a limit-preserved property (in the sense of Prop. 5.1.1) and ex-boxes are simple nets, it follows (in view of the results proved in Chap. 5) that solving recursive equations in the process of giving the box semantics of recursion variables always yields simple boxes.

There are no other possible sources of non-simpleness. □

Corollary 9.5.5. If P is closed, then all the transitions of $\mathcal{B}_P$ are labelled by $\emptyset$.

Proof. This follows syntactically, since every occurrence of a variable has a corresponding declaration, and hence any non-$\emptyset$ transition is enclosed by the scoping construct arising from the block in which the corresponding declaration occurs. □

Thus, in the translation given above, the notion of program closedness (i.e., the property that all variables are declared) implies nonobservability (in the sense that all transitions are 'silent').

Exercise 9.5.1. Find a *Razor* program which is not closed, yet its box contains only silent transitions.

9.6 Proofs of Distributed Algorithms

The main purpose of a concurrent programming language such as *Razor* is to express parallel programs and distributed algorithms. It is still acknowledged to be a difficult problem, in general, to state and prove the correctness of such programs and algorithms formally. An important goal of a compositional translation such as that defined by the mapping $\mathsf{box}_{\mathsf{PBC}}(\mathsf{pbc}_{Razor}(\cdot))$ is to create a usable framework for making Petri net specific methods readily available — ideally, hand in hand with other methods such as the Owicki-Gries method [85] — in order to solve this problem.

In the remainder of this section we discuss briefly some (automatised) net-based correctness proofs of three mutual exclusion algorithms due to Peterson [87], Dekker [33], and Morris [82].

9.6.1 A Final Set of Petri-Net-Related Definitions

We now define some basic net-theoretic notions which will be helpful in the correctness proofs. They extend the definitions given in Sects. 2.2, 4.2.3, and 6.1.1, and will be applied to nets (S,T,W) (as in Sect. 2.2) that are simple. Corollary 9.5.4 guarantees that this is not a restriction in the present context. From the notions introduced in Sect. 4.2.3, we will use only those that are independent of transition labellings, λ, and which can therefore be applied to an unmarked net (S,T,W) or to a marked net (S,T,W,M). We will write

$$M_0 \,[\, t_1 \ldots t_m \,\rangle\, M_1$$

in order to indicate that marking M_1 is reachable from marking M_0 by the sequence of singleton steps $\{t_1\}\ldots\{t_m\}$, where $t_j \in T$ for $1 \le j \le m$. We will call $t_1 \ldots t_m$ a *firing sequence*. If $m = 0$, then the firing sequence is empty and $M_0 = M_1$.

We define the *Parikh vector* of a firing sequence $t_1 \ldots t_m$ to be the (well-defined) function from T to $\mathbf{N}$ which assigns to every $t \in T$ the number of its occurrences in $t_1 \ldots t_m$. By writing

$$M_0 \,[\, t_1 t_2 t_3 \ldots \rangle$$

we indicate that $t_1 t_2 t_3 \ldots$ is an *infinite firing sequence* starting from M_0, i.e., $t_1 \ldots t_i$ is a valid firing sequence, for every $i \ge 0$. Such a sequence models infinite behaviour, but note that for an infinite sequence, the 'marking reached after executing it' is not a well-defined notion, and the 'Parikh vector' (with the range $\mathbf{N}$) may not be well defined. Sometimes we write

$$M_0\, t_1\, M_1\, t_2\, M_2 \ldots$$

in order to emphasise the intermediate markings of an infinite firing sequence, i.e., $M_0\ [t_1 \ldots t_i\rangle\ M_i$, for every $i \geq 0$.

In what follows, we will use S-invariants and S-components, but without taking into account the labellings, and hence the issues related to, e.g., 1-conservativeness, will be disregarded. The basic behavioural property of an S-invariant $\theta : (S \cup T) \to \Re^+$ is the (weighted) marking preservation captured in Prop. 6.1.1, which is a direct consequence of the balancing equation (6.6). Notice that by the latter, we may as well restrict θ to be a function $\theta : S \to \Re^+$ because the values of θ at transitions can be reconstructed from the values of θ at places. We next define a notion similar to an S-invariant, but relating to transitions rather than places.

A *T-invariant* of (S, T, W) is a mapping $\zeta : T \to \Re^+$ such that, for every place $s \in S$, the following balancing equation is satisfied:

$$\sum_{t \in T} W(t, s) \cdot \zeta(t) = \sum_{t \in T} W(s, t) \cdot \zeta(t) \in \mathbf{N}. \tag{9.2}$$

The notion of a T-invariant as defined here is a restricted version of the more general notion where the values are allowed to be negative. Similarly as for S-invariants, ζ is a *T-aggregate* or *T-component* if its range is $\mathbf{N}$ or $\{0, 1\}$, respectively. A transition t is *covered* by ζ if $\zeta(t) > 0$. The *support* of a T-invariant is the set of transitions it covers. The basic behavioural property of a T-invariant is stated in the next proposition which follows directly from (9.2).

Proposition 9.6.1. Let M_0 and M_1 be two markings of N such that $M_0\ [t_1 \ldots t_m\rangle\ M_1$. Then $M_0 = M_1$ if and only if the Parikh vector of $t_1 \ldots t_m$ is a T-aggregate. □

But a T-aggregate may fail to be the Parikh vector of an execution sequence, unless we start from a large enough marking. For *finite* nets, we have the following result that ensures the existence of a T-aggregate whenever there is some infinite behaviour for a net with *bounded* markings (cf. Sect. 4.2.3).

Proposition 9.6.2. Let $N = (S, T, W, M)$ be a bounded finite net. If $t_1 t_2 \ldots$ is an infinite firing sequence starting with M, then there exists a T-aggregate ζ whose support is the set of all transitions occurring infinitely often in that firing sequence.

Proof. Let T' be the set of transitions occurring infinitely often in the firing sequence $t_1 t_2 \ldots$. Because N is finite and bounded, only a finite number of markings are reachable from M. Hence one of them must occur infinitely often in the sequence $M\ t_1\ M_1\ t_2\ M_2 \ldots$, and between any two such occurrences, M_i and M_j, a T-invariant can be found by means of Prop. 9.6.1($\Rightarrow$). Moreover, it is always possible to choose the first occurrence M_i in such a way that no transition in the (finite) set $T \setminus T'$ occurred after it, and the second occurrence

M_j in such a way that all the transitions in T' appear at least once between M_i and M_j. Hence the property of the support. □

A set of places $\mathcal{F}$ of a net will be called a *trap* if $\mathcal{F}^\bullet \subseteq {}^\bullet\mathcal{F}$. The basic property enjoyed by traps is the preservation of nonempty markings, as formalised in the next proposition which follows directly from the definition.

Proposition 9.6.3. Let $\mathcal{F}$ be a trap and M be a marking such that $M(s) > 0$, for some $s \in \mathcal{F}$. Then, for every marking $M' \in [M\rangle$, there is at least one $s' \in \mathcal{F}$ such that $M'(s') > 0$. □

The final notion we shall need is deadlock-freeness. It is the only one among those defined in this section that takes the labelling of a net into account (more precisely, the labelling of places). Let $\Sigma = (S, T, W, \lambda, M)$ be a plain box. A transition $t \in T$ is called *dead* at M if there is no $M' \in [M\rangle$ such that M' enables t. If M is a marking such that $M \neq \Sigma^\circ$ and M enables no transition, then it is called a *deadlock.* A static box Σ will be called *deadlock-free* if there is no marking $M \in [{}^\circ\Sigma\rangle \setminus \{\Sigma^\circ\}$ such that M is a deadlock. Usually, deadlocks should be avoided.

Exercise 9.6.1. Define the support of an S-invariant and show that it is always a trap. Then show that Prop. 9.6.3 is a weakening of Prop. 6.1.1.

9.6.2 Peterson's Mutual Exclusion Algorithm

Figure 9.11 shows a parallel *Razor* program implementing Peterson's mutual exclusion protocol for two processes [87].

```
begin
var in1, in2: {0,1} (init 0) ; hold: {1,2} (init 1);
do                                  ||  do
    ⟦ in1:=1 ⟧;                 a1  ||      ⟦ in2:=1 ⟧;                 b1
    ⟦ hold:=1 ⟧;                a2  ||      ⟦ hold:=2 ⟧;                b2
    if ⟦ in2=0 ⟧                    ||      if ⟦ in1=0 ⟧
        □ ⟦ (hold ≠ 1) ⟧ fi;    a3  ||          □ ⟦ (hold ≠ 2) ⟧ fi;    b3
    CritSct1;                       ||      CritSct2;
    ⟦ in1:=0 ⟧                  a4  ||      ⟦ in2:=0 ⟧                  b4
od                                  ||  od
end
```

Fig. 9.11. Peterson's algorithm for two processes

For action a_1 we use the abbreviation ⟦ in_1:=1 ⟧ to denote ⟦ **true**; in_1:=1 ⟧ (see also Sect. 9.2.4), and similarly for b_1. It is claimed that if no *CritSct*$_i$ internally deadlocks or loops indefinitely, then:

- It is not possible for both processes to execute their respective critical sections *CritSct*$_i$ concurrently.
- There are no deadlocks.
- The program is conditionally *starvation-free*, in the sense that if the first process has succeeded in executing a_1, the second one cannot prevent it from entering its critical section; and, *vice versa*, if the second process has executed b_1, the first one cannot prevent it from entering its critical section.

The translation of this *Razor* program into a box is shown, in a slightly simplified form, in Fig. 9.12. The simplification concerns the fact that initialisations and terminations (hence also the e- and x-places), as well as dead transitions have been omitted. Referring to Figs. 9.11 and 9.12, transition u_1 corresponds to a_1, transitions u_2 and u_3 correspond to a_2, transitions u_4 and u_5 correspond to a_3, and transition u_6 corresponds to a_4. Place p_4 corresponds to *CritSct*$_1$. Transitions r_j correspond to the b_i's, and q_4 to *CritSct*$_2$, in a symmetric way. However, note that the initial marking is asymmetric since *hold* is initialised to 1 rather than to 2. This asymmetry is insignificant for the three properties we wish to prove, i.e., *hold* might have been initialised to 2. Note that, by Cor. 9.5.1, the net in Fig. 9.12 is safe.

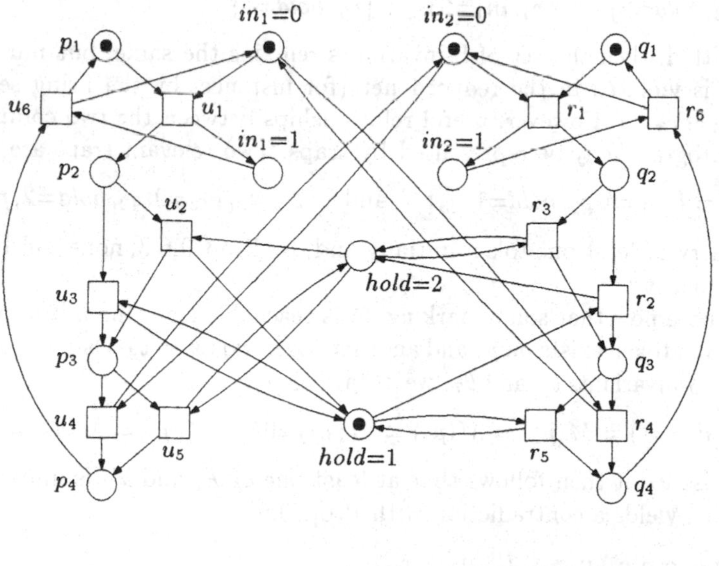

Fig. 9.12. Translation of Peterson's algorithm into a safe Petri net

A Petri-net-based analysis of this algorithm employs three of the results mentioned in Sect. 9.6.1, namely Props. 6.1.1, 9.6.2, and 9.6.3.

Safety analysis: the critical section property is satisfied. The net shown in Fig. 9.12 is decomposable into five S-components, two for the sequential components (generated by the four leftmost places and the four rightmost places, respectively) and three for the variables *hold*, in_1, and in_2 (generated pairwise by the six remaining places):

$$\theta_1 = \{p_1, p_2, p_3, p_4\} \qquad \theta_2 = \{q_1, q_2, q_3, q_4\}$$
$$\theta_3 = \{hold{=}1, hold{=}2\} \qquad \theta_4 = \{in_1{=}0, in_1{=}1\}$$
$$\theta_5 = \{in_2{=}0, in_2{=}1\} .$$

Recall from Sect. 6.1.2 that an S-component θ can be treated as a special subnet; moreover, it is possible to identify θ with the set of places s such that $\theta(s) = 1$. These S-components are initially marked with one token each. Notice, however, that the net has more S-components than these five, e.g.,

$$\theta_6 = \{in_1{=}0, p_2, p_3, p_4\} \quad \text{and} \quad \theta_7 = \{in_2{=}0, q_2, q_3, q_4\} .$$

The places p_4 and q_4 represent the two critical sections. To prove the property of mutual exclusion, it has to be shown that $M(p_4){=}0$ or $M(q_4){=}0$, for every reachable marking M. The S-invariants of the net in Fig. 9.12 alone lack the power to prove this property because if the arrows making up the six side-loops:

$$\{u_3, hold{=}1\} \qquad \{u_4, in_2{=}0\} \qquad \{u_5, hold{=}2\}$$
$$\{r_3, hold{=}2\} \qquad \{r_4, in_1{=}0\} \qquad \{r_5, hold{=}1\}$$

are omitted, then the set of S-invariants remains the same, but mutual exclusion is violated in the reduced net (for instance, by the firing sequence $u_1\, u_3\, u_4\, r_1\, r_3\, r_4$). However, useful relationships between the two components of the program may be represented by traps. Two relevant traps are

$$\mathcal{F}_1 = \{in_1{=}0, p_2, hold{=}1, q_3\} \quad \text{and} \quad \mathcal{F}_2 = \{in_2{=}0, q_2, hold{=}2, p_3\} .$$

Both carry at least one token initially and, by Prop. 9.6.3, none can be emptied of tokens.

Suppose now that some marking M is reachable from the initial marking such that a token exists on p_4 and another token exists on q_4. Using Prop. 6.1.1 and the S-invariants θ_6 and θ_7, we obtain that:

$$M(in_1{=}0) = M(p_2) = M(p_3) = M(in_2{=}0) = M(q_2) = M(q_3) = 0 .$$

Using also θ_3, it then follows that at least one of $\mathcal{F}_1$ and $\mathcal{F}_2$ is unmarked at M, which yields a contradiction with Prop. 9.6.3.

Exercise 9.6.2. Prove deadlock-freeness.

Progress analysis: the program is conditionally starvation-free. The program is not unconditionally starvation-free because the process on the left-hand side can be executed indefinitely without the process on the right-hand side having to be executed at all, and *vice versa*. Thus, both processes can be 'starved'. However, if transition r_1 has just been executed, then the other process may no longer be executed indefinitely, and conversely, if u_1 has just been executed, then the process on the right-hand side may not be executed indefinitely.

In order to demonstrate this property with respect to u_1, it needs to be shown that there is no infinite execution which, from a certain point on, continues to have a token on place p_2:

$$M_0\, t_1\, M_1\, t_2\, M_2 \ldots \underbrace{M_{j-1}\, t_j\, M_j\, t_{j+1} \quad \ldots}_{1=M_{j-1}(p_2)=M_j(p_2)=\ldots} \quad . \tag{9.3}$$

We may do this net-theoretically by using a method which was introduced in [17]. We prove the nonexistence of the execution (9.3) by contradiction, assuming that $M_0\, t_1\, M_1\, t_2\, M_2 \ldots$ as above does exist.

Let T' be the set of transitions which occur infinitely often in

$M_0\, t_1\, M_1\, t_2\, M_2 \ldots$.

From the assumption that p_2 carries a token from M_{j-1} onwards, it follows that $u_2 \notin T'$ and $u_3 \notin T'$ since both belong to $p_2^\bullet \setminus {}^\bullet p_2$ and the net is safe. By Prop. 9.6.2, there is a T-aggregate $\zeta\colon T \to \mathbf{N}$ such that T' is the support of ζ. The following list enumerates eight T-components that are candidates for ζ. Note that it is possible to identify a T-component with the set of transitions for which it returns 1, since its range is $\{0,1\}$. All other T-aggregates are linear combinations of these eight T-components (see Exc. 9.6.3).

$$\begin{aligned} \zeta_1 &= \{u_1, u_3, u_4, u_6\} & \zeta_2 &= \{u_1, u_3, u_5, u_6\} \\ \zeta_3 &= \{u_1, u_2, u_4, u_6, r_1, r_2, r_4, r_6\} & \zeta_4 &= \{u_1, u_2, u_5, u_6, r_1, r_2, r_4, r_6\} \\ \zeta_5 &= \{u_1, u_2, u_4, u_6, r_1, r_2, r_5, r_6\} & \zeta_6 &= \{u_1, u_2, u_5, u_6, r_1, r_2, r_5, r_6\} \\ \zeta_7 &= \{r_1, r_3, r_4, r_6\} & \zeta_8 &= \{r_1, r_3, r_5, r_6\}\ . \end{aligned}$$

We now check which of these eight T-invariants can possibly be such that its transitions occur in the sequence $M_0\, t_1\, M_1\, t_2\, M_2 \ldots$ after the index j. ζ_1–ζ_6 are immediately excluded since they contain transitions u_2 and u_3. ζ_7 is excluded because, from M_{j-1} onwards, p_2 is marked and hence *in*$_1$=0 is unmarked, by the S-component θ_6 (and its initial marking and Prop. 6.1.1) but *in*$_1$=0 is necessary to execute r_4. Finally, in order to execute infinitely many times r_3 and r_5 in ζ_8, it must be possible, by the S-component θ_3 (and its initial marking and Prop. 6.1.1), to move cyclically the token between *hold*=1 and *hold*=2. However, *hold*=1 can only be filled by u_2 which is never executed after M_{j-1}.

Hence the existence of $M_0\, t_1\, M_1\, t_2\, M_2 \ldots$ as in (9.3) would lead to a contradiction, proving conditional starvation-freeness.

Exercise 9.6.3. Show that, for the net in Fig. 9.12, all T-aggregates can be computed additively from the T-components $\zeta_1, \ldots, \zeta_8$.

Automatic verification of the mutual exclusion property. The program shown in Fig. 9.11 can be fed (after a syntactic translation into the tool's input language $B(PN)^2$) into the PEP partial-order-based model checker [4, 86]. This model checker translates any such program automatically into a net, calculates the McMillan finite prefix of that net [78], and executes Esparza's model checker [37] (with corrections by Graves [51] and improvements by Esparza, Römer, and Vogler [38]) with a given input formula of the associated temporal logic. In this case, the formula to be checked is $\neg(\Diamond(p_4 \wedge q_4))$. This formula uses the temporal logic operator $\Diamond$, which means, intuitively: 'possibly in the future'. Thus, informally, the formula reads: 'it is not the case that, possibly in the future, both p_4 and q_4 hold a token at the same time'. On a SUN Ultra-1/140 with normal load, the execution times shown in Fig. 9.13 were measured. For any given program and net, the prefix has to be constructed only once, so that only the third line is relevant for any given formula (except for the very first formula to be checked).

Task	Peterson (2 processes)	Dekker (2 processes)	Morris (3 processes)
Calculating the net	0.07 seconds	0.05 seconds	0.04 seconds
Building the finite prefix	0.01 seconds	0.05 seconds	1.19 seconds
Checking the formula	0.02 seconds	0.01 seconds	0.12 seconds

Fig. 9.13. Execution times for automatic verification by PEP [86]

9.6.3 Dekker's and Morris's Mutual Exclusion Algorithms

Dekker's algorithm for mutual exclusion of two processes is defined as follows:

begin **var** in_1, in_2 : {**true**, **false**} (**init false**);
  var *hold* : $\{1, 2\}$ (**init** 1);
  $c_1 \parallel c_2$
end ,

where c_1 is shown in Fig. 9.14 and c_2 arises from c_1 by swapping all the 1's with 2's (also in the indices). Initialisation of *hold* is inessential (it could be 2 instead of 1, or a nondeterministic choice between these two values).

Morris's algorithm for mutual exclusion of n processes is defined as follows:

begin **var** $a, b : \{0, 1\}$ (**init** 1);
  $m : \{0, 1\}$ (**init** 0);
  $na, nm : \mathbf{N}$ (**init** 0);
  $c_1 \parallel \ldots \parallel c_n$
end,

```
c1 : do [ in1:=true ];
       do [ in2 ] ; if [ hold = 2 ] ; [ in1:=false ];
                                      if [ hold = 1 ] fi;
                                      [ in1:=true ]
                    □ [ hold ≠ 2 ]
                    fi
       od;
       CritSct 1;
       [ hold:=2 ];
       [ in1:=false ]
     od
```

Fig. 9.14. The first component of Dekker's algorithm

where c_i is shown in Fig. 9.15. It holds for any number of parallel processes, but uses atomic semaphore operations.

```
ci : P(b); [ na:=na + 1 ]; V(b);
     P(a); [ nm:=nm + 1 ];
     P(b); [ na:=na - 1 ];
     if [ na = 0 ]; V(b); V(m) □ [ na ≠ 0 ]; V(b); V(a) fi;
     P(m); [ nm:=nm - 1 ];
     CritSct ;
     if [ nm = 0 ]; V(a) □ [ nm ≠ 0 ]; V(m) fi
```

Fig. 9.15. The i-th component of Morris's algorithm

Table 9.13 shows the execution times of automatic Petri-net-based verification of the mutual exclusion property for Dekker's algorithm with two processes and Morris's algorithm with three processes and a correspondingly adapted temporal logic formula. These figures show how useful and efficient our approach may be, in analysing the semantical correctness of an algorithm through an automatic translation of the latter into Petri boxes and using an associated automated model-checking tool.

9.7 Literature and Background

In [10], it is shown how a more powerful concurrent language can be treated in a similar way, and also how high-level Petri nets [63] (rather than place/transition nets) can be used as a target domain of the semantic function. Both translations have been optimised [50, 93] and are incorporated in the PEP tool [4, 86].

It is shown in [60] how parameter mechanisms can be added to the procedure semantics given in Sect. 9.4. Actually, this is rather straightforward. Reference parameters correspond to syntactic substitution, while value parameters can be modelled (in the usual way, see, e.g., [52]) by additional assignments preceding invocations.

Dekker's algorithm was the first (correct) protocol for the mutual exclusion between two processes based only on atomic reads and writes to a common memory variable. Morris's algorithm is often cited as an example for two unusual properties: first, that there is no syntactic difference between the processes, and second, that the ordering of two V actions (in the first **if fi** command) matters for correctness.

For the trap method, see [3, 7, 17, 91, 92]. 'By-hand' verification of Dekker's and Morris's algorithms using traps can be found in [17, 92] and [96], respectively. However, it should be noted that the trap method is not universal because there exist examples in which neither S-invariants nor traps are strong enough to prove a given safety property.

The T-invariant method was made more efficient in [79]. [92] describes an alternative, though still Petri-net-based, method for proving progress properties.

10. Conclusion

In the introduction of this book, we had indicated that we aimed at forging as tight and strong links between process algebras and Petri nets as possible. In retrospect, it is possible to say that we have achieved this aim both at the structural and at the behavioural levels. Moreover, we have carried out this investigation not only for a specific process algebra, but rather in a generic way, for the box algebra. The latter encompasses an infinity of concrete process algebras, of which the existing ones can be viewed as suitably adapted derivatives or instances. It is also worth mentioning that we have strived to allow infinity in the definitions and formulations of the results wherever it was possible; in particular, we considered operators with infinitely many arguments.

In the process of proving the isomorphism between the box algebra and its associated Petri net model, we have identified a series of properties whose judicious interplay was instrumental for the validity of this result such as safeness, x-directedness and ex-exclusiveness of operands for non-pure operators. These investigations were often triggered by the needs in the translation of a realistic concurrent programming language. They also often generalise results published elsewhere, so that this book is full of essentially new results and mostly unpublished proofs.

A main driving concept behind the theory described in this book has been compositionality. In the first instance our aim was to match the operators of a (specific) process algebra one-to-one with operators on nets. This idea has developed into matching the operators of process algebras with (operator) nets themselves, thus resulting in using nets both as objects and as functions.

Compositionality has several other incarnations in this book, for instance, concerning Petri net indigenous techniques. We have developed general methods for deriving S-invariants, S-components, and S-aggregates for nets which are constructed modularly, from the corresponding invariants of their components. Moreover, this has been done even after allowing infinite nets, unguarded recursion, and non-pure operator boxes. It has been shown how to use these constructions to obtain coverability results, and thus to derive important behavioural characteristics of whole families of nets. This makes it unnecessary to consider separately each relevant behavioural property (such as safeness or cleanness) and to prove its validity by structural induction.

Compositionality also played an important role in our treatment of recursion, in the sense that properties of a fixpoint are derived from those of its approximations. Since we have developed the whole theory for unrestricted (in particular, unguarded) recursion, we could not rely on any of the standard techniques of denotational semantics (such as uniqueness of minimal fixpoints). We have discussed this phenomenon and shown how standard fixpoint techniques can nevertheless be exploited.

Another example of compositionality is that of behaviour, i.e., whether or not the behaviour of a composed object is the composition of the behaviour of its components. We have discussed this question and shown under which conditions it can be answered affirmatively.

Finally, we have shown that the denotational semantics of a full programming language (including shared variables, channels, parallelism, and recursion) can be given compositionally. We have also made the point that this can help in proving properties about programs written in such a language (though we would not go so far as to claim that such proofs can always be done completely compositionally).

The theory presented in this book can be extended in several ways. One of these extensions consists in allowing nets to be high-level nets (rather than place-transition nets), but still consider the same types of process algebra. The M-net and A-net theories initiated by the group around Elisabeth Pelz and Hanna Klaudel in Paris [67, 65, 66] serve this purpose. process algebra domain.

Having thus reached the end of the book, it may seem right to recall some of the events which had led to its conception. It was during the three years of the running of the European Union funded Basic Research Action Project DEMON (Design Methods based on Nets) that the initial idea of developing a model encompassing Petri nets and process algebras was put forward and then developed in a series of articles. This book has given an extensive account of the resulting PBC, which can be thought of as a foundational framework based on which other, related investigations have later been carried out.

These various strands of PBC-related work were continued during the lifetime of the successor of the DEMON Project, the Basic Research Action Working Group CALIBAN (Causal Calculi based on Nets). The idea of writing a book on the PBC was first considered in 1992. However, at that time its scope was much narrower, and moreover, the PBC theory was not yet developed to the extent it is now.

Thus, as it is often the case in life, our initial plans were only a guideline, and the real work on the book has transformed into something rather different. The main reason for this was an early realisation that the box algebra lends itself easily to a range of generalisations and modifications, which could result both in an increased expressive power and more powerful results. This has eventually developed into the piece of work of which you are just reading the last sentence.

Appendix: Solutions of Selected Exercises

Solution of Exercise 3.2.3 (the commutativity part)

We need to show that, for every pair of dynamic PBC expressions, G and H, the transition systems $\mathbf{ts}_{G\|H} = (V_1, L, A_1, v_0^1)$ and $\mathbf{ts}_{H\|G} = (V_2, L, A_2, v_0^2)$ are isomorphic. We first observe that, for all dynamic expressions C, D, C', and D',

$$C\|D \equiv C'\|D' \implies C \equiv C' \wedge D \equiv D' . \tag{.1}$$

Moreover, for every dynamic expression J such that $J \equiv C\|D$, we have

$$J = K\|L \text{ or } J = \overline{E\|F} \text{ or } J = \underline{E\|F} \text{ or } J = \overline{X} \text{ or } J = \underline{X} , \tag{.2}$$

where K, L are dynamic PBC expressions, E, F are static PBC expressions, and X is a process variable such that $X \stackrel{\mathrm{df}}{=} E'\|F'$, for some E' and F'. Both (.1) and (.2) can be shown by considering the various inaction rules (as well as action rules with empty steps) defined in Chap. 3. We can now proceed with a sketch of the main part of the proof.

Let Q be a relation comprising all pairs

$$\left([C\|D]_\equiv \, , \, [D\|C]_\equiv\right)$$

where C, D are dynamic PBC expressions, and let $\mathit{isom} = (V_1 \times V_2) \cap Q$. We will now argue that isom is an isomorphism for $\mathbf{ts}_{G\|H}$ and $\mathbf{ts}_{H\|G}$.

We first observe that $V_1 = \{v \mid (v,w) \in \mathit{isom}\}$ and $V_2 = \{w \mid (v,w) \in \mathit{isom}\}$ which follows from (.2), the action rules PAR, SKP, RDO, the fact that no other action rules are applicable at the topmost level for expressions in (.2), and, finally, IREC, IPAR1, and IPAR2. To show that isom is injective, we take $(v,w), (v',w) \in \mathit{isom}$. Then there are C, D, C', and D' such that $D\|C \in v$, $D'\|C' \in v'$, and $C\|D \equiv C'\|D' \in w$. By (.1), $C \equiv C'$ and $D \equiv D'$. Hence $D\|C \equiv D'\|C'$, and so $v = v'$. Thus isom is a bijection.

Clearly, $(v_0^1, v_0^2) \in \mathit{isom}$. Thus (by symmetry) it suffices to show that if $(v, \Gamma, w) \in A_1$, then $(v', \Gamma, w') \in A_2$, where $(v,v'), (w,w') \in \mathit{isom}$. By (.2), we can consider the following three cases.

Case 1: $C \xrightarrow{\Gamma'} C'$, $D \xrightarrow{\Gamma''} D'$, $\Gamma = \Gamma' + \Gamma''$, $C\|D \in v$, and $C'\|D' \in w$. Since $(v,v') \in \mathit{isom}$, there are K, L such that $K\|L \in v$ and

$L\|K \in v'$. By (.1), $C \equiv K$ and $D \equiv L$. Hence $L\|K \xrightarrow{\Gamma} D'\|C'$. Moreover, $([C'\|D']_{\equiv}, [D'\|C']_{\equiv}) \in \mathit{isom}$.

Case 2: $\Gamma = \{\mathsf{skip}\}$, $\overline{M} \xrightarrow{\Gamma} \underline{M}$, $\overline{M} \in v$, and $\underline{M} \in w$. Since $(v, v') \in \mathit{isom}$, there are K, L such that $K\|L \in v$ and $L\|K \in v'$. By (.2), we may assume that $M = \overline{E\|F} \equiv \overline{E}\|\overline{F}$ for some E, F. Hence, by (.1), $\overline{E} \equiv K$ and $\overline{F} \equiv L$. We then proceed similarly as in Case 1.

Case 3: $\Gamma = \{\mathsf{redo}\}$, $\underline{M} \xrightarrow{\Gamma} \overline{M}$, $\underline{M} \in v$, and $\overline{M} \in w$. We then proceed as in Case 2.

Solution of Exercise 3.2.9

This exercise can be dealt with similarly as the previous one, with Q being defined as the set of all pairs:

$$\Big([E;(F;G)]_{\equiv} \ , \ [(E;F);G]_{\equiv} \Big)$$
$$\Big([E;(G;F)]_{\equiv} \ , \ [(E;G);F]_{\equiv} \Big)$$
$$\Big([G;(E;F)]_{\equiv} \ , \ [(G;E);F]_{\equiv} \Big)$$

where E, F are static and G dynamic PBC expressions.

Solution of Exercise 3.2.12

In what follows a *cluster* is a finite multiset of multiactions Δ which can be decomposed as $\Delta = \sum_{j=1}^{m} \{\alpha_j + h_j \cdot \{a\} + k_j \cdot \{\widehat{a}\}\}$ in such a way that $m \geq 2$,

$$\sum_{j=1}^{m} h_j = \sum_{j=1}^{m} k_j = m - 1 \ \text{ and } \ \prod_{j=1}^{m} (h_j + k_j) > 0 \, .$$

We also denote $\mathit{synch}(\Delta) = \sum_{j=1}^{m} \alpha_j$. Then SY12 can be rewritten to another format which will be used in the proof:

$$\frac{G \xrightarrow{\Gamma + \sum_{i=1}^{n} \Delta_i} G'}{G \ \mathsf{sy}\ a \xrightarrow{\Gamma + \sum_{i=1}^{n} \{\mathit{synch}(\Delta_i)\}} G' \ \mathsf{sy}\ a} \quad n \geq 0 \text{ and each } \Delta_i \text{ is a cluster.} \qquad (.3)$$

We will first show that a move obtained using SY1 and SY2 can be obtained by a single application of SY12
To begin with, we notice that any derivation for a move of G sy a with only SY1 and SY2, uses one application of SY1 followed by a series of k ($k \geq 0$) applications of SY2. We then proceed by induction on k.

Base case ($k = 0$): Immediate if we observe that SY12 with $n = 0$ is nothing but SY1.

Induction step: Assume that the result holds for $k \geq 0$, and that we have

$$G \text{ sy } a \xrightarrow{\{\alpha+\{a\}\}+\{\beta+\{\widehat{a}\}\}+\Gamma'} G' \text{ sy } a \quad \text{and} \quad G \text{ sy } a \xrightarrow{\{\alpha+\beta\}+\Gamma'} G' \text{ sy } a \;,$$

where the first move has been derived with k applications of SY2, and the second move has been derived from the first move by a $(k+1)$-st application of SY2. By the induction hypothesis, it is the case that the first move can be derived using a single application of SY12, i.e., we have

$$G \xrightarrow{\Gamma+\sum_{i=1}^{n}\Delta_i} G' \quad \text{and} \quad G \text{ sy } a \xrightarrow{\Gamma+\sum_{i=1}^{n}\{synch(\Delta_i)\}} G' \text{ sy } a \;,$$

where $n \geq 0$, each Δ_i is a cluster, and

$$\Gamma + \sum_{i=1}^{n}\{synch(\Delta_i)\} = \{\alpha + \{a\}\} + \{\beta + \{\widehat{a}\}\} + \Gamma' \;.$$

We can therefore consider three cases depending on whether $\alpha + \{a\}$ and $\beta + \{\widehat{a}\}$ belong to Γ (the fourth case is symmetric to Case 2).

Case 1: $\alpha + \{a\} \notin \Gamma$ and $\beta + \{\widehat{a}\} \notin \Gamma$. Then $n \geq 2$ and, without loss of generality, we have

$$\Gamma' = \Gamma + \sum_{i=1}^{n-2}\{synch(\Delta_i)\} \;,$$

$\alpha + \{a\} = synch(\Delta_{n-1})$, and $\beta + \{\widehat{a}\} = synch(\Delta_n)$. We will show that $\Delta = \Delta_{n-1} + \Delta_n$ is a cluster such that $synch(\Delta) = \alpha + \beta$. Let

$$\Delta_{n-1} = \sum_{j=1}^{m}\{\alpha_j + h_j \cdot \{a\} + k_j \cdot \{\widehat{a}\}\}$$

$$\Delta_n = \sum_{j=m+1}^{m+l}\{\alpha_j + h_j \cdot \{a\} + k_j \cdot \{\widehat{a}\}\}$$

be decompositions of Δ_{n-1} and Δ_n satisfying the definition of a cluster. From

$$a \in synch(\Delta_{n-1}) = \sum_{j=1}^{m}\alpha_j \quad \text{and} \quad \widehat{a} \in synch(\Delta_n) = \sum_{j=m+1}^{m+l}\alpha_j$$

it follows (without loss of generality) that $a \in \alpha_1$ and $\widehat{a} \in \alpha_{m+1}$. Then we define $\alpha'_1 = \alpha_1 - \{a\}$, $\alpha'_{m+1} = \alpha_{m+1} - \{\widehat{a}\}$, $h'_1 = h_1 + 1$, $k'_1 = k_1$, $h'_{m+1} = h_{m+1}$, and $k'_{m+1} = k_{m+1} + 1$. Moreover, for all other indices j, $\alpha'_j = \alpha_j$, $h'_j = h_j$, and $k'_j = k_j$. Thus

$$\Delta = \sum_{j=1}^{m+l}\{\alpha'_j + h'_j \cdot \{a\} + k'_j \cdot \{\widehat{a}\}\}$$

and it is easy to see that Δ is indeed a cluster; in particular, we have

$$\sum_{j=1}^{m+l} h'_j = (1+h_1) + \sum_{j=2}^{m+l} h_j = 1 + \sum_{j=1}^{m} h_j + \sum_{j=m+1}^{m+l} h_j$$
$$= 1 + (m-1) + (l-1) = (m+l) - 1$$

and, likewise, $\sum_{j=1}^{m+l} k'_j = (m+l) - 1$. Also,

$$synch(\Delta) = \sum_{j=1}^{m+l} \alpha'_j = \sum_{j=1}^{m+l} \alpha_j - \{a, \widehat{a}\}$$
$$= synch(\Delta_{n-1}) + synch(\Delta_n) - \{a, \widehat{a}\} = \alpha + \beta\,.$$

Thus we have

$$G \xrightarrow{\Gamma + \sum_{i=1}^{n-2} \Delta_i + \Delta} G'$$

and, by SY12, we can deduce in one step that

$$G \text{ sy } a \xrightarrow{\Gamma + \sum_{i=1}^{n-2} \{synch(\Delta_i)\} + \{synch(\Delta)\}} G' \text{ sy } a\,.$$

Moreover,

$$\Gamma + \sum_{i=1}^{n-2} \{synch(\Delta_i)\} + \{synch(\Delta)\} = \Gamma' + \{\alpha + \beta\}$$

which completes Case 1.

Case 2: $\alpha + \{a\} \in \Gamma$ and $\beta + \{\widehat{a}\} \notin \Gamma$. Then $n \geq 1$ and, without loss of generality, we have $\Gamma = \Gamma'' + \{\alpha + \{a\}\}$,

$$\Gamma' = \Gamma'' + \sum_{i=1}^{n-1} \{synch(\Delta_i)\}\,,$$

and $\beta + \{\widehat{a}\} = synch(\Delta_n)$. We will show that $\Delta = \Delta_n + \{\alpha + \{a\}\}$ is a cluster such that $synch(\Delta) = \alpha + \beta$.

Let

$$\Delta_n = \sum_{j=1}^{m} \{\alpha_j + h_j \cdot \{a\} + k_j \cdot \{\widehat{a}\}\}$$

be a decomposition of Δ_n satisfying the definition of a cluster. From

$$\widehat{a} \in synch(\Delta_n) = \sum_{j=1}^{m} \alpha_j$$

it follows (without loss of generality) that $\widehat{a} \in \alpha_1$. Then we define $\alpha'_1 = \alpha_1 - \{\widehat{a}\}$, $\alpha'_{m+1} = \alpha$, $h'_1 = h_1$, $k'_1 = k_1 + 1$, $h'_{m+1} = 1$, and $k'_{m+1} = 0$. Moreover, for all other indices j, $\alpha'_j = \alpha_j$, $h'_j = h_j$, and $k'_j = k_j$. Thus

$$\Delta = \sum_{j=1}^{m+1} \{\alpha'_j + h'_j \cdot \{a\} + k'_j \cdot \{\widehat{a}\}\}$$

and it is easy to see that it is indeed a cluster satisfying

$$\begin{aligned} synch(\Delta) &= \sum_{j=1}^{m+1} \alpha'_j = \sum_{j=1}^{m} \alpha_j - \{\widehat{a}\} + \alpha = synch(\Delta_n) - \{\widehat{a}\} + \alpha \\ &= \beta + \{\widehat{a}\} - \{\widehat{a}\} + \alpha = \alpha + \beta \,. \end{aligned}$$

Thus we have

$$G \xrightarrow{\Gamma'' + \{\alpha + \{a\}\} + \sum_{i=1}^{n-1} \Delta_i + \Delta_n} G'$$

which means that

$$G \xrightarrow{\Gamma'' + \sum_{i=1}^{n-1} \Delta_i + \Delta} G'$$

and, by SY12, we can deduce in one step that

$$G \text{ sy } a \xrightarrow{\Gamma'' + \sum_{i=1}^{n-1} \{synch(\Delta_i)\} + \{synch(\Delta)\}} G' \text{ sy } a \,.$$

Moreover,

$$\Gamma'' + \sum_{i=1}^{n-1} \{synch(\Delta_i)\} + \{synch(\Delta)\} = \Gamma' + \{\alpha + \beta\}$$

which completes Case 2.

Case 3: $\alpha + \{a\} \in \Gamma$ and $\beta + \{\widehat{a}\} \in \Gamma - \{\alpha + \{a\}\}$. Then $n \geq 0$, $\Gamma = \Gamma'' + \{\alpha + \{a\}, \beta + \{\widehat{a}\}\}$, and

$$\Gamma' = \Gamma'' + \sum_{i=1}^{n} \{synch(\Delta_i)\} \,.$$

It is straightforward to see that $\Delta = \{\alpha + \{a\}, \beta + \{\widehat{a}\}\}$ is a cluster such that $synch(\Delta) = \alpha + \beta$. We then proceed similarly as in the two previous cases.

We will now show, proceeding by induction on $l = \sum_{i=1}^{n} |\Delta_i|$, that a move obtained using SY12 can also be obtained using SY1 and SY2.

Base case (l=0): Immediate since $n = 0$ and we have already noted that SY12 with $n = 0$ is nothing but SY1.

Inductive step: Suppose that the result holds for $l \geq 0$ and, moreover, that we have $\sum_{i=1}^{n} |\Delta_i| = l + 1$ in (.3). Then $n \geq 1$ and we can assume that

$$\Delta_n = \sum_{j=1}^{m} \{\alpha_j + h_j \cdot \{a\} + k_j \cdot \{\widehat{a}\}\}$$

is a decomposition of Δ_n satisfying the definition of a cluster. We consider two cases.

Case 1: $m = 2$. Then, since $h_1 + h_2 = 1 = k_1 + k_2$ and $h_1 + k_1 \neq 0 \neq h_2 + k_2$, without loss of generality we may assume that $h_1 = k_2 = 1$ and $k_1 = h_2 = 0$. By the induction hypothesis, we may derive

$$G \text{ sy } a \xrightarrow{\Gamma+\{\alpha_1+\{a\}\}+\{\alpha_2+\{\widehat{a}\}\}+\sum_{i=1}^{n-1}\{synch(\Delta_i)\}} G' \text{ sy } a$$

using only SY1 and SY2. Hence one more application of SY2 yields

$$G \text{ sy } a \xrightarrow{\Gamma+\{\alpha_1+\alpha_2\}+\sum_{i=1}^{n-1}\{synch(\Delta_i)\}} G' \text{ sy } a\ .$$

But $\alpha_1 + \alpha_2 = synch(\Delta_n)$, so we have

$$G \text{ sy } a \xrightarrow{\Gamma+\sum_{i=1}^{n}\{synch(\Delta_i)\}} G' \text{ sy } a$$

using only SY1 and SY2.

Case 2: $m > 2$. Then, since $\sum_{j=1}^{m}(h_j + k_j) = 2m - 2$ and $\prod_{j=1}^{m}(h_j + k_j) > 0$, by the pigeon hole principle, we have (without loss of generality) that $h_m = 1$ and $k_m = 0$. Thus, since $\prod_{j=1}^{m-1}(k_j + h_j) > 0$, $\sum_{j=1}^{m-1} k_j = m - 1$, and $\sum_{j=1}^{m-1} h_j = m - 2 > 0$, we may assume (again without loss of generality) that $k_{m-1} \geq 1$ and $h_{m-1} + k_{m-1} \geq 2$. Then we define $\alpha'_{m-1} = \alpha_{m-1} + \{\widehat{a}\}$, $h'_{m-1} = h_{m-1}$, and $k'_{m-1} = k_{m-1} - 1$. Moreover, for all other indices $j \leq m-1$, $\alpha'_j = \alpha_j$, $h'_j = h_j$, and $k'_j = k_j$.

It is easy to see that $\Delta = \sum_{j=1}^{m-1}\{\alpha'_j + h'_j \cdot \{a\} + k'_j \cdot \{\widehat{a}\}\}$ is a cluster such that $synch(\Delta) = \sum_{j=1}^{m-1} \alpha_j + \{\widehat{a}\}$. We thus can rewrite the premise of (.3) in the following way:

$$G \xrightarrow{\Gamma+\{\alpha_m+\{a\}\}+\sum_{i=1}^{n-1}\Delta_i+\Delta} G' \text{ sy } a\ .$$

By the induction hypothesis (note that $\sum_{i=1}^{n-1}|\Delta_i| + |\Delta| = l$), we may derive

$$G \text{ sy } a \xrightarrow{\Gamma+\{\alpha_m+\{a\}\}+\sum_{i=1}^{n-1}\{synch(\Delta_i)\}+\{synch(\Delta)\}} G' \text{ sy } a$$

using only SY1 and SY2, which is nothing but

$$G \text{ sy } a \xrightarrow{\Gamma+\sum_{i=1}^{n-1}\{synch(\Delta_i)\}+\{\alpha_m+\{a\}\}+\{\sum_{j=1}^{m-1}\alpha_j+\{\widehat{a}\}\}} G' \text{ sy } a\ .$$

Hence one more application of SY2 yields

$$G \text{ sy } a \xrightarrow{\Gamma+\sum_{i=1}^{n-1}\{synch(\Delta_i)\}+\{\sum_{j=1}^{m}\alpha_j\}} G' \text{ sy } a$$

which completes Case 2 since

$$\Gamma + \sum_{i=1}^{n-1}\{synch(\Delta_i)\} + \{\sum_{j=1}^{m}\alpha_j\} = \Gamma + \sum_{i=1}^{n}\{synch(\Delta_i)\}.$$

Solution of Exercise 7.1.2

Consider the following SOS-operator box Ω

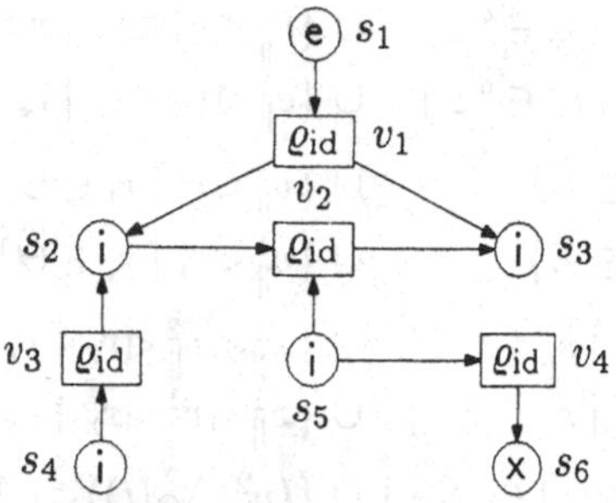

whose complex markings (M, Q) reachable from ${}^\circ\Omega$ and Ω° and the corresponding factorisations $(\mu_e, \mu_d, \mu_x, \mu_s)$ are as follows:

M	Q	μ_e	μ_d	μ_x	μ_s
$\{s_1\}$	$\emptyset$	$\{v_1\}$	$\emptyset$	$\emptyset$	$\{v_2, v_3, v_4\}$
$\emptyset$	$\{v_1\}$	$\emptyset$	$\{v_1\}$	$\emptyset$	$\{v_2, v_3, v_4\}$
$\{s_2, s_3\}$	$\emptyset$	$\emptyset$	$\emptyset$	$\{v_1\}$	$\{v_2, v_3, v_4\}$
		$\emptyset$	$\emptyset$	$\{v_2, v_3\}$	$\{v_1, v_4\}$
$\{s_6\}$	$\emptyset$	$\emptyset$	$\emptyset$	$\{v_4\}$	$\{v_1, v_2, v_3\}$

It may be observed that for the factorisation $\mu = (\emptyset, \emptyset, \{v_2, v_3\}, \{v_1, v_4\})$, we have $s_2 \in {}^\bullet\mu_x \cap \mu_x^\bullet$ which is sufficient to construct an example illustrating the property in question; for example, one can take the refinement defined for the nets $\Sigma = (\mathsf{N}_\alpha, \underline{\mathsf{N}_\alpha}, \underline{\mathsf{N}_\alpha}, \mathsf{N}_\alpha)$.

Solution of Exercise 7.3.14

Consider the following SOS-operator box Ω

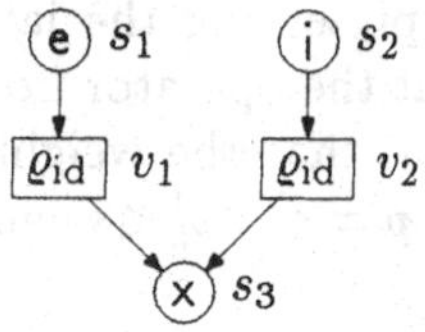

and two expressions, $\Omega(a,\overline{b})$ and $\Omega(a,\underline{b})$. According to the rules of operational semantics, we have $\Omega(a,\overline{b}) \xrightarrow{\{b\}} \Omega(a,\underline{b})$. We then observe that $\Omega(a,\underline{b})$ is a box expression since the corresponding factorisation, $\mu = (\emptyset, \emptyset, \{v_2\}, \{v_1\})$, is a factorisation of the exit marking of Ω. However, $\Omega(a,\overline{b})$ is not a box expression since the corresponding $\mu = (\{v_2\}, \emptyset, \emptyset, \{v_1\})$ is a factorisation of the complex marking $(M, Q) = (\{s_2\}, \emptyset)$ which is not reachable from the entry or exit marking of Ω.

Solution of Exercise 4.4.1 (the commutativity part)

Let $\Delta = \Sigma \| \Theta$ and $\Xi = \Theta \| \Sigma$. We first observe that

$$\begin{aligned}
{}^\circ\Delta &= \{e_{\|}^1 \lhd v_{\|}^1 \lhd s_{\mathrm{e}} \mid s_{\mathrm{e}} \in {}^\circ\Sigma\} &\cup \{e_{\|}^2 \lhd v_{\|}^2 \lhd r_{\mathrm{e}} \mid r_{\mathrm{e}} \in {}^\circ\Theta\}\\
{}^\circ\Xi &= \{e_{\|}^2 \lhd v_{\|}^2 \lhd s_{\mathrm{e}} \mid s_{\mathrm{e}} \in {}^\circ\Sigma\} &\cup \{e_{\|}^1 \lhd v_{\|}^1 \lhd r_{\mathrm{e}} \mid r_{\mathrm{e}} \in {}^\circ\Theta\}\\
\ddot{\Delta} &= \{v_{\|}^1 \lhd s_{\mathrm{i}} \mid s_{\mathrm{i}} \in \ddot{\Sigma}\} &\cup \{v_{\|}^2 \lhd r_{\mathrm{i}} \mid r_{\mathrm{i}} \in \ddot{\Theta}\}\\
\ddot{\Xi} &= \{v_{\|}^2 \lhd s_{\mathrm{i}} \mid s_{\mathrm{i}} \in \ddot{\Sigma}\} &\cup \{v_{\|}^1 \lhd r_{\mathrm{i}} \mid r_{\mathrm{i}} \in \ddot{\Theta}\}\\
\Delta^\circ &= \{x_{\|}^1 \lhd v_{\|}^1 \lhd s_{\mathrm{x}} \mid s_{\mathrm{x}} \in \Sigma^\circ\} &\cup \{x_{\|}^2 \lhd v_{\|}^2 \lhd r_{\mathrm{x}} \mid r_{\mathrm{x}} \in \Theta^\circ\}\\
\Xi^\circ &= \{x_{\|}^2 \lhd v_{\|}^2 \lhd s_{\mathrm{x}} \mid s_{\mathrm{x}} \in \Sigma^\circ\} &\cup \{x_{\|}^1 \lhd v_{\|}^1 \lhd r_{\mathrm{x}} \mid r_{\mathrm{x}} \in \Theta^\circ\}\\
T_\Delta &= \{(v_{\|}^1, \lambda_\Sigma(t)) \lhd t \mid t \in T_\Sigma\} \cup \{(v_{\|}^2, \lambda_\Theta(t)) \lhd u \mid u \in T_\Theta\}\\
T_\Xi &= \{(v_{\|}^2, \lambda_\Sigma(t)) \lhd t \mid t \in T_\Sigma\} \cup \{(v_{\|}^1, \lambda_\Theta(u)) \lhd u \mid u \in T_\Theta\}\,.
\end{aligned}$$

An isomorphism *isom* between Δ and Ξ can be defined as follows (where $s_{\mathrm{e}} \in {}^\circ\Sigma$, $u \in T_\Theta$, etc, as in the above formulae):

$$\begin{aligned}
e_{\|}^1 \lhd v_{\|}^1 \lhd s_{\mathrm{e}} &\mapsto e_{\|}^2 \lhd v_{\|}^2 \lhd s_{\mathrm{e}}\\
e_{\|}^2 \lhd v_{\|}^2 \lhd r_{\mathrm{e}} &\mapsto e_{\|}^1 \lhd v_{\|}^1 \lhd r_{\mathrm{e}}\\
v_{\|}^1 \lhd s_{\mathrm{i}} &\mapsto v_{\|}^2 \lhd s_{\mathrm{i}}\\
v_{\|}^2 \lhd r_{\mathrm{i}} &\mapsto v_{\|}^1 \lhd r_{\mathrm{i}}\\
x_{\|}^1 \lhd v_{\|}^1 \lhd s_{\mathrm{x}} &\mapsto x_{\|}^2 \lhd v_{\|}^2 \lhd s_{\mathrm{x}}\\
x_{\|}^2 \lhd v_{\|}^2 \lhd r_{\mathrm{x}} &\mapsto x_{\|}^1 \lhd v_{\|}^1 \lhd r_{\mathrm{x}}\\
(v_{\|}^1, \lambda_\Sigma(t)) \lhd t &\mapsto (v_{\|}^2, \lambda_\Sigma(t)) \lhd t\\
(v_{\|}^2, \lambda_\Theta(u)) \lhd u &\mapsto (v_{\|}^1, \lambda_\Theta(u)) \lhd u\,.
\end{aligned}$$

Clearly, *isom* is a bijection preserving the labelling of nodes as well as the markings of places (note that the operator box $\Omega_{\|}$ has the empty marking). What remains to be shown is that the weight functions are also preserved. We show that this holds for $p = e_{\|}^1 \lhd v_{\|}^1 \lhd s_{\mathrm{e}}$ and $y = (v_{\|}^1, \lambda_\Sigma(t)) \lhd t$. We then have:

$$W_\Delta(p, y) =_{(4.7)} \sum_{z \in \{s_{\mathrm{e}}\}} \sum_{w \in \{t\}} W_\Sigma(z, w) = W_\Sigma(s_{\mathrm{e}}, t)\,.$$

We next observe that $p' = isom(p) = e_{\|}^2 \lhd v_{\|}^2 \lhd s_{\mathrm{e}}$ and $y' = isom(y) = (v_{\|}^2, \lambda_\Sigma(t)) \lhd t$, and that we have

$$W_\Xi(p, y) =_{(4.7)} \sum_{z \in \{s_{\mathrm{e}}\}} \sum_{w \in \{t\}} W_\Sigma(z, w) = W_\Sigma(s_{\mathrm{e}}, t)\,.$$

Thus $W_\Delta(p,y) = W_\Xi(p',y')$. As another case, we take $p = v^1_{\|} \lhd s_\mathrm{i}$ and $y = (v^2_{\|}, \lambda_\Theta(u)) \lhd u$. Then

$$W_\Delta(p,y) =_{(4.7)} \sum_{z\in\emptyset} \sum_{w\in\{u\}} W_\Theta(z,w) = 0 \,.$$

We next observe that $p' = isom(p) = v^2_{\|} \lhd s_\mathrm{i}$ and $y' = isom(y) = (v^1_{\|}, \lambda_\Theta(u)) \lhd u$, and that we have

$$W_\Delta(p',y') =_{(4.7)} \sum_{z\in\emptyset} \sum_{w\in\{u\}} W_\Theta(z,w) = 0 \,.$$

Thus $W_\Delta(p,y) = W_\Xi(p',y')$. Other cases can be dealt with in a similar way.

Solution of Exercise 4.4.3

Let $\Delta = \Sigma;(\Theta;\Psi)$ and $\Xi = (\Sigma;\Theta);\Psi$. We first observe that

$$
\begin{aligned}
{}^\circ\Delta &= \{e_; \lhd v^1_; \lhd s_\mathrm{e} \mid s_\mathrm{e} \in {}^\circ\Sigma\}\\
{}^\circ\Xi &= \{e_; \lhd v^1_; \lhd e_; \lhd v^1_; \lhd s_\mathrm{e} \mid s_\mathrm{e} \in {}^\circ\Sigma\}
\end{aligned}
$$

$$
\begin{aligned}
\ddot{\Delta} = {} & \{v^1_; \lhd s_\mathrm{i} \mid s_\mathrm{i} \in \ddot{\Sigma}\}\\
& \cup \{v^2_; \lhd v^1_; \lhd r_\mathrm{i} \mid r_\mathrm{i} \in \ddot{\Theta}\}\\
& \cup \{v^2_; \lhd v^2_; \lhd r_\mathrm{i} \mid r_\mathrm{i} \in \ddot{\Theta}\}\\
& \cup \{i_; \lhd \{v^1_; \lhd s_\mathrm{x}, v^2_; \lhd e_; \lhd v^1_; \lhd r_\mathrm{e}\} \mid s_\mathrm{x} \in \Sigma^\circ \ \wedge\ r_\mathrm{e} \in {}^\circ\Theta\}\\
& \cup \{v^2_; \lhd i_; \lhd \{v^1_; \lhd r_\mathrm{x}, v^2_; \lhd p_\mathrm{e}\} \mid r_\mathrm{x} \in \Sigma^\circ \ \wedge\ p_\mathrm{e} \in {}^\circ\Psi\}\\
\ddot{\Xi} = {} & \{v^1_; \lhd v^1_; \lhd s_\mathrm{i} \mid s_\mathrm{i} \in \ddot{\Sigma}\}\\
& \cup \{v^1_; \lhd v^2_; \lhd r_\mathrm{i} \mid r_\mathrm{i} \in \ddot{\Theta}\}\\
& \cup \{v^1_; \lhd r_\mathrm{i} \mid r_\mathrm{i} \in \ddot{\Theta}\}\\
& \cup \{v^1_; \lhd i_; \lhd \{v^1_; \lhd s_\mathrm{x}, v^2_; \lhd r_\mathrm{e}\} \mid s_\mathrm{x} \in \Sigma^\circ \ \wedge\ r_\mathrm{e} \in {}^\circ\Theta\}\\
& \cup \{i_; \lhd \{v^1_; \lhd v^2_; \lhd r_\mathrm{x}, v^2_; \lhd p_\mathrm{e}\} \mid r_\mathrm{x} \in \Sigma^\circ \ \wedge\ p_\mathrm{e} \in {}^\circ\Psi\}
\end{aligned}
$$

$$
\begin{aligned}
\Delta^\circ &= \{x;\lhd v^2_; \lhd x_; \lhd v^2_; \lhd p_\mathrm{e} \mid p_\mathrm{e} \in \Psi^\circ\}\\
\Xi^\circ &= \{x_; \lhd v^2_; \lhd p_\mathrm{e} \mid p_\mathrm{e} \in \Sigma^\circ\}
\end{aligned}
$$

$$
\begin{aligned}
T_\Delta = {} & \{(v^1_;, \lambda_\Sigma(t)) \lhd t \mid t \in T_\Sigma\}\\
& \cup \{(v^2_;, \lambda_\Theta(u)) \lhd (v^1_;, \lambda_\Theta(u)) \lhd u \mid u \in T_\Theta\}\\
& \cup \{(v^2_;, \lambda_\Psi(v)) \lhd (v^2_;, \lambda_\Psi(v)) \lhd v \mid v \in T_\Psi\}\\
T_\Xi = {} & \{(v^1_;, \lambda_\Sigma(t)) \lhd (v^1_;, \lambda_\Sigma(t)) \lhd t \mid t \in T_\Sigma\}\\
& \cup \{(v^1_;, \lambda_\Theta(u)) \lhd (v^2_;, \lambda_\Theta(u)) \lhd u \mid u \in T_\Theta\}\\
& \cup \{(v^2_;, \lambda_\Psi(v)) \lhd v \mid v \in T_\Psi\} \,.
\end{aligned}
$$

An isomorphism *isom* between Δ and Ξ can be defined as follows (where $s_e \in {}^\circ\Sigma$, $u \in T_\Theta$, etc, as in the above formulae):

$$\begin{array}{rcl}
e_; \lhd v^1_; \lhd s_e & \mapsto & e_; \lhd v^1_; \lhd e_; \lhd v^1_; \lhd s_e \\
v^1_; \lhd s_i & \mapsto & v^1_; \lhd v^1_; \lhd s_i \\
v^2_; \lhd v^1_; \lhd r_i & \mapsto & v^1_; \lhd v^2_; \lhd r_i \\
v^2_; \lhd v^2_; \lhd r_i & \mapsto & v^1_; \lhd r_i \\
i_; \lhd \{v^1_; \lhd s_x, v^2_; \lhd e_; \lhd v^1_; \lhd r_e\} & \mapsto & v^1_; \lhd i_; \lhd \{v^1_; \lhd s_x, v^2_; \lhd r_e\} \\
v^2_; \lhd i_; \lhd \{v^1_; \lhd r_x, v^2_; \lhd p_e\} & \mapsto & \{i_; \lhd \{v^1_; \lhd v^2_; \lhd r_x, v^2_; \lhd p_e\} \\
x_; \lhd v^2_; \lhd x_; \lhd v^2_; \lhd p_e & \mapsto & x_; \lhd v^2_; \lhd p_e \\
(v^1_;, \lambda_\Sigma(t)) \lhd t & \mapsto & (v^1_;, \lambda_\Sigma(t)) \lhd (v^1_;, \lambda_\Sigma(t)) \lhd t \\
(v^2_;, \lambda_\Theta(u)) \lhd (v^1_;, \lambda_\Theta(u)) \lhd u & \mapsto & (v^1_;, \lambda_\Theta(u)) \lhd (v^2_;, \lambda_\Theta(u)) \lhd u \\
(v^2_;, \lambda_\Psi(v)) \lhd (v^2_;, \lambda_\Psi(v)) \lhd v & \mapsto & (v^2_;, \lambda_\Psi(v)) \lhd v \; .
\end{array}$$

Clearly, *isom* is a bijection preserving the labelling of nodes as well as the markings of places (note that the operator box $\Omega_;$ has the empty marking). What remains to be shown is that the weight functions are also preserved. We show that this indeed holds for the case of $p = i_; \lhd \{v^1_; \lhd s_x, v^2_; \lhd e_; \lhd v^1_; \lhd r_e\}$ and $y = (v^2_;, \lambda_\Theta(u)) \lhd (v^1_;, \lambda_\Theta(u)) \lhd u$. Then we have:

$$\begin{array}{rcl}
W_\Delta(p, y) & =_{(4.7)} & \displaystyle\sum_{z \in \{e_; \lhd v^1_; \lhd r_e\}} \; \sum_{w \in \{(v^1_;, \lambda_\Theta(u)) \lhd u\}} W_{\Theta;\Psi}(z, w) \\
& = & W_{\Theta;\Psi}\left(e_; \lhd v^1_; \lhd r_e, (v^1_;, \lambda_\Theta(u)) \lhd u\right) \\
& =_{(4.7)} & \displaystyle\sum_{z \in \{r_e\}} \sum_{w \in \{u\}} W_\Theta(z, w) \\
& = & W_\Theta(r_e, u) \; .
\end{array}$$

We next observe that $p' = isom(p) = v^1_; \lhd i_; \lhd \{v^1_; \lhd s_x, v^2_; \lhd r_e\}$ and $y' = isom(y) = (v^1_;, \lambda_\Theta(u)) \lhd (v^2_;, \lambda_\Theta(u)) \lhd u$ and that we have

$$\begin{array}{rcl}
W_\Xi(p', y') & =_{(4.7)} & \displaystyle\sum_{z \in \{i_; \lhd \{v^1_; \lhd s_x, v^2_; \lhd r_e\}\}} \; \sum_{w \in \{(v^2_;, \lambda_\Theta(u)) \lhd u\}} W_{\Sigma;\Theta}(z, w) \\
& = & W_{\Sigma;\Theta}\left(i_; \lhd \{v^1_; \lhd s_x, v^2_; \lhd r_e\}, (v^2_;, \lambda_\Theta(u)) \lhd u\right) \\
& =_{(4.7)} & \displaystyle\sum_{z \in \{r_e\}} \sum_{w \in \{u\}} W_\Theta(z, w) \\
& = & W_\Theta(r_e, u) \; .
\end{array}$$

Thus $W_{\Delta}(p,y) = W_{\Xi}(p',y')$. Other cases can be dealt with in a similar way.

Solution of Exercise 8.1.5

From the results presented in Chaps. 7 and 8, we obtain the following PBC syntax with the two binary loops and the ternary one, modelled with side conditions, incorporating both (Dom1)—(Dom2) and (Beh1)—(Beh5).

$$
\begin{array}{lllllll}
E^{\mathrm{wf}} & ::= & X^{\mathrm{wf}} & \mid E^{\mathrm{wf}}[f] & \mid E^{\mathrm{e}} & \mid [E^{\mathrm{x}} * E^{\mathrm{e:x:xcl}}\rangle & \mid \\
 & & E^{\mathrm{xcl}} & \mid E^{\mathrm{wf}} \| E^{\mathrm{wf}} & \mid E^{\mathrm{x}} & \mid \langle E^{\mathrm{e:x:xcl}} * E^{\mathrm{e}}] & \\
E^{\mathrm{e}} & ::= & X^{\mathrm{e}} & \mid E^{\mathrm{e}}[f] & \mid E^{\mathrm{e:xcl}} & \mid E^{\mathrm{e:x}} & \mid \\
 & & E^{\mathrm{e}} \| E^{\mathrm{e}} & \mid [E^{\mathrm{e:x}} * E^{\mathrm{e:x:xcl}}\rangle & & & \\
E^{\mathrm{x}} & ::= & X^{\mathrm{x}} & \mid E^{\mathrm{x}}[f] & \mid E^{\mathrm{x:xcl}} & \mid E^{\mathrm{e:x}} & \mid \\
 & & E^{\mathrm{x}} \| E^{\mathrm{x}} & \mid \langle E^{\mathrm{e:x:xcl}} * E^{\mathrm{e:x}}] & & & \\
E^{\mathrm{xcl}} & ::= & X^{\mathrm{xcl}} & \mid E^{\mathrm{xcl}}[f] & \mid E^{\mathrm{e:xcl}} & \mid \langle E^{\mathrm{e:x:xcl}} * E^{\mathrm{e:xcl}}] & \mid \\
 & & E^{\mathrm{x}}; E^{\mathrm{wf}} & \mid E^{\mathrm{wf}}; E^{\mathrm{e}} & \mid E^{\mathrm{x:xcl}} & \mid [E^{\mathrm{x:xcl}} * E^{\mathrm{e:x:xcl}}\rangle & \mid \\
 & & [E^{\mathrm{x}} * E^{\mathrm{e:x:xcl}} * E^{\mathrm{e}}] & & & & \\
E^{\mathrm{e:x}} & ::= & X^{\mathrm{e:x}} & \mid E^{\mathrm{e:x}}[f] & \mid E^{\mathrm{e:x:xcl}} & \mid E^{\mathrm{e:x}} \| E^{\mathrm{e:x}} & \mid \\
 & & E^{\mathrm{e:x}} \Box E^{\mathrm{e:x}} & & & & \\
E^{\mathrm{e:xcl}} & ::= & X^{\mathrm{e:xcl}} & \mid E^{\mathrm{e:xcl}}[f] & \mid E^{\mathrm{e:x:xcl}} & \mid [E^{\mathrm{e:x:xcl}} * E^{\mathrm{e:x:xcl}}\rangle & \mid \\
 & & E^{\mathrm{e}}; E^{\mathrm{e}} & \mid E^{\mathrm{e:x}}; E^{\mathrm{wf}} & \mid [E^{\mathrm{e:x}} * E^{\mathrm{e:x:xcl}} * E^{\mathrm{e}}] & & \\
E^{\mathrm{x:xcl}} & ::= & X^{\mathrm{x:xcl}} & \mid E^{\mathrm{x:xcl}}[f] & \mid E^{\mathrm{e:x:xcl}} & \mid \langle E^{\mathrm{e:x:xcl}} * E^{\mathrm{e:x:xcl}}] & \mid \\
 & & E^{\mathrm{x}}; E^{\mathrm{x}} & \mid E^{\mathrm{wf}}; E^{\mathrm{e:x}} & \mid [E^{\mathrm{x}} * E^{\mathrm{e:x:xcl}} * E^{\mathrm{e:x}}] & & \\
E^{\mathrm{e:x:xcl}} & ::= & X^{\mathrm{e:x:xcl}} & \mid E^{\mathrm{e:x:xcl}}[f] & \mid \alpha & \mid E^{\mathrm{e:x:xcl}} \Box E^{\mathrm{e:x:xcl}} & \mid \\
 & & E^{\mathrm{e:x}}; E^{\mathrm{x}} & \mid E^{\mathrm{e}}; E^{\mathrm{e:x}} & \mid [E^{\mathrm{e:x}} * E^{\mathrm{e:x:xcl}} * E^{\mathrm{e:x}}] & &
\end{array}
$$

References

1. van der Aalst W.M.P. (1998) The Application of Petri Nets to Workflow Management. The Journal of Circuits, Systems and Computers 8: 21–66
2. Baeten J.C.M., Weijland W.P. (1990) Process Algebra. Cambridge Tracts in Theoretical Computer Science 18, Cambridge University Press
3. Best E. (1996) Semantics of Sequential and Parallel Programs. Prentice Hall
4. Best E. (1997) Partial Order Verification with PEP. In: Holzmann G., Peled D., Pratt V. (Eds.) POMIV'96: Partial Order Methods in Verification. American Mathematical Society, 305–328
5. Best E., Devillers R. (1988) Sequential and Concurrent Behaviour in Petri Net Theory. Theoretical Computer Science 55: 87–136
6. Best E., Devillers R., Esparza J. (1993) General Refinement and Recursion Operators for the Petri Box Calculus. In: Enjalbert P., Finkel A., Wagner K.W. (Eds.) STACS 93, 10th Annual Symposium on Theoretical Aspects of Computer Science. Lecture Notes in Computer Science 665, Springer-Verlag, 130–140
7. Best E., Devillers R., and Hall J.G. (1992) The Petri Box Calculus: a New Causal Algebra with Multilabel Communication. In: Advances in Petri Nets 1992, Rozenberg G. (Ed.). Lecture Notes in Computer Science 609, Springer-Verlag, 21–69
8. Best E., Devillers R., Koutny M. (1998) Petri Nets, Process Algebras and Concurrent Programming Languages. In: Advances in Petri Nets. Lectures on Petri Nets II: Applications, Reisig R., Rozenberg G. (Eds.). Lecture Notes in Computer Science 1492, Springer-Verlag, 1–84
9. Best E., Devillers R., Koutny M. (1999) The Box Algebra —a Model of Nets and Process Expressions. In: Donatelli S., Kleijn J. (Eds.) ICATPN'99. Lecture Notes in Computer Science 1639, Springer-Verlag, 344–363
10. Best E., Frączak W., Hopkins R.P., Klaudel H., Pelz E. (1998) M-nets: an Algebra of High-level Petri Nets, with an Application to the Semantics of Concurrent Programming Languages. Acta Informatica 53: 813-857
11. Best E., Koutny M. (1995) A Refined View of the Box Algebra. In: De Michelis G., Diaz M. (Eds.) ICATPN'95. Lecture Notes in Computer Science 935, Springer-Verlag, 1–20
12. Best E., Koutny M. (1995) Solving Recursive Net Equations. In: Fülöp Z., Gécseg F.(Eds.) Automata, Languages and Programming, 22nd International Colloquium, ICALP 95. Lecture Notes in Computer Science 944, Springer-Verlag, 605–623
13. Boudol G., Castellani I. (1994) Flow Models of Distributed Computations: Three Equivalent Semantics for CCS. Information and Computation 114(2): 247–314
14. Boussinot F., de Simone R. (1991) The Esterel Language. Technical Report RR-1487, INRIA, Sophia Antipolis

15. Brams G.W. (nom collectif de André Ch., Berthelot G., Girault C., Memmi G., Roucairol G., Sifakis J., Valette R., Vidal-Naquet G.) (1985) Résaux de Petri: Théorie et Pratique. Two volumes. Editions Masson
16. Brookes S.D. (1996) The Essence of Parallel Algol. In: LICS'96. IEEE Computer Society Press, 164–173
17. Bruns G., Esparza J. (1995) Trapping Mutual Exclusion in the Box Calculus. Theoretical Computer Science 153(1-2): 95–128
18. Davey B.A., Priestley H.A. (1990) Introduction to Lattices and Order. Cambridge Mathematical Textbook
19. Degano P., De Nicola R., Montanari U. (1988) A Distributed Operational Semantics for CCS Based on C/E Systems. Acta Informatica 26(1-2): 59–91
20. Desel J., Esparza J. (1995) Free Choice Petri Nets. Cambridge Tracts in Theoretical Computer Science 40, Cambridge University Press
21. Devillers R. (1985) The Semantics of Capacities in P/T Nets: a First Look. In: ICATPN'85. Espoo (Finland), 171–190
22. Devillers R. (1990) The Semantics of Capacities in P/T Nets. In: Advances in Petri Nets 1989, Rozenberg G. (Ed.). Lecture Notes in Computer Science 424, Springer-Verlag, 128–150
23. Devillers R. (1992) Maximality Preservation and the ST-idea for Action Refinements. In: Advances in Petri Nets 1992, Rozenberg G. (Ed.). Lecture Notes in Computer Science 609, Springer-Verlag, 108–151
24. Devillers R. (1992) Tree Interfaces in the Petri Box Calculus. Technical Report TR-LIT-265, Laboratoire d'Informatique Théorique, Université Libre de Bruxelles
25. Devillers R. (1993) Construction of S-invariants and S-components for Refined Petri Boxes. In: Ajmone Marsan M. (Ed.) ICATPN'93. Lecture Notes in Computer Science 691, Springer-Verlag, 242–261
26. Devillers R. (1993) S-invariant Analysis of Petri Boxes. Technical Report TR-LIT 273, Laboratoire d'Informatique Théorique, Université Libre de Bruxelles
27. Devillers R. (1993) Analysis of General Refined Petri Boxes. In: Baeza-Yates R. (Ed.) XIIIth International Conference of the Chilean Society of Computer Science. Sociedad Chilena de Ciencia de la Computación, 419–434
Also appeared as: Devillers R. (1994) Analysis of General Refined Petri Boxes. Computer Science 2: Research and Application, Plenum Publishing, 411–428
28. Devillers R. (1996) Petri Boxes and Finite Precedence. In: Montanari U., Sassone V. (Eds.) CONCUR 96, Concurrency Theory, 7th International Conference. Lecture Notes in Computer Science 1119, Springer-Verlag, 465–480
29. Devillers R. (1993) On a More Liberal Synchronisation Operator for the Petri Box Calculus. Technical Report LIT-281, Laboratoire d'Informatique Théorique, Université Libre de Bruxelles
30. Devillers R. (1994) The Synchronisation Operator Revisited for the Petri Box Calculus. Technical Report LIT-290, Laboratoire d'Informatique Théorique, Université Libre de Bruxelles
31. Devillers R. (1995) S-invariant Analysis of Recursive Petri Boxes. Acta Informatica 32(4): 313–345
32. Devillers R., Koutny M. (1998) Recursive Nets in the Box Algebra. In: CSD'98: International Conference on Application of Concurrency to System Design. IEEE Press, 239–249
33. Cited in: [35]
34. Dijkstra E.W. (1976) A Discipline of Programming. Prentice Hall
35. Dijkstra E.W. (1968) Cooperating Sequential Processes. In: Programming Languages, Genuys F. (Ed.). Academic Press, 43–112

36. Esparza J. (1991) Fixpoint Definition of Recursion in the Box Algebra. Memorandum, Universität Hildesheim
37. Esparza J. (1994) Model Checking Using Unfoldings. Science of Computer Programming 23(2-3): 151–195
38. Esparza J., Römer S., Vogler W. (1996) An Improvement of McMillan's Unfolding Algorithm. In: Margaria T., Steffen B. (Eds.) TACAS'96. Lecture Notes in Computer Science 1055, Springer-Verlag, 87–106
39. Fleischhack H., Grahlmann B. (1996) A Petri Net Semantics for $B(PN)^2$ with Procedures which Allows Verification. Technical Report 21/96, Hildesheimer Informatikbericht
40. Frączak W., Klaudel H. (1994) A Multi-action Synchronisation Scheme and its Application to the Petri Box Calculus. In: ESDA: Engineering Systems Design and Analysis Conference. London, 91–100
41. Frączak W. (1995) Multi-action Process Algebra. In: Kanchanasut K., Lévy J.-J. (Eds.) Algorithms, Concurrency and Knowledge. Lecture Notes in Computer Science 1023, Springer-Verlag, 126–140
42. Genrich H.J., Lautenbach K., Thiagarajan P.S. (1980) Elements of General Net Theory. In: Net Theory and Applications, Proc. of the Advanced Course on General Net Theory of Processes and Systems, Brauer W. (Ed.). Lecture Notes in Computer Science 84, Springer-Verlag, 21–163
43. van Glabbeek R., Vaandrager F.W. (1987) Petri Net Models for Algebraic Theories of Concurrency (Extended Abstract). In: de Bakker J.W., Nijman A.J., Treleaven P.C., (Eds.) PARLE, Parallel Architectures and Languages Europe, Volume II. Lecture Notes in Computer Science 259, Springer-Verlag, 224–242
44. Goltz U. (1988) On Representing CCS Programs by Finite Petri Nets. In: Chytil M.P., Janiga L., Koubek V. (Eds.) MFCS'88. Lecture Notes in Computer Science 324, Springer-Verlag, 339–350
45. Goltz U. (1988) Über die Darstellung von CCS-Programmen durch Petrinetze. Technical Report 172, GMD-Bericht, Verlag R. Oldenbourg
46. Goltz U., van Glabbeek R. (1990) Refinement of Actions in Causality Based Models. In: de Bakker J.W., de Roever W.-P., Rozenberg G. (Eds.) Stepwise Refinement of Distributed Systems: Models, Formalism, Correctness, Proceedings REX Workshop. Lecture Notes in Computer Science 430, Springer-Verlag, 267–300
47. Goltz U., Loogen R. (1991) A Non-interleaving Semantic Model for Nondeterministic Concurrent Processes. Fundamenta Informaticae 14(1): 39–73
48. Goltz U., Reisig W. (1983) The Non-sequential Behaviour of Petri Nets. Information and Control 57: 125–147
49. Grabowski J. (1981) On Partial Languages. Fundamenta Informaticae 4: 427–498
50. Grahlmann B. (1998) Parallel Programs as Petri Nets. PhD Thesis, Universität Hildesheim
51. Graves B. (1997) Computing Reachability Properties Hidden in Finite Net Unfoldings. In: Ramesh S., Sivakumar G. (Eds.) FST-TCS'97. Lecture Notes in Computer Science 1346, Springer-Verlag, 327–341
52. Gries D. (1981) Science of Computer Programming. Springer-Verlag
53. Hall J.G. (1996) An Algebra of High-level Petri Nets. PhD Thesis, University of Newcastle upon Tyne
54. Hennessy M.B. (1988) Algebraic Theory of Processes. MIT Press
55. Hesketh M., Koutny M. (1998) An Axiomatisation of Duplication Equivalence in the Petri Box Calculus. In: Desel J., Silva M. (Eds.) ICATPN'98. Lecture Notes in Computer Science 1420, Springer-Verlag, 165–184

56. Hesketh M. (1998) Synthesis and Axiomatisation for Structural Equivalences in the Petri Box Calculus. PhD Thesis, University of Newcastle upon Tyne
57. Hesselink W.H. (1992) Programs, Recursion and Unbounded Choice. Cambridge Tracts in Theoretical Computer Science 27, Cambridge University Press
58. Hoare C.A.R. (1978) Communicating Sequential Processes. Communications of the ACM 21: 666–677
59. Hoare C.A.R. (1985) Communicating Sequential Processes. Prentice Hall
60. Jenner L. (1994) Ein Prozedurkonzept für die parallele Hochsprache $B(PN)^2$. Diplomarbeit, Universität Hildesheim
61. Janicki R., Koutny M. (1995) Semantics of Inhibitor Nets. Information and Computation 123(1): 1–16
62. Janicki R., Lauer P.E. (1992) Specification and Analysis of Concurrent Systems - the COSY Approach. EATCS Monographs on Theoretical Computer Science, Springer-Verlag
63. Jensen K. (1997) Coloured Petri Nets. Basic Concepts, Analysis Methods and Practical Use Vol. 1, 2nd ed., 2nd corr. printing, Vol. 2, 2nd corr. printing, and Vol. 3. Monographs in Theoretical Computer Science – An EATCS Series, Springer-Verlag
64. Kiehn A. (1990) Petri Net Systems and their Closure Properties. In: Rozenberg G. (Ed.) Advances in Petri Nets 1989. Lecture Notes in Computer Science 424, Springer-Verlag, 306–328
65. Klaudel H., Pelz E. (1994) Handling Abstract Data Types in the Petri Box Calculus. In: Czaja L., Burkhard H.-D., Starke P.H. (Eds.) CS&P'94. Humboldt-Universität zu Berlin, Informatik-Bericht Nr. 36, 47–57
66. Klaudel H., Pelz E. (1995) Communication as Unification in the Petri Box Calculus. In: Reichel H. (Ed.) Fundamentals of Computation Theory. Lecture Notes in Computer Science 965, Springer-Verlag, 303–312
67. Klaudel H. (1995) Modèles algébriques, basés sur les réseaux de Petri, pour la sémantique des langages de programmation concurrents. PhD Thesis, Université Paris XI Orsay
68. Koutny M., Best E. (1999) Fundamental Study: Operational and Denotational Semantics for the Box Algebra. Theoretical Computer Science 211(1-2): 1–83
69. Koutny M., Esparza J., Best E. (1994) Operational Semantics for the Petri Box Calculus. In: Jonsson B., Parrow J. (Eds.) CONCUR 94, Concurrency Theory, 5th International Conference. Lecture Notes in Computer Science 836, Springer-Verlag, 210–225
70. Koutny M. (1994) Partial Order Semantics of Box Expressions. In: Valette R. (Ed.) ICATPN'94. Lecture Notes in Computer Science 815, Springer-Verlag, 318–337
71. Lauer P.E. (1974) Path Expressions as Petri Nets, or Petri Nets with Fewer Tears. Technical Report MRM/70, Computing Laboratory, University of Newcastle upon Tyne
72. Lautenbach K. (1975) Liveness in Petri Nets. Technical Report ISF/75-02-1, GMD, Bonn
73. Levy A. (1979) Basic Set Theory. Springer-Verlag
74. Lilius J., Pelz E. (1996) An M-net Semantics for $B(PN)^2$ with Procedures. In: Atalay V., Halici U., Inan K., Yalabik N., Yazici A. (Eds.) ISCIS, 11th International Symposium on Computer and Information Science. Middle East Technical University, Antalya, 365–374

75. LOTOS: International Standard ISO 8807 (1988) In: Brinksma E. (Ed.) Information Processing Systems - Open Systems Interconnection - LOTOS - A formal description technique based on the temporal ordering of observational behaviour
76. May D. (1983) Occam. ACM SIGPLAN Notices 18: 69–79
77. Mazurkiewicz A. (1987) Trace Theory. In: Advances in Petri Nets 1986, Petri Nets: Applications and Relationships to Other Models of Concurrency, Part II, Brauer W., Reisig W., Rozenberg G. (Eds.). Lecture Notes in Computer Science 255, Springer-Verlag, 279–324
78. McMillan K.L. (1992) Using Unfoldings to Avoid the State Explosion Problem in the Verification of Asynchronous Circuits. In: van Bochmann G., Probst D.K. (Eds.) 4th Workshop on Computer Aided Verification. Lecture Notes in Computer Science 663, Springer-Verlag, 164–174
79. Melzer S., Esparza J. (1996) Checking System Properties via Integer Programming. In: Nielson H.R.(Ed.) ESOP'96: Programming Languages and Systems. Lecture Notes in Computer Science 1058, Springer-Verlag, 250–264
80. Milner R. (1980) A Calculus of Communicating Systems. Lecture Notes in Computer Science 92, Springer-Verlag
81. Milner R. (1989) Communication and Concurrency. Prentice Hall
82. Morris J.H. (1979) A Starvation-free Solution to the Mutual Exclusion Problem. Information Processing Letters 8(2): 76–80
83. Murata T. (1989) Petri Nets: Properties, Analysis and Applications. Proc. of IEEE 77(4): 541–580
84. Olderog E.R. (1991) Nets, Terms and Formulas. Cambridge Tracts in Theoretical Computer Science 23, Cambridge University Press
85. Owicki S.S., Gries D. (1976) An Axiomatic Proof Technique for Parallel Programs. Acta Informatica 6: 319–340
86. PEP (a Programming Environment Based of Petri Nets) can be found at: http://theoretica.informatik.uni-oldenburg.de/~pep/HomePage.html.
87. Peterson G.L. (1981) Myths About the Mutual Exclusion Problem. Information Processing Letters 12(3): 115–116
88. Peterson J.L. (1981) Petri Net Theory and the Modeling of Systems. Prentice Hall
89. Plotkin G.D. (1981) A Structural Approach to Operational Semantics. Technical Report FN-19, Computer Science Department, University of Aarhus
90. Reisig W. (1985) Petri Nets. An Introduction. EATCS Monographs on Theoretical Computer Science, vol. 4, Springer-Verlag
91. Reisig W. (1996) Modelling and Verification of Distributed Algorithms. In: Montanari U., Sassone V. (Eds.) CONCUR 96, Concurrency Theory, 7th International Conference. Lecture Notes in Computer Science 1119, Springer-Verlag, 579–595
92. Reisig W. (1998) Elements of Distributed Algorithms. Springer-Verlag
93. Riemann R.-Ch. (1999) Modelling of Concurrent Systems: Structural and Semantical Methods in the High Level Petri Net Calculus. PhD Thesis, Université Paris-Sud and Universität Hildesheim
94. Starke P.H. (1981) Processes in Petri Nets. Elektronische Informationsverarbeitung und Kybernetik 17: 389–416
95. Taubner D. (1989) Finite Representation of CCS and TCSP Programs by Automata and Petri Nets. Lecture Notes in Computer Science 369, Springer-Verlag
96. Teßmer B. (1996) S- und T-Invarianten zum Nachweis von Eigenschaften paralleler Algorithmen. Diplomarbeit, Universität Hildesheim

97. van Glabbeek R. (1990) The Refinement Theorem for ST-bisimulation Semantics. In: Broy M., Jones C. (Eds.) IFIP Working Conference on Programming Concepts and Methods. IFIP, Elsevier Science Publishers B.V. (North-Holland), 27–52
98. Vaught R.L. (1985) Set Theory. An Introduction. Birkhäuser
99. Vogler W. (1991) Bisimulation and Action Refinement. In: Choffrut C., Jantzen M. (Eds.) STACS 91, 8th Annual Symposium on Theoretical Aspects of Computer Science. Lecture Notes in Computer Science 480, Springer-Verlag, 309–321
100. Vogler W. (1992) Partial Words versus Processes: a Short Comparison. In: Advances in Petri Nets 1992, Rozenberg G. (Ed.). Lecture Notes in Computer Science 609, Springer-Verlag, 292–303
101. Wirth N. (1971) The Programming Language Pascal. Acta Informatica 1(1): 35–64

Index

Monographs in Theoretical Computer Science · An EATCS Series

C. Calude
Information and Randomness
An Algorithmic Perspective

K. Jensen
Coloured Petri Nets
***Basic Concepts,* Analysis Methods**
and Practical Use, Vol. 1
2nd ed.

K. Jensen
Coloured Petri Nets
Basic Concepts, *Analysis Methods*
and Practical Use, Vol. 2

K. Jensen
Coloured Petri Nets
Basic Concepts, Analysis Methods
and *Practical Use,* Vol. 3

A. Nait Abdallah
The Logic of Partial Information

Z. Fülöp, H. Vogler
Syntax-Directed Semantics
Formal Models
Based on Tree Transducers

A. de Luca, S. Varricchio
Finiteness and Regularity
in Semigroups
and Formal Languages

E. Best, R. Devillers, M. Koutny
Petri Net Algebra

Texts in Theoretical Computer Science · An EATCS Series

J. L. Balcázar, J. Díaz, J. Gabarró
Structural Complexity I
2nd ed. (see also overleaf, Vol. 22)

M. Garzon
Models of Massive Parallelism
Analysis of Cellular Automata
and Neural Networks

J. Hromkovič
Communication Complexity
and Parallel Computing

A. Leitsch
The Resolution Calculus

G. Păun, G. Rozenberg, A. Salomaa
DNA Computing
New Computing Paradigms

A. Salomaa
Public-Key Cryptography
2nd ed.

K. Sikkel
Parsing Schemata
A Framework for Specification
and Analysis of Parsing Algorithms

H. Vollmer
Introduction to Circuit Complexity
A Uniform Approach

W. Fokkink
Introduction to Process Algebra

K. Weihrauch
Computable Analysis
An Introduction

S. Jukna
Extremal Combinatorics
With Applications in Computer Science

Former volumes appeared as EATCS Monographs on Theoretical Computer Science

Vol. 5: W. Kuich, A. Salomaa
Semirings, Automata, Languages

Vol. 6: H. Ehrig, B. Mahr
Fundamentals of Algebraic Specification 1
Equations and Initial Semantics

Vol. 7: F. Gécseg
Products of Automata

Vol. 8: F. Kröger
Temporal Logic of Programs

Vol. 9: K. Weihrauch
Computability

Vol. 10: H. Edelsbrunner
Algorithms in Combinatorial Geometry

Vol. 12: J. Berstel, C. Reutenauer
Rational Series and Their Languages

Vol. 13: E. Best, C. Fernández C.
Nonsequential Processes
A Petri Net View

Vol. 14: M. Jantzen
Confluent String Rewriting

Vol. 15: S. Sippu, E. Soisalon-Soininen
Parsing Theory
Volume I: Languages and Parsing

Vol. 16: P. Padawitz
Computing in Horn Clause Theories

Vol. 17: J. Paredaens, P. DeBra, M. Gyssens, D. Van Gucht
The Structure of the Relational Database Model

Vol. 18: J. Dassow, G. Páun
Regulated Rewriting in Formal Language Theory

Vol. 19: M. Tofte
Compiler Generators
What they can do, what they might do, and what they will probably never do

Vol. 20: S. Sippu, E. Soisalon-Soininen
Parsing Theory
Volume II: LR(k) and LL(k) Parsing

Vol. 21: H. Ehrig, B. Mahr
Fundamentals of Algebraic Specification 2
Module Specifications and Constraints

Vol. 22: J. L. Balcázar, J. Díaz, J. Gabarró
Structural Complexity II

Vol. 24: T. Gergely, L. Úry
First-Order Programming Theories

R. Janicki, P. E. Lauer
Specification and Analysis of Concurrent Systems
The COSY Approach

O. Watanabe (Ed.)
Kolmogorov Complexity and Computational Complexity

G. Schmidt, Th. Ströhlein
Relations and Graphs
Discrete Mathematics for Computer Scientists

S. L. Bloom, Z. Ésik
Iteration Theories
The Equational Logic of Iterative Processes